Eukaryotes

	DIVISION	SUBDIVISION	CLASS	SUBCLASS	ORDER
Slime Molds	Myxomycota		Myxomycetes		Liceales
					Trichiales
					Echinosteliales
					Stemonitales
					Physarales
			Plasmodiophoromycetes		Plasmodiophorales
			Dictyosteliomycetes		Dictyosteliales
			Labyrinthulomycetes		Labyrinthulales
					Thraustochytriales
True Fungi	Oomycota		Oomycetes		Saprolegniales
					Leptomitales
					Peronosporales
	Hyphochytridiomy-cota		Hyphochytridiomycetes		Hyphochytridiales
	Chytridiomycota		Chytridiomycetes		Chytridiales
					Blastocladiales
					Monoblepharidales
	Zygomycota		Zygomycetes		Mucorales
					Entomophthorales
			Trichomycetes		Eccrinales
	Eumycota	Ascomycotina	Hemiascomycetes		Endomycetales
					Taphrinales
			Euascomycetes		
			Series Plectomycetes		Eurotiales
			Series Pyrenomycetes		Erysiphales
					Chaetomiales
					Xylariales
					Clavicepitales
			Series Discomycetes		Helotiales
					Lecanorales
					Pezizales
					Tuberales
			Series Loculoascomycetes		Pleosporales
					Dothidiales
					Pyrenulales
		Basidiomycotina	Series Laboulbeniomycetes		Laboulbeniales
			Heterobasidiomycetes		Tremellales
					Uredinales
					Ustilaginales
			Homobasidiomycetes		
			Series Hymenomycetes		Aphyllophorales
					Agaricales
			Series Gasteromycetes		Lycoperdales
					Phallales
					Nidulariales
	Form Class Fungi Imperfecti				Sphaeropsidales
					Melanconiales
					Moniliales
					Mycelia Sterilia
Bryophytes	Bryophyta		Hepaticae (Hepatopsida)	Jungermanniae	Calobryales
					Jungermanniales
					Metzgeriales
				Marchantiae	Sphaerocarpales
					Monocleales
					Marchantiales
			Anthocerotae (Anthoceropsida)		Anthocerotales
			Musci (Bryopsida)	Sphagnidae	Sphagnales
				Andreaeidae	Andreaeales
				Tetraphidae	Tetraphidales
				Polytrichidae	Polytrichales
				Buxbaumiidae	Buxbaumiales
				Bryidae	Dicranales
					Archidiales
					Fissidentales
					Pottiales
					Grimmiales
					Funariales
					Eubryales
					Isobryales
					Hookeriales
					Hypnobryales

NONVASCULAR PLANTS
An Evolutionary Survey

R. F. Scagel

R. J. Bandoni

J. R. Maze

G. E. Rouse

W. B. Schofield

J. R. Stein

Department of Botany
The University of British Columbia

Wadsworth Publishing Company
Belmont, California
A Division of Wadsworth, Inc.

Biology Editor: Jack Carey

Production: Greg Hubit Bookworks

Copy Editor: Nick Murray

Cover photo: Coral fungus, Hocking County, Ohio
Photo by Roger Burnard

Printed in the United States of America

1 2 3 4 5 6 7 8 9 10—86 85 84 83 82

Library of Congress Cataloging in Publication Data
Main entry under title:

Nonvascular plants.

"A comprehensive revision and expansion of the nonvascular plant section (including bryophytes) of An evolutionary survey of the plant kingdom" —Pref.
    Includes bibliographies and index.
    1. Cryptogams.   2. Cryptogams—Evolution.
3. Phylogeny (Botany)   I. Scagel, Robert Francis, 1921–
QK505.N66   1982   586   81-24066
ISBN 0-534-01029-6   AACR2

Books in the Wadsworth Biology Series

Biology: The Unity and Diversity of Life, 2nd, Starr and Taggart

Energy and Environment: Four Energy Crises, Miller

Living in the Environment, 3rd, Miller

Replenish the Earth: A Primer in Human Ecology, Miller

Oceanography: An Introduction, 2nd, Ingmanson and Wallace

Biology books under the editorship of William A. Jensen, University of California, Berkeley

Biology, Jensen, Heinrich, Wake, Wake

Biology: The Foundations, Wolfe

Botany: An Ecological Approach, Jensen and Salisbury

Plant Physiology, 2nd, Salisbury and Ross

Plant Physiology Laboratory Manual, Ross

Nonvascular Plants, Scagel et al.

Plant Diversity: An Evolutionary Approach, Scagel et al.

An Evolutionary Survey of the Plant Kingdom, Scagel et al.

Plants and the Ecosystem, 3rd, Billings

Plants and Civilization, 3rd, Baker

PREFACE

In 1965, the first edition of *An Evolutionary Survey of the Plant Kingdom* appeared. It surveyed the whole plant kingdom, with emphasis on comparative morphology and evolution. The wide use of *An Evolutionary Survey of the Plant Kingdom* by second-year students, as well as in more advanced undergraduate courses and as a reference, has confirmed our philosophy that a collaboration of specialists in one text can make a unique contribution to students of the plant kingdom, especially at a more advanced level, providing a more satisfactory coverage of each group than is often achieved in a book written by one author. *Nonvascular Plants: An Evolutionary Survey* is a comprehensive revision and expansion of the nonvascular plant section (including bryophytes) of *An Evolutionary Survey of the Plant Kingdom.*

Since 1965, there have been significant advances in botanical research, which have necessitated some major changes in classification systems. An outline of the classification system followed in this text appears on the end pages of the book. In addition to these major changes, we have updated all chapters to take into account much new information that has been reported in the literature appearing since 1965.

The text provides additional references for the more advanced student and teacher. A list of pertinent references appears at the end of each chapter. The text assumes that the student has already had a first-year university course in botany or the equivalent.

We have standardized our terminology throughout the text as far as is practical. We have also tried to maintain continuity in treatment so that the overall trends in phylogeny and evolution are not lost or interrupted. By the same token, we have attempted to apply the findings of paleobotany directly to living representatives in the various plant groups in order to gain perspective on these two complementary aspects of botanical science. A comprehensive glossary, including boldface terms used throughout the text, is included at the end of the text.

Individual contributions, which have been primarily in our respective fields of specialization, are as follows: bacteria, slime molds, and fungi (including lichens) (Robert J. Bandoni); bryophytes (W. B. Schofield); algae (Robert F. Scagel and Janet R. Stein); introduction and general aspects of evolution and classification (J. R. Maze); paleobotany (Glenn E. Rouse). However, since we have consolidated our efforts and consulted closely with each other in the interests of uniformity and continuity, it is impossible to define completely the limits of each person's contribution.

The quality of the drawings is a feature that has been universally well received by students and teachers using the earlier text. Many of the original drawings prepared by Margaret (Dean) Jensen, Ernani Menez, and P. Drukker Brammall for *An Evolutionary Survey of the Plant Kingdom* and *Plant Diversity: An Evolutionary Approach* have been reproduced in *Nonvascular Plants: An Evolutionary Survey;* some of their drawings have been regrouped or redrawn. In addition, some new drawings have been added to the figures; these were prepared by Lesley R. Bohm, P. Drukker Brammall, and Muriel Schofield. The following figures were prepared by Lesley R. Bohm: 1-1, 2; 3-1, 3, 15E, 17, 19, 22D, 27; 4-25; 5-8, 16, 18, 20, 24, 25, 28, 35, 37, 39, 40, 53; 6-10, 13, 14, 16, 18, 20, 21, 23, 25, 28, 34, 35, 40, 42, 44, 46, 58, 62, 79, 80, 83; 11-13, 14, 15, 29, 31, 33; 13-4, 5, 6, 7, 8, 9, 10, 11, 15, 20, 22, 24, 26. P. Drukker Brammall prepared the following figures: 2-3, 4G, H, 5, 6, 7A; 3-49, 50A, B, F, 51A; 7-3C, 4, 7, 8C, E, F, 15, 16A-C, 23B, C, 28 (in part), 38, 39; 8-5A, 12, 14C; 9-4A, E, 7A, B. Muriel Schofield prepared the following figures: 12-1, 12, 25, 29D, G, J, K, 30B, C, 32, 39.

Many new photographs, including transmission and scanning electron micrographs, have been added to this revised material to complement the drawings.

We are indebted to individuals and publishers who loaned original illustrations and photographs, and who gave permission to copy or redraw figures; these are acknowledged in the appropriate captions.

Acknowledgments

For constructive criticism and valuable recommendations concerning preparation of this text, and for reading and commenting on an early draft of the manuscript, or portions thereof, we are indebted to the following reviewers:

Chester Bosworth, Southern Connecticut State University; Dennis Clark, Arizona State University; Martha Christensen, University of Wyoming; Lois J. Cutter, University of North Carolina, Greensboro; William E. Dietrich, Indiana University of Pennsylvania; Patricia Gensel, University of North Carolina, Raleigh; Allen Graham, Kent State University; Robert B. Kaul, University of Nebraska, Lincoln; P. W. Kirk, Old Dominion University; Norma Lang, University of California, Davis; Estelle Levetin, University of Tulsa; Davis McLaughlin, University of Minnesota; M. W. Miller, University of South Alabama; Florence Neely, Augusta College; Nancy Nicholson, Miami University of Ohio; Paul Nighswonger, Northwestern State College; Harry K. Phinney, Oregon State University; Don R. Reynolds, Natural History Museum; Judith E. Skog, George Mason University; Shirley R. Sparling, California Polytechnic University; Anne S. Susalla, St. Mary's College; Sanford S. Tepfer, University of Oregon; Henry S. Webert, Nicholl's State University; Jonathon J. Westfall, University of Georgia, Athens; Richard A. White, Duke University.

Portions of the manuscript were also read and constructively criticized by Drs. Fred R. Ganders, J. W. D. Garbary, Lynda J. Goff, Gayle I. Hansen, M. Neushul, R. E. Norris, F. J. R. Taylor, Nancy J. Turner, and M. J. Wynne; to each of these persons we express our sincere appreciation. Although these comments and criticisms have been most helpful and have contributed significantly to the text, the authors take full responsibility for the final product.

We are grateful to the following for providing reference literature or specimens for study: Drs. R. M. Schuster, G. H. M. Scott, W. C. Steere, Jane Taylor, and D. H. Vitt. Finally, we wish to acknowledge the assistance provided by the Department of Botany, The University of British Columbia, and especially to thank Mr. J. Reid for photographic assistance and Miss J. Celestino Oliveira for technical assistance.

CONTENTS

The plant world exhibits a tremendous range of diversity, from unicellular bacteria and algae to immense trees. Plants occur in every conceivable environment, from aquatic (both marine and freshwater) to deserts, from tropical to alpine and arctic environments, and from hot springs to snow banks. Nutritionally, plants can be **autotrophs** or **heterotrophs,** and the latter **saprophytes** or **parasites.** Despite this great range of diversity, there is a natural order that exists, and the primary impetus behind much biological research has been to elucidate, describe, and understand this order.

Plant morphology, the theme of this book, is concerned with one aspect of biological order—the relationships between plant groups and between **tissues,** structures, and **organs** of plants exhibiting structural complexity. However, an appreciation of biological **evolution,** the mechanism that most biologists consider to be the cause of natural order, adds to the study of morphology. The purpose of this chapter is to offer a brief introduction to that mechanism and to the study of natural order.

The Process of Evolution

There are many ways to describe the process of evolution, but practically any description will include the following points: (1) evolution is a **genetic** phenomenon; (2) **populations** are the units of evolution; (3) populations exhibit **variation;** (4) variants are not equally successful in contributing progeny to the next generation—i.e., they are subject to **selection;** and (5) selection produces genetic change in populations over time.

Population Genetics

Population is a general term that is used in several contexts. In evolutionary biology, a population is a group of potentially interbreeding individuals occupying a specific habitat. Because a population consists of more than one individual, its genetics is slightly different from the genetics of an individual. All traits seen in organisms are under control of units called **genes.** Genes occur on

1

INTRODUCTION

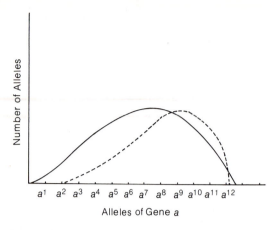

Figure 1-1 Change in gene frequency as a result of selection. Dotted line represents the change in a population as a result of selection for a_7 to a_{11}.

chromosomes at positions called **loci** (sing. **locus**). In diploid individuals, there are two **alleles** for each gene locus. Thus, in a population, the total number of alleles for any one locus is twice the number of individuals—e.g., in a population of 100 plants, there will be 200 alleles for any gene locus. The two alleles at one locus in any one plant need not be the same as the two alleles for the same locus in another plant. For example, one plant can have alleles a_1 and a_2, in which case it is **heterozygous** at the a locus. Another plant can have different alleles, a_3 and a_4. For this reason many different alleles for one gene locus can exist in a population. The relative number of alleles for any gene in a population is usually called gene frequency, although, strictly speaking, it is allele frequency. For example, in a population of 100 plants there would be 200 alleles for the a locus. If 100 were a_1, 50 were a_2, 40 were a_3, and 10 were a_4, the gene frequencies would be 0.5 for a_1 (100/200), 0.25 for a_2, 0.20 for a_3, and 0.05 for a_4. Allele frequencies always add up to 1.00. A change in gene frequency is an integral part of evolution (see Fig. 1-1).

Variation

Variation means detectable differences that exist between members of a population. It can be due either to environmental or to genetic differences. Environmentally induced variation results from the response of plants with similar genotypes to different environments. An example of this might be plant size, as illustrated by the brown alga *Laminaria digitata*, which grows in northern European waters. In areas where there is rapid water movement, its blades are narrow and often split longitudinally. When these plants are moved to calm waters, the blades become wider (up to 8 times as wide), and there is very little splitting. Another example is the phenomenon of heterophylly, the formation of different leaves. On some trees, such as bur oak (*Quercus macrocarpa*), leaves formed in the shade are much larger, thinner, and less deeply lobed than leaves formed in the sun. Morphological responses to environmental differences such as these are often called **phenotypic plasticity.**

Genotypic differences are due to variation in the genetic makeup of plants. Such differences will usually be expressed regardless of environment, although environment may modify the degree of difference expressed. Genetic differences are most important in evolution.

Genetic Sources of Variation

Our concern here is with genotypic variation. There are two sources of genetic variation—**mutation** and **recombination.** Mutation is a change in base pair sequence in DNA so as to give new (or different) alleles of genes. Recombination generates variation as a result of bringing together new combinations of alleles.

There are two sources of recombination: (a) **independent assortment** of chromosomes at **meiosis** with subsequent sexual reproduction; and (b) **crossing-over** between **homologous chromosomes** during meiosis. The former phenomenon results in new and unique allele combinations in the offspring. For example, if two homozygous individuals with the genetic constitutions $a_1a_1b_1b_1c_1c_1$ and $a_2a_2b_2b_2c_2c_2$ are crossed, the result is $a_1a_2b_1b_2c_1c_2$ in the first generation (F_1). If the F_1, which is heterozygous for these genes, is selfed (crossed with itself), there is the possibility of 27 different genotypes appearing in the second generation, the F_2. Of these 27, 3 will be reconstituted parental and F_1 genotypes, and 24 will be new. This is assuming the genes are on different chromosomes and thus, like chromosomes, show independent assortment. Recombination, as a result of independent assortment and subsequent sexual reproduction, can produce many genotypes. If a diploid plant that is heterozygous for ten independently assorting genes is selfed, the number of possible genotypes is 59,049 (3^n, where $n =$ the number of independently assorting genes). Recombination within a sexually reproducing population can produce an even greater amount of variation. In a population, it is theoretically possible for a gamete from any individual to fertilize a gamete from any other individual. Thus, any allelic combination from all the alleles available in the population can be produced. The presence of **multiple alleles** in a population increases the already great potential for producing variants as a result of recombination. The following formula gives the possible number of recombinations where more than two alleles exist for a gene:

$$R = \left[\frac{r(r + 1)}{2} \right]^n$$

where $R =$ number of recombinations possible
$r =$ number of alleles for each gene
$n =$ number of independently assorting genes

For example, a population with ten independently assorting genes, each with four alleles, can give rise to 10 billion variants as a result of recombination. A comparison of this with the number of recombinants possible when only two alleles exist for ten independently assorting genes shows that doubling in the allele number increases the recombinants by about 170,000 times.

Crossing-over results in the exchange of chromosome parts between homologous chromosomes when they pair early in meiosis. It is significant in recombination, for it increases the number of independently assorting genes by breaking the genes linked together on one chromosome. The higher the rate of crossing-over, the greater the number of independently assorting genes.

Both mutation and genetic recombination are significant in evolution, but at different levels. Recombination is the immediate source of variation for **natural selection,** whereas mutation is the ultimate source for new alleles, which can be brought into new combinations by recombination.

Natural Selection

Natural populations of plants produce more progeny than can be supported by the environment. Hence, only some survive, the survivors being those better adapted to the environment. Of those that survive, not all grow equally well. Some will grow more vigorously and produce a greater proportion of progeny for the following generation (i.e., will have higher fitness). This differential reproduction by individuals in a population means that certain genotypes increase and others decrease, resulting in a change in gene frequency. Figure 1-1 presents a diagram of a hypothetical situation in which there is a change in gene frequency in a population.

Although natural selection results in genetic change, it does not act directly on the genotype. The genotype of an organism interacts with the environment, resulting in plants as we observe them growing. The expression of this environmentally modified genotype is the **phenotype.** When selection acts, it does so on the phenotype.

Selection results in populations that are adapted to the habitat in which they live, and this adaptation is genetically based. Populations of one species that are genetically adapted to different environments are **ecotypes.**

There are several examples of experimental proof of adaptation resulting from natural selection. One is from a series of experiments performed on ecotypes at the Carnegie Institute at Stanford University by J. Clausen, D. D. Keck, and W. M. Hiesey (1948). These experiments involved populations of a perennial herb, yarrow (*Achillea millefolium*), growing along an altitudinal gradient from coastal California to alpine areas in the Sierra Nevada. Plants from different altitudes could be ordered into a hierarchy and had traits that formed obvious adaptations to different habitats. Clausen, Keck, and Hiesey performed reciprocal transplants with alpine and lowland plants, transplanting alpine plants to gardens at Stanford and lowland plants to gardens at timberline in the Sierra Nevada. Hence, they were able to assess the performance of plants in habitats different from their native ones. In habitats to which they had been transplanted, plants maintained certain features exhibited in their native environments. That the transplants did not "adapt" to their new environments indicates features correlated with native environments of the plants that were under genetic control.

Plants from lowland and alpine habitats were crossed, and in the F_2 generation there was extreme variability. Plants of the F_2 generation were then transplanted to alpine and lowland stations, and the results were studied over a period of time. In time, some individuals were eliminated at various transplant sites as the environment selected out those that were not genetically adapted to it.

Similar adaptation has been demonstrated for a kelp species, *Laminaria groenlandica* (Druehl 1967). This species is restricted to a narrow range of temperature and salinity and occurs in either heavy or moderate surf conditions. Two morphological forms can be distinguished: a long-stipe form, 10–30 cm long with a wedge-shaped base to the blade (where it is attached to the stipe); and a short-stipe form, usually less than 10 cm with a flat or heart-shaped blade base. The long-stipe forms occur in areas exposed to heavy surf with cool (6–13 °C), high salinity (28–32 °/oo) waters having little seasonal variation. This habitat occurs on the outer coast of the west coast of North America (Washington and British Columbia). In contrast, the short-stipe form occurs in areas having moderate seasonal variation of temperature (6–17 °C) and salinity (24–32°/oo). Such areas are not subject to heavy surf and occur in the protected waters of Juan de Fuca Strait and Puget Sound (Washington) and Georgia Strait (British Columbia). Transplant studies of both forms in quiet waters show growth only in winter when the temperature is low (about 6–8 °C) and salinity high (26°/oo). In spring, an increase in temperature and a drop in salinity due to freshwater runoff cause death of the plants.

Other selection experiments have been performed on bacteria. Such organisms are conven-

Table 1-1 Various Mechanisms That May Prevent Hybridization between Different Plants

1. Mechanisms acting before fertilization
   A. Ecological—plants adapted to different habitats
   B. Seasonal—plants flower at different times of year
   C. Mechanical—pollination prevented by structural differences in the flowers of different species
   D. Gametic—gametes of different plants incompatible
   E. Physiological—inability of pollen to germinate on style

2. Mechanisms acting after fertilization
   A. Hybrid inviability or weakness
   B. Hybrid sterility
   C. F_2 breakdown—F_1's viable and fertile but F_2's weak or sterile

ient for studies on selection because many individuals can be grown in a small area, and there is very rapid generation turnover. In these experiments, bacteria were subjected to a medium with high concentration of antibiotics. Such an environment would select for individuals tolerant to antibiotics and select against those intolerant to antibiotics. The result was production of strains of bacteria more resistant to antibiotics than normal strains. The same phenomenon is seen in many disease-causing microorganisms in which strains have evolved that are resistant to antibiotics. In insects, strains have also evolved that are resistant to pesticides, such as DDT. Genetic studies of both bacteria and insects demonstrate that resistance to antibiotics and pesticides is inherited.

Populations of one species face different selection pressures in different environments, which result in unlike changes in their genetic makeup. This leads to an increase in the number of genetic differences between populations. With continuing selection, genetic dissimilarity between populations increases. In early stages of the development of differences between populations, the populations will not be distinct, since intermediates between all types will exist. These intermediates may reflect intermediate environments, crossing between different populations, or the presence of some genetic similarities in the populations. Eventually, populations become morphologically and physiologically distinct, and intermediates between the populations disappear. Populations such as these are considered to be separate species. As well, different species are usually reproductively isolated (individuals of the different species cannot cross and produce viable offspring). In most instances, reproductive isolation seems to be the result of an accumulation of genetic differences in physically isolated populations. There are many different kinds of isolating mechanisms. A brief outline of them is presented in Table 1-1.

Evolution is a gradual and continuing process (but see Polyploidy, below). Therefore, it is possible to find populations of plants that have not diverged sufficiently to be called separate species, yet have accumulated enough genetic differences so that it is difficult to consider them as one species. Examples of this occur in the flowering plants *Stipa* (needle grass) and *Clarkia* (clarkia), and in the green algae *Chara* (stonewort), and the microscopic *Gonium*. In *Stipa*, there are species that are morphologically and physiologically somewhat distinct yet interfertile and, in some instances, connected by series of intermediates. In *Clarkia*, there are species that are similar morphologically and physiologically yet completely intersterile. In both *Gonium* and *Chara*, there are species that are reproductively isolated in nature but can produce some viable zygotes when grown together under artificial conditions. Problematic species such as these are the focal point for many taxonomic studies; they also make it absolutely impossible to give one precise species definition that will fit all plants. The definition of a species will vary from plant group to plant group but will usually contain the following points: (1) species are morphologically, ecologically, and physiologically distinct; and (2) they usually do not hybridize in nature.

Polyploids

The gradual nature of evolution is a constant background, and against that background there can be periodic episodes of rapid evolution. **Polyploidy** is such an episode. Polyploidy, or chromosome doubling, is a means of evolution that is common in plants. Unlike gradual evolution, as discussed in the preceding section, evolution through polyploidy is abrupt and can result in almost instantaneous evolution of new species. There are different types of polyploids: two basic types, determined by parentage of the polyploid, and some intermediate types.

Allopolyploidy is one basic type of polyploidy in which two different processes are involved. The first step is the formation of a sterile hybrid; the second is doubling of chromosomes in that hybrid. Sterile hybrids are often a result of crossing between different species of plants, and hybrid sterility is due to the inability of the chromosomes to pair during meiosis (i.e., homologous chromosomes are lacking). The lack of chromosome pairing during meiosis prevents normal meiosis and results in **inviable gametes.** However, in sterile hybrids it is possible that unreduced gametes will form. Should such gametes fuse, they produce a zygote in which chromosomes are doubled as compared with the parent. Should this zygote develop into a plant, each chromosome will have an identical partner, and homologous chromosomes will exist. Such a plant would be a fertile offspring from a sterile hybrid. It is also possible for somatic chromosome doubling to occur in a sterile hybrid, which would make it fertile. Either way, the newly fertile plant is capable of reproducing its "type," and a new species may be created.

Autopolyploidy, the other basic type of polyploidy, is a result of the chromosomes of one fertile plant doubling to give a plant with twice the original chromosome number. This type of polyploidy tends to produce less fertile individuals than allopolyploidy, since each chromosome will have three (instead of one) homologous partners. During meiosis, more than two chromosomes can pair, so that instead of homologous pairs on a metaphase plate there can be groups of three to four chromosomes. Meiosis under such conditions often results in at least some genetic imbalance in the gametes—a cause of gamete inviability.

The intermediates between auto- and allopolyploids are a result of chromosome doubling in hybrid plants that have some pairing of homologous chromosomes but not enough to be completely fertile. There are two ways in which such hybrids can be created. One is the crossing between two slightly different individuals of the same species, or between two very closely related species (A_1A_1AA, Fig. 1-2). Polyploids formed from such hybrids are called *segmental allopolyploids*. The other means whereby intermedi-

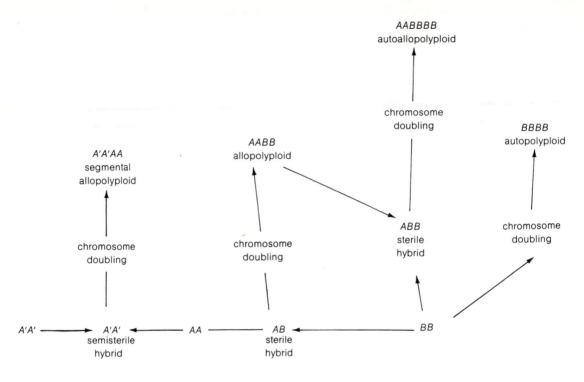

Figure 1-2 Kinds of polyploids and their origins

ate types of polyploids arise is by crossing between an allopolyploid species and one of its parental species, followed by chromosome doubling. Interrelationships between the different kinds of polyploid species are shown in Figure 1-2.

Hybridization and subsequent formation of polyploids is a very common phenomenon in plants. In the green alga *Nitella*, there are chromosome numbers of $n = 6, 9, 12, 18,$ and 24, indicating evolution through polyploidy. One species of *Nitella*, *N. furcata*, with $n = 12$, has been interpreted as being an autopolyploid. Polyploid series have also been demonstrated for *Chara*, with numbers of $n = 14, 28,$ and 42, and for *Tolypella*, with chromosome numbers of $n = 8, 11,$ and 33 reported. In the grass genus *Bromus* (brome grass), there has been extensive polyploidy. Figure 1-2 is based on a study of *Bromus* by Stebbins. As a result of hybridization and polyploid formation, complex aggregations of species can develop (Fig. 1-3).

Our concern here has been a brief summary of the process of biological evolution. Many examples are known from vascular plants, especially in seed plants. More detailed information for many organisms is present in the works of G. L. Stebbins (1950, 1966) and T. Dobzhansky (1951).

Recent and current studies are elucidating similar processes in nonvascular plants. Evolution of species in bryophytes probably results from phenomena similar to those described; however, as very little research has been done with these plants, there is a tremendous potential for investigation. Many fungi, algae, and bacteria do not undergo sexual reproduction as a regular feature. Thus, although the general mode of evolution (variation acted upon by natural selection) is similar, details of how this occurs may be different. These are discussed in following chapters.

To this point, we have considered the process of biological evolution; we now discuss the results of evolution, natural order.

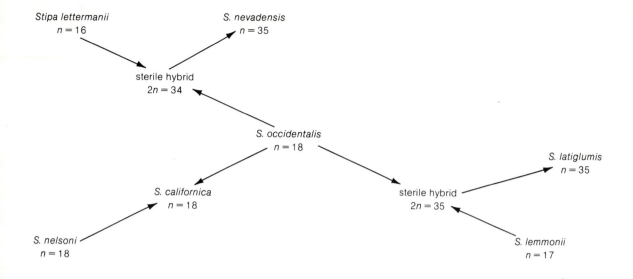

Figure 1-3 Formation of species, sterile hybrids, and polyploid species in *Stipa* (needle grass)

Natural Order

The biological world is not a random aggregation of individuals but consists of groups of individuals that are distinct from each other, are readily recognizable, and can be ordered into a hierarchy (see p. 8). The existence of these groups of individuals is called natural order, and an understanding of this natural order is one of the primary goals of biology. In attempting to understand it, biologists erect classifications. Thus, this section begins with a discussion of classification.

General Comments on Classification

Before discussing classification, a number of terms should be defined to avoid any ambiguity.

Class forms the base of several words. Class itself denotes a group of objects that are related to each other in some manner. This relationship is usually one of similarity. **Classify,** a cognate of class, means to arrange objects into classes. It is also used to denote the procedure involved in identification of a plant. In this instance, a plant is being placed in a previously defined group rather than arranged with others into new groups. **Classification,** another cognate of class, is used in two different senses. One is the act of arranging objects into classes; the other is the end result of this act.

Although classification is commonly associated with scientific endeavors, libraries, or telephone companies, it is, in fact, one of the most general aspects of our lives. Classification is the basis for language. If we had a unique word for every individual object we perceive, language would degenerate into chaos. For example, the word *tree* applies to all objects that fit the concept of tree. If each tree in a forest had a different term, communication about that forest would be impossible.

Another term that must be understood in order to discuss classification is the general term **taxon** (pl., **taxa**), a group of related organisms. The basis for relationship can be anything: aquatic plants represent a taxon based on habitat; *Cannabis* (marihuana or hemp) represents a taxon based on many chemical, morphological, and cytological similarities.

Biological Classification

A scientist concerned with classification is called a *taxonomist* or *systematist*, and the actual practice of classifying is usually called *taxonomy* or *systematics*.

Why Classify Organisms?   The primary reason for classifying is to understand natural order. However, a classification produced as an attempt to understand natural order also has many other uses. One very valuable use of classification is that names are attached to groups of plants that are recognized. These names become a means of communication about plants and about meaningful facts and theories concerning them. Thus, one can speak of "plastids of *Porphyra perforata*," a phrase which has far greater precision than the phrase "plastids of an aquatic plant." Also, when names are attached to objects of research, they add the element of repeatability. To duplicate a piece of research, the same kind of organism must be studied.

What Is Classification?   In a general sense, a classification is an **information storage and retrieval system.** The storage of information occurs when an individual is placed in a certain class; information retrieval occurs when a biologist wants an individual with certain attributes and goes to a class that is based on the attributes desired. Because of the relationship between information and classification, there are different kinds of classification depending on the amount of information they contain. A classification containing little information is very different from one containing much. For example, a classification based on whether or not plants are aquatic is one of very low information content. All that is known about individuals in the class "aquatic plants," is that they grow in water. The plants in such a class would be those that grow in marine, fresh, and brackish water, and would include representatives from most plant groups. On the other hand, a classification based on many features is one of higher information content. Much is known about the individuals in such a class. Take, for example, the angiosperm family Solanaceae: many Solanaceae contain chemicals called *alkaloids*. If a biologist wants to find new plants with alkaloids, the Solanaceae would be an appropriate place to look. In studying natural order, systematists want to produce the classification with maximum information content. Such a classification is presented as a hierarchy—a classification in which closely related classes are grouped together in higher taxa that are mutually exclusive. A classification of this kind is presented in Figure 1-4. In taxonomy, this hierarchy is referred to as a **Linnaean hierarchy** after Carolus Linnaeus, founder of biological taxonomy.

There are some important points that must be made here regarding the nature of taxa. Taxa of any one category (e.g., species or genera) will not necessarily be of the same size, exhibit the same amount of variation, or show the same degree of distinctness from other taxa of the same category. The differences in numbers of individuals in taxa are readily obvious. For example, the number of individuals of the kelp *Macrocystis angustifolia* (bladder kelp), which occurs chiefly in southern Australia and South Africa, is obviously much lower than that of *Macrocystis pyrifera* (giant kelp), which occurs mainly along the Pacific Coast between lat. 35°N–23°N and lat. 14°S–55°S and is circumpolar in the Southern Hemisphere. The differences in absolute amount of variation in taxa are likewise apparent. *Macrocystis angustifolia* will express far more absolute variability than will a species of a filamentous alga. As a large kelp, it has far more cell types than a filamentous alga. The amount of relative variation also differs in taxa of the same rank. Certain species are notorious for their inherent variability. An excellent example is the green alga *Enteromorpha intestinalis*, which varies from broad tubular plants to narrow, very slender, elongate forms. Other species are not so variable, an example being the brown alga *Postelsia palmaeformis* (sea palm). The amount of relative variation can also differ in higher taxa. If one genus contains 200 species and another has only 2, the relative amount of variation in the genera will be different. This aspect of relative variation can be carried to fairly high levels. Witness the gymnosperm order Ginkgoales, with the single species, *Ginkgo biloba*.

Kinds of Biological Classification   Generally speaking, there are three different kinds of bio-

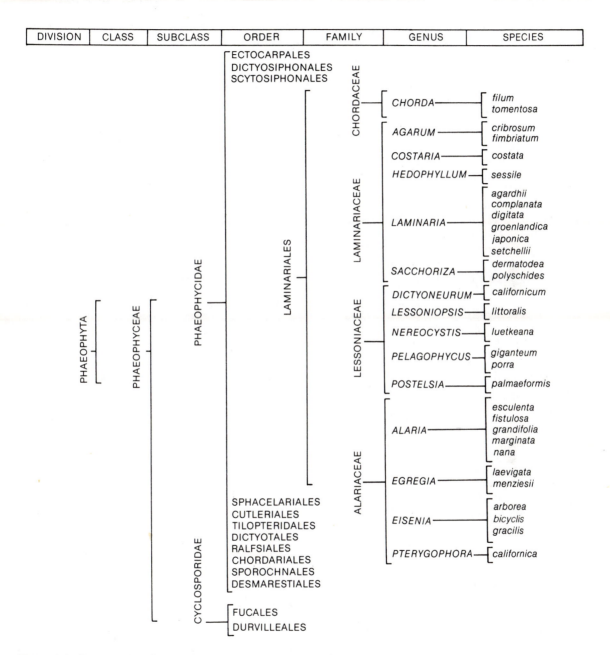

DIVISION	CLASS	SUBCLASS	ORDER	FAMILY	GENUS	SPECIES

ECTOCARPALES
DICTYOSIPHONALES
SCYTOSIPHONALES

CHORDACEAE
CHORDA — *filum* / *tomentosa*

LAMINARIACEAE
AGARUM — *cribrosum* / *fimbriatum*
COSTARIA — *costata*
HEDOPHYLLUM — *sessile*
LAMINARIA — *agardhii* / *complanata* / *digitata* / *groenlandica* / *japonica* / *setchellii*
SACCHORIZA — *dermatodea* / *polyschides*

LESSONIACEAE
DICTYONEURUM — *californicum*
LESSONIOPSIS — *littoralis*
NEREOCYSTIS — *luetkeana*
PELAGOPHYCUS — *giganteum* / *porra*
POSTELSIA — *palmaeformis*

ALARIACEAE
ALARIA — *esculenta* / *fistulosa* / *grandifolia* / *marginata* / *nana*
EGREGIA — *laevigata* / *menziesii*
EISENIA — *arborea* / *bicyclis* / *gracilis*
PTERYGOPHORA — *californica*

LAMINARIALES

PHAEOPHYTA

PHAEOPHYCEAE

PHAEOPHYCIDAE

SPHACELARIALES
CUTLERIALES
TILOPTERIDALES
DICTYOTALES
RALFSIALES
CHORDARIALES
SPOROCHNALES
DESMARESTIALES

CYCLOSPORIDAE

FUCALES
DURVILLEALES

Figure 1-4 Representation of a classification of brown algae (Phaeophyta), inclusive of all orders, but including subordinal categories (families, genera, and species) only in one order, the Laminariales

logical classification: **special purpose, key,** and **general purpose.** These are considered below.

Special-Purpose Classification. Classifications based on similarities only in certain features of the organisms are known as **special-purpose classifications**; they are low information classifications. Examples of such classifications would be those based on the physiogomy or growth habit of the plant, on the soil in which the plants grow, on leaf type, or on time of flowering. There are thousands of special purpose classifications.

Key Classification. A classification that is used to identify objects is a **key classification.** Like the previous classification, it has low information content. This kind of classification, usually referred to as a **key,** aids in identification by erecting a series of mutually exclusive classes. In doing so, a key classification (or key) eliminates choices until the object in question is identified.

An example of a key classification is a key to some motile, green algae (see Fig. 7-7A):

1. Cells arranged as a slightly curved plate . . . . *Gonium*

1. Cells arranged in a sphere; size variable . . 2

    2. Cells tightly packed together, each assuming triangular shape . . . . *Pandorina*

    2. Cells some distance apart, each round in form 3

3. Colony large, over 500 μm; composed of many cells; usually a few very large cells present . . . . *Volvox*

3. Colony less than 500 μm (average 100–200 μm); composed of (32–) 64–128 (–256) cells; cells the same size throughout . . . . *Eudorina*

General-Purpose Classification. The kind of classification usually associated with biology is a **general-purpose classification.** Such a classification is (1) of interest to the greatest number of biologists; (2) concerned with names (**nomenclature**) and Linnaean hierarchy; (3) the classification that taxonomists try to produce; and (4) the classification with the highest information content. We refer to such a general-purpose classification as a **natural classification.**

A natural classification is, at least theoretically, based on information derived from all aspects or characteristics of the organism: its biochemistry, parasites, genetics, physiology, cytology, life history, anatomy, development, embryology, morphology, ecology, reproductive features, and distribution. However, it must be pointed out that a natural classification is rarely, if ever, based on data derived from all these fields of study. The primary reason for this is that there is simply not enough time to gather all the information available.

Because all the available information is not actually used in erecting a classification, there is an experimental aspect that is an integral part of classifications. A taxon (group of related plants) is an hypothesis that is based on the evidence used. This hypothesis (taxon) is tested by gathering new information. For example, some taxa (hypotheses) have been erected in the Poaceae (grass family) on the basis of cell type in the leaf epidermis. Data derived from other fields of study, such as leaf anatomy, physiology, root anatomy, chromosome number, and floral morphology have also indicated the existence of the same taxa.

The Making of a Natural Classification

What we are concerned with here are the mechanics of producing a natural classification. In discussing the erection of a classification, we will consider two different methodologies: intuitive and numerically assisted. A systematist who erects a classification using an intuitive methodology will study museum specimens, specimens growing in nature, and the appropriate literature. Characteristics may or may not be recorded. Then, relying on the impression gained from the specimens studied, a classification will be erected. This method is that most commonly used in systematic studies. Although intuitive classification may appear to be highly subjective and perhaps even suspect as a scientific methodology, it has produced remarkably stable classifications. Different systematists who use this method on a common group of plants often arrive at very similar classifications. We suspect the efficacy of this method is a demonstration of the resolving power of the human mind.

When using intuitive methodology, a systematist often follows a certain approach. Some try to erect a classification based on overall similarities between individuals in a taxon. Others erect a classification that stresses certain types of similarities—shared derived traits. Individuals in a taxon share the same derived traits. A derived trait is a change in a trait from the condition seen in the closest relative to the group being studied. The two approaches presented are often identified by name. Erecting a classification based on overall similarities is called **phenetics;** erecting a classification based on derived traits is called **phylogenetic systematics** or **cladistics.** Despite the different approaches, phenetics and cladistics can give identical classifications.

The second methodology used in erecting a classification, numerical assistance, is the mathematical formalization of an intuitive classification; it involves checking and modifying the position in each taxon by using a formula. This often leads to a more detailed understanding of relationships.

An initial step in mathematical formalization of a classification is to measure carefully and record the characteristics of the organisms under study. The data derived from the organisms is called a data matrix, wherein the state of all characteristics has been recorded for all organisms. The *data matrix* is subjected to mathematical analysis. In phenetic classifications, the mathematical analysis used is called *cluster analysis*—a numerical tool designed to demonstrate the existence of groups of similar individuals amongst the organisms in the data matrix. The philosophy and mathematical methods involved in erecting a phenetic classification have been referred to as **numerical taxonomy.** In phylogenetic systematics, numerical methods, referred to here as *cladistic analysis,* have been developed by Farris (see Wiley 1981) to identify shared derived traits. The numerical methods associated with cladistic analysis are not as complicated as cluster analysis. Thus, computer assistance, although very handy, is not as necessary for cladistics as it is for cluster analysis.

Because of the mathematical precision associated with formalization of classification, it has been possible to compare phenetic and cladistic classifications. According to these comparisons, cladistic classifications seem to be superior.

Farris and Brooks have demonstrated that for any data matrix, cladistic analysis results in classifications with information content greater than or equal to those resulting from phenetic analysis. As well, the proponents of cladistics have demonstrated the compatibility between cladistics and the philosophy of science (see Wiley 1981). For these reasons, the remainder of this discussion is on concepts taken from cladistics.

Figure 1-5 is a cladogram, which shows the results of cladistic analysis. Using this figure, we will briefly trace certain concepts of significance in classification.

Monophyly, Polyphyly, and Paraphyly The terms *monophyly, polyphyly,* and *paraphyly* are adjectives applied to taxa in a classification. A **monophyletic taxon** is one in which all members have the same derived traits. For example, in Figure 1-5, a taxon consisting of *H, I,* and *J* is monophyletic; all share the derived trait *11. F* and *G* likewise make up a monophyletic taxon having the unique derived trait *8.* Other examples of monophyletic taxa in Figure 1-5 are *FGHIJ, BC, ABC, EFGHIJ.*

A **polyphyletic taxon** is one that includes individuals that do not share the same derived traits. For example, in Figure 1-5, a taxon consisting of *BC* and *D* would be polyphyletic. The derived traits for *BC* are *24;* for *D* they are *1, 2,* and *3.*

A **paraphyletic taxon** is one that does not include all individuals that share the same derived traits. For example, a taxon consisting only of *EF* and *G* is paraphyletic in that it does not include all individuals that have the derived trait *5.*

It is the goal of systematics to produce classifications that do not have paraphyletic and polyphyletic taxa, since such classifications contain less information than those with only monophyletic taxa. As a corollary of the lower information content in paraphyletic or polyphyletic taxa, knowledge of one member of that taxon does not necessarily mean there is knowledge about other members of the same taxon.

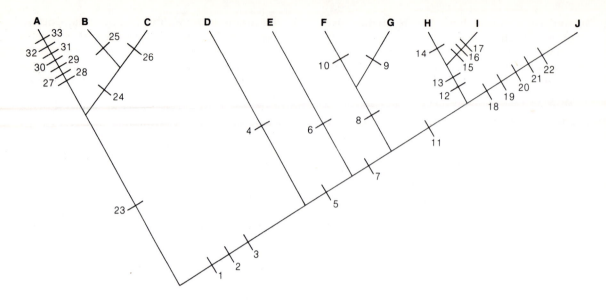

Figure 1-5 Cladogram of ten taxa. The ten taxa classified are called terminal taxa. Each vertical bar on the stem of the cladogram represents shared derived traits.

There are other names applied to monophyletic, paraphyletic, and polyphyletic taxa. Monophyletic taxa are sometimes called **natural;** polyphyletic and paraphyletic taxa are called **unnatural** or **artificial.**

Homology and Analogy The terms **homology** and **analogy** refer to types of similarities between traits. Homologous similarities are those seen in shared derived traits. Analogous similarities are not shared derived traits; they are similarities between taxa that are not classed together. It is possible that a systematist may erect taxa based on analogous traits. These are a type of artificial taxon and, in the past, were rather common in plants. The old group Thallophyta, which combined together all fungi, bacteria and algae, was one such taxon; the common analagous characteristic was a very simple plant body.

Primitive versus Advanced **Primitive** refers to a trait like that seen in the ancestor of a taxon;

advanced refers to a trait different from that in the ancestor of a taxon. A derived trait would be advanced. The terms *primitive* and *advanced* are also used by some to refer to taxa; we prefer not to. It is confusing to see the same words in different contexts. As well, the different rate of evolution that different characteristics undergo means that any one extant plant can have both primitive and advanced traits. When referring to taxa, those that have fewer unique derived traits we call generalized; those having more unique derived traits we call specialized.

Sympleisiomorphy, Synapomorphy, and Homoplasy Cladists use special terms to refer to certain types of similarities. These terms have a distinct advantage in that they are less ambiguous than other terms.

Sympleisiomorphy refers to generalized traits in a group. **Synapomorphy** is a shared derived trait shared by all members of a taxon. **Homoplasy** is a convergent characteristic, a similarity that does not identify a natural or monophyletic taxon. Synapomorphies are those traits that confer the greatest information content on a classification; homoplasies give interesting insight into the biology of the organisms that share them. The reason why homoplasies do so is that they often represent adaptations to particular environments.

Anagenesis and Cladogenesis There are two more terms that should be understood: *anagenesis* and *cladogenesis*. **Anagenesis** refers to evolutionary change within a lineage. **Cladogenesis** refers to evolutionary splitting of a lineage to give two taxa where one existed before.

Interpretation of a Classification

We now discuss briefly the interpretation that can be put on taxa in a classification. The most common interpretation is that each monophyletic taxon in a classification is the result of evolution. As such, it has a unique history. Some biologists (see Wiley 1981) would argue that the existence of natural order that can be arranged in a hierarchy is sufficient proof of evolution.

Features Used in Classification

Characteristics are the source of information used in erecting a classification. There are certain criteria that characteristics should meet. First, they should show variation within the group being studied. Second, the characteristics should not show greater variation within an individual than they do between individuals. Third, they should be readily measurable. The use of certain features when erecting a classification at one level does not mean that these same features will be useful when classifying organisms at a higher taxonomic level. The presence of chlorophyll, wood, tracheids, and phloem may well be worthwhile data to use when classifying pines at the level of class or division, but they are of little value when classifying pines at the species level.

In choosing characteristics, taxonomists depend on their knowledge of the organisms to know *what* features should be analyzed. Furthermore, it is an axiom of taxonomy that the quality of the characteristics is the sole determinant of the quality of the resulting classification. Neither the most brilliant intellect nor the most sophisticated computer program can do anything to compensate for poor information. Thus, the better taxonomists know the organisms, and the more they know about organisms in general, the better taxonomists they will be.

What we have presented here in summary form is what seems to us the means whereby taxonomists work and think. We hasten to point out that such theoretical aspects of taxonomy have been the point of much debate, both in the literature and in various scientific meetings. Anyone who desires to follow such arguments is referred to the journal *Systematic Zoology*, or to the recent book by Wiley (1981).

References

Items marked with an asterisk (*) in the following list are of broad interest.

Anderson, E. 1949. *Introgressive hybridization*. John Wiley & Sons, New York.

*Clausen, J. 1951. *Stages in the evolution of plant species.* Cornell Univ. Press, Ithaca, N.Y.

Clausen, J., and Heisey, W. M. 1958. Experimental studies on the nature of species. IV. *Genetic structure of ecological races*. Carnegie Institution of Washington Publication 615.

Clausen, J.; Keck, D. D.; and Heisey, W. M. 1940. Experimental studies on the nature of species. I. *The effect of varied environment on western North American plants*. Carnegie Institution of Washington Publication 520.

———.1945. Experimental studies on the nature of species. II. *Plant evolution through amphiploidy and autoploidy with examples from the Madiinae*. Carnegie Institution of Washington Publication 564.

———.1948. Experimental studies on the nature of plant species. III. *Environmental responses of climatic races of Achillea*. Carnegie Institution of Washington Publication 581.

Darwin, C. 1872. *The origin of species*. 6th ed. London.

*Davis, P. H., and Heywood, V. H. 1963. *Principles of*

angiosperm taxonomy. D. Van Nostrand Co., Princeton, N. J.

*Dobzhansky, T. 1951. *Genetics and the origin of species.* Columbia Univ. Press, New York. (First published in 1937).

Dobzhansky, T.; Ayala, F. J.; Stebbins, G. L.; and Valentine, J. W. 1977. *Evolution.* W. H. Freeman, San Francisco.

Druehl, L. D. 1967. Distribution of two species of *Laminaria* as related to some environmental factors. *J. Phycology* 3:103–8.

*Hennig, W. 1966. *Phylogenetic systematics.* Univ. of Illinois Press, Urbana, Ill.

Heslop-Harrison, J. 1953. *New concepts in flowering plant taxonomy.* Wm. Heinemann, London.

Johnson, L. A. S. 1968. Rainbow's end: The quest for an optimal taxonomy. *Proc. Linn. Soc. New South Wales* 93:8–45.

Lewis, H. 1953. The mechanism of evolution in the genus *Clarkia. Evolution* 7:1–20.

Mayr, E. 1963. *Animal species and evolution.* Harvard Univ. Press, Belknap Press, Cambridge, Mass.

*Simpson, G. G. 1961. *Principles of animal taxonomy.* Columbia Univ. Press, New York.

Sneath, P. H. A., and Sokal, R. R. 1973. *Principles and practices of numerical taxonomy.* W. H. Freeman, San Francisco.

*Stebbins, G. L. 1950. *Variation and evolution in plants.* Columbia Univ. Press, New York.

———. 1966. *Processes of organic evolution.* Prentice-Hall, Englewood Cliffs, N.J.

*Wiley, F. O. 1981. *Phylogenetics: The theory and practice of phylogenetic systematics.* John Wiley & Sons, New York.

In this text, plants are separated into two groups on the basis of ultrastructural cellular detail, as clearly shown by the electron microscope (Fig. 2-1). The first, the **prokaryotes** (*pro* = before; *kary* = nut, referring to nucleus), lacks typical membrane-bound organelles, such as a nucleus, mitochondria, chloroplasts, and vacuoles (Fig. 2-1A). The prokaryotic group consists of only two divisions, both containing photosynthetic and nonphotosynthetic representatives. The second is a very large group of plants that has organelles separated from the cytoplasm by a unit membrane (Fig. 2-1B). These are the **eukaryotes** (*eu* = true) (Fig. 2-1C), which have photosynthetic and nonphotosynthetic representatives. Most living organisms (plants and animals) are eukaryotic, as are most nonvascular plants. Many biologists now classify living organisms into five kingdoms. Following this system, the prokaryotes would be in the Monera and the eukaryotic plants would include some of the Protista, all the Plantae, and the Fungi. In this five-kingdom system, only the kingdom Animalia is excluded from the concept of nonvascular plants.

The term *nonvascular plants* is used in this text in the broadest sense to include a diverse assemblage of organisms in which regular water- and food-conducting systems are generally lacking. Many are composed of cells that are almost identical throughout the plant body, or **thallus** (*thall* = young twig or shoot). Often there is little differentiation in these morphologically simple plants. Thus, every cell can potentially function either as a vegetative or reproductive cell. Because of this basic morphological simplicity, biochemical and physiological characteristics are of fundamental importance in classifying many of these plants.

Nonvascular plants can be separated into several basic groups: bacteria, algae, fungi, slime molds, and bryophytes (Table 2-1). However, this simple separation does not acknowledge some fundamental differences within these groups. Bacteria are morphologically simple and are distinguished from other nonvascular plants by details of cell structure (prokaryotic) and physiology. The term *algae* has been used for centuries to include those photosynthetic organisms with relatively simple morphology and unprotected reproductive structures. Because of information provided by the electron microscope, prokaryotic

2

NONVASCULAR PLANTS

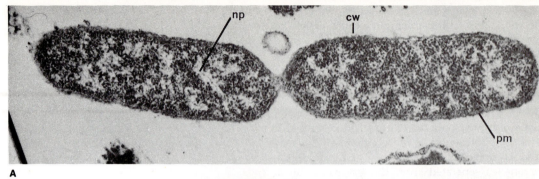

A

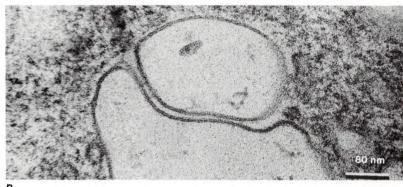

B

Figure 2-1 Comparison of ultrastructure of prokaryotic and eukaryotic cells. **A,** prokaryotic cell showing only nucleoplasm (*np*) and plasma membrane (*pm*) with cell wall (*cw*); note lack of distinct organelles, ×90,000. **B,** eukaryotic membranes (from two adjacent cells) showing nature of membrane, 10 mm = 80 nm, × 150,000. **C,** eukaryotic cells showing nucleus (*n*), chloroplasts (*c*), mitochondria (*m*), cell wall (*cw*), × 15,000. (**B,** courtesy of L. Goff; **C,** courtesy of T. Bisalputra.)

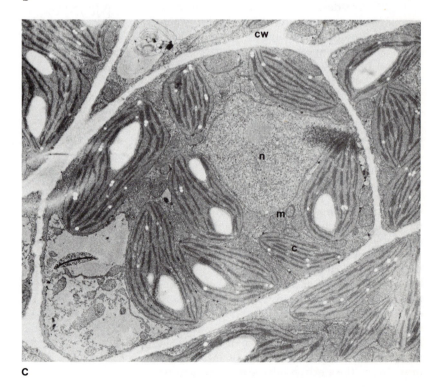

C

Table 2-1 General Groups of Nonvascular Plants

Common Name	Motile Cell Assimilative	Motile Cell Reproductive	Vegetative Form	Habitat	Energy Source	Average Size (range)
Prokaryotes						
Bacteria	gliding (flagella)*	none	unicell, multicell	ubiquitous	autotrophic, heterotrophic	1–5 μm (0.5–500 μm)
Bluegreen algae	(gliding)	none	unicell, multicell	aquatic, terrestrial	autotrophic	5–100 μm (1–500 μm)
Eukaryotes						
Fungi	none	(flagella)	unicell, multicell	aquatic, terrestrial	heterotrophic	10 μm–1 m (1 μm–∞)
Slime molds	none	(flagella)	multicell	terrestrial	heterotrophic, ingest particles	5 μm–10 mm
Algae	flagella	flagella	unicell, multicell	aquatic (terrestrial)	autotrophic	5 μm–1 m (1 μm–75 m)
Bryophytes	none	flagella (male gamete only)	multicell	terrestrial (aquatic)	autotrophic	10–30 mm (0.1–60.0 mm)

*Parentheses indicate feature occurs in some representatives, but not the majority.

bluegreen algae have now been separated from other algae which are eukaryotic. **Autotrophic** (*auto* = self; *troph* = nourish) algae, with their photosynthetic pigments, differ physiologically from the **heterotrophic** (*heter* = different) fungi and slime molds. This difference in mode of nutrition is of fundamental importance in characterizing groups of nonvascular plants. In addition, characteristics such as reproduction, ultrastructure, and biochemistry separate fungi and slime molds from algae. In fact, with more knowledge of these diverse groups, it is becoming increasingly apparent that fungi and slime molds should not be considered as plants, but as a separate kingdom of organisms distinct from plants and animals. Bryophytes are easily distinguished from algae and fungi by several morphological and anatomical features. They are structurally and reproductively more complex than either.

Habitat

Nonvascular plants are ubiquitous. Although most occur in moist situations, they are not restricted to an aquatic habitat (see Table 2-1). However, liquid water, often in very minute amounts, is required for sexual reproduction for most algae and for bryophytes. Fungi, with their different type of nutrition, also have diverse sexual mechanisms; liquid water is not an absolute requirement for many species. The most widespread and adaptable organisms are the bacteria. Different species are able to live in aerobic or anaerobic conditions; to utilize a vast number of different energy sources, including light and both inorganic and organic compounds; and to endure extremes of both temperature and humidity. In contrast, algae are limited generally to aerobic environments and utilize light as their energy source. Eukaryotic algae grow in a limited temperature range as compared with some bluegreen algae (prokaryotic), which have been collected from temperatures over 70 °C. Fungi are also adaptable to temperature; some grow at extremes as high as 60 °C or as low as below 0 °C. Fungi are limited to organic substrates for their source of energy, but otherwise they are almost as adaptable a group as bacteria. Bryophytes, which are photosynthetic, are more restricted in

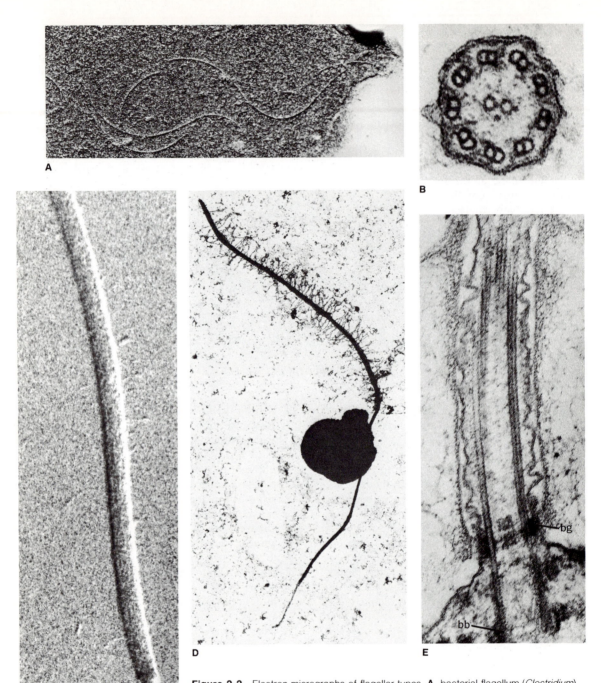

Figure 2-2 Electron micrographs of flagellar types. **A,** bacterial flagellum (*Clostridium*), shadowed, ×17,600. **B,** transverse section of eukaryotic flagellum (*Volvox*), showing nine peripheral pairs of microtubules surrounding two central microtubules, ×90,000. **C,** whip-lash flagellum (*Chlamydomonas*), shadowed, ×17,600. **D,** tinsel (forward *t*) and whiplash (posterior *w*) flagella (*Chorda* zoospore), ×4100. **E,** longitudinal section (*Volvox*) showing continuous membrane of flagellum and protoplast, ×90,000. (**A,** courtesy of L. Veto; **B, E,** courtesy of T. Bisalputra; **C,** courtesy of J. F. Gerrath; **D,** courtesy of R. Toth.)

their habitat, probably as a result of their more complex morphology and type of reproduction.

Cell Structure

As noted, prokaryotes generally lack internal membranes that separate nuclear material, respiratory enzymes, and photosynthetic apparatus (when present) from the rest of the protoplast (Fig. 2-1A). Nevertheless, the entire protoplast is circumscribed by the typical lipoprotein plasma membrane. The eukaryotic cell is easily recognized by the presence of discrete membrane-enclosed organelles (Fig. 2-1C).

Some of the nonvascular plants are motile in the vegetative, or assimilative, stage (in particular, some bacteria, and algae; see Table 2-1). Most nonvascular plants have motile reproductive cells with one or more **flagella** (*flag* = whip) for propulsion. In prokaryotes, the flagellar structure (Fig. 2-2A) differs from that of eukaryotes, which is composed of **microtubules** bounded by the plasma membrane (Fig. 2-2B).

The nature of the outer cell covering (outside the plasma membrane) is almost as diverse as the organisms to be considered. Some nonvascular plants lack a cell wall but have rigid scales or a firm outer proteinaceous membrane. Others have a wall, and still others have a wall for only a part of their life history. Polysaccharides, sometimes containing additional nitrogenous compounds, are the usual components of cell coverings. In addition to or in place of the organic compounds, there may be present such inorganic materials as iron, silica, manganese, magnesium, or calcium carbonate. The nature of the cell covering, like the flagellation, is specific for a given species or stage in the life history of a species. Discussion of the variety of these materials is presented later in this chapter.

Food reserves, or storage products, of nonvascular plants are also diverse. These reserves are carbohydrates, proteins, or fats and oils, and can vary depending upon environmental conditions. There is more information available concerning carbohydrate storage products than others. These polysaccharides are sometimes similar to those in cell coverings.

Reproduction

In many nonvascular plants reproduction is vegetative, or asexual. This involves simple **fission** (*fiss* = cleft; Fig. 2-3A), budding (Fig. 2-3B), fragmentation, development of specialized **deciduous** (*decid* = falling off) parts, or production of special reproductive structures termed **spores** (*spor* = seed; Fig. 2-4). In many nonvascular plants (excluding bryophytes), spores are produced by either mitosis (equational division) or meiosis (reduction division). In bryophytes and vascular plants, spores are produced by meiosis and can be termed meiospores. Spores, which may have a cell wall, are generally produced in **sporangia** (*spor* = seed; *ang* = box) that are usually unicellular (Fig. 2-4A–E) but may be multicellular (Fig. 2-4H). Spores that move by means of flagella are **planospores** (*plano* = wandering) or **zoospores** (*zoo* = animal; Fig. 2-4C, D). Nonmotile spores are **aplanospores** (*a* = without) or **sporangiospores** (Fig. 2-4A, B, E). In some organisms a heavy wall surrounds the protoplast, and such spores serve as resting stages (Fig. 2-4F, G).

Nonvascular plants reproduce extensively by asexual means, and for many this is the only type of reproduction. Sexual reproduction is unknown for some, whereas for others it is the rule. This is discussed further with the specific groups of nonvascular plants.

Classification

Nonvascular plants are classed in two subkingdoms, each with more than one division (or phylum) as shown in Table 2-2. The Eukaryota (as noted previously) is the larger subkingdom, and 12 divisions of nonvascular plants are recognized. There is some debate by botanists concerning some of these divisions which appear to be heterogeneous and should probably be further subdivided. The system here is selected for ease in understanding relationships among various groups. The emphasis is not on the absoluteness of each division but on an understanding of the diverse organisms that are known.

Table 2-2 Classification of Nonvascular Plants

Subkingdom	Division	Common Name
Prokaryota	Schizomycophyta	bacteria
	Cyanophyta	bluegreen algae
Eukaryota	Myxomycota	slime molds
	Chytridiomycota	chytrids
	Hyphochytridiomycota	hyphochytrids
	Oomycota	oomycetes
	Zygomycota	zygomycetes
	Eumycota	true fungi
	Chlorophyta	green algae
	Chrysophyta	chrysophytes
	Pyrrhophyta	dinoflagellates
	Cryptophyta	cryptophytes
	Euglenophyta	euglenoids
	Phaeophyta	brown algae
	Rhodophyta	red algae
	Bryophyta	bryophytes

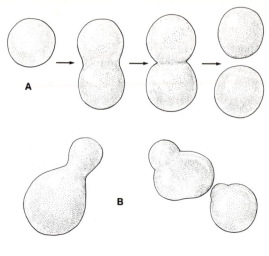

Figure 2-3 Asexual, or vegetative, reproduction in nonvascular plants. **A,** fission. **B,** budding.

Subkingdom Prokaryota (Prokaryote Divisions)

Schizomycophyta (bacteria) and Cyanophyta (bluegreen algae) are the main prokaryotic organisms. The cells are characterized by a lack of internal differentiation (Fig. 2-1A). The nuclear material present, deoxyribonucleic acid (DNA), is in the form of very small fibrils often located near the center of the cell. Respiratory enzymes are scattered throughout the cell, although they can be concentrated on the plasma membrane. Photosynthetic pigments are located on and within the individual photosynthetic lamellae, or **thylakoids** (*thylac* = sac, pouch), distributed peripherally in the cytoplasm. Some bacterial cells have flagella. The cell wall (**peptidoglycan**) of prokaryotes is unique in containing large amounts of nitrogen present as amino acids and amino sugars.

Reproduction is primarily asexual, but in some there is a type of sexual reproduction involving a unidirectional transfer of genetic material between cells. This results in genetic recombination but does not involve fusion of sexual partners.

Schizomycophyta are primarily nonphoto-synthetic, but there are a few photosynthetic forms. In contrast, Cyanophyta are primarily photosynthetic, with very few nonphotosynthetic representatives. Both bacteria and bluegreen algae have representatives that fix atmospheric nitrogen into cellular material. Some prokaryotes live at high temperatures (45–95 °C); others colonize bare rock or soil surfaces or live in conditions of extreme cold and dryness. Extant Cyanophyta and Schizomycophyta are considered descendants of the earliest forms of life, recognizable in rocks approximately 3 billion years old, with some reports from even older rocks.

Schizomycophyta (Bacteria) There are few morphological features used for identification within the bacteria (see Chapter 3). They are unicellular or colonial, ranging in size from 0.5 to 500 μm but with most 1–5 μm. Most reproduce vegetatively, but genetic recombination is known for some. Physiologically, bacteria are complex and diverse. They are important in nutrient cycles (nitrogen, sulfur, iron), decay, and disease production. Most bacteria are heterotrophic, requiring external organic compounds for both energy and carbon assimilation. However, some bacteria utilize either inorganic compounds or light as an energy source and carbon

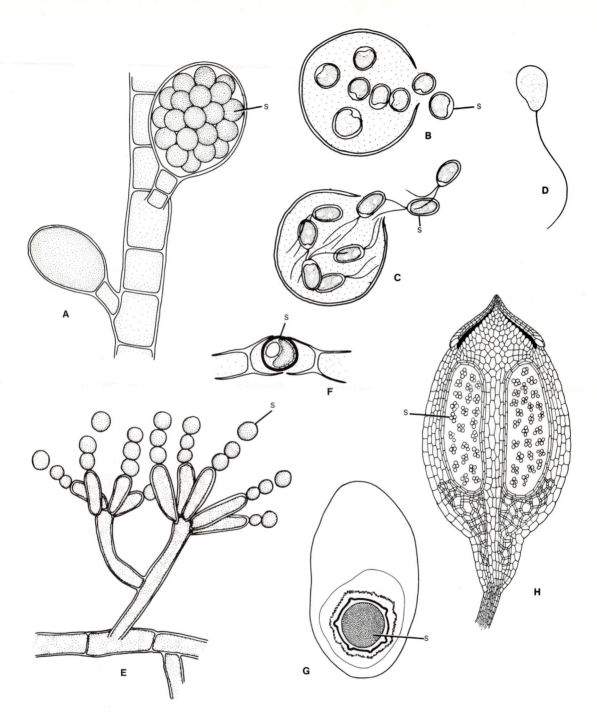

Figure 2-4 Variation in types of sporangia and spores (*s*) in nonvascular plants. **A–E,** unicellular sporangia; **A, B,** aplanosporangia with aplanospores; **C, D,** zoosporangium and zoospores (planospores); **E,** sporangiospores (aplanospores). **F, G,** resting spores; **F,** akinete; **G,** bacterial endospore; **H,** multicellular sporangium.

dioxide for a carbon source. Photosynthetic bacteria contain photosynthetic pigments similar to those in other plants but photosynthesize only in anaerobic conditions. Oxygen is neither released nor used in bacterial photosynthesis. Reserve foods are often stored as poly-β-hydroxybutyric acid.

Bacteria are classified according to morphology and physiology as outlined in Chapter 3. Most bacteria are included in the class Schizomycetes, which is divided into several orders based mainly on morphology. A second class includes rickettsias, viruses, and mycoplasma (pleuropneumonia) organisms, which generally require a living host.

Cyanophyta (Bluegreen Algae) Morphologically, bluegreen algae resemble bacteria and can be classified as bluegreen bacteria (Cyanobacteria). Many cells are larger than the "average" bacterial cell, and there are many more filamentous bluegreen representatives. All photosynthetic bluegreen algae contain the photosynthetic pigment chlorophyll *a*, which occurs in all photosynthetic plants but bacteria. Photosynthesis occurs in aerobic conditions with oxygen released. Reserve products are carbohydrates and proteins, with a glucose-containing carbohydrate (cyanophyte starch) being predominant (see Table 2-4, p. 28).

Classification of Cyanophyta by botanists is based on morphological features. Current research by microbiologists incorporates physiological and biochemical characteristics as well, since many morphological features are subject to environmental changes. There is only one class, Cyanophyceae, with several orders delimited by morphological and reproductive features (Chapter 3).

**Subkingdom Eukaryota
(Eukaryote Divisions)**

The remaining nonvascular plants contain cells with discrete organelles (Fig. 2-1C). The DNA is in chromosomes within a well-defined nuclear envelope, and division is regularly by mitosis. Some respiratory enzymes are within mitochondria and, when present, photosynthetic pigments and enzymes are in chloroplasts. Motile cells are propelled by flagella composed of nine pairs of microtubules surrounding one or two central microtubules, which are surrounded by a membrane continuous with the plasma membrane (Fig. 2-2B, E). If the flagellar membrane is smooth, the flagellum is referred to as a **whiplash flagellum** (Fig. 2-2C–E). However, if the membrane has tubular thin hairs along its length, it is a **tinsel flagellum** (Fig. 2-2D). In addition, some organisms have flagella bearing very delicate "hairs" and/or fine scales. As the form, location, and number of flagella on a cell are constant for a given species, they are used to characterize some nonvascular groups (see Fig. 2-5). Flagella will continue to beat if isolated from the cell when a chemical energy supply is available.

In addition to asexual reproduction, which commonly occurs in eukaryotic nonvascular plants, sexual reproduction is present (although not universal). In sexual reproduction there is diversity in morphology of **gametes** (*gam* = marriage; Fig. 2-6). Gametes are either identical in appearance (**isogamous;** *iso* = alike) or dissimilar (**anisogamous;** *an* = not). Both gametes can be flagellated, only one, or neither. If one gamete is large and nonmotile and the other markedly smaller (motile or not), sexual reproduction is **oogamous** (*oo* = egg). The larger gamete is regarded as female (egg) and the smaller as male (sperm). When gametes are almost identical in appearance, they are not referred to as male and female; they are termed *plus* and *minus* and are known as **mating types.**

Gametes are produced in structures, **gametangia** (*gam* = marriage; *ang* = box; Fig. 2-7), which in fungi and algae are typically unicellular (Fig. 2-7B, C). In bryophytes, however, gametangia are more specialized and are multicellular (Fig. 2-7A). In some nonvascular plants, gametes as distinct units are not present, but nuclei function as gametes. In some of these, nuclei that function as gametes move through a fertilization tube that develops after gametangial contact (Fig. 2-7C). Another modification of sexual reproduction occurs through fusion of whole protoplasts and is termed **conjugation** (*conjug* = joined together; Fig. 2-7D, E). In this instance, male and female gametes are usually not dis-

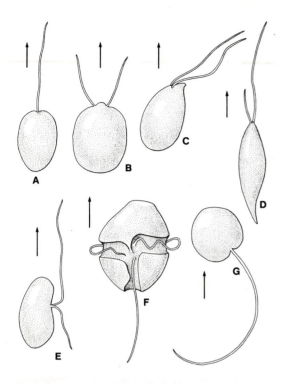

	ISOGAMY	ANISOGAMY	OOGAMY
A	♂ ♀	B ♂ ♀	C ♂ ♀
D	♂ ♀	E ♂ ♀	F ♂ ♀

Figure 2-6 Variation in types of sexual reproduction in nonvascular plants. **A–C,** motile gametes; **D–F,** nonmotile gametes. **A, D,** isogamy; **B, E,** anisogamy; **C, F,** oogamy.

Figure 2-5 Types of flagellar arrangement in nonvascular plants. **A–D,** anterior (apical or subapical) attachment. **E, F,** lateral attachment. **G,** posterior attachment. (*Arrows* indicate direction of movement.)

tinguished; however, if one protoplast moves, it is considered the male gamete.

Another feature is the variation in relative time and location of gamete fusion, known as **syngamy** (*syn* = with, together; *gam* = marriage), and meiosis (reduction division) in the life history (Figs. 2-8 and 2-9). The time of syngamy relative to meiosis determines the type of life history as well as whether the predominant phase is **haploid** (*hap* = single; *oid* = form; Fig. 2-8A) or **diploid** (*di* = double; Fig. 2-8B), or whether both phases are distinguishable (Fig. 2-9A). These types of life histories have been given different names by researchers, and there is no agreement (even among the authors of this text). If the haploid (*n*) phase is dominant (Fig. 2-8A), the zygote is the only diploid (*2n*) stage. This life history is restricted to some algae (pri-

marily freshwater) and some fungi. In a second type of life history, the plant is diploid (*2n*) and gametes are the only haploid (*n*) stage (Fig. 2-8B). This is restricted in the plant kingdom to only a few algae (mainly marine) and some fungi; however, it is typical of animals. The third life history in which both haploid and diploid plants are present is considered the "typical plant life history" (Fig. 2-9A). In these plants the haploid phase produces gametes and is termed the **gametophyte** (*phyte* = plant), whereas the plant producing meiospores is the **sporophyte.** The two plants can be identical in appearance (**isomorphic:** *iso* = alike; *morph* = form) or markedly different (**heteromorphic:** *heter* = different). This last type of life history is that found in bryophytes and all embryo-producing plants. However, as noted, it is only one of several life-history types occurring in algae and fungi. A fourth type of life history is restricted to some fungi and involves a delay in nuclear fusion after protoplast fusion (Fig. 2-9B).

Nonphotosynthetic Eukaryotes The nonphotosynthetic eukaryotic organisms include slime

Nonvascular Plants **23**

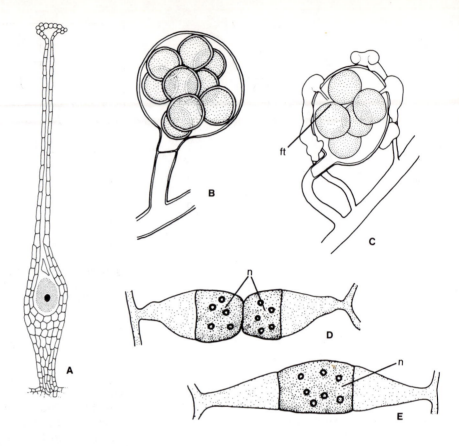

Figure 2-7 Variation in gametangia in nonvascular plants. **A,** multicellular; **B,** unicellular. **C,** fertilization tubes (*ft*). **D, E,** gametangial conjugation (*n*, nuclei).

molds and fungi. Some biologists consider these a single division, Mycota. However, differences in morphology, cell structure, and method of obtaining food separate these into several distinct groups (see Chapters 4–6). All slime molds are characterized by wall-less (naked) assimilative stages, and most are capable of feeding by ingesting particulate food. Fungi are primarily filamentous, although some are unicellular, and generally have a rigid cell wall during most of their life history. Fungi cannot ingest food; instead they secrete exoenzymes (*exo* = outside) capable of digesting materials. Subsequent breakdown products are taken into the cells. Food reserves of fungi are generally the polysaccharide glycogen, sugar-alcohols, and lipids. The walls can contain

chitin, which is similar to cellulose, the main component of cell walls.

Myxomycota (Slime Molds). There are two very distinct types of slime molds: the true slime molds and cellular slime molds (see Chapter 4). Cellular slime molds do not appear closely related to each other or to true slime molds. All Myxomycota have naked assimilative stages and reproduce by spores. The assimilative body is multinucleate in true slime molds and uninucleate in cellular slime molds.

Fungi. The group of organisms known as fungi is a large heterogeneous assemblage that presents

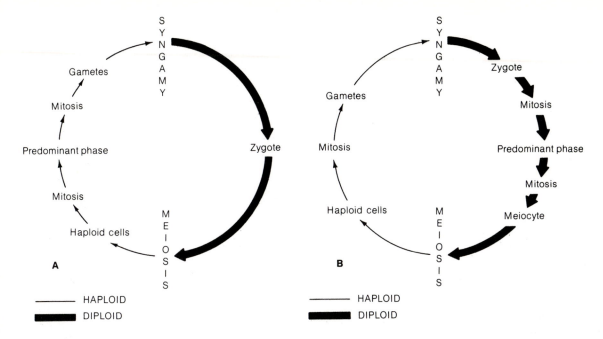

Figure 2-8 Types of alternation of life history or phases (generations) restricted to non-vascular plants. **A,** haploid stage predominant with meiosis in zygote (only 2*n* stage). **B,** diploid stage predominant with gametes produced by meiosis (only *n* stage); also typical of animals.

formidable problems in classification; no system is completely acceptable to all **mycologists** (*myco* = fungus; *ology* = study). In this text several divisions are recognized (Chapters 5 and 6). Some mycologists treat these divisions as separate classes in a single, large division. However, differences in cell wall composition, flagellated reproductive cells (when present), and morphological characteristics separate the divisions (see Table 2-3). Fungi range in size from discrete units of a few micrometers to masses of material of indeterminate size. Many fungi are microscopic but produce reproductive structures (sometimes termed *fruiting structures*) that are easily observed. Some fungi appear closely related to some algal groups; others are biochemically more similar to animals.

In many fungi sexual reproduction occurs regularly and is easily recognized; in others there is a modified life history. In these fungi with modified life histories, syngamy is divided into two distinct phases: **plasmogamy** (*plasmo* =

form, shape; *gam* = marriage), in which there is fusion of only cytoplasm; and **karyogamy** (*kary* = nucleus), in which there is nuclear fusion (Fig. 2-9B). Plasmogamy is followed by a period of time in which paired nuclei divide synchronously and cells are binucleate (*bi* = two). After plasmogamy, paired nuclei from two gametes divide mitotically to form binucleate cells that are either of limited duration or persist indefinitely, producing the **dikaryon** (*di* = two; *kary* = nucleus), or *n* + *n*, stage (Fig. 2-9B). Eventually many nuclear pairs fuse, with meiosis usually following immediately to restore the haploid (*n*) phase.

Included with fungi are lichens, which in the past have been treated as a separate division. The lichen association involves algal (**phyco-biont:** *phyco* = alga; *bio* = life) and fungal (**my-cobiont:** *myco* = fungus) partners. Lichen fungi differ little from other parasitic fungi, except that fungal filaments and algal cells associate to produce a single "plant." As the lichen reproductive

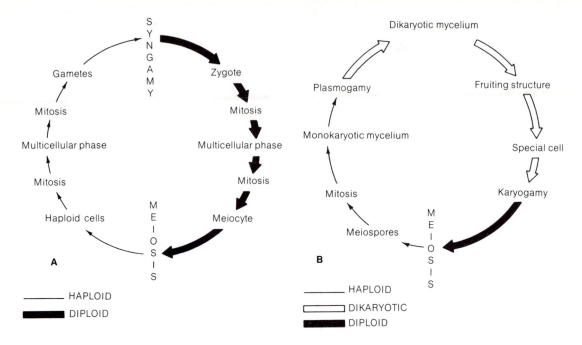

Figure 2-9 Types of alternation of life history or phases (generations). **A,** alternation of haploid and diploid phases with meiosis occuring in diploid plant (typical of plants). **B,** delayed nuclear fusion (dikaryon) occurring in many fungi.

structures are fungal in form, classification is essentially that of the fungi (see Chapter 6).

Photosynthetic Eukaryotes   All photosynthetic eukaryotes contain chlorophyll *a* in their chloroplasts and release oxygen as a by-product. Other pigments present absorb light energy that is passed to chlorophyll *a* for photosynthesis. Included in this group are eukaryotic algae as well as embryo-producing plants, which include the bryophytes.

Algae.   Until the twentieth century, organisms known as algae were grouped in one taxonomic category. With increasing knowledge resulting from studies of biochemistry, physiology, and morphology, it became apparent that such a grouping comprised a number of distinct units. At present algae are classed in several divisions (see Chapters 7–11). Often chlorophyll *a* is

masked by other pigments contained on or within the chloroplast thylakoids. In this text the term *chloroplast* is used for all chlorophyll-bearing organelles with membrane-limited, lamellar structure.*

The other pigments in the algae that are important in photosynthesis and are used for classification of algae are other chlorophylls (chlorophyll *b*, c_1, c_2, *d*), carotenoids (*carot* = carrot), which include carotenes and xanthophylls (*xantho* = yellow; *phyll* = leaf), and the phycobiliproteins (*phyco* = alga; *bili* = bile) phyco-

---

*Some biologists use the term chromatophore for nongreen, photosynthetic organelles; however, in current usage by biologists in general, the term chromatophore applies to both nonphotosynthetic and nonlamellar structures. Strictly speaking, photosynthetic organelles should be named by their color. As this is not practical due to the variety of colors resulting from different pigments, this text uses chloroplast for all phytosynthetic plastids regardless of color.

Table 2-3 Major Distinguishing Characteristics of the Nonphotosynthetic Nonvascular Eukaryotes (Slime Molds and Fungal Divisions)

Division (Common Name)	Major Wall Component* (polysaccharides)	Asexual Reproduction (including flagellation)	Habitat
Myxomycota (slime molds)	lacking (in assimilative phase)	zoospores and/or myxamoebae anterior; 1–2; whiplash	terrestrial aquatic
Hyphochytridiomycota (hyphochytrids)	cellulose-chitin	zoospores anterior; 1; tinsel	aquatic
Chytridiomycota (chytrids)	chitin-glucan	zoospores posterior; 1; whiplash	aquatic
Oomycota (oomycetes)	cellulose-glucan	zoospores anterior; 2; whiplash and tinsel	aquatic
Zygomycota (zygomycetes)	chitin-chitosan chitin-glucan	aplanospores; sporangiospores	terrestrial
Eumycota (true fungi) includes: ascomycetes basidiomycetes imperfect fungi	mannan-glucan; mannan-chitin chitin-glucan	aplanospores; budding; conidia	terrestrial (aquatic)

*Glucan and cellulose are composed of glucose units; chitin and chitosan contain nitrogen; mannan is composed of mannose units.

cyanin (*cyan* = blue) and phycoerythrin (*erythr* = red).

Potential chemical energy may be retained in organic compounds termed *photosynthate*, or storage (reserve) products. Generally they are carbohydrates, although fats, oils, and proteins are present in varied amounts. More information is available about the nature of carbohydrates, which are polysaccharides composed mainly of the monosaccharide glucose. The way in which the glucose molecules are joined together determines the exact carbohydrate. These are classed either as starchlike or laminaranlike compounds.

Cell covering is varied among the algal groups. Often a definite wall is present at some stage in the life history of an alga. This wall can be cellulose, as in bryophytes and vascular plants, or some other polysaccharide. Cellulose is also composed of glucose but the linkage and configuration of the units are different than in the storage polysaccharides (i.e., starchlike compounds). Other wall polysaccharides contain different monosaccharides. In the red and brown

algae, other wall materials termed phycocolloids (*phyco* = alga; *coll* = glue) occur. Additionally, inorganic compounds (iron, silicon) are deposited in the wall. Some algae have a covering of inorganic scales (silicon, calcium) or organic scales similar to the cell wall materials.

Separation of algal groups is based primarily on biochemical features such as pigments, storage products, and cell coverings, as well as on the nature and location of flagella. Table 2-4 shows the use of these features in the classification of algae and indicates the extent of their variety.

Bryophyta (Bryophytes). The last group of nonvascular plants, Bryophyta, encompasses mosses and moss allies (liverworts, hornworts) and is discussed in Chapter 12. Pigments. cell walls, and storage products are the same as for most green algae (division Chlorophyta; see Table 2-4) and vascular plants. Many botanists consider bryophytes as part of the large group of green terrestrial plants because of their habitat, the

Table 2-4 Major Distinguishing Characteristics of the Photosynthetic Eukaryotes (Algal Divisions, Bryophytes) as Compared to the Bluegreen Algae (Prokaryotes)

Division (Common Name)	Major Pigments	Major Storage Products	Major Cell Coverings	Flagellation	Habitat (Approximate %)
Prokaryota					
Cyanophyta (bluegreen algae)	chlorophyll *a* (carotenoids)* phycobiliproteins	starchlike	peptidoglycan	none	freshwater—80 marine—20
Eukaryota					
Chlorophyta (green algae)	chlorophyll *a, b* (carotenoids)	starch	cellulose (xylan, mannan)	apical; 2, 4; equal; whiplash (scales)	freshwater—60 marine—40
Chrysophyta† (chrysophytes)	chlorophyll *a, c* carotenoids	laminaranlike	(cellulose) organic, CaCO₃, silicon	apical; 1, 2; unequal (equal); tinsel and/ or whiplash	freshwater—60 marine—40
Pyrrhophyta (dinoflagellates)	chlorophyll *a, c* carotenoids	starch	cellulose	lateral; 2; whiplash	freshwater—40 marine—60
Cryptophyta (cryptophytes)	chlorophyll *a, c* (carotenoids) phycobiliproteins	starch	protein	lateral; 2; unequal; tinsel	freshwater—90 marine—10
Euglenophyta (euglenoids)	chlorophyll *a, b* (carotenoids)	laminaranlike	protein	apical; 1, 2; unequal; tinsel	freshwater—90 marine—10
Phaeophyta (brown algae)	chlorophyll *a, c* carotenoids	laminaran	cellulose phycocolloids	lateral; 2; unequal; tinsel and whiplash	freshwater—0.1 marine—99.9
Rhodophyta (red algae)	chlorophyll *a, (d)* (carotenoids) phycobiliproteins	starchlike	cellulose phycocolloids	none	freshwater—10 marine—90
Bryophyta (bryophytes)	chlorophyll *a, b* (carotenoids)	starch	cellulose	apical; 2; equal; whiplash (scales)	aquatic—5 terrestrial—95

*Parentheses indicate may occur in some representatives, but not the majority.
†Constituting a diverse assemblage in this text (see Chapter 8)

multicellular complexity of their reproductive structures, the dependence of the young diploid embryo on the parent plant, their life history, and their gross morphology. In both morphology and anatomy, bryophytes are structurally simpler than vascular plants; both gamete-producing (gametophyte) and meiospore-producing (sporophyte) phases are relatively conspicuous, each with distinctive features. In a morphological and physiological sense, bryophytes are a link between those photosynthetic organisms occupying aquatic and terrestrial habitats, although they probably are not a link in an evolutionary sense.

The smallest bryophytes are a few millimeters or nearly microscopic; the largest erect forms are 60 cm high, but some creeping forms are several meters in length. Most bryophytes are strictly terrestrial and grow in humid environments, although some grow in arid sites with little humidity, and a few are aquatic.

References

For references on slime molds and fungi, see Chapters 4–6; on bryophytes, see Chapter 12.

Bell, P. R., and Coombe, D. E., trans. 1976. *Strasburger's textbook of botany.* Longmans, Green & Company, London.

Bell, P. R., and Woodcock, C. L. F. 1968. *The diversity of green plants.* Edward Arnold, London.

Bold, H. C. 1977. *The plant kingdom.* 4th ed. Prentice-Hall, Englewood Cliffs, N.J.

Bold, H. C.; Alexopoulos, C. J.; and Delevoryas, T. 1980. *Morphology of plants and fungi.* 4th ed. Harper & Row, New York.

Bold, H. C., and Wynne, M. J. 1978. *Introduction to the algae.* Prentice-Hall, Englewood Cliffs, N.J.

Brown, W. H. 1935. *The plant kingdom.* Ginn & Company, Boston.

Chadefaud, M. 1960. *Les végétaux non vasculaires.* In Chadefaud, M., and Emberger, L., *Traité de botanique,* vol. 1. Masson & Cie Ed., Paris.

Corner, E. J. H. 1968. *The life of plants.* Mentor, New American Library, New York.

Cronquist, A. 1981. *Basic botany.* 2nd ed. Harper & Row, New York.

Dittmer, H. C. 1964. *Phylogeny and form in the plant kingdom.* D. Van Nostrand, Princeton, N.J.

Dodd, J. D. 1962. *Form and function in plants.* Iowa State Univ. Press, Ames, Iowa.

Dodge, J. D. 1979. The phytoflagellates: fine structure and phylogeny. *Biochemistry and physiology of protozoa,* vol. 1. 2nd ed. Academic Press, New York.

Duddington, C. L. 1967. *Flora of the sea.* Thomas Y. Crowell, New York. (Also as *Seaweeds and other algae,* Faber & Faber, London.)

Fott, B., and Glenk, H. O. 1971. *Algenkunde.* 2nd ed. G. Fischer Verlag, Jena.

Fritsch, F. E. 1935, 1945. *The structure and reproduction of the algae,* vols. 1, 2. Cambridge Univ. Press, London.

van den Hoek, C. 1978. *Algen.* George Thieme Verlag, Stuttgart.

Jensen, W. A., and Salisbury, F. B. 1972. *Botany: An ecological approach.* Wadsworth Publishing Company, Belmont, Calif.

Lee, R. E. 1980. *Phycology.* Cambridge Univ. Press, Cambridge.

McLean, R. C., and Ivimey-Cook, W. R. 1951. *Textbook of theoretical botany,* vol. 1. Longmans, Green & Company, London.

Norstog, K., and Long, R. W. 1976. *Plant biology.* W. B. Saunders Company, Philadelphia.

Raven, P. H.; Evert, R. F.; and Curtis, H. 1981. *Biology of plants.* 3rd ed. Worth Publishers, New York.

Rayle, D. L., and Wedberg, W. D. 1980. *Botany: A human concern.* W. B. Saunders Company, Philadelphia.

Round, F. E. 1969. *Introduction to the lower plants.* Butterworth, London.

———. 1973. *Biology of the algae.* 2nd ed. Edward Arnold, London.

Rushforth, S. R. 1976. *The plant kingdom: Evolution and form.* Prentice-Hall, Englewood Cliffs, N.J.

Smith, G. M. 1955. *Cryptogamic botany.* In *Algae and fungi,* 2nd ed. Vol. 1, McGraw-Hill Book Company, New York.

Tiffany, L. H. 1958. *Algae, the grass of many waters.* 2nd ed. Charles C Thomas, Springfield, Mo.

Trainor, F. R. 1978. *Introductory phycology.* John Wiley & Sons, New York.

The subkingdom Prokaryota comprises three groups of **prokaryotic** organisms, Schizomycophyta, Cyanophyta, and Archaebacteria. The Archaebacteria, not treated here, is a small group which appears to be only remotely related to Schizomycophyta and Cyanophyta. Structural and chemical similarities indicate a close relationship between the main group of bacteria, Schizomycophyta, and bluegreen algae (Cyanophyta or Cyanobacteria). In these prokaryotes, the nuclear material is not surrounded by a nuclear envelope, and they lack both nucleoli and the chromosomal organization of eukaryotic organisms. Prokaryotic cells also lack plastids, mitochondria, golgi, and other unit-membrane organelles characteristic of eukaryotes. All photoautotrophic prokaryotes that possess chlorophyll *a* and carry on aerobic photosynthesis are classified in the division Cyanophyta (= Cyanobacteria), or bluegreen algae. The remaining prokaryotic autotrophs and heterotrophs constitute the diverse assemblage designated as the division Schizomycophyta, or bacteria.

3

PROKARYOTA (PROKARYOTES—BACTERIA AND BLUEGREEN ALGAE)

Division Schizomycophyta (Bacteria)

Bacteria are structurally simple organisms, but they are biochemically complex and display a greater variety of nutritional modes than is found in any other group. Bacterial cells are small and their large surface/volume ratio is accompanied by exceptionally high metabolic and growth rates. In terms of numbers of individuals, bacteria undoubtedly outnumber any other group of organisms in any particular environment. Thus, although they are not conspicuous organisms in our surroundings, bacteria have an enormous influence on the environment and its inhabitants.

Most bacteria are heterotrophs and obtain nutrients from other living or dead organisms. They can absorb small molecules directly from their surroundings and, in addition, secrete exoenzymes that digest complex molecules such as those of proteins and polysaccharides. **Saprobic** heterotrophs—i.e., those living on dead plant and animal materials—are common in foods, water, sewage, soil, dung, and decaying materials of all types. Parasitic bacteria attack plants and animals of all types. Paradoxically, some bacteria produce antibiotics that are useful in the treatment of certain bacterial diseases.

Many heterotrophic bacteria form mutualistic associations with other organisms. Some are of relatively great economic importance to man—e.g., **nitrogen-fixing** bacteria in root nodules and bacteria present in the digestive tracts of ruminants.

Autotrophic bacteria include both photosynthetic and chemosynthetic forms. **Photoautotrophic** bacteria mainly inhabit mud and water and most require only inorganic nutrients—i.e., carbon dioxide, mineral nutrients, and energy in the form of sunlight.

Chemoautotrophic bacteria obtain energy through various oxidation-reduction reactions, utilizing carbon dioxide as a carbon source. Chemoautotrophic bacteria occur in soil, mud, and water, where oxidizable CO, Fe^{++}, NH_3, H_2S, or similar substances are present.

Most bacteria require oxygen—i.e., they are aerobic organisms. In contrast, obligately anaerobic bacteria can grow only in the absence of oxygen. Some species, the facultative anaerobes, grow either in the presence or absence of oxygen. Bacteria show similar variations with respect to certain other environmental factors. For example, some "heat-loving" bacteria grow only in hot springs where water temperatures exceed 90 °C; others can grow at temperatures below 0 °C. Some bacteria are adapted to life in concentrated salt or sugar solutions that prevent the growth of other forms.

Cell Form and Structure

Most bacteria are unicellular, and the growth of cells generally is determinate. Commonly encountered forms are spheres and straight or curved rods (Fig. 3-1). Spherical bacterial cells are called **cocci** (*sing.* **coccus**); straight rod-shaped cells are designated as **bacilli** (*sing.* **bacillus**) or simply "rods." Helically curved bacterial cells are referred to as spirils, or **spirilla,** and very short, curved rods are termed **vibrios.** Some bacteria have cellular extensions, as in *Caulobacter* and *Ancalomicrobium* (Figs. 3-30 and 3-31), and thus differ from the above categories. In addition to these

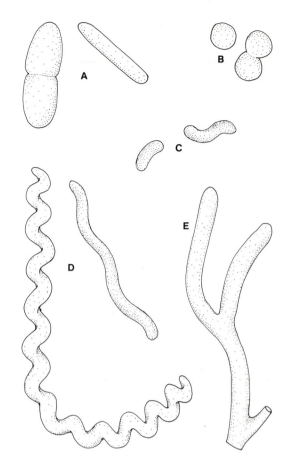

Figure 3-1 Bacterial form. **A,** bacillus; **B,** coccus; **C,** vibrio; **D,** spirillum; **E,** coenocytic filaments.

cell types, continuous tubular filaments are produced by some members of the Actinomycetales.

Cell shape is relatively constant in most bacterial species but can vary with age or environmental conditions. Thus, *Rhizobium* cells in root nodules are rod- to X- or Y-shaped (Fig. 3-2), but are rod-shaped in culture.

Size Bacilli of some species measure less than 0.5 μm in diameter and 1 μm in length. The longest of single-celled bacteria are certain of the Spirochaetales (Figs. 3-38 and 3-39), which can exceed 500 μm in length. Most bacilli fall into the range 1–5 μm in length and 0.5–1 μm in diameter; cocci usually attain a diameter of 0.5–1 μm.

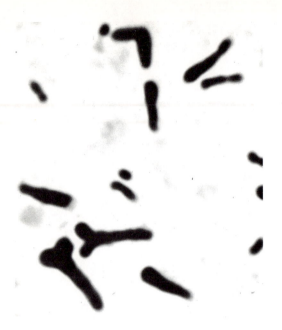

Figure 3-2 Cells of *Rhizobium* from a root nodule, ×2000. Note variation in form.

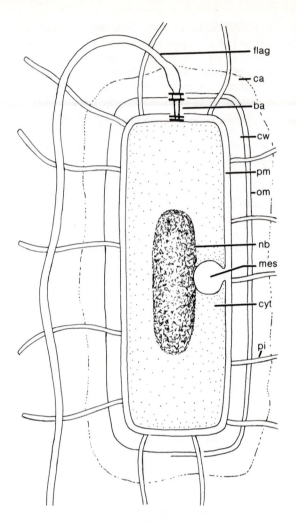

Figure 3-3 Bacterial structure: *flag*, flagellum; *ba*, basal apparatus; *cap*, capsule; *om*, outer membrane (gram-negative bacteria only); *cw*, cell wall; *pm*, plasma membrane; *nb*, nuclear body; *mes*, mesosome (gram-positive bacteria); *cyt*, cytoplasm; *pi*, pilus (gram-negative bacteria).

Cell Structure Structures common to many bacterial groups are shown in a composite diagram of a bacilliform cell in Figure 3-3. The bacterial protoplast usually is surrounded by several layers. These include (1) a **capsule** or slime layer, (2) a rigid wall layer, (3) a plasma membrane, and (4) a membrane outside the cell wall in some types.

The outermost layer can be of relatively constant thickness and sharply defined, in which case it is called a **capsule** (Fig. 3-4A). In some species, this capsule has been shown to consist of vertically oriented polysaccharide fibrils, as shown in Figure 3-4B. The capsular material of some bacteria is soft, and it dissolves or is lost into the surrounding medium. Such material is sometimes called a slime layer, or simply slime. Capsules or slimes are of relatively simple chemical composition, often consisting of a single polysaccharide or polypeptide. Slimes of some bacteria are composed of dextrans, levans, cellulose or alginatelike substances. Presence or absence of capsules and slimes can be strongly influenced by nutritional conditions in culture. In those bacteria regularly producing capsules, mutation to nonencapsulated forms often occurs.

Since mutants of this type survive and continue to reproduce, the capsule is not considered essential to the life of the bacterium. However, the presence of a capsule does play an important role in parasitic species, since it protects cells in the animal body from **phagocytosis** by **phagocytes.** Thus, strains of pneumococcus, *Diplococcus pneumoniae*, that lack capsules are avirulent or less virulent than **encapsulated** strains.

Cell wall chemistry and structure vary

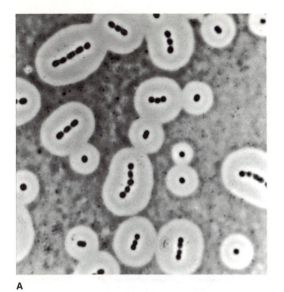

A

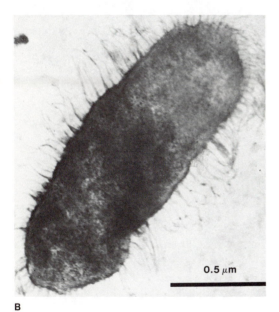

B

Figure 3-4 A, photograph of encapsulated bacteria mounted in India ink (*scale line* shown in 10 μm). **B,** photomicrograph showing fibrillar nature of capsular material in *Klebsiella pneumoniae* (*scale line* shown is 0.5 μm). (**A,** photograph courtesy of E. Juni and W. H. Taylor, with permission of *Journal of Bacteriology;* **B,** from E. L. Springer and I. L. Roth, "The ultrastructure of the capsules of *Diplococcus pneumoniae* and *Klebsiella pneumoniae* stained with ruthenium red," *Journal of General Microbiology* 74:21–31, 1973, courtesy of the authors and with permission of Cambridge University Press.)

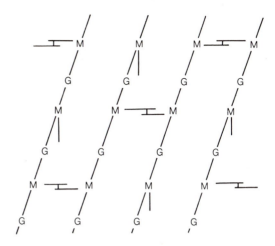

Figure 3-5 Diagrammatic representation of cell wall composition in *Escherichia coli.* Alternating N-acetylglucosamine (*G*) and N-acetylmuramic acid (*M*) units form chains; peptide subunits are attached to the N-acetylmuramic acid units and, in some instances, cross-link adjacent chains.

among different groups of bacteria. The most noteworthy of variations are those that distinguish gram-positive and gram-negative bacteria. The gram strain, originally devised by Christian Gram in 1884 as a method of staining bacterial cells in animal tissues, has long been used as an aid to bacterial identification. In this procedure, bacterial smears are stained with crystal violet and then **mordanted** with an iodine-potassium iodide solution. The smear is next treated with an organic solvent such as ethanol. Gram-positive bacteria retain the stain-mordant complex; all stain is quickly removed by ethanol from gram-negative bacteria. Although some bacterial species are gram-indeterminate, and there are sometimes inconsistencies in staining, most bacterial species fall into one or the other of the above categories.

One type of cell wall substance, called peptidoglycan, occurs in both gram-positive and gram-negative cells as well as in bluegreen algae. These complex molecules, occurring only in cell walls of prokaryotic organisms, are composed of sugar-amines and amino acids. Peptidoglycans consist of chains of alternating units of sugar-amines, N-acetylglucosamine and N-acetylmuramic acid, to which are attached peptide subunits (Fig. 3-5). Some peptides attached to

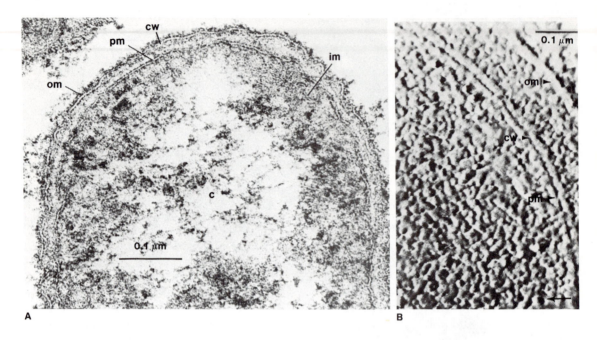

A

B

Figure 3-6 **A,** longitudinal section of *Acetobacter suboxydans* cell, showing membranous outer layer of cell wall (*scale line* shown is 0.1 μm): *om,* outer membrane; *cw,* cell wall; *pm,* plasma membrane; *im,* intracytoplasmic membranes. **B,** freeze-etched *Rhizobium* cell (*scale line* shown is 0.1 μm): *om,* outer membrane; *pm,* plasma membrane. **C,** electron micrograph of cell wall preparation, showing surface pattern in *Spirillum* (*scale line* shown is 0.1 μm). (**A,** photograph courtesy of B. L. Batzing and G. W. Claus, with permission of *Journal of Bacteriology;* **B,** photograph courtesy of C. R. MacKenzie, W. J. Vail, and D. C. Jordan, with permission of *Journal of Bacteriology;* **C,** photograph courtesy T. A. Beveridge and R. G. E. Murray, with permission of *Journal of Bacteriology.*)

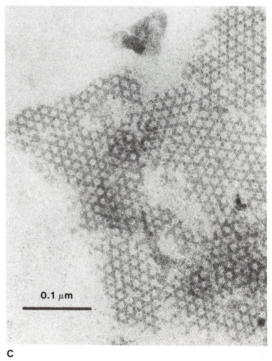

C

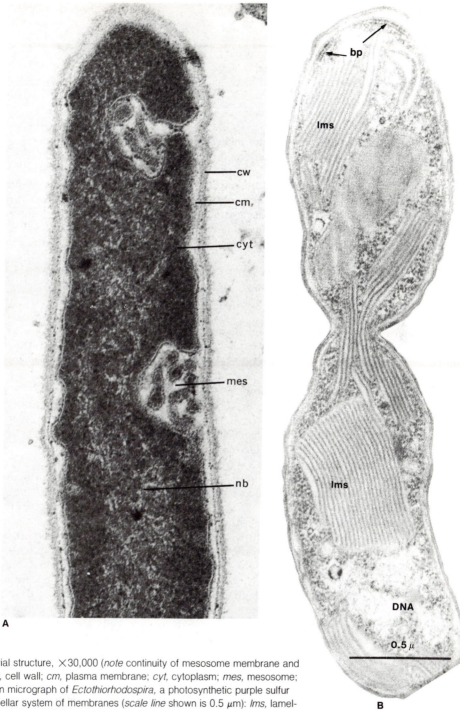

Figure 3-7 **A,** bacterial structure, ×30,000 (*note* continuity of mesosome membrane and plasma membrane): *cw,* cell wall; *cm,* plasma membrane; *cyt,* cytoplasm; *mes,* mesosome; *nb,* nucleoid. **B,** electron micrograph of *Ectothiorhodospira,* a photosynthetic purple sulfur bacterium, showing lamellar system of membranes (*scale line* shown is 0.5 μm): *lms,* lamellar stacks; *bp,* basal structure of flagellar tuft; *DNA,* nucleoid. (**A,** photograph courtesy of T. Bisalputra; **B,** photograph courtesy of H. G. Trüper, with permission of *Journal of Bacteriology.*)

adjacent sugar-amine chains are cross-linked, the degree of linkage varying with species. The peptide subunits usually are composed of D-alanine, L-alanine, D-glutamic acid, and a dibasic acid, L-lysine or meso-diaminopimelic acid. Peptidoglycans constitute from as little as 5–10% to as much as 80–90% of the cell wall.

Cell walls of gram-negative bacteria consist of a thin layer of peptidoglycan sandwiched between the plasma membrane and an outer membrane. The latter resembles the plasma membrane in electron micrographs (Fig. 3-6), and both resemble the unit membranes of other organisms. However, their chemical composition does not include the sterols found in the membranes of eukaryotes. Additionally, the outer membrane of gram-negative cells is not chemically identical to the plasma membrane.

In addition to peptidoglycan, cell walls of gram-positive bacteria contain complex substances, called *teichoic acids*, linked to the peptidoglycan.

High magnification of the surface of the wall has revealed various surface patterns in both gram-positive and gram-negative bacteria. The patterns (Fig. 3-6C) described as hexagonal, tetragonal, "holey," etc. are often covered by amorphous material.

The cell wall protects bacterial cells from osmotic damage, maintains cell form, and is necessary to motility. Bacilliform cells treated with enzymes that digest the cell wall assume a spherical form. These delicate protoplasts must be maintained in isotonic solutions to prevent osmotic rupture.

Differences in the susceptibility of gram-positive and gram-negative bacteria to antibiotics is sometimes attributable to structural differences in their cell walls. Incorporation of newly synthesized peptidoglycan into the cell wall involves cross-linking to existing peptidoglycan. This cross-linking process is strongly inhibited by small concentrations of penicillin. Perhaps because of the higher percentage of peptidoglycan in walls of gram-positive bacteria, these forms are more subject to damage by penicillin than are gram-negative species.

The Protoplast A plasma membrane similar to the unit membranes of other organisms sur-

rounds the bacterial protoplast. The plasma membrane can consist of a simple peripheral layer, or it can have one or more internal extensions that protrude into the cytoplasm. The intrusions in gram-positive bacteria often are in the form of complex vesicular structures called **mesosomes** (Fig. 3-7). Mesosomes are lacking in gram-negative bacteria, but complex infolding of the membrane can occur (Fig. 3-7B, *Ectothiorhodospira*), or irregular intracytoplasmic membrane systems can be present.

The bacterial plasma membrane is composed of proteins and phospholipids. Enzymes functioning in respiration and, in photoautotrophic bacteria, photosynthetic pigments can also be associated with the cytoplasmic membrane. The plasma membrane can therefore have other functions than controlling the movement of materials into or out of the cell; these functions include some of those carried out by unit-membrane organelles in eukaryotes.

Mesosomes are associated with the formation of new cell walls during cell division and **endospore** formation. Where present, mesosomes also appear to play a role in insuring that daughter nucleoids go into separate cells during cell division.

Bacterial cytoplasm is devoid of unit-membrane organelles such as mitochondria, endoplasmic reticulum, golgi, and (generally) vacuoles. Membrane (nonunit membrane)-enclosed gas vacuoles are present in certain marine bacteria and are similar to those found in some blue-green algae. Because of the lack of organelles, bacterial cytoplasm is of relatively homogeneous appearance. However, granular materials can be present and are visible either with or without staining (Fig. 3-8). Refractile globules of sulfur and of poly-beta-hydroxybutyric acid are often visible without special staining procedures. Food reserve materials other than poly-beta-hydroxybutyric acid such as glycogen and starch, can be seen if cells are treated with iodine-potassium-iodide solution.

In electron micrographs, numerous **ribosomes** are visible in bacterial cytoplasm. These structures have a diameter of about 10 nm and are called 70-S ribosomes in contrast to the larger 80-S ribosomes found in the eukaryotic cytoplasm. The 70-S ribosomes also are produced in eukaryotic cells, but they occur only in mitochondria and chloroplasts.

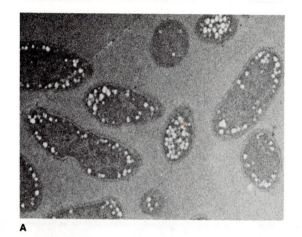

A

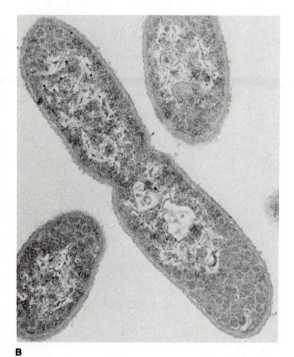

B

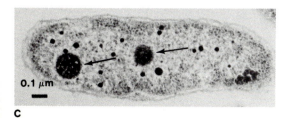

0.1 μm

C

The Nucleoid Bacterial cells each contain one of more nuclear bodies, or *nucleoids* (Figs. 3-7 and 3-9). These lack the nuclear envelope, nucleolus, and chromosomal organization of eukaryotic nuclei. The nucleoid consists of a single strand of DNA that is up to 1.4 μm in length and is continuous—i.e., it is a ring. The strand is folded and forms a compact mass that is centrally located in the cell; it is attached to the plasma membrane. In gram-positive bacteria, attachment is to a mesosome; in gram-negative bacteria, attachment is at the cell periphery.

DNA of the nucleoid bears the main genome of the bacterial cell. In addition to this DNA, bacterial cells often contain independent, self-replicating rings of DNA called **plasmids.** Plasmids do not have overall genetic homology with the nucleoid genome; genes borne on plasmids differ from those on the chromosome. Among cell structures or properties known to be determined by plasmid genes are formation of certain **pili** (p. 40), resistance to many antibiotics and toxins, and synthesis of some amino acids. Plasmids, in addition to replicating within the cell, can determine conjugation and their own transfer.

Flagella Many bacteria exhibit either swimming or gliding motion. Swimming motion depends upon the action of one or more flagella. Flagella can occur singly or in tufts and are variously positioned on cells (Fig. 3-10A–D). If tufted, they are aggregated into one large strand and function as a unit. There can be a single flagellum at one pole of the cell, single flagella at both poles, tufts of flagella at one or both poles, or numerous flagella inserted over the entire cell surface.

Bacterial flagella consist of a single strand, in contrast to the 9 + 2 arrangement found in all eukaryotic flagellated cells. Because of their small diameter, single flagella can be seen by light mi-

Figure 3-8 Granular materials in bacterial cells. **A, B,** glycogenlike granules in *Escherichia coli* (**A,** ×14,500; **B,** ×36,000); **C,** metachromatic granules (polyphosphate, *arrows*) in *Thiobacillus novellus* (*scale line* shown is 0.1 μm). (Figures courtesy of J. M. Shively; **C,** with permission of *Annual Review of Microbiology.*)

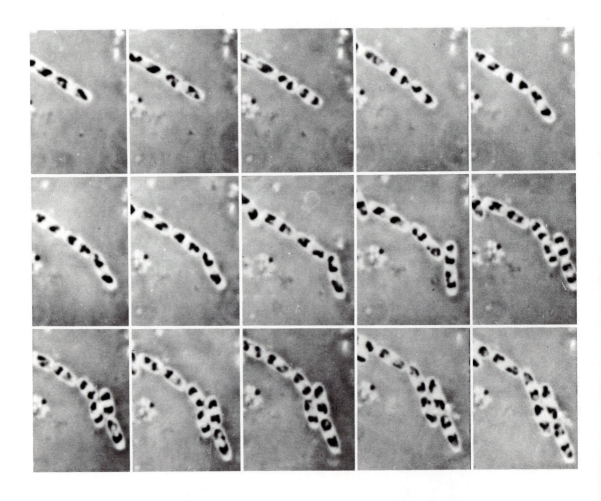

Figure 3-9 Bacterial nucleoid or nuclear body. Division of nucleoid and growth in a single group of living cells of *Escherichia coli*. Sequence taken over a period of 78 minutes, ×1800. (Photographs courtesy of D. J. Mason and D. M. Powelson, with permission of *Journal of Bacteriology*.)

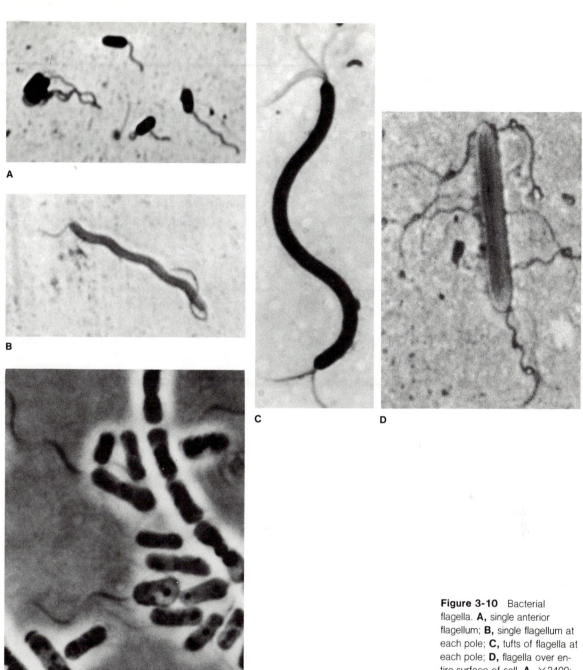

Figure 3-10 Bacterial flagella. **A,** single anterior flagellum; **B,** single flagellum at each pole; **C,** tufts of flagella at each pole; **D,** flagella over entire surface of cell. **A,** ×2400; **B,** ×2300; **C,** ×2000; **D,** ×2000. **E,** phase-contrast photograph of living *Sphaerotillus* cells showing flagellar tufts, ×3400. (Photograph courtesy of J. L. Stokes, with permission of *Journal of Bacteriology*.)

croscopy only if they are first mordanted and stained. Tufts of flagella that are aggregated can be seen by means of dark-field or phase-contrast microscopy (Fig. 3-10E) in unstained cells.

Pure preparations of bacterial flagella can be obtained by shaking cells with minute glass beads. Sheared flagella can then be collected and examined chemically. Bacterial flagella consist almost entirely of proteins called **flagellins.** If flagella are suspended in acidic solutions (pH 3–4), they disintegrate; the proteins will reunite to form flagellar filaments if the pH is adjusted to 5.5–6.

Flagella of bacteria penetrate only through the cell wall and are not membrane-ensheathed as in eukaryotes. Each flagellum is bent slightly, or "hooked," at the point where its base approaches the cell wall, and each is provided with a basal apparatus (Fig. 3-11). The basal apparatus includes one or two sets of spoollike structures adjacent to the plasma membrane and, if present, the outer membrane.

Bacterial flagella have a length 2 to 3 times that of the cell, and they are spirally curved (Fig. 3-12). Characteristics of the spiral or helix are constant for a given species, and the flagellum is known to be stiff. The mechanism of flagellar action is still uncertain, although some observations indicate that they rotate. If spirilla are sandwiched between glass slides to restrict flagellar movement (Fig. 3-12), the cells rotate. Flagellar action in such spirilla appears to involve rotation of the flagellum, rather than contraction or flexing.

It has been established that bacterial movements are directed responses; they occur in response to stimuli such as light intensity and nutrients. These **tactic** responses can be induced by placing cells in habitats where diffusional gradients of nutrients, oxygen, and so on, occur. In pure culture, flagella may be lost as cultures age; nutrients can also determine the presence or absence of flagella.

Many motile and nonmotile gram-negative bacteria possess filamentous appendages called *pili* (Fig. 3-13). Pili are shorter than flagella, are of smaller diameter, and are often more numerous than flagella. Like flagella, pili can be restricted to the poles of the cell, or they can be on other surfaces as well. They are composed of single proteins, called *pilins.*

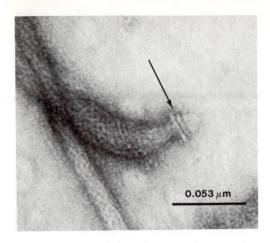

0.053 μm

Figure 3-11 Basal apparatus of *Clostridium flagellum*, including the basal hook and spoollike structure (*arrow*) (*scale line* shown is 0.053 μm). (From R. C. Hamilton and H. M. Chandler, "Ultrastructure of the basal organelles of flagella of *Clostridium chauvoei*," *Journal of General Microbiology* 89:191–194, 1975. Courtesy of the authors and with permission of Cambridge University Press.)

Several types of pili are known. One, designated as *F*-pilus, occurs only on donor cells in conjugation (p. 44). In some soil-inhabiting bacteria, polar pili are responsible for starlike aggregates of cells; these pili are contractile. Another function of pili is in the adhesion of cells to one another, to inert objects, or possibly to host cells.

Reproduction

Binary Fission Reproduction in most bacteria is by **binary fission** (Figs. 3-14 and 3-15), a process in which the cell and its contents are equally divided into two. Up to 20–30 min before actual cell cleavage, the nucleoid divides, and separation of daughter nucleoids probably results through their attachment to the mesosome or plasma membrane. More than two nucleoids are often present at the time of cell division—i.e., nucleoid division is not necessarily immediately followed by transverse wall formation.

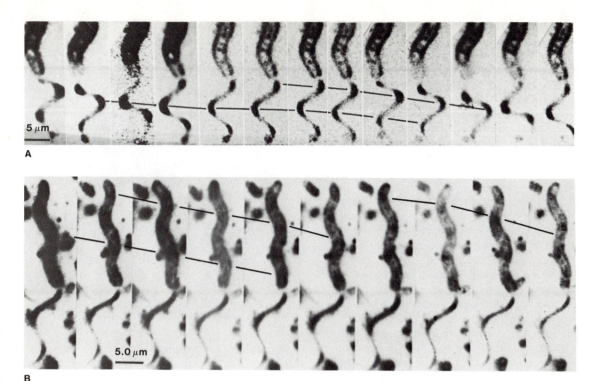

A

B

Figure 3-12 A, B, *Spirillum volutans*. **A,** the body is motionless and the flagellum rotates. **B,** the body rotates when the cells are tightly sandwiched between glass plates that prevent flagellar rotation (the flagellum is rigid and its gyre width is greater than that of the cell) (*scale line* shown is 5 μm). (Photographs courtesy of M. Mussill and R. Jarosch, with permission of *Protoplasma*.)

In many gram-positive bacteria, a thickened band is visible in the cell wall prior to cell division (Fig. 3-14). A doughnut-shaped mesosome forms at the point of intrusion or infolding of the plasma membrane (Fig. 3-15) and gradually decreases in diameter. As the mesosome closes in, new cell wall material is deposited outside the membrane at the mesosome base. When transverse wall formation is complete, the two new cells snap apart forcefully in some species; in others, the cells remain attached forming filaments, or separation is gentle.

Following cell division, new wall material must be formed as young cells increase in size.

Under ideal conditions, fission occurs in many bacteria as frequently as once in 20–30 min. Theoretically, this rate of reproduction could result in the production of immense masses of bacterial cells within a few hours. Fortunately various natural factors prevent such increases. Among these factors are limitation of nutrients, accumulation of metabolic products produced by the bacteria, and competition from other organisms. In culture, growth of a bacterial population follows a predictable pattern, or growth curve (Fig. 3-16). This growth curve has four distinct phases: (1) *initial* "lag" phase, during which cell numbers increase slowly; (2) period of extremely rapid increase, called the *logarithmic* or *exponential* phase; (3) phase in which the population neither increases nor decreases, called the *stationary* phase; and (4) *senescent* phase, in which cell division, if any, is exceeded by rate of cell death.

Cell Aggregates and Colonies Cells of some coccoid species adhere to one another after fission occurs, and aggregates of cells so produced have a relatively constant form. These aggregates may be comparable to the simple colonies of some algal groups. If division of cocci occurs consistently on a single plane, and cells adhere, pairs or chains result. Pairs or chains of adhering cells are produced in *Streptococcus* (Fig. 3-17A, B). In other cocci, divisions occur regularly in two or three opposing planes, and packets of cells, as in *Sarcina*, are the result (Fig. 3-17C). In *Staphylococcus*, divisions occur randomly (Fig. 3-17D), and adhesion of cells results in grapelike clusters. Many bacilli produce chains of cells or filaments as a result of transverse division and adhesion.

Masses of cells that develop on agar media (Fig. 3-18) are generally referred to as **colonies.** The appearance of these colonies varies among species and with different culture conditions. If standardized media, temperature, and other conditions are used, colony characteristics are of some value in identifying bacteria. Colony features that are used are profile and nature of margin (Fig. 3-18), texture and surface appearance, and pigmentation. Pigmentation is generally consistent for a given species, and pigments vary in their nature. Some species form water-soluble pigments that diffuse out into the medium; in others, pigments remain within the cells, and only the colony is colored.

Endospores and Cysts Many bacteria form resistant **endospores** or **cysts.** Endospores are thick-walled structures generally produced singly within parent cells (Fig. 3-19). Almost all endospore-forming bacteria are bacilli, and most are species of *Clostridium* or *Bacillus.*

In the initial stage of endospore formation (Fig. 3-20), the nucleoid forms an elongate axial filament with associated granules of reserve materials. The nucleoid divides, and a membranous septum develops between the mother cell and the somewhat smaller endospore protoplast. Unidirectional growth of the mother cell plasma membrane then results in the "engulfment" of the endospore protoplast or *forespore.* As a result of this engulfment, the forespore becomes surrounded by two membranes, between which the endospore wall is deposited. As in fission, depo-

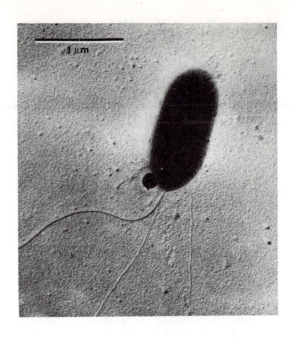

Figure 3-13 *Pseudomonas.* Metal-shadowed cell, showing single polar flagellum and bipolar pili (*scale line* shown is 1 μm). (Photograph courtesy of R. L. Weiss and H. D. Raj, with permission of *Australian Journal of Experimental Biology and Medical Science.*)

sition of new wall material is associated with the presence of mesosomes. As wall material is deposited, the forespore becomes refractile and, at maturity, the endospore is surrounded by a thick wall, or **cortex,** two coat layers, and an outermost layer called the **exosporium** (Fig. 3-21A). During maturation, refractility increases, resistance to heat increases, and there is an uptake of calcium and other substances. Finally, the endospore is liberated through the breakdown of the mother cell wall.

Endospores of many bacteria have appendages or other ornamentation of the outer wall layer (Fig. 3-21). Endospore size, shape, and position in the mother cell vary with the species.

Endospores are extremely resistant to heat, chemicals, and desiccation. In some species, they have been reported to withstand boiling for more than two hours and the action of chemicals that rapidly kill vegetative cells. They are also capable of retaining their viability for more than 50 years.

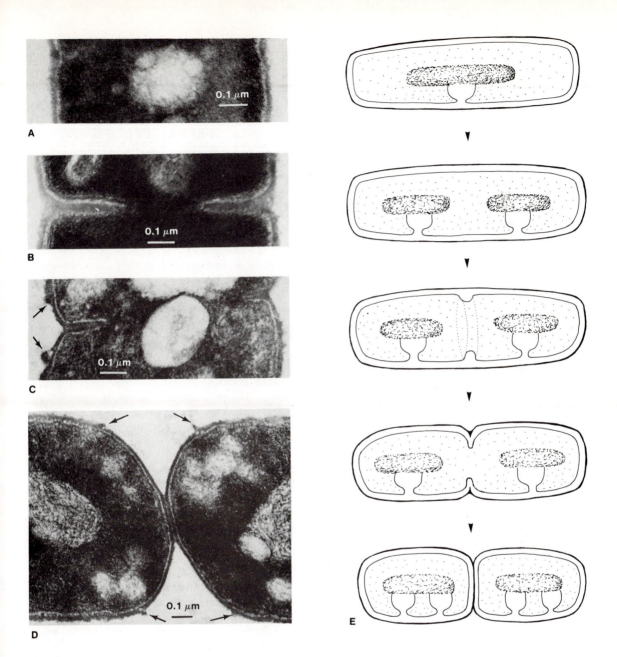

Figure 3-14 **A–D,** cell division in *Bacillus cereus*. **A,** initial band has developed; **B,** developing transverse wall; **C,** rupture of the outer wall layer; **D,** final stage in division (*arrows* indicate peripheral wall scars) (*scale lines* shown are 0.1 µm). **E,** diagram of bacterial fission. Before cell division occurs, a mesosome attached to the nucleoid splits; the two parts separate with attached daughter "chromosomes." A thickened band then develops on the inner surface of the cell wall and grows centripetally until division is complete. The outer wall layer splits before transverse wall formation is completed. (Photographs **A–D,** courtesy of K. L. Chung, reproduced by permission of the National Research Council of Canada from *Canadian Journal of Microbiology* 19:217–221, 1973.)

When endospores germinate (Fig. 3-21B, C), each gives rise to a single cell. Thus, endospores do not generally lead to an increase in the number of cells; they are a device for surviving unfavorable environmental conditions. In culture, the nature of the medium and other factors are known to influence endospore production.

Cysts are produced by cells of *Azotobacter* and by species of Myxobacteriales. In cyst formation, the entire protoplast of a vegetative cell becomes surrounded by a thickened wall (Fig. 3-22). Cyst formation is often preceded by a rounding up of the protoplast.

Sexual reproduction of the type found in eukaryotes is not known in bacteria, but several types of phenomena occur that lead to genetic recombination. All these phenomena involve unilateral transfer of DNA from one cell, the donor, to a second cell, the recipient. Generally, only a fraction of the genome is transferred.

Conjugation Bacterial **conjugation** (Fig. 3-23) superficially resembles conjugation in some yeasts and other eukaryotic organisms. It involves the transfer of genetic material from donor cell to recipient cell through a temporary fusion bridge (Fig. 3-24). In *Escherichia coli*, the species in which conjugation was first discovered and has since been most studied, two types of donor cells are known. The two types are designated as *F+* and *Hfr* cells; recipient cells are designated as *F−*. When a suspension of either *Hfr* or *F+* cells is mixed with a suspension of *F−* recipients, pairing of some individuals of the two types will occur. When such mating pairs form, DNA moves through an *F*-pilus to the recipient cell. The amount of DNA transferred depends upon the length of time that contact is maintained between cells and upon the nature of the donor cells.

F+ cells usually transfer only a small amount of their genome; frequently it is only the *F* factor (a plasmid) that is involved. Conjugation between *Hfr* and *F−* cells leads to the transfer of part or all of a donor chromosome. Using genetic markers, it can be shown that markers are transferred in a regular sequence. The proportion of the chromosome transferred depends upon how long the bridge between donor and recipient is maintained. Transfer of the entire chromosome requires approximately 1½ to 1¾ hours.

The number and length of donor-cell pili vary with the strain and with environmental conditions. The number and length of these structures are correlated with faster initiation of chromosome transfer. It has also been suggested that *F*-pili, through contractile action, might aid in drawing mating pairs into close contact.

After conjugation, daughter cells of the recipient line can exhibit combinations of characteristics of both parental lines. Such genetic changes are the result of recombination between the transferred segment of donor chromosome and chromosome of the recipient cell.

Conjugation is known to occur within species of *Escherichia*, *Vibrio*, *Salmonella*, *Enterobacter*, and others. Intergeneric mating will also occur between strains of *Escherichia* and *Salmonella*, *Shigella*, *Serratia*, *Pasteurella*, and *Proteus*.

Transformation In **transformation,** soluble DNA from donor cells is taken up by recipient cells. Transformation was first observed in *Streptococcus pneumoniae*, causing one type of pneumonia. In this species, virulent strains are characterized by the presence of capsules, and they produce smooth, glistening colonies. These smooth *S*-forms regularly give rise to capsuleless *R*-forms; *R*-forms produce rough colonies and are avirulent. In early experiments, dead *S*-forms and living *R*-forms were mixed and inoculated into mice; living *S*-forms were recovered from the mice. It is now known that reversion of *R*- to *S*-forms in the mice involved transformation of these cells by DNA from the dead *S*-forms.

Generally, in transformation, the amount of DNA participating is no more than 5% of the total genome. Within a few minutes of uptake by recipient cells, donor DNA forms a complex with recipient DNA.

Transformation has been observed in species of many genera, including *Streptococcus*, *Hemophilus*, *Neisseria*, *Escherichia*, *Agrobacterium*, and *Rhizobium*.

Transduction In **transduction** (Fig. 3-25), fragments of DNA are carried from one cell to another by bacterial viruses, called **bacteriophages** or simply *phages*. During the formation of virus particles within one bacterial cell, a fragment of

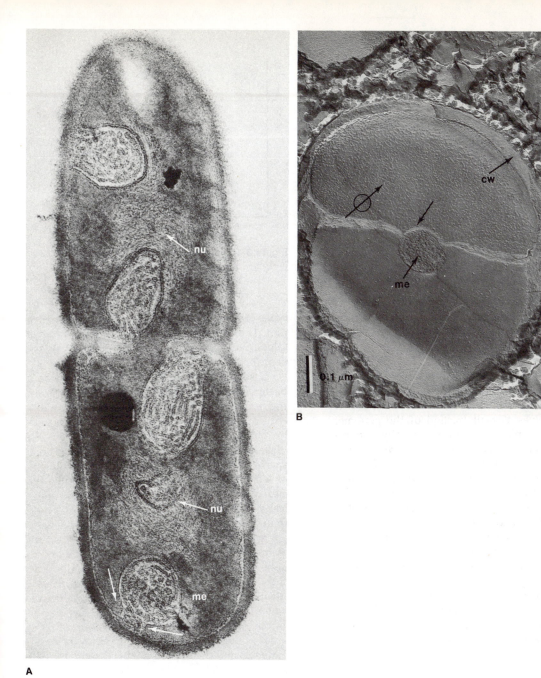

Figure 3-15 **A,** thin section of dividing *Lactobacillus* cell, showing develop-ing transverse wall and mesosome, ×65,800; *nu,* nucleoid; *me,* mesosome; *arrow* indicates continuity of mesosome membrane and plasma membrane. **B,** replica of freeze-etched, dividing *Staphyloccus* showing developing transverse wall and mesosome (*scale line* shown is 0.1 μm): *cw,* cell wall; *me,* mesosome. (**A,** photograph courtesy of T. Kakefuda, J. T. Holden, and N. M. Utech, with permission of *Journal of Bacteriology;* **B,** photograph cour-tesy of J. E. Gilchrist and I. W. De Voe, reproduced by permission of the National Research Council of Canada from *Canadian Journal of Microbiology* 19:294–295, 1973.)

bacterial DNA becomes enclosed within the protein coat of some virus particles. Upon dissolution (*lysis*) of the host cell, or donor, defective virus particles carrying bacterial DNA are released. When such a defective particle infects another cell, bacterial DNA is injected into the new cell together with viral DNA. Because of the incorporated bacterial DNA, the defective phage can neither replicate in nor lyse the recipient cell.

Transduction is mediated through a special kind of bacteriophage called a temperate phage. The DNA of such viruses can exist within their host cells as inactive particles called **prophages,** the prophage DNA being in association with and replicating at the same time as the host DNA. In contrast, virulent phages multiply rapidly within the host cell, killing the host and causing its lysis. Prophages can be transformed into virulent phages by exposure to ultraviolet light.

DNA introduced into recipient cells in conjugation, transformation, or transduction, can result in heritable changes in offspring of the recipient. The recipient is "diploid" only for that portion of the donor DNA received, and this often involves a small fraction of the genome. The partial diploid condition is usually ephemeral; heritable changes depend upon recombination between the donor and recipient DNAs.

Genetic recombination in bacteria differs from sexual reproduction in eukaryotes, in that there is no union of gametes or gametic nuclei, and meiosis does not occur.

Phenotypic changes in bacteria can also occur by viral conversion, caused by the presence of viral DNA in bacterial cells. For example, strains of *Corynebacterium diphtheriae* infected by temperate phages produce diphtheria toxin; uninfected strains of the bacterium do not. The production of extremely potent toxins by some *Clostridium* species is also mediated through viral conversion.

Nutritional and Biochemical Characteristics

Morphology and major nutritional categories— e.g., heterotrophy and autotrophy—can be used to characterize families and orders of bacteria. However, the morphological similarity of taxa at lower levels is often so great that many other

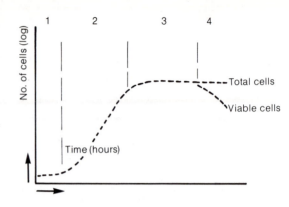

Figure 3-16 Typical growth curve of bacterial population in culture. **1,** lag phase; **2,** exponential growth phase; **3,** stationary phase; **4,** senescent phase.

types of characteristics must be used in classification. The characteristics used are primarily physiological or biochemical; only a few types will be described here.

Use of Gram's stain is a routine procedure in bacterial identification, as are several other staining procedures. Bacteria that can be grown in culture, whether parasites or saprobes, can also be characterized by their nutritional and environmental requirements. For example, some species are obligate anaerobes; others are obligate or facultative anaerobes. Most heterotrophic species can be grown on relatively simple, chemically defined media. They can require no more than a single organic compound to serve as an energy and carbon source, mineral salts, and water. However, others are more exacting and require one of more amino acids, vitamins, or other substances. Additionally, the ability to utilize a substance and the products of its use are both of value in characterizing bacterial species. For example, many bacteria are capable of utilizing glucose, but the manner in which they attack the molecule and the byproducts of dissimilation differ widely.

Antigenic Specificity If certain foreign substances are introduced into tissues of living vertebrates, the animal body will respond by producing protective substances called **antibodies.** The inducing substance, or **antigen,** can be a polysaccharide, a protein, or a polypeptide;

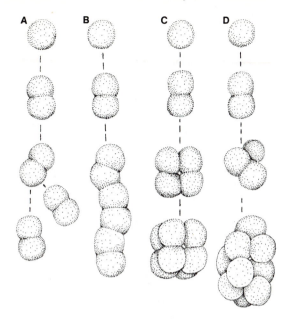

Figure 3-17 Diagrammatic illustration of pairs, chains, packets, and irregular clusters of cells characteristic of **A, B,** *Streptococcus;* **C,** *Sarcena* and **D,** *Staphylococcus.*

into laboratory animals, inducing antibody formation. The serum containing antibodies can then be used in testing unidentified bacteria; a positive reaction occurs only if the homologous antigen is present. Bacterial flagellins, capsular polysaccharides, pili, cell walls and their associated proteins, and other cellular proteins are all antigenic. Related species can have a number of common antigens; a given antigenic substance can also occur in more than one genus. However, relatively minor differences in the capsular polysaccharides of different strains of a single species can elicit the formation of different specific antibodies. At least 75 serological strains of the pneumoccus, *Streptococcus pneumoniae,* can be differentiated on this basis. Antibodies can also be labelled (dyed) with appropriate fluorescent dyes. If the labelled antibody is then applied to a mixture of bacteria, such as might be obtained with a throat swab, it will become attached only to cells carrying specific antigens for the labelled antibody. Such cells will fluoresce, when viewed by fluorescence microscopy, allowing immediate identification of the pathogen.

each antigen induces the formation of a specific antibody. Smallpox vaccine contains viral particles, and vaccination with this substance induces the formation of specific antibodies. The virulence of the virus has been reduced, so that vaccination does not cause smallpox in the vaccinated individual. However, the localized infection that results is sufficient to stimulate antibody production and thereby provide immunity to smallpox.

Antibodies are serum proteins that react or unite with inducing or homologous antigens. In agglutination reactions, when serum containing an antibody is mixed with a suspension of the homologous antigen, each antibody molecule can combine with two molecules of antigen. A network, or lattice, is formed, resulting in the production of a visible precipitate.

The specificity of antigen-antibody reaction and the fact that reaction between the two is often macroscopically visible have made this a useful tool in bacterial taxonomy. Killed whole bacterial cells, or portions thereof, can be injected

DNA and Classification Perhaps the most fundamental of the biochemical characteristics used in classifying bacteria are those concerned with the genetic material. Two types of DNA analyses are currently in use: (1) base-composition, or relative percentage of complementary guanine + cytosine pairs (% G + C); and (2) degree of genetic homology between DNA strands of different species. Both kinds of analysis require extraction and purification of DNA from the organism being examined.

The G + C values that have been obtained for bacterial species range from about 25% to almost 80% G + C. The range within a single large genus (e.g., *Bacillus*) may be as great as 20% G + C, but it commonly is not that great. Strains of a single species that differ by as much as 10% G + C cannot be considered closely related and, in most instances, probably have been incorrectly identified or classified. Morphologically similar genera, such as *Micrococcus* and *Staphylococcus,* can have very different values that would indicate only remote relationship. The genera *Actinoplanes, Mycobacterium, Streptomyces, Nocardia, Micromonospora* (all members of the Actinomyce-

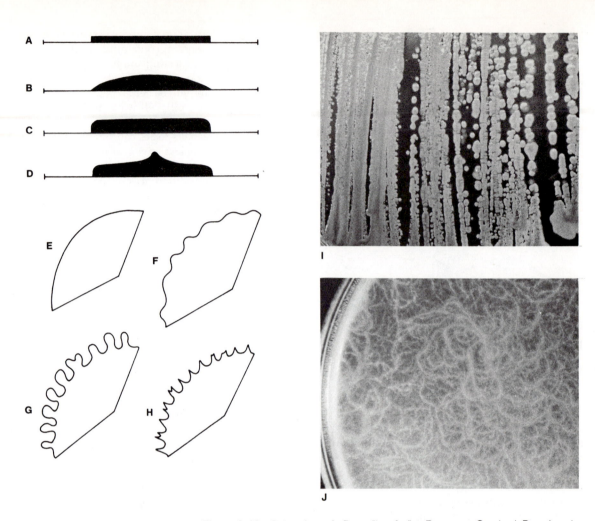

Figure 3-18 Colony form. **A–D,** profiles: **A,** flat; **B,** convex; **C,** raised, **D,** umbonate. **E–H,** margins: **E,** entire; **F,** undulate; **G,** lobate; **H,** erose. **I, J,** two types of bacterial colonies showing difference in general appearance, both ×1.2.

tales) have relatively close percentages of G + C, although they differ greatly morphologically.

Although greatly differing DNA base-composition values indicate a lack of relatedness between two species, and close values are expected for related species, close or identical values need not indicate close relationship. For example, many true fungi have G + C percentages that are close, or even identical, to those of certain bacteria.

The degree of genetic homology in the DNA from two taxa does provide for an estimate of relatedness. In these analyses, isotopically labelled

DNA is obtained by growing a test species on a medium containing a suitable isotopically labelled substance. A DNA preparation from the test species is mixed with a similar preparation of unlabelled DNA from the second species. The preparation is then heated to cause separation of complementary strands of DNA. Upon cooling, a number of unlabelled strands will recombine with labelled strands. The percentage of labelled-unlabelled strands that have combined can be determined and is considered to be an accurate indicator of relatedness. Two samples of DNA from a single strain of bacterium give a "hybridization" value arbitrarily set at 100%. Closely related strains will yield high recombination percentages; conversely, the more distantly related two taxa are, the lower will be the DNA hybridization value.

Orders of Bacteria

The orders of bacteria treated in the following discussion are those commonly recognized in traditional treatments of the group. Some orders now recognized (e.g., Caryophanales, Spirochaetales, and Rickettsiales) appear to be relatively coherent groups; others encompass rather heterogeneous assemblages of bacteria. The abundant studies currently underway in bacterial classification should eventually result in a more natural scheme.

The general introductory section of this chapter, covering structure and reproduction, applies primarily to the two orders Eubacteriales and Pseudomonadales. Because of their economic importance, these two groups have been intensively studied, and our knowledge of them is much more complete than that of other orders. These bacteria include many economically important pathogenic species, nitrogen-fixing bacteria, and industrially important species.

Order Eubacteriales (True Bacteria) Eubacteria are common in soil, water, sewage, decaying plant and animal remains, food, and the atmosphere. Many live as symbionts in association with higher plants or animals, either mutualistically or as parasites. All have simple coccoid or bacilliform cells with rigid walls. If the cells are motile, flagella are inserted over the entire

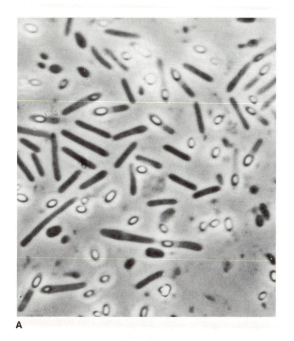

A

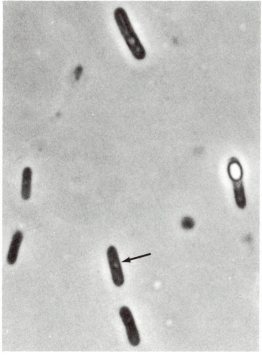

B

Figure 3-19 Bacterial endospores. **A,** heavily sporulating *Bacillus* culture, showing numerous mature and released endospores, ×2000. **B,** single cell with mature endospore, and a second cell with a clear area (*arrow*) indicating endospore formation, ×2000.

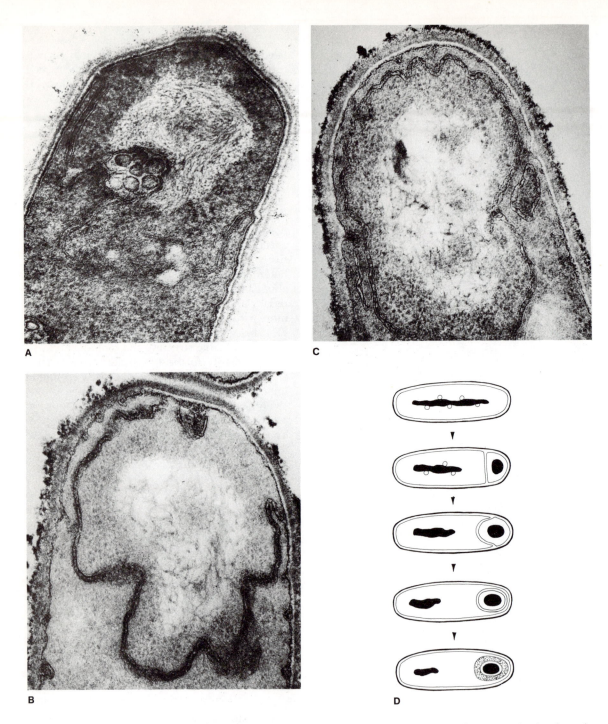

Figure 3-20 Endospore formation in *Bacillus*. **A,** initial membranous septum has formed, ×107,065; **B,** "engulfment" almost completed, ×101,100; **C,** completed forespore membrane, ×64,520; **D,** diagrams showing endospore formation. (Photographs courtesy of P. D. Walker, with permission of *Journal of Applied Bacteriology*.)

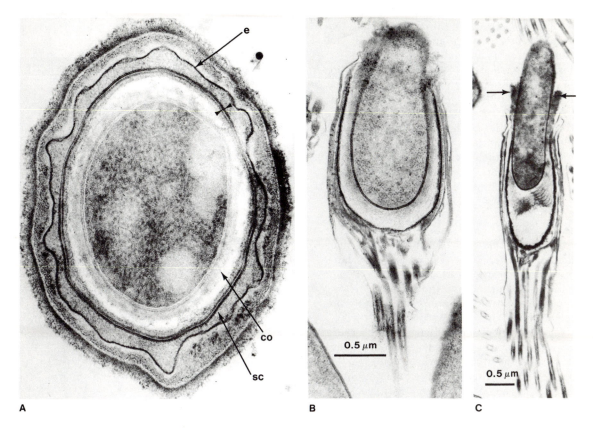

A   **B**   **C**

Figure 3-21 Bacterial endospores. **A,** *Bacillus.* Electron micrograph of thin section of mature endospore, ×95,100: *co,* cortex; *e,* exosporium; *sc,* spore coat layers. **B, C,** *Clostridium.* Germination and outgrowth of an appendaged endospore (*scale lines* shown are 0.5 μm). (**A,** photograph courtesy of P. D. Walker, with permission of *Journal of Applied Bacteriology;* **B, C,** photographs courtesy of W. A. Samsonoff, T. Hashimoto, and S. F. Conti, with permission of *Journal of Bacteriology.*)

cell surface. The order includes both gram-positive and gram-negative species.

Because some species are parasites, especially of man and domesticated animals, eubacteria have been intensively studied. Such diseases as anthrax (*Bacillus anthracis*), diphtheria (*Corynebacterium diphtheriae*), bubonic and pneumonic plague (*Pasteurella pestis*), lobar pneumonia (*Streptococcus pneumoniae*), and typhoid fever (*Salmonella typhi*), have resulted in death for thousands of humans. Botulism, caused by *Clostridium botulinum*, and other types of food poisoning or food infections (*Salmonella enteritidis*), have added to the toll. Whooping cough, undulant fever, gangrene, tetanus, gonorrhea, and meningitis are additional examples of diseases caused by eubacteria. Our domesticated animals are subject to the same or similar infections, and domesticated plants are also attacked by bacteria. Among the latter are bacterial canker of tomato (*Corynebacterium michiganense*), bacterial wilts (*Corynebacterium* and *Erwinia* spp.), crown gall of alfalfa (*Agrobacterium tumefaciens*), and others.

Among the many bacteria commonly associated with humans are coliform bacteria, including species of *Escherichia, Klebsiella, Aerobacter, Proteus,* and others. *Escherichia coli* is invariably present in the normal intestinal tract, as are certain species of *Clostridium* and *Streptococcus.* The coliform bacteria generally are abundant in water contaminated by fecal material and are used as indicator species in tests for contamination of water supplies.

Escherichia coli (Fig. 3-9) has been the subject of countless studies of many types and may rank as the most commonly used biological "tool." The ease with which it can be cultured, its common occurrence, and its rapid growth all make it a valuable organism for study.

The endospore-forming bacteria virtually all belong to the genera *Clostridium* and *Bacillus.* *Bacillus* species are mainly soil-inhabiting saprobes, but *B. anthracis* causes anthrax of man and animals; other species cause diseases of honeybees and other insects. Most *Bacillus* species are aerobic, in contrast to the closely related *Clostridium* species, which are predominantly obligate anaerobes.

Clostridia are economically important because of both detrimental and beneficial activities. Several species, the most notorious being *C. botulinum,* produce toxins that cause food poisoning. The toxins, called botulins, are proteinaceous and are the most toxic substances known. They result in numerous deaths annually. Species of *Clostridium* are common in soil, and their endospores are extremely resistant to heat. When vegetables or meats are canned, heating must be sufficient to kill all vegetative cells of bacteria and to destroy the much more resistant endospores. If contaminating endospores are not destroyed, they can germinate and reproduce in the anaerobic environment that occurs in sealed cans. Some instances of botulism have also been attributed to fish packaged in sealed plastic packages that exclude oxygen. Boiling of foods containing botulins for ten minutes will destroy the toxins.

Some *Clostridium* species are important free-living, nitrogen-fixing organisms; others produce a variety of interesting substances, such as butyric and acetic acids, *n*-butanol, acetone, isopropyl alcohol, ethanol, and butyl alcohol. Butyl alcohol is produced commercially by fermentation using *C. acetobutylicum.*

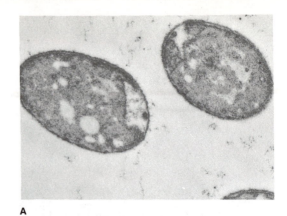

A

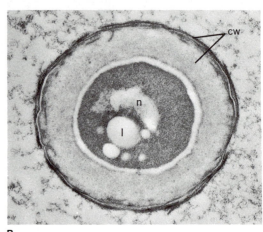

B

Figure 3-22 *Azotobacter.* **A,** vegetative cells, ×24,500. **B,** thin section of cyst, ×27,500: *cw,* cyst wall layers; *l,* lipid globule; *n,* nuclear material. (After M. D. Scolofsky and D. Wyss, with permission of *Journal of Bacteriology.*)

Nitrogen Fixation. Conversion of atmospheric N_2 into organic nitrogen, or **nitrogen fixation,** is carried out by many prokaryotes. Species of *Azotobacter, Bacillus, Beijerinckia, Clostridium,* and others are capable of fixing nitrogen as free-living forms. Other bacteria—e.g., *Rhizobium* species and certain actinomycetes—live symbiotically in root nodules and carry out nitrogen fixation in cooperation with higher plants. Species of *Clostridium* and other anaerobic, nitrogen-fixing species can contribute to the fertility of water-

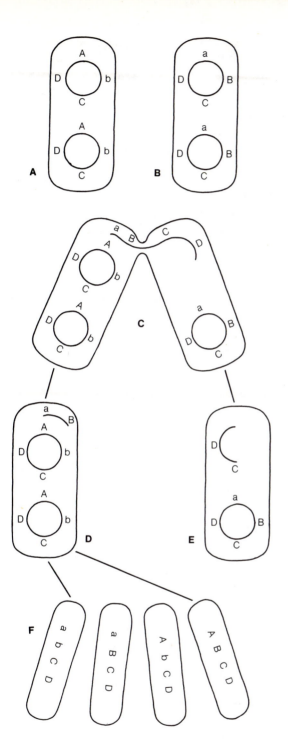

logged soils or mud where oxygen is low. *Azotobacter* (Fig. 3-22) is abundant in well-aerated agricultural soils in many regions of the world, but its contribution to soil fertility has not yet been accurately assessed. Nitrogen in a usable form is a major requirement of green plants and is often present in soils in limited amounts.

Symbiotic nitrogen-fixing bacteria occur in association with leguminous plants and with species of *Alnus, Eleagnus,* and others. By far the best-known and the most important symbiotic nitrogen-fixing bacteria are species of *Rhizobium* (Figs. 3-2 and 3-26) associated with legume roots. *Rhizobium* species exist as free-living forms in soil and are easily grown in culture. However, they will fix nitrogen in culture only if provided with nutrients normally obtained from their plant associate. Such nutrients include arabinose or galactose as a carbon source, nitrogen in the form of glutamine, and a citric acid cycle intermediate, such as succinate or fumarate.

Infection of legume roots occurs through root hairs. Root exudates are known to stimulate growth of the bacteria, and substances produced by the bacteria (e.g., indole acetic acid) affect the development of root hairs. *Rhizobium* cells become trapped in folds of root hair walls, and a filament of *Rhizobium* cells, called an infection thread, develops inside the root hair. The infection thread penetrates to the root cortex, and nodule formation ensues (Fig. 3-26). Within the nodule cells (Fig. 3-26B), multiplication of the *Rhizobium* cells occurs. Healthy root nodules are pink because of presence of a haeme-type pigment, leghaemoglobin. Leghaemoglobin does not have a direct role in nitrogen fixation, but probably transports oxygen to the nodule bacteria; it may also protect the critical enzyme, nitrogenase, from destructive contact with oxygen.

Figure 3-23 Bacterial conjugation. **A, B,** recipient and donor cells, each with two nuclear bodies. **C,** transfer of genetic material from donor to recipient via temporary fusion bridge. **D, E,** donor and recipient cells after conjugation. **F,** possible types of descendants of recipient cell **E** after conjugation.

The legume and *Rhizobium* association can fix more nitrogen than is required for their individual needs. Maximum estimates for a given field are as high as 500 pounds of fixed nitrogen per acre per year; smaller but still significant amounts are generally involved. Such fixation is of great importance agriculturally, and is responsible for the agricultural practice of planting legumes in depleted soils or alternating such plants with other crops. The quantity of nitrogen fixed in the symbiotic association depends upon the plant species and strain or species of *Rhizobium* involved. In some combinations, no nitrogen fixation occurs; the bacteria in such root nodules are essentially parasites. To insure colonization by the best strains, it is common agricultural practice to dust seeds with preparations of appropriate nitrogen-fixing bacteria.

The mechanism of nitrogen fixation is not yet completely known, although its practical importance makes this a subject of great interest. Reduction of N_2 to N_3 is energy-requiring, and bacteria generate the necessary ATP through metabolism of pyruvic acid. Ferredoxin, an iron-containing electron carrier, and the enzyme nitrogenase also are involved in nitrogen fixation. Nitrogenase, a critical enzyme in the process, is rapidly destroyed by contact with oxygen. Hence, all nitrogen-fixing systems involve some sort of protective mechanism to prevent such contact.

Food Bacteria. Many eubacteria grow in milk and other foods; their presence there may lead to spoilage, disease in those who consume the food, or preservation of the food, depending upon the bacteria involved. Several species of *Streptococcus*, *Lactobacillus*, and *Leuconostoc* are regularly present in milk; they are important in the manufacture of cottage cheese, cheese, yogurt, and similar fermented milk products. In cheese manufacture, such bacteria are added to milk and, together with the enzyme rennin, bring about curdling of the milk and coagulation of casein to form curd. The curd is salted and then ripened under conditions that favor the bacteria responsible for the given type of cheese. *Lactobacillus*, *Streptococcus*, *Propionibacterium*, and *Brevibacterium* species are important in the ripening process, as are molds for some cheeses. Microorganisms break down protein or fatty substance present in curd and, in

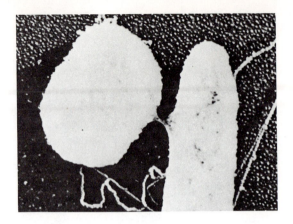

Figure 3-24 Conjugation in *Escherichia coli*, ×8300. The two mating strains are distinguishable on the basis of form, the Hfr strain (*right*) having elongate, narrow cells in contrast to the short broad cells of the F⁻ strain. The Hfr strain shown is motile, but the F⁻ strain lacks flagella. (Photograph courtesy of T. F. Anderson, E. L. Wollman, and F. Jacob, "Sur les processus de conjugaison et de recombinaison chez Escherichia coli III.—aspects morphologiques en microscopie électronique." *Annales de L'Institut Pasteur* 93:450–455, 1957.)

so doing, are responsible for the specific flavors of different cheeses.

Order Pseudomonadales (The Pseudomonads) Pseudomonadales include mostly gram-negative, motile bacilli, spirilla, vibrios, or short, ovoid forms. Flagella are polar and can occur either singly or in tufts at one or both poles of the cells. Pseudomonads are widely distributed in fresh water, sea water, mud, soils, and other habitats; the order contains both autotrophic and heterotrophic types. Some plant and animal pathogens are classified in Pseudomonadales, but these are much fewer than in Eubacteriales.

Photoautotrophic Pseudomonads. Three families of photoautotrophic bacteria are included in Pseudomonadales: purple sulfur bacteria (Thiorhodaceae); purple nonsulfur bacteria (Athiorhodaceae); and green sulfur bacteria (Chlorobacteriaceae). The three are differentiated on the

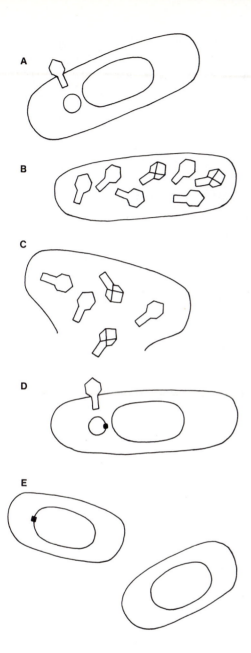

basis of their photosynthetic pigments and processes. Purple sulfur bacteria and purple non-sulfur bacteria have bacteriochlorophylls *a* and *b* and carotenoids; the latter obscure the green color of the chlorophyll. Green sulfur bacteria are characterized by the presence of bacteriochlorophylls *a* and *c, d,* or *e.* Although carotenoids are present, they do not mask the chlorophyll. Bacterial chlorophylls are similar to higher plant chlorophylls in their chemical structure. The photosynthetic sulfur bacteria utilize inorganic sulfur-containing compounds (e.g., H_2S) as hydrogen donors. Free sulfur results and is deposited as globules within the cells of most purple sulfur bacteria and externally in the case of green sulfur bacteria. Purple nonsulfur bacteria can utilize organic hydrogen donors, require an external source of B vitamins, and can live heterotrophically in the dark. The general overall reaction for photosynthesis in bacteria is similar to that in other plants:

$$CO_2 + 2\,H_2A \xrightarrow[\text{bacteriochlorophyll}]{\text{light}} CH_2O + H_2O + 2\,A$$

In higher plants, *A* would represent oxygen; in bacteria, it is sulfur or an organic substance. For example, if the hydrogen donor is H_2S the reaction is

$$CO_2 + 2\,H_2S \xrightarrow[\text{bacteriochlorophyll}]{\text{light}} CH_2O + H_2O + 2\,S$$

The free sulfur shown in this reaction, or that from other sources, can be further oxidized as follows:

$$2\,CO_2 + S^{--} + 2\,H_2O \xrightarrow[\text{bacteriochlorophyll}]{\text{light}} 2\,CH_2O + SO^{--}$$

Molecular hydrogen can replace that from H_2S in bacterial photosynthesis. Purple nonsulfur bacteria can use either molecular hydrogen or H derived from organic molecules to convert organic substances photosynthetically. For example, acetic acid can be converted to poly-beta-hydroxybutyric acid, the common intracellular reserve. One way in which this occurs is as follows:

Figure 3-25 Bacterial transduction mediated by bacterial virus or ''phage.'' **A,** virus injects DNA into bacterial cell; **B,** numerous new phage particles are produced. **C,** phage particles are released and defective particle (**D**) injects DNA strand, carrying small fragment of bacterial DNA into new bacterial cell. In **E,** the fragment transmitted by the phage is shown attached to bacterial chromosome.

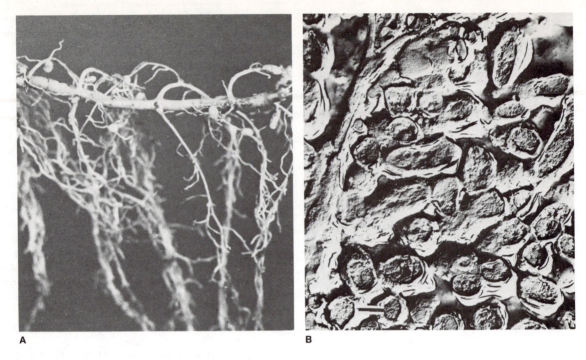

A   B

Figure 3-26   A, B, *Rhizobium.* **A,** nodules on roots of common white clover, ×1.5; **B,** cells within nodule cell of alfalfa (*Medicago*) (*scale line* shown is 1 μm). (**B,** photograph courtesy of C. R. MacKenzie, W. J. Vail, and D. C. Jordan, with permission of *Journal of Bacteriology.*)

$$2\,C_4H_4O_2 + H_2 \xrightarrow[\text{bacteriochlorophyll}]{\text{light}} (C_4H_6O_2) + 2\,H_2O$$

The purple nonsulfur bacteria can also reduce CO_2 photosynthetically, but they commonly convert preformed organic substances, as above. This photosynthetic process is remarkably efficient, and most substrate molecules are converted to cellular substances.

Bacterial photosynthesis occurs only in the absence of oxygen; molecular oxygen is neither liberated nor consumed in the process. Because of this feature and the light-absorption range of their photosynthetic pigments, bacteria carry on photosynthesis where other plants cannot. Photosynthetic bacteria often cause massive "blooms" in water where oxygen is limited, H_2S or certain organic substances are abundant, and sufficient light penetrates for photosynthesis.

Pigments of photosynthetic bacteria are not located in chloroplasts but are complexed with proteins to form reaction centers. These complexes are associated with vesicular or lamellar invaginations of the plasma membrane (Fig. 3-7B and 3-28) in purple bacteria; in the green sulfur bacteria, bacteriochlorophyll is associated with special flattened vesicles arranged around the periphery of the cytoplasm.

In addition to their ability to fix CO_2 through photosynthesis, several species of *Chromatium*, *Rhodospirillum*, and *Chlorobium* can fix nitrogen.

Among the more commonly encountered photosynthetic bacteria is the genus *Chromatium*. Its cells are coccoid to bacilliform or curved, and the diameter may be as great as 10 μm. These large, polarly flagellated cells typically contain refractile globules of sulfur (Fig. 3-27). *Chro-*

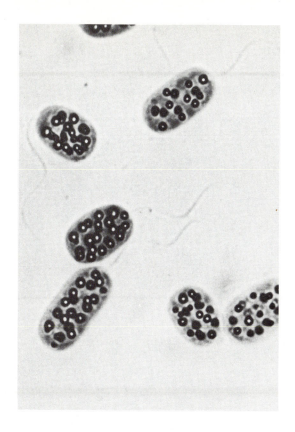

Figure 3-27 Phase-contrast photograph of living *Chromatium* cells, × 1900. *Note* flagella and refractile sulfur granules. (Photograph courtesy of H. G. Schlegel and N. Pfennig, with permission of *Archiv für Mikrobiologie.*)

matium occurs in marine or stagnant water. *Rhodospirillum* (Fig. 3-28A) and *Rhodopseudomonas* lack sulfur granules but are polarly flagellated. They require organic substances and grow in water or mud in which the requisite substances are present. *Chlorobium* species (Fig. 3-28B) occur in fresh and salt water or muds, especially under conditions of high sulfide and low pH. Cells of *Chlorobium* are nonmotile, coccoid to rod-shaped, and rather variable in form.

The H_2S used by photosynthetic sulfur bacteria (and by some chemoautotrophs) is produced through bacterial reduction of elemental sulfur or degradation of organic sulfur-containing substances.

H_2S is unstable under aerobic conditions but can accumulate in anaerobic environments such as marine black muds or sediments of estuaries and lakes.

Nitrifying Bacteria. An important part of the nitrogen cycle in soils involves the conversion of ammonium nitrogen to nitrite and nitrate. The genera *Nitrosomonas* (Fig. 3-29), *Nitrosococcus,* and *Nitrobacter* are important in these conversions. *Nitrosomonas* and *Nitrobacter* occur in neutral or alkaline soils and in manure. They are strictly aerobic and, in culture, are extremely sensitive to many types of organic substances. In nature, these organisms form colonies on mineral soil particles that are free of organic matter.

Nitrification by *Nitrosomonas* and *Nitrobacter* involves only the oxidation of ammonia to nitrite and nitrite to nitrate. The source of ammonia is proteins, amino acids, or other N-containing organic substances, and conversion of organic N to ammonia is carried out by many different organisms. Oxidation of ammonia to nitrite and nitrate can be diagrammed as follows:

Nitrosomonas
$$2\,NH_4 + 3\,O_2 \longrightarrow 2\,NO_2^- + H^+ + 2\,H_2O$$

Nitrobacter
$$2\,NO_2^- + O_2 \longrightarrow 2\,NO_3^-$$

The oxidation reactions are energy-yielding processes; the energy obtained is used in reduction of CO_2 to form organic substances. *Nitrosomonas* and *Nitrobacter* are therefore chemoautotrophs.

Denitrification—that is, reduction of nitrate to nitrite and nitrite to N_2—is common in poorly aerated habitats. Conversion of NO_3 to NO_2 is carried out by species of *Achromobacter, Pseudomonas,* and *Bacillus,* as well as by some fungi. Reduction of nitrite to free N_2 is carried out by only a few species of bacteria (e.g., *Thiobacillus denitrificans*).

Other Autotrophic Pseudomonads. Some chemoautotrophic pseudomonads obtain energy through oxidation of hydrogen, carbon monoxide, methane, or sulfur. Such organisms occur in mud or water in which decaying materials provide requisite oxidizable substrates. *Thiobacillus*

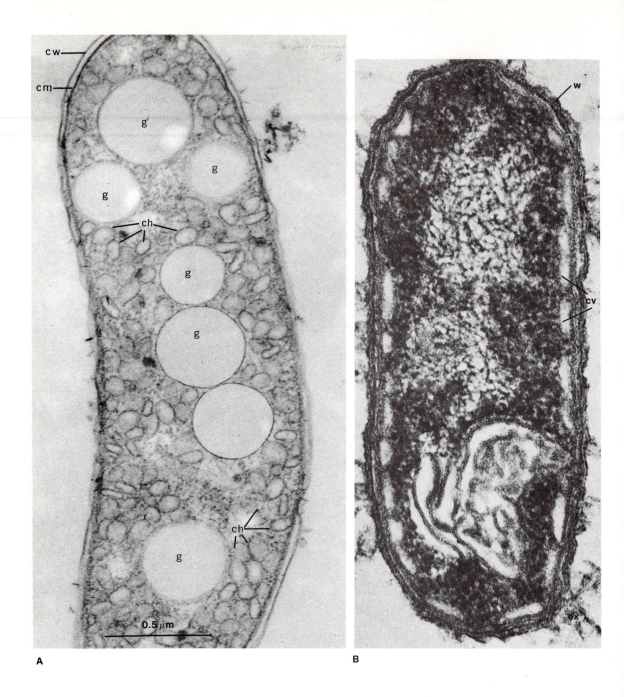

Figure 3-28 A, *Rhodospirillum rubrum*. Electron micrograph showing vesicular chromatophores: *cw,* cell wall; *cm,* cytoplasmic membrane; *ch,* chromatophore; *g,* granule. **B,** thin section of *Chlorobium* cell, × 131,800: *cv,* cortical vesicles; *w,* cell wall; *ex,* capsular material. (**A,** photograph courtesy of E. S. Boatman, "Observations on the fine structure of *Rhodospirillum rubrum," Journal of Cell Biology* 20:297–311, 1964, with permission of Rockefeller University Press; **B,** photograph courtesy G. Cohen-Bazire.)

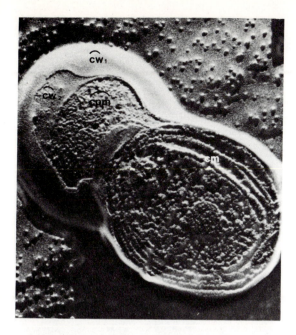

Figure 3-29 *Nitrosomonas.* Cellular structures as revealed by a replica of freeze-etched preparation, ×60,300: *cw*₁, *cw*₂, wall layers; *cpm*, cytoplasmic membrane; *cm*, intracytoplasmic membranes. (Photograph courtesy of A. P. Van Gool, with permission of *Archiv für Mikrobiologie.*)

species, among the most interesting of those oxidizing sulfur, include strains capable of growing either autotrophically or heterotrophically. These bacteria occur in acidic environments and, since they can oxidize H_2S to sulfur or sulfuric acid, their activities increase the acidity of their environment. The formation of sulfuric acid and tolerance to strongly acid conditions are important factors in the ability of this organism to compete with other microorganisms.

Heterotrophic Pseudomonads. In culture, many of the approximately 150 described species of *Pseudomonas* (Fig. 3-13) produce water soluble, fluorescent pigments. These predominantly aerobic, polarly flagellated bacilli occur in soil or water. More than half the described species are pathogenic to plants. Bacterial wilt of such plants as tomato and tobacco, bacterial stripe and halo blight of oats, and many other plant diseases are caused by pseudomonads. Few *Pseudomonas* spe-

cies are parasites of animals, but the specific epithet of *P. aeruginosa* is derived from the fact that this species produces bluish pus in infected wounds in man and many other animals.

Pseudomonas species are versatile scavengers that attack and degrade a great variety of organic substances. The latter include not only simple sugars, polysaccharides (such as starch and cellulose), and proteins, but various hydrocarbons (e.g., kerosene and gasoline) as well.

Stalked Pseudomonads. The stalked bacteria, sometimes classified as the Caulobacteriales, include a heterogeneous group of aquatic organisms. *Caulobacter* (Fig. 3-30), the best known genus, has straight or curved cells attached to submerged objects or to the surface film of water. Stalks are sometimes attached to other microorganisms, but this does not seem to harm the latter. The stalk is tipped with an adhesive holdfast; the stalk wall is continuous with that of the cell. Reproduction in *Caulobacter* involves transverse fission of the cell (Fig. 3-30B), the distal "daughter" cell being stalkless and motile by means of a polar flagellum. When released, the new cell swims for a time and eventually attaches to some object and produces a stalk. Attachment is at the flagellar pole, and reproduction does not occur until a new stalk has been formed. Although the stalk wall and cell wall are continuous, the stalk of *Caulobacter* is equipped with a series of transverse "bulkheads." Some recently described bacteria—e.g., *Ancalomicrobium* (Fig. 3-31)—have from two to eight cellular extensions, but none serves in attachment. The cellular extensions of this group are of unknown function but might serve to increase the surface area of the cells.

A second genus of stalked bacteria, *Gallionella* (Fig. 3-32), is very unlike *Caulobacter*. *Gallionella* occurs in iron-containing fresh water and in sea water. The cells are short, curved, or bean shaped and are attached to a ribbonlike structure, the stalk, by their concave sides. For many years, the view was held that the stalklike structure consisted of a nonliving secretion. However, more recent investigations indicate that the stalk consists of minute fibrils embedded in an organic matrix, and that it has some reproductive capacity. Small budding cells and sporangiumlike sacs have been found in association with the stalk.

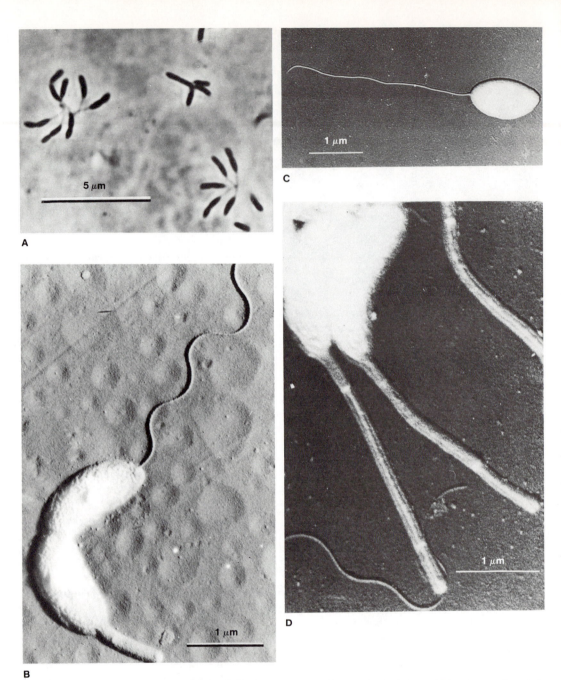

Figure 3-30 A–D, *Caulobacter.* **A,** phase-contrast photograph showing rosettes of at-tached cells (*scale line* shown is 5 μm); **B,** dividing cell, the daughter cell (*right*) with single polar flagellum; **C,** swarmer after release from parent cell; **D,** detail of stalk (note that the stalk has a distinct central core and septa, *arrows*) (*scale lines* shown in **B–D** are 1 μm). (Photographs from J. S. Poindexter, ''Biological properties and classification of the *Caulobacter* group.'' *Bacteriological Reviews* 28:231–295, 1964, with permission of *Bac-teriological Reviews.*)

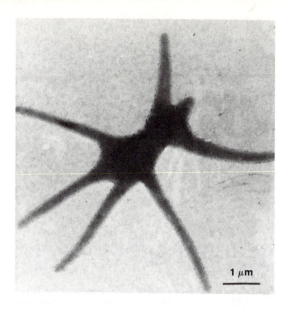

Figure 3-31 *Ancalomicrobium* cell with several cellular extensions (*scale line* shown is 1 μm). (Photograph courtesy of J. F. Staley, with permission of *Journal of Bacteriology*.)

Thin sections and electron microscopy of the bean-shaped cells have shown the presence of an organelle apparently responsible for stalk formation. The stalk is secreted continuously, is twisted, and is impregnated with iron oxides.

Curved Pseudomonads. Many bacterial cells of the spirilla and vibrio type belong to species of Pseudomonadales. *Vibrio* (Fig. 3-33) species include a number of human and animal parasites, the most important being *V. comma*, the cause of Asiatic cholera. The recorded cases of this disease over the past twenty years have ranged from a low of 20,000 to highs of around 100,000 annually. The disease, frequently fatal, is transmitted by contact, or by water or food contaminated with fecal material from diseased individuals or carriers.

Spirillum species are common in stagnant water, especially if there is an abundance of decaying matter present. They are motile by means of polar flagella (Fig. 3-12) and have rigid walls. Reproduction is by binary fission, but some spe-

cies produce coccoid bodies in aging cultures. These coccoid bodies germinate when placed in fresh medium and give rise to new spirilla.

One of the most interesting parasitic pseudomonad genera is *Bdellovibrio* (Fig. 3-34), whose species parasitize other bacteria. The very small, curved rods swim at an astonishing speed estimated at 100 cell lengths per second. Attachment occurs upon contact or collision with a suitable host, and a pore is formed in the host cell wall. The *Bdellovibrio* cell passes through the pore; it does not penetrate the protoplast, but the latter is consumed. Elongation of the parasite then occurs and a C- or spiral-shaped filament up to ten times the usual cell length is formed. This filament undergoes segmentation into new cells, which are then released.

Order Chlamydobacteriales (Sheathed Bacteria) The sheathed bacteria are aquatic organisms that form motile cells called **swarmers.** The bacilliform swarmers of *Sphaerotillus* (Fig. 3-35) are equipped with a subpolar tuft of flagella. After swimming for a time, motile cells become fixed to solid objects. Adhesion occurs at one pole by means of a gummy substance secreted by the cells. Growth of the cell and fission results in a multicellular filament, or **trichome,** with a distinct holdfast. During growth, a capsular substance is secreted; a sheath is present beneath the capsular material. In the presence of organic iron compounds, iron is deposited in the sheath and it becomes yellowish-brown or brown. Sulfur accumulates within the cells in the presence of H_2S, and granules of poly-beta-hydroxybutyrate also are present within the cell. Swarmers are released through the open tip of the filament or at points where it ruptures. The motile cells often attach to sheaths of *Sphaerotillus* and thus produce false branches. Sheaths of *Sphaerotillus* are resistant to decay and empty sheaths are common in many freshwater habitats.

Order Spirochaetales (Spirochaetes) Spirochaetes are spirilla with extremely thin, flexible walls and polar flagella. Flagella occur in a tuft at either end of the cell and are morphologically and chemically similar to other bacterial flagella. However, they do not extend out from the cell of

A

B

Figure 3-32 A, B, *Gallionella ferruginea.* Electron micrographs showing stalks and bean-shaped terminal cells: **A,** ×75,000; **B,** ×65,000. Note that each stalk consists of a number of strands; the very large stalk in **A** is tipped by two cells. (Photographs courtesy of A. E. Vatter and R. S. Wolfe, with permission of *Journal of Bacteriology.*)

the bacterium but are spirally wound around the body (Fig. 3-36). If transverse sections of the cell are examined under the electron microscope, posterior and anterior bundles of flagella can be seen to overlap in the midsection of the body; the bundles are enclosed within a thin outer membrane.

Spirochaetes are actively motile (Fig. 3-36A, B), but their motility is very different from that of other spirilla (p. 138). The mechanism involved is as yet unclear. Flexing of cells often occurs, and when the cell is in contact with a solid substrate, creeping movement is visible.

Cells of Spirochaetales are narrow; the diameter for most species does not exceed 0.75 μm, but the length often exceeds 300 μm. The transparent cells do not stain readily with many common bacterial stains, making observations difficult.

All spirochaetes are heterotrophic; some are free-living, mud- or water-inhabiting species, and some are symbionts associated with animals. *Treponema pallidum,* causing syphilis, and *Leptospira icterohaemorrhagiae,* causing spirochaetal jaundice, are among the more important species parasitizing humans.

Order Hyphomicrobiales (Budding Bacteria)

Budding bacteria are characterized by a type of reproduction in which division results in unequal cells. In bacterial **budding,** a bleb develops either at the pole of a cell or at the tip of a tubular extension developed from the pole. The bleb increases in length, becomes walled off from the parent cell, and generally is released before attaining the size of the parent cell. A number of budding bacteria have complex membranous systems internally, and budding is seen as a method of reproduction that does not require disruption or reorganization of this system of parallel membranes. In *Rhodomicrobium* (Fig. 3-37A), photosynthetic pigments are associated with the membranous lamellae.

Rhodomicrobium and *Hyphomicrobium* are morphologically similar, but nutritionally distinct genera of budding bacteria. *Rhodomicrobium* (Fig. 3-37A–C) cells are photoautotrophic and contain carotenoids and bacteriochlorophyll. Their anaerobic photosynthetic process, like that of the Athiorhodaceae, requires organic hydrogen donors. *Hyphomicrobium* (Fig. 3-37D), by contrast, is a colorless heterotroph.

Rhodomicrobium has top-shaped or broadly oval cells (Fig. 3-37) interconnected by narrow, tubular filaments. A terminal cell can become a flagellated swarmer, or it may remain attached and produce a filament and bud. The swarmers have a number of polar flagella and, after a motile period, become fixed to a solid substrate. A tubular filament then develops opposite the point of attachment, and budding occurs. The tubular filaments are provided with **septa** between cells, and they sometimes branch.

Hyphomicrobium differs from *Rhodomicrobium*

in its nutrition and in having motile cells with a single polar flagellum. It occurs in soil and aquatic habitats; it is a denitrifier.

Order Myxobacteriales (Slime Bacteria) Slime bacteria have extremely thin-walled, flexible bacilliform cells that are capable of gliding and flexing movements. The cells lack flagella, and their motility requires contact with a solid substratum. Colonies of these bacteria produce slimy materials and, during migration of cells, a visible trail of slime is often present in cultures. Slime bacteria are common in soil, decaying vegetation, dung, and on tree bark, occurring wherever other bacteria are abundant.

Myxobacteria "prey" upon other bacteria and small microorganisms. They produce antibiotics that are especially active against gram-positive bacteria and possibly assist in their mode of nutrition. They secrete enzymes in addition to the antibiotics, and these extracellular enzymes are capable of digesting the cell walls and protoplasts of other bacteria and probably fungi. Cells of the myxobacteria divide by binary fission; they also form resistant cysts.

The most striking feature of myxobacteria is their production of communal, macroscopically visible "fruiting" bodies (Figs. 3-38 and 3-39). In fruiting body formation, vegetative cells heap together and form **sessile** or stalked structures of varied degrees of complexity. In *Myxococcus* (Fig. 3-38), the **fructification** is little more than a colored drop of slime in which the cells have become encysted. In *Chondromyces* (Fig. 3-39), the fruiting body is stalked, branched, and shrublike. Branch tips bear large **macrocysts,** each of which contains many rod-shaped cysts. Upon germinating, each cyst gives rise to a single bacilliform cell.

Order Beggiatoales (Gliding Bacteria) Gliding bacteria are colorless, predominantly filamentous organisms; the cells lack flagella, but gliding and flexing movements of entire filaments occur in most genera. Gliding bacteria are found in fresh or salt water, and in moist terrestrial habitats in decaying material.

The cells of some species contain refractive globules of sulfur and granules of poly-beta-hy-

Figure 3-33 *Vibrio.* Shadowed cells showing characteristic curved forms, × 19,450. (Photograph courtesy of A. E. Ritchie, R. F. Keeler, and J. H. Bryner, from "Anatomical features of *Vibrio fetus,*" *Journal of General Microbiology* 43:427–438, 1966, with permission of Cambridge University Press.)

droxybutyrate. Reproduction of the cells is by fission. The numbers of filaments increase through fragmentation and by production of short spore-like segments resembling hormogonia of blue-green algae (see p. 77).

Beggiatoa species (Fig. 3-40A–C) are known to be heterotrophic, but in the presence of H_2S, the cells deposit sulfur internally. Filaments of *Beggiatoa* remain unattached and are capable of both gliding and flexing movements. The species of this genus bear a remarkable similarity, morphologically and in their movement, to those of *Oscillatoria* (see p. 157 and Figs. 3-48C, 3-51A).

Thiothrix (Fig. 3-40D) is similar to *Beggiatoa*, but its filaments become attached to solid objects and are nonmotile. Each filament has a basal holdfast, and groups of filaments are commonly arranged in rosettes. Reproduction is by fission and by sporelike segments as in *Beggiatoa*. Segments can glide when released, but soon become fixed to nearby objects.

Order Rickettsiales (The Rickettsias) Rickettsias are obligate intracellular symbionts found primarily in the cells of arthropods—e.g., lice, fleas, ticks, and mites. Some occur in the cytoplasm of infected cells; others are found in or associated with the nucleus. A number of species occurring in arthropods infect humans and cause such diseases as epidemic and murine typhus, Rocky Mountain spotted fever, Q-fever, tsutsugamushi (scrub typhus), and rickettsialpox. Other species cause a number of important livestock diseases. Arthropod vectors of these pathogenic species are apparently unharmed by the bacteria they transmit.

Rickettsias, because of their small size and obligate intracellular parasitism, once were considered intermediate between viruses and true bacteria. However, electron microscopy has revealed that their structure is essentially like that of eubacteria; the chemistry of their cell walls also is the same as in that group, and they are sensitive to similar antibiotics. The cells are mostly rod- to coccus-shaped and nomotile. They reproduce by binary fission.

The most important rickettsial species is *Rickettsia prowazekii* (Fig. 3-41), the cause of epidemic typhus. Epidemics of this disease have regularly accompanied wars, famine, and other disasters during the history of man. In many early wars, typhus was responsible for many more deaths than occurred in fighting. In the siege of Granada in 1489 and that of Naples in 1528, 17,000 Spanish and 30,000 French soldiers, respectively, were struck down by typhus. Recent wars have seen a reduction in cases of typhus, although about 30,000 cases were reported in Japan and Korea in 1946–47. The disease is transmitted by lice.

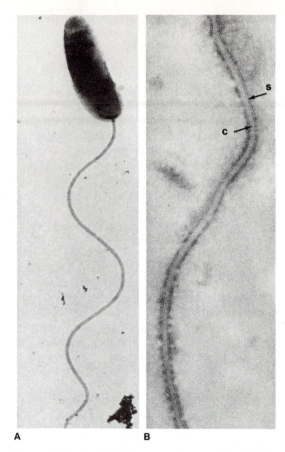

A   **B**

Figure 3-34 **A,** *Bdellovibrio bacteriovorus* cell, ×27,775; **B,** enlarged flagellum, ×71,760: *c*, core; *s*, sheath. (Photograph courtesy of R. J. Seidler and M. P. Starr, with permission of *Journal of Bacteriology*.)

Rickettsia prowazekii is named after H.T. Ricketts and S. van Prowazek, pioneers in the study of rickettsial diseases. Both men died of typhus contracted during laboratory studies of the disease.

Order Actinomycetales (Actinomycetes) Actinomycetales include a number of moldlike bacteria, most of which produce tubular filaments called **hyphae.** In some, hyphae are short, infrequently branched, and soon fragment into rodlike segments. In others, hyphae are extensive, nonfragmenting, and produce moldlike

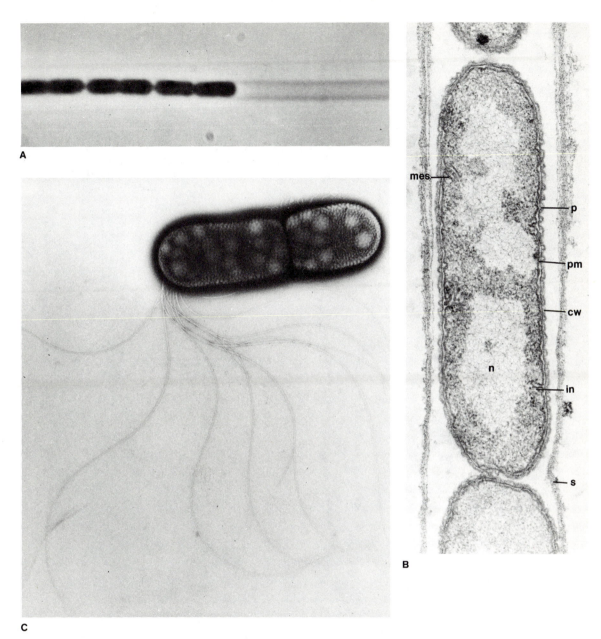

Figure 3-35 A–C, *Sphaerotillus natans.* **A,** sheathed filament, the cells of which contain numerous storage granules, ×3600. **B,** electron micrograph of thin section of filament, ×13,500: *mes,* mesosome; *p,* peptidoglycan wall layer; *pm,* plasma membrane; *in,* intrusion of plasma membrane; *n,* nucleoid, **C,** negative stained preparation of swarm cell after first division, ×4600 (note subpolar tuft of flagella characteristic of this species; granules visible in **C** are polyhydroxybutyric acid). (Photographs courtesy of J. F. M. Hoeniger, H.-D. Tauschel, and J. L. Stokes, reproduced by permission of the National Research Council of Canada, from the *Canadian Journal of Microbiology* 19:309-313, 1973).

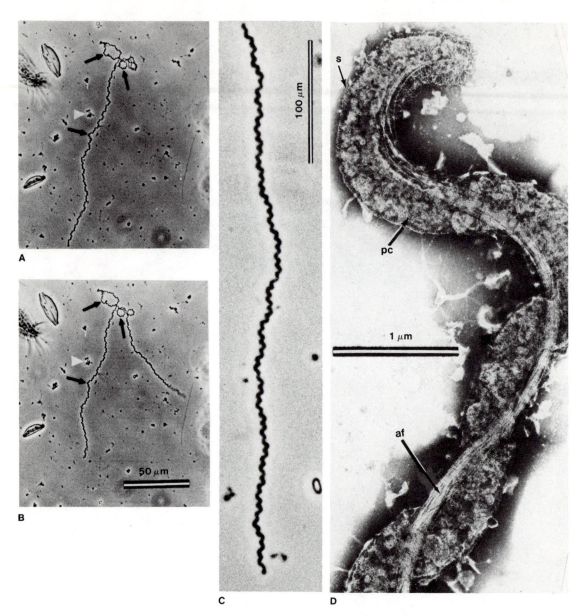

Figure 3-36 *Spirochaeta plicatilis.* **A, B.** living cell, the two photographs taken several seconds apart, illustrating the creeping motion common to this group (*scale line* shown is 50 μm). Note that net displacement to the right has occurred, as indicated by the position of the cell ends and white marker. However, the cell has maintained a fixed pattern of coils (*arrows*). The extreme length of the cell can be judged by comparing with the diatom cells visible in the photographs. **C,** phase-contrast photomicrograph of extended cell (*scale line* shown is 50 μm). **D,** electron micrograph of one end of cell (*scale line* shown is 1 μm); *af,* axial fibrils; *pc,* protplasmic cylinder; *s,* outer sheath. (Photographs courtesy of R. P. Blakemore and E. Canale-Parola, with permission of *Archiv für Mikrobiologie.*)

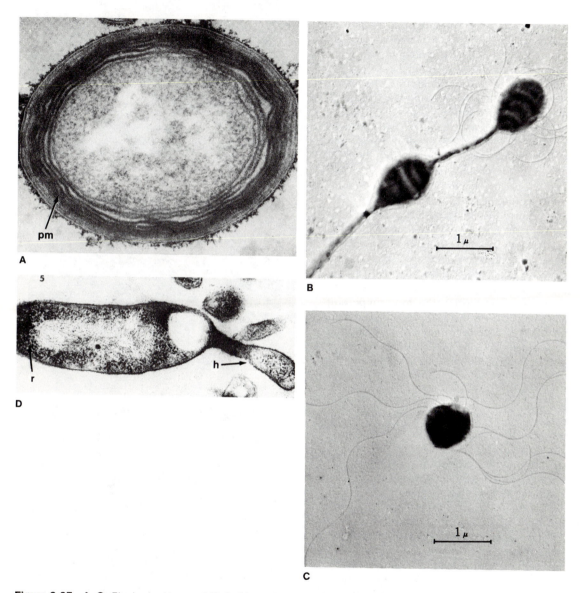

Figure 3-37 **A–C,** *Rhodomicrobium vaniellii.* **A,** thin section of cell showing membrane system, ×67,700 **B,** cells attached to tubular filament; **C,** flagellated swarmer (*scale lines* in **B** and **C** are 1 μm). Note flagella on terminal cell in **B,** a swarmer that has not yet been released. **D,** *Hyphomicrobium.* Thin section of cell and portion of tubular process, ×47,065; *h,* hypha or tubular process; *r,* ribosomes. (**A, D,** photographs courtesy of S. F. Conti and P. Hirsch, with permission of *Journal of Bacteriology;* **B, C,** courtesy of H. C. Douglas and R. S. Wolfe, with permission of *Journal of Bacteriology.*)

colonies. The hyphae have a typical prokaryotic structure with numerous nuclear bodies, mesosomes, and so forth; they may lack septa, but these are generally formed either in association with branching, with age and fragmentation, or with spore production. These hyphae are mostly under 1.5 μm in diameter and thus are smaller than most fungal hyphae. All actinomycetes are heterotrophic; the order is divided into several families based upon peptidoglycan chemistry, the extent of hyphal development, and the method of reproduction.

The actinomycetes are abundant in soil, fresh water, salt water, decaying material, and dung. The characteristic odor of ploughed soil and the foul taste of some drinking water are attributed to substances produced by some members of this group.

Morphologically, the simplest actinomycetes are members of the genus *Mycobacterium*. These form only short, infrequently branched filaments that soon fragment into rod-shaped segments (Fig. 3-42). Cell walls of *Mycobacterium* contain large amounts of lipid or waxlike substances. *Mycobacterium tuberculosum*, the most important pathogenic species, is the cause of human tuberculosis and is reported to cause 2–3 million deaths annually. Bovine tuberculosis, caused by *M. bovis*, is an important disease of cattle, and the species sometimes infects man. *M. leprae* is the cause of leprosy, and other species are responsible for a number of tuberculosis-like diseases of animals.

Actinomyces produces a more extensive **mycelium** than that of *Mycobacterium*, but it, too, soon undergoes fragmentation. Fragments may be rod-shaped or branched. Species of *Actinomyces* sometimes infect humans and animals; lumpy jaw of cattle is an example of the latter infections.

In *Nocardia* (Fig. 3-42B, C) as in *Actinomyces* and *Mycobacterium*, fragmentation generally is the method of reproduction. Rod- or coccus-shaped cells are produced; the former have been reported to be polarly flagellated, with one or more flagella in some species.

Species of *Nocardia* are predominantly soil saprobes, but some cause diseases of humans and animals.

Species of *Streptomyces* have extensive, nonfragmenting hyphae, masses of which resemble

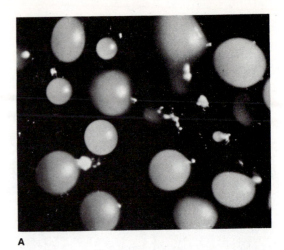

A

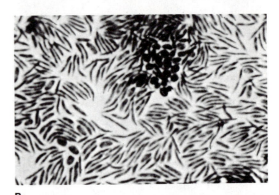

B

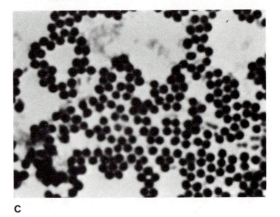

C

Figure 3-38 *Myxococcus.* **A,** mature fruiting bodies, ×30; **B,** cells fixed and stained at time of cyst formation, ×1100; **C,** stained, mature cysts, ×1500. (Photographs **A** and **B** by N. A. Woods, after Henrici and Ordal, with permission of D. C. Heath and Company.)

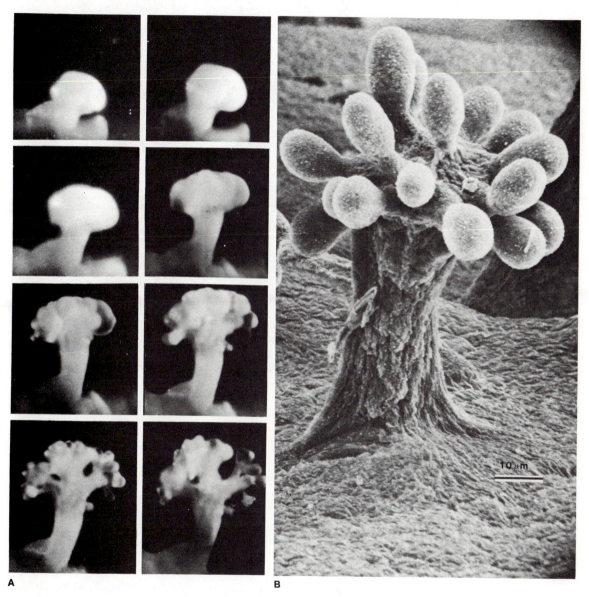

Figure 3-39 *Chondromyces.* **A,** successive stages in fruiting body formation, × 13.8. **B,** scanning electron micrograph of mature fruiting body (*scale line* shown is 10 μm). (**A,** photographs courtesy of J. T. Bonner, from John Tyler Bonner, *Morphogenesis* (Princeton, N.J.: Princeton Univ. Press. 1952. Reprinted by permission of Princeton University Press; **B,** photograph courtesy of P. L. Grilione and J. Pangborn, with permission of *Journal of Bacteriology.*)

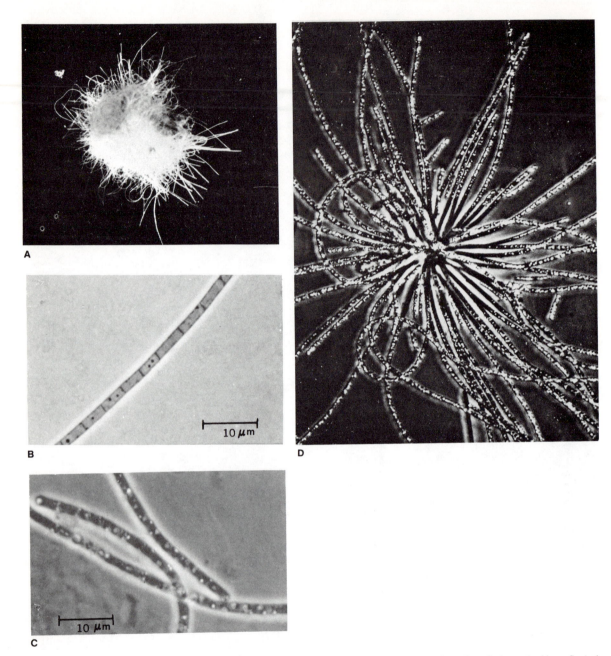

Figure 3-40   A–C, *Beggiatoa.* **A,** colony of filaments in water, photographed by reflected light; **B, C,** details of filaments, phase-contrast (*scale lines* shown are 10 μm). Transverse walls are visible in the dying filament in **B** and sulfur granules are present in **C.** The latter filaments were exposed to an atmosphere containing hydrogen sulfide. **D,** *Thiothrix* filaments attached to a common object, ×650. (**A–C,** photographs courtesy of L. Faust and R. S. Wolfe, with permission of *Journal of Bacteriology.* **D,** photograph courtesy of E. J. Ordal and F. E. Palmer.)

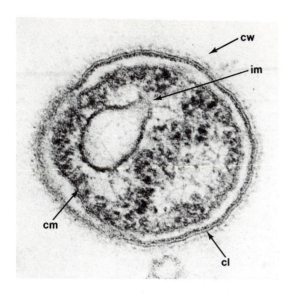

Figure 3-41 *Rickettsia prowazekii.* Thin section from cell cultured in chick yolk sac, ×135,650: *cw,* cell wall; *im,* intracytoplasmic membrane; *cl,* capsule; *cm,* cytoplasmic membrane. (Photograph courtesy of R. L. Anacker, E. G. Pickens, and D. B. Lackman, with permission of *Journal of Bacteriology.*)

mold colonies (Fig. 3-43A, B). Aerial hyphal branches produce chains of **conidia.** Conidia, developing in straight or coiled hyphal tips (Fig. 3-43C), are smooth or spiny and dry. They are predominantly airborne structures and are not so resistant to heat or other adverse conditions as are the endospores of true bacteria.

Some species of *Streptomyces* have been isolated from human and animal infections, and one, *S. scabies,* causes a potato scab disease. *Streptomyces* species, mainly soil inhabitants, are a major source of useful antibiotics, e.g., streptomycin, nystatin, amphotericin B, chloramphenicol, aureomycin, and tetracycline.

Members of the aquatic genus *Actinoplanes* (Fig. 3-44) produce extensive mycelium and motile sporangiospores. Polarly flagellated spores are produced in vesicular swellings or sporangia.

Order Mycoplasmatales (Mycoplasmas or PPLOs) Most Mycoplasmatales are included in the genus *Mycoplasma.* The first species dis-

covered, *M. mycoides,* is the cause of bovine pleuropneumonia; it is often referred to as the *pleuropneumonia organism,* or *PPO.* Related or morphologically similar forms often are called *pleuropneumonialike organisms,* or *PPLOs.*

The cells of *Mycoplasma* species lack cell walls and are enclosed only in a cytoplasmic membrane. Cells can be coccoid, filamentous, branched, or irregular in form (Fig. 3-45). Filamentous cells reach 2–50 µm in length, but they are generally less than 0.5 µm wide. Reproduction has been reported to be both by a form of fragmentation and by the production of small coccoid bodies, which are only 0.1–0.3 µm in diameter and can pass through bacterial filters. They develop in very large numbers within large, older cells and, upon release, give rise to new cells.

Mycoplasmas require complex organic media for growth in culture, and a sterol must be added to the medium. They grow slowly, many producing small, "fried egg" colonies with raised, granular centers and translucent, flat peripheral areas.

In addition to bovine pleuropneumonia, *Mycoplasma* species cause *Mycoplasma* pneumonia in humans—especially in young adults and children—and infections of the respiratory and genital tracts of humans and animals. Some species have been found to cause plant diseases; others are saprobes found in sewage, soil, and other habitats.

Division Cyanophyta (Bluegreen Algae)

The division Cyanophyta (= Cyanobacteria) comprises a single class, the Cyanophyceae. The organisms are generally bluegreen but are sometimes rust-red to brown or black. These photoautotrophic prokaryotes are ubiquitous, as are bacteria, and are also represented by some of the oldest fossils. Bluegreen algae have many similarities and interrelationships with the eukaryotic algae and are considered as algae rather than cyanobacteria in this text. Cells of bluegreen algae range from less than 1 µm to several micrometers and are rarely the size of eukaryotic algal cells.

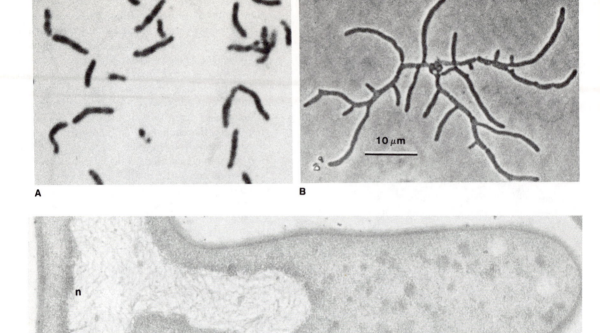

Figure 3-42 A, *Mycobacterium.* Photomicrograph of cells grown in culture, ×3000. **B, C,** *Nocardia.* **B,** phase contrast micrograph of filaments (*scale line* shown is 10 μm); **C,** thin section electron micrograph showing nucleoid (*n*) (*scale line* shown is 1 μm). (**B, C,** photographs courtesy of B. L. Beaman and D. M. Shankel, with permission of *Journal of Bacteriology.*)

Cell Structure

The cell is composed of a layered cell wall—generally of four layers—with the innermost one consisting of peptidoglycans (also known as *mucopeptides* or *glycoprotein*), as in bacteria. The cell wall is most similar to that of gram-negative bacteria (see p. 33). External to the wall may be a fibrous sheath (Figs. 3-46A and 3-47E), which appears to originate from the outer wall layer. Sometimes the sheath is very noticeable with the light microscope, being highly colored and/or stratified. It is considered to be composed of mixed polysaccharides, although a great deal more investigation is needed to elucidate its chemical nature.

Internal to the wall is the plasma membrane, which has the typical tripartite structure of membranes. In some bluegreen algae, invaginations of this membrane are present; these are considered comparable to the mesosomes of gram-positive bacteria (see p. 36).

The pigments in photosynthetic lamellae (the thylakoids) are distributed evenly throughout the cell (Fig. 3-46A). The thylakoids, which are closed discs, contain chlorophyll and carotenoids. Associated with the thylakoids are red and blue pigments, the **phycobiliproteins** (*phyco* = alga; *bili* = bile), in granules, the **phycobilisomes** (*some* = body) (Fig. 3-47B). The phycobiliproteins are present in other algae (see Chapter 9, Cryptophyta; Chapter 11, Rhodophyta).

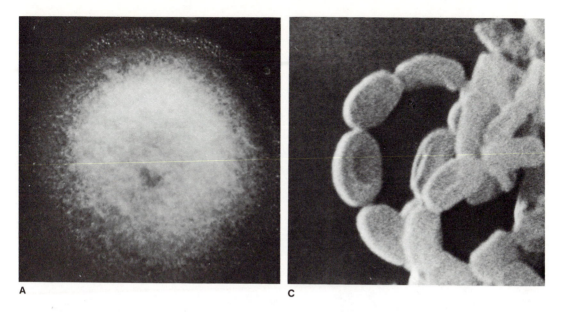

A

C

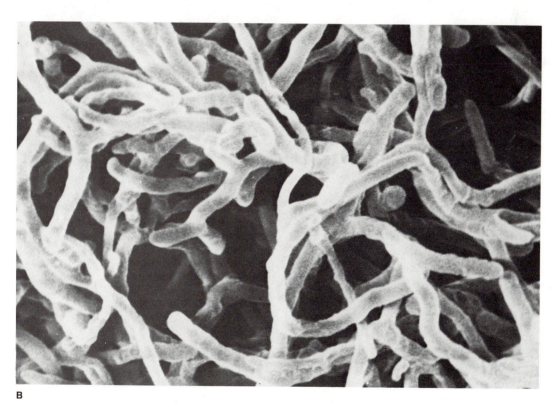

B

Figure 3-43 *Streptomyces.* **A,** young colony, ×50; **B,** scanning electron micrograph of colony showing typical filamentous growth, ×5000; **C,** scanning electron micrograph of conidial chain, ×8000.

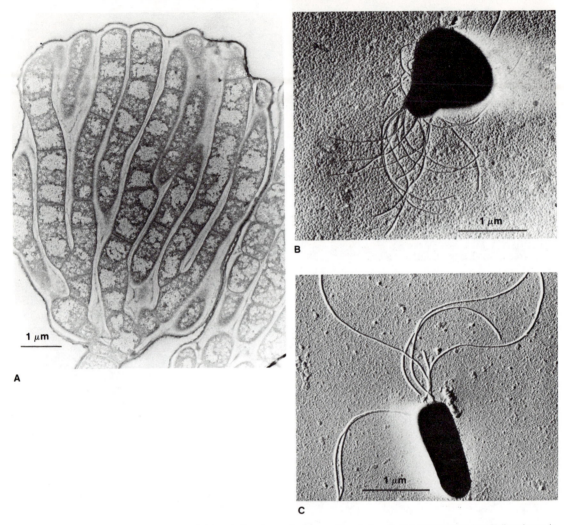

Figure 3-44 *Actinoplanes.* **A,** thin section of developing sporangium (*scale line* shown is 1 μm) (note septation in individual filaments, the cells of which will be transformed into swarmers); **B, C,** swarmers of two species of *Actinoplanes* (*scale lines* shown are 1 μm). (**A,** courtesy of H. A. Lechevalier, M. P. Lechevalier, and P. E. Holbert; **B, C,** courtesy of M. L. Higgins, M. P. Lechevalier, and H. A. Lechevalier; **A, B, C,** with permission of *Journal of Bacteriology.*)

The bluegreen algae do not have a nucleoid comparable to that in bacteria (see p. 37), and there is no condensed state of DNA or histone proteins (Fig. 3-46A). Genetic studies lag well behind those with bacteria, although genetic transfer and transformation do occur (see p. 44). Bluegreen algal DNA is like that of eukaryotic chloroplasts and mitochondria but different from that of eukaryotic nuclei. The RNA, like the DNA, is also similar to that in the chloroplasts of eukaryotic cells.

Cellular inclusions are of varied appearance and function. Some are just large enough to be resolved with the light microscope (ca. 0.2–0.5

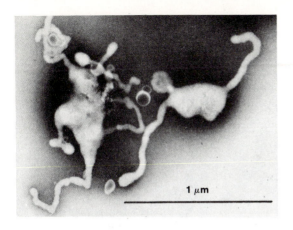

Figure 3-45 *Mycoplasma*. Electron micrograph of negative stained preparation showing pleomorphism (*scale line* shown is 1 μm). (Photograph courtesy of K. Reuss, with permission of *Journal of Bacteriology*.)

μm diam.); however, it is not possible to resolve structural details. These inclusions comprise **polyhedral bodies** as well as **polyglucan granules, cyanophycin granules**, and **polyphosphate granules** (Figs. 3-46 and 3-47). The names indicate the chemical nature or the structure of the inclusion.

The crystalline polyhedral bodies (*poly* = many; *hedr* = seat) are 200–300 nm (0.2–0.3 μm) in diameter, contain enzymes involved in carbon dioxide fixation, and should be called **carboxysomes**. Polyglucan granules (*glucan* = sweet; also known as α granules) are around 250 nm (or 0.25 μm) and associated with the thylakoids (Fig. 3-46B). They are considered carbohydrate storage products, similar to the glycogen generally stored by animals. The cyanophycin granules, previously known as structured granules, often have a substructure of undulating flattened sacs (Figs. 3-46A and 3-47C). They are irregularly shaped, are 500–1000 nm (or 0.5–1 μm) in size, and contain amino acids. The polyphosphate granules (Fig. 3-46) lack membranes, are often porous when young, and occupy a central position in the cell. In addition to the larger granules, dense **osmophilic bodies** (osmium positive) of a lipid nature are present near the thylakoids (Figs. 3-46A and 3-47C).

One other inclusion easily observed with the

light microscope as clear areas in the cell are the **gas vacuoles** (Figs. 3-47A and 3-48D). Structurally, these are very different from the vacuoles of eukaryotic plants. They do not contain water but are permeable to certain gases and susceptible to pressure changes. With the electron microscope, gas vacuoles are resolved into packed arrays of cylindrical vesicles 0.2–1.0 μm long and 70–75 nm in diameter (Fig. 3-47A). Each of the cylindrical units has conical ends and consists of a thin membrane of helically-wound ribs approximately 4 nm apart. The vesicles themselves are proteinaceous and almost completely lacking in lipids, which occur in other cell membranes.

Cell Division

In cell division the plasma membrane and the inner two wall layers invaginate into the cytoplasm from outside the cell to form a broad **septum** (*sept* = fence) (Fig. 3-46B). Additional wall material is formed, and the septum continues to grow toward the center of the cell, separating the two new cells. Prior to septum initiation, there may or may not be an invagination of the thylakoids. In multicellular forms, the outer wall layers and materials secreted by the cells hold the cells together. Submicroscopic (15–20 nm diam.) pores connect the plasma membranes of adjoining cells. In some of the filamentous forms, the separating septum appears to be penetrated by large protoplasmic connections (visible with the light microscope, 0.5 μm diam.), but these do not connect the protoplasts of adjacent cells (Fig. 3-51F).

Movement

In contrast to some bacteria, bluegreen algae do not have any flagellated cells. However, some filamentous forms are capable of independent movements seen usually as a gliding, spiralling, or gentle waving of the loose end of the filament. One species of *Oscillatoria* (Figs. 3-48C and 3-51A) has a maximum gliding rate of 11.1 μm/sec^{-1}. At present, no single theory explains the movement.

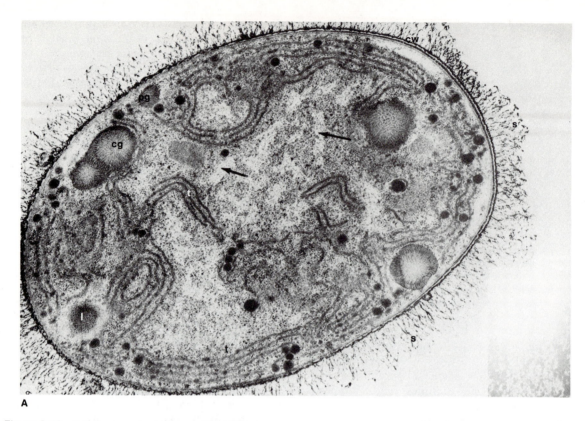

A

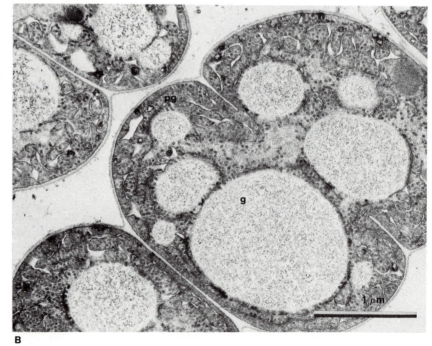

B

Figure 3-46 Cell structure of Cyanophyceae (*Anabaena* spp.), electron microscope. **A,** section through cell, showing fibrous sheath (*s;* stained with ruthenium red), wall layers (*cw*), peripheral single thylakoids (*t*), nucleoplasm (*arrow*), cyanophycin granules (*cg*), polyphosphate granules (*pg*), and osmophilic or lipid bodies (*l*), ×340. **B,** dividing cell containing large polyphosphate granules and numerous polyglucan granules (*g*) on peripheral thyalkoids, ×28,000. (**A,** courtesy of L. V. Leak and *Journal of Ultrastructure Research,* from L. V. Leak, "Studies on the preservation and organization of DNA-containing regions in a blue-green alga, a cytochemical and ultrastructural study," *J. Ultrastruct. Res.* 20:190–205, 1967; **B,** courtesy of N. J. Lang.)

Reproduction

Reproduction in Cyanophyceae is restricted to vegetative, or asexual, reproduction, and gamete production as known for eukaryotic organisms does not exist. The main method is by **fragmentation,** which is a simple breaking apart of the thallus into two or more units that grow into new organisms. In filamentous forms, small fragments, termed **hormogonia** (*horm* = chain; *gon* = offspring) are released and glide away from the parent filament. Hormogonia often result from the death of an **intercalary** (*inter* = between; *cala* = insert) cell with a biconcave area produced by the decrease in pressure of adjacent cells.

In addition to fragmentation and hormogonia production, various types of nonflagellated spores or sporelike cells can be produced. These can also serve as dormant, or resting, structures as well as reproductive units. True spore formation occurs in some genera, especially some nonfilamentous **epiphytes** (*epi* = upon); for example, in *Dermocarpa* the protoplast divides internally to produce a number of spores (Fig. 3-50C, D), and in *Chamaesiphon* a linear series of spores is produced (Fig. 3-50E).

The most commonly produced spore is the **akinete** (*a* = without; *kine* = movement) (Figs. 3-47C and 3-51C), characteristic of filamentous cyanophytes. It is larger than a vegetative cell and has a thicker wall. Differentiation involves not only cell enlargement and deposition of several envelope layers, but an increase in the number of cyanophycin granules (Fig. 3-47C). In akinete germination, cyanophycin granules disappear and numerous polyglycan granules are present. Akinetes serve as resting stages able to withstand unfavorable environmental conditions.

Another sporelike cell is the **heterocyst** (*hetero* = different; *cyst* = bag) (Figs. 3-47D, 3-48F, I, and 3-51C, E). With the light microscope, the heterocyst appears as a translucent or shiny cell, yellow to bluegreen in color. It has wall thickenings at the ends where it is attached to other cells (Figs. 3-47D, E, and 3-51E, G). The electron microscope shows the heterocyst to contain a complex reticular pattern of membranes within a heavy wall and envelope (Fig. 3-47D, E). Heterocysts serve as a point of weakness in a filament, thus promoting fragmentation. A few heterocysts have been seen to produce new filaments; however, the main function is in nitrogen fixation (see p. 124).

Sexual reproduction, as observed in eukaryotic algae, is not present in cyanophytes. However, **anastomoses** (*anastomos* = coming together) of cells in filaments do occur in some strains of *Nostoc* (Fig. 3-48H) under certain growth conditions. There is evidence for genetic transfer in bluegreen algae similar to that in bacteria, but more research is needed to approach our knowledge of the bacteria.

Morphological Variation and Classification

Morphologically some Cyanophyceae are as simple as bacteria; however, most are somewhat more complex. Bluegreens are unicellular or colonial, free-floating or attached, filamentous or nonfilamentous (Figs. 3-50 and 3-51). Cell shape is variable and is used (with varied success) for classification. A combination of morphological and biochemical attributes (as used with bacteria) are presently being used to establish a more workable system. Bacteriologists have proposed that by 1985 all bluegreen algae be classified as Cyanobacteria, using features of bacterial classification. In this text, a morphological system with emphasis primarily on external appearance will be used to accent the algallike features of Cyanophyceae.

Two distinctive morphological lines of development (and possibly evolution) are evident: the nonfilamentous (Fig. 3-49A) and filamentous (Fig. 3-49B) types. Both lines consist of more than one order (see Table 3-1). Morphologically, the simplest member is a unicellular, spherical, free-floating form (*Synechocystis*; Fig. 3-50A), from which two morphological lines can be derived. The nonfilamentous line includes unicellular, attached forms like *Dermocarpa* and *Chamaesiphon* (Fig. 3-50C–E), as well as free-floating, unicellular, but nonspherical algae like *Synechococcus* (Fig. 3-50B). From the unicellular form, a colony can develop. The simplest type of colony results

Table 3-1 Characteristics of Orders of Cyanophyceae (Bluegreen Algae)

Order	Cell Arrangement	Branching	Heterocyst/Akinete
Chroococcales	unicellular or colonial; free-floating	none	none
Chamaesiphonales	unicellular; attached	none	none
Pleurocapsales	colonial or filamentous; attached, sessile	none	none
Nostocales	filamentous	false, if present	none, either, or both
Stigonematales	filamentous or parenchymatous	true	both

After Bourrelly (1970).

when the cells remain fastened after division and are embedded in a common sheath (*Gloeothece, Chroococcus, Gloeocapsa*; Figs. 3-48B and 3-50F, G). The colony usually fragments before reaching more than four to eight cells in size. In *Merismopedia* (Fig. 3-50H; referred to as *Agmenellum* by some workers), cell divisions occur regularly in only two planes, forming a single-layered sheet of cells. When there are no regular planes of cell division, an irregular colony, as in *Microcystis* (Figs. 3-48A and 3-50I), is formed.

The filamentous line (Fig. 3-49B) can be derived directly from unicellular, free-floating forms or through unicellular, attached forms. Cell divisions in this series occur in only one plane in unbranched forms, such as *Oscillatoria* (Figs. 3-48C and 3-51A) and *Anabaena* (Fig. 3-48D, E). Most filamentous genera are straight, although moving in a helical manner; but some genera, such as *Spirulina* (Fig. 3-51B), are helical-shaped. A cyanophyte filament generally consists of a row of cells, the **trichome** (*trich* = hair) and a sheath (Fig. 3-51D). The number of trichomes in a sheath (when present) is used to separate genera (see *Schizothrix*, Fig. 3-51D). The distinction between trichome and filament is used only for bluegreen algae because of the nature of the sheath. The filaments can be free or aggregated into various macroscopic, mucilaginous masses of regular or irregular form, as

Figure 3-47 Cell structure of Cyanophyceae, electron microscope. **A,** gas vacuoles (*gv*) in *Nostoc*, showing cylindrical shape (*upper*) and packed arrays in transverse section, ×2500. **B,** section (*Fremyella*) showing phycobilisomes (*arrows*) on thylakoids (*t*), ×37,800. **C,** akinete (*Anabaenopsis*), showing large cyanophycin granules (*cg*) and osmophilic bodies (*l*), ×24,800. **D,** heterocyst (*Anabaena*) attached to vegetative filament by pore (*po*), showing thick envelope (*e*), and contorted thylakoids (*t*), especially near poles, ×16,500. **E,** heterocyst (*top*) and adjacent vegetative cell (*bottom*) (*Anabaena*) with layered heterocyst envelope (*e*) appearing as space, and vegetative cell sheath (*sh*) present, ×24,000. (**A,** courtesy of T. Bisalputra and *Journal of Ultrastructure Research*, from T. Bisalputra, D. L. Brown, and T. E. Weier, "Possible respiratory sites in a blue-green alga *Nostoc sphaericum* as demonstrated by potassium tellurite and tetranitro-blue tetrazolium reduction," *J. Ultrastruc. Res.*, 27:182–197, 1969; **B,** courtesy of E. Gantt and *Journal of Bacteriology*; **C,** courtesy of N. J. Lang and Blackwell Scientific Publications; **D,** courtesy of Norma J. Lang and The Royal Society (London); **E,** courtesy of Norma J. Lang.)

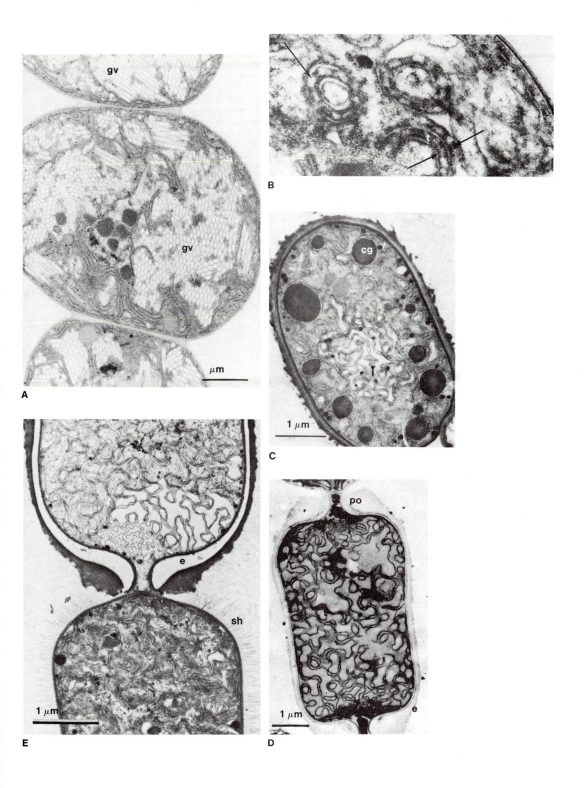

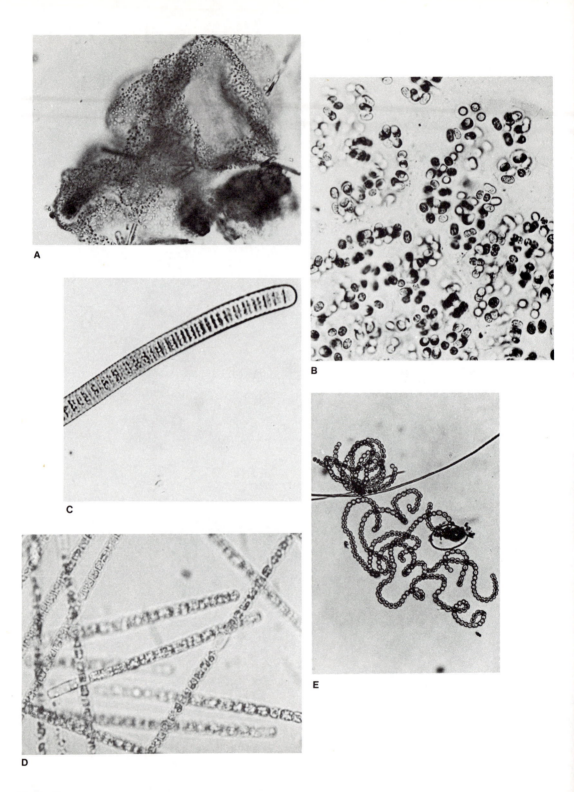

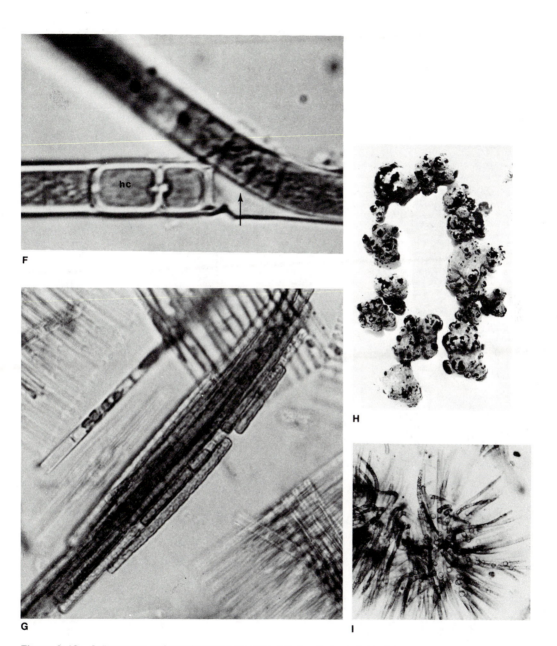

Figure 3-48 Cell structure and morphological diversity of Cyanophyceae, light micro-scope. **A, B,** colonial, nonfilamentous forms; **A,** *Microcystis,* ×200; **B,** *Gloeothece,* ×2000. **C–I,** filamentous forms: **C,** *Oscillatoria,* filament characteristically lacking sheath, ×700; **D, E,** *Anabaena* spp., showing gas vacuoles in **D,** ×350; **E,** ×150; **F,** *Tolypothrix* with false branching (*arrow*) and heterocyst (*hc*), ×700; **G,** *Aphanizomenon* with parallel trichomes without sheath, ×300; **H,** *Nostoc,* macroscopic colonies, ×0.4; **I,** *Rivularia* with tapering trichomes and basal heterocysts (*arrow*), ×150.

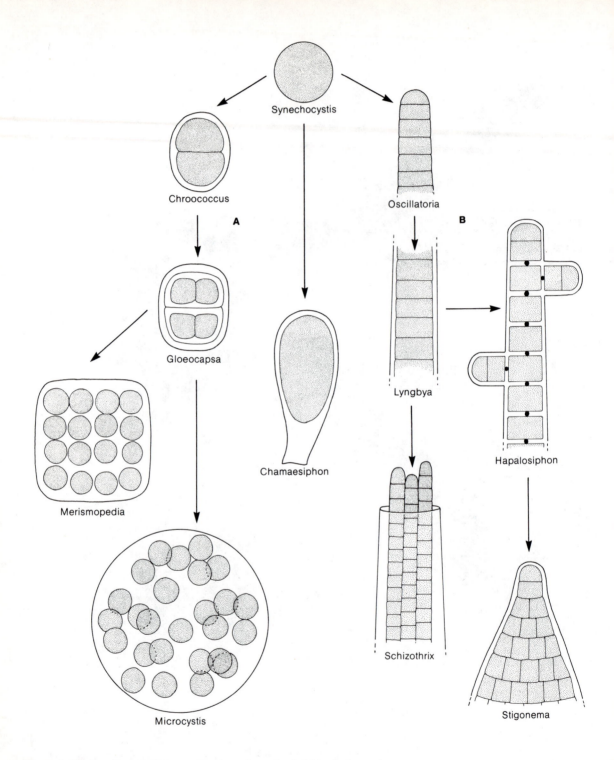

Figure 3-49 Developmental lines resulting from plane of cell division in Cyanophyceae. **A,** nonfilamentous lines, including attached forms (left, center). **B,** filamentous line (right).

in *Anabaena* and *Nostoc* (Figs. 3-48D, E, H and 3-51C). In filamentous forms, a branch occurs when cell division is in another plane, as in *Hapalosiphon* (Fig. 3-51F). This is referred to as **true branching** and occurs regularly in many eukaryotic algae.

In some filamentous genera, there is a break in the trichome, and one end (or both) breaks through the sheath at the point of fragmentation, as in *Tolypothrix* (Fig. 3-48F). This gives the appearance of a "branch," but since it is not attached to the rest of the trichome and does not result from a change in the plane of cell division, it is termed a **false branch.** The break may also result from the death of an intercalary cell or extensive growth of a trichome within its sheath.

Most filamentous forms have cell division throughout the trichome; however, in some, division and growth may be restricted. The cells of a trichome can be uniform in size or have a gradual tapering with the apical (free) end possessing long and narrower cells (as in *Rivularia*, Figs. 3-48I and 3-51E).

Nonfilamentous forms are separated into orders according to whether they are attached or not, whereas filamentous forms are separated according to the occurrence of specific cells and appearance of the filament (Table 3-1). The order Chroococcales contains most nonfilamentous forms and includes the unicellular *Synechocystis*, *Synechococcus*, and colonial *Chroococcus*, *Gloeocapsa*, *Merismopedia* (*Agmenellum*), and *Microcystis* (see Figs. 3-48A, B and 3-50A, B, F–I). Because of their small size and morphological simplicity, some are difficult to distinguish from bacteria.

The largest order is Nostocales, which includes most well-known genera of cyanophytes. The basic structure for this order is the trichome, which may have little sheath, as in *Oscillatoria* and *Aphanizomenon* (Figs. 3-48C, G and 3-51A). The cells of the trichome are in a single, linear series, forming a **uniseriate** (*uni* = one; *seri* = row) structure. The occurrence of heterocysts and akinetes and the relative dimensions of the cells of a trichome are used to separate families of the order. The commonly occurring and important Nostocales are *Anabaena*, *Nostoc*, *Tolypothrix*, and Yas well as *Oscillatoria* and *Aphanizomenon* (Figs. 3-48C–I and 3-51A–E).

The order Stigonematales is characterized by true branching and is considered to comprise the most complex of Cyanophyceae. Some members have several linear rows of cells (**multiseriate; multi** = many), such as *Hapalosiphon* (Fig. 3-51F); in others, random cell divisions form **parenchymatous** plants, as in *Stigonema* (Fig. 3-51G). There are connections of cell wall material between cells, although protoplasts are not connected (Fig. 3-51F). Some Stigonematales morphologically resemble certain red algae (genera of Bangiophyceae in Rhodophyta, Chapter 11), thus providing good morphological examples of evolutionary relationships between the red and bluegreen algae.

Physiology and Biochemistry

The photoautotrophic nature of bluegreen algae separates them from bacteria. Cyanophyte photosynthesis is an aerobic process in which water is the hydrogen donor and molecular oxygen is released. This is identical to photosynthesis in eukaryotic, photoautotrophic plants and involves cyclic and noncyclic photophosphorylation, the presence of two photosystems in photosynthesis, and chlorophyll *a*. In contrast, bacterial photosynthesis is anaerobic, using other compounds as hydrogen donors, other chlorophyll pigments, and so on, as discussed previously (p. 55). Some bluegreen algae can photosynthesize in anaerobic conditions, but this occurs only for short periods of time and is not a "way of life" for many.

In addition to chlorophyll *a*, carotenoids and phycobiliproteins are present, giving bluegreen algae their characteristic colors. These accessory pigments, having different absorption spectra, pass light energy of different wavelengths to chlorophyll *a*, which is the photoreactive pigment.

Some bluegreen algae share with bacteria the ability to fix atmospheric nitrogen (N_2). As far as is known, all nitrogen-fixing bluegreen algae are photosynthetic, although some have symbiotic relationships with eukaryotic plants. Less is known about nitrogen fixation in cyanophytes than in bacteria; however, the processes are very similar (see p. 57 for details of bacterial nitrogen fixation). The heterocyst (see p. 77, and Fig. 3-47D, E) is the main (but not exclusive) site of nitrogen fixation. Energy from photo-

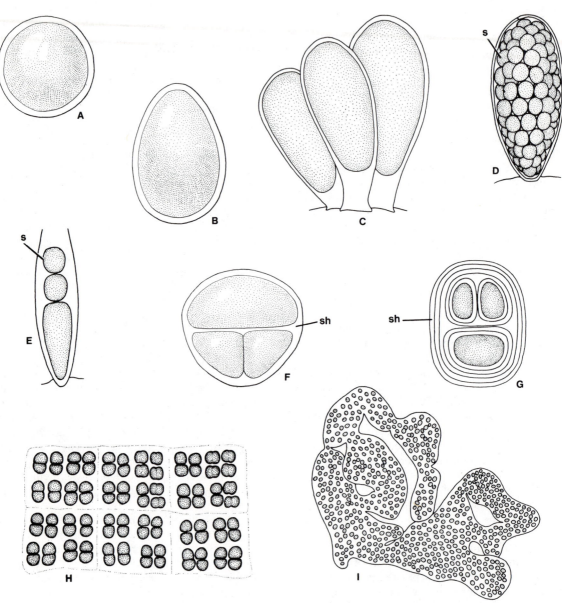

Figure 3-50 Morphological diversity of Cyanophyceae, nonfilamentous forms. **A,** *Synechocystis,* ×3500. **B,** *Synechococcus,* ×3500. **C, D,** *Dermocarpa, vegetative* (**C**) and containing spores (*s*) (**D**), ×1000. **E,** *Chamaesiphon,* with spores (*s*), ×3000. **F,** *Chroococcus,* with sheath (*sh*), ×3500. **G,** *Gloeocapsa,* ×2250. **H,** *Merismopedia,* ×1175. **I,** *Microcystis,* ×325.

synthesis is required, yet the main enzyme (nitrogenase) is oxygen-sensitive. Thus, the process must occur in a low-oxygen environment or under reducing conditions. In the heterocyst, this is achieved by a high respiratory rate, reduction in available oxygen, and virtual absence of photosynthesis, since only photosystem I of photosynthesis is present (no oxygen evolution). The extensive membrane system (Fig. 3-47E) contains approximately 45% less chlorophyll than vegetative cells and very few phycobiliproteins. Forms lacking heterocysts fix nitrogen in vegetative cells under anaerobic or almost anaerobic conditions. A significant amount of the nitrogen fixed is secreted into the surrounding environment.

Some Cyanophyceae produce toxins, as do some bacteria (see p. 52). The toxins often cause the death of cattle or birds that drink water containing the algae (Fig. 3-52B, C). The nature of the toxins varies, but they must be released from the cell to be effective. The genera involved are the colonial *Microcystis* (Figs. 3-48A and 3-50I) and filamentous *Aphanizomenon* and *Anabaena* (Fig. 3-48D, E, G). Toxin production generally occurs in warm water, but the temperature for optimum toxin production, growth, and photosynthesis differs for each process as well as each species, which partly explains the difficulty in predicting when a given cyanophycean growth will be troublesome.

Ecology

The Cyanophyceae are probably the most ubiquitous photosynthetic plants. They can be free-floating, sessile, or attached in the aquatic habitat; they can be macroscopic cushions or layers on rocks, soil, or trees. Their ability to occupy habitats with extreme environmental conditions, such as heat, cold, and moisture, is unsurpassed by any other algae or even by higher green plants (bryophytes, tracheophytes). This success at such extremes probably results from the prokaryotic nature of the cell. In fact, only bacteria are able to live in more extreme and diversified environments. The lack of membrane-bound organelles and small cell size appear to be the factors allowing survival.

The ability of bluegreen algae to colonize bare surfaces has been recognized as important in the establishment of vegetation in some areas, such as the volcanic island Surtsey off Iceland, formed first in 1963. Bluegreen algae will form bluish-black patches on lawns, if growth of the grass is poor or if bare areas are present. Some cyanophytes comprise the algal partner in lichens (see Chapter 6), which are themselves successful colonizers of bare areas. Bluegreen algae are also known in hot springs and grow at temperatures as high as 74 °C.

Since cyanophytes generally have higher temperature tolerances than eukaryotes, one problem with thermal pollution may be the promotion of bluegreen algal growths. In any habitat, massive amounts of material are termed **blooms** because masses of algae suddenly appear almost overnight in surface waters (Fig. 3-52A). Actually the sudden increase in numbers may be caused by the rising of algae from deeper waters. The bluegreens grow at depths where light intensity is low, oxygen is low, and available nitrogen may be low, although other nutrients are plentiful. During growth, gas vacuoles (Figs. 3-47A and 3-48D) develop, and the algae rise in the water column. Thus the "sudden" appearance of masses of bluegreen algae results from the surfacing of these deep-water organisms.

Another factor with regard to bluegreen algal blooms is that abundant growth can occur when inorganic nitrogen is low. Thus, if the other required nutrients are available, only cyanophytes can utilize atmospheric nitrogen. Bluegreens secrete a large amount of fixed nitrogen and carbon. In so doing, they may produce substances (including toxins) that both inhibit the growth of eukaryotic algae and provide necessary nutrients for their own growth or for that of eukaryotes. Bluegreen algae seem particularly resistant to grazing by zooplankton; thus, the standing crop is not diminished and does not enter into the normal food chain of the aquatic ecosystem.

The tolerance of bluegreen algae to extreme environments does not seem to include acid habitats where the pH is less than 4. Many large bodies of water are near neutral or slightly alkaline (pH ca. 7–8.5); thus, they are susceptible to algal blooms. In areas of high pH (above 9), bluegreens are often the dominant algae. Their ability to use carbonate and bicarbonate, instead of

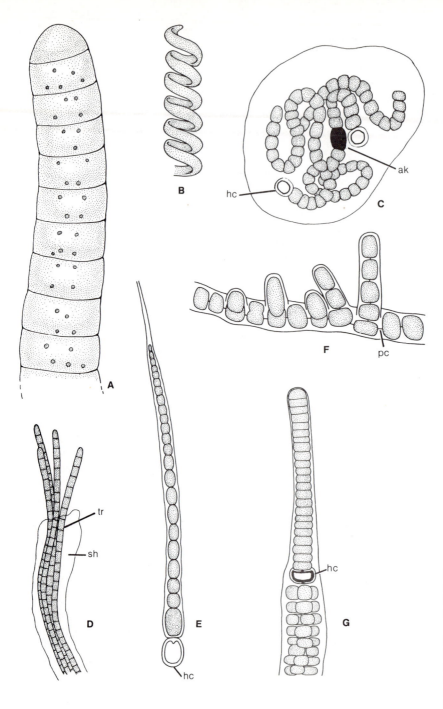

Figure 3-51 Morphological diversity of Cyanophyceae, filamentous forms. **A,** *Oscillatoria,* ×1500. **B,** *Spirulina,* ×1800. **C,** *Nostoc,* showing heterocyst (*hc*) and akinete (*ak*), ×1000. **D,** *Schizothrix,* with several trichomes (*tr*) within a sheath (*sh*), ×1000. **E,** *Rivularia,* tapering filament with basal heterocyst, ×1095. **F,** *Hapalosiphon,* with true branching showing cell connections (*pc*), ×1180. **G,** *Stigonema,* showing parenchymatous nature, ×500.

A

B

C

Figure 3-52 Blooms of blue-green algae and effect of it on cattle. **A,** bloom on surface of pasture pond. **B,** cattle drinking from pond with bluegreen bloom on surface, especially lower left. **C,** result of cattle drinking water containing toxin-producing bluegreen algae. (Courtesy of W. Carmichael.)

Prokaryota (Prokaryotes—Bacteria and Bluegreen Algae)   **87**

carbon dioxide, for a carbon source partially explains their success in these environments.

Some bluegreen algae live in intimate association with other organisms, including eukaryotic algae. Bluegreens also occur in diatoms, dinoflagellates, bryophytes, and organs of various tracheophytes. The alga probably contributes photosynthate and nitrogenous compounds to the host.

Relationships of the Prokaryota

The oldest fossils known are Precambrian prokaryotes from rocks over 3 billion years old (see Fig. 13-1). These are in the geologic series termed Swaziland Supergroup, in northeastern South Africa. Morphological and biochemical evidence indicates that anaerobic, heterotrophic, and photosynthetic prokaryotes were present at least 3.1 billion years ago. These include minute bacteria and bluegreen algae. Possible filamentous bluegreen algae in finely layered, calcareous formations (**stromatolites**) are also known in rocks some 2.5 billion years old. Present-day stromatolites are known in tropical and subtropical environments. More diverse prokaryotic fossils are present in Precambrian sediments of the massive Canadian Shield of eastern North America (north of the Great Lakes). The formation, which is between 2.0 and 1.6 billion years old, contains a variety of **microfossils** (*micro* = small) resembling bluegreen algae. The geochemistry of this formation supports the occurrence of aerobic photosynthetic organisms.

Younger sediments (1.4–1.2 billion years old) in southeastern California and the Northern Territory of Australia indicate an even greater diversity of prokaryotic fossils. Generally, bacterial photosynthesis does not liberate oxygen, whereas bluegreen algae, like other photosynthetic plants, do liberate oxygen. However, at least one species of *Oscillatoria* can carry out both types of photosynthetic reactions. In anaerobic waters, it utilizes hydrogen sulfide as the hydrogen donor, and in aerobic waters it uses water. This suggests a possible origin of oxygen-releasing photosynthesis.

The greatest evolutionary discontinuity among living organisms is that separating prokaryotes from eukaryotes. Although it is generally believed that eukaryotes arose from a prokaryotic ancestor, neither fossil nor living intermediates are known. One widely accepted theory is that the eukaryotic cell, as we know it, evolved from incorporation of prokaryotes into other prokaryote cells. An example of this **endosymbiotic** (*endo* = within) origin, which created the organelles in eukaryote cells, involves a cyanophyte cell becoming the chloroplast of a red algal cell. The pigments of red algae and bluegreen algae are similar and in some instances are identical. Both contain the phycobiliproteins in phycobilisomes on the thylakoids. Conceivably, red algae may have developed through the incorporation of bluegreen endosymbionts in primitive eukaryotic cells. The endosymbiont theory is considered further in Chapter 13.

References

Prokaryota

Echlin, P., and Morris, I. 1965. The relationship between bluegreen algae and bacteria. *Biol. Rev.* 40:143–187.

Hall, G. 1971. Evolution of the prokaryotes. *J. Theoret. Biol.* 30:429–454.

Starr, M. P., Stolp, H., Trüper, H. G., Balows, A., and Schlegel, H. G., eds. 1981. *The prokaryotes.* 2 vols. Springer-Verlag, Berlin.

Schizomycophyta

Ashworth, J. M., and Smith, J. E., eds. 1973. *Microbial differentiation.* Twenty-third Symposium of the Society for General Microbiology. Cambridge Univ. Press, Cambridge.

Bisset, K. A. 1970. *The cytology and life-history of bacteria.* 3rd ed. E. & S. Livingstone, London.

Brock, C. D. 1974. *Biology of microorganisms.* 2nd ed. Prentice-Hall, Englewood Cliffs, N.J.

Buchanan, R. E., and Gibbons, N. E., eds. 1974. *Bergey's manual of determinative bacteriology.* Williams & Wilkins, Baltimore.

Bull, A. T., and Meadow, P., eds. 1978. *Companion to microbiology.* Longman, New York.

Burrows, W. 1973. *Textbook of microbiology.* 20th ed. W. B. Saunders, Philadelphia.

Campbell, R. 1977. *Microbial ecology.* John Wiley & Sons, New York.

Doetsch, R. N., and Cook, T. M. 1973. *Introduction to bacteria and their ecology.* University Park Press, Baltimore.

Frobisher, M.; Hinsdill, R. D.; Crabtree, K. T.; and Goodheart, C. R. 1974. *Fundamentals of microbiology.* 9th ed. W. B. Saunders, Philadelphia.

Hawker, L. E., Linton, A. H. 1979. *Micro-organisms: Function, form and environment.* 2nd ed. Edward Arnold, London.

Jay, J. M. 1970. *Modern food microbiology.* Van Nostrand Reinhold, New York.

Lascelles, J., ed. 1973. *Microbial photosynthesis.* Dowden, Hutchinson & Ross, Stroudsburg, Penn.

Peberdy, J. F. 1980. *Developmental microbiology.* Halsted Press, New York.

Pelczar, M. J., Jr., and Reid, R. D. 1977. *Microbiology.* 4th ed. McGraw-Hill, New York.

Skerman, V. B. D. 1967. *A guide to the identification of the genera of bacteria with methods and digests of generic characteristics.* 2nd ed. Williams & Wilkins, Baltimore.

Stanier, R. Y.; Adelberg, E. A.; Ingraham, J. L. (1976). *The microbial world.* 3rd ed. Prentice-Hall, Englewood Cliffs, N.J.

Stanier, R. Y.; Adelberg, E. A.; Ingraham J. L.; and Wheelis, M. L. 1979. *Introduction to the microbial world.* Prentice-Hall, Englewood Cliffs, N.J.

Cyanophyta

Bourrelly, P. 1970. *Les algues d'eau douce.* Vol. 3, *Les algues bleues et rouges, eugléniens, peridiniens et cryptomonadines,* pp. 285–453. N. Boubée & Cie, Paris.

Carr, N. G., and Whitton, B. A., eds. 1973. *The biology of blue-green algae.* Univ. of California Press. Berkeley and Los Angeles.

Fogg, G. E.; Stewart, W. D. P.; Fay, P.; and Walsby, A. E. 1973. *The blue-green algae.* Academic Press, New York.

Geitler, L. 1932. *Cyanophyceae.* In Rabenhorst, L., ed., *Kryptogamen-flora von Deutschland, Österreich und der Schweiz,* vol. 14. Akademische Verlagsgesellschaft, Leipzig.

Humm, H. J., and Wicks, S. R. 1978. *Introduction and guide to the marine bluegreen algae.* Wiley-Interscience, New York.

Wolk, P. 1973. Physiology and cytological chemistry of blue-green algae. *Bact. Rev.* 37:32–101.

The division Myxomycota, as treated in this chapter, includes a heterogeneous assemblage of four groups: Myxomycetes, Dictyosteliomycetes, Plasmodiophoromycetes, and Labyrinthulomycetes. The relationships of these groups, either to one another or to other organisms, are unclear. Historically, the groups have been classified together at times because of superficial similarities and misconceptions concerning their structure. All have naked, wall-less, assimilative phases with some animallike characteristics. All have life histories involving spores or similar structures, and in this feature they resemble fungi. However, each class seems distinctly different from the others; a comparative summary appears in Table 4-1.

4

MYXOMYCOTA

Class Myxomycetes (True Slime Molds or Acellular Slime Molds)

About 450 species of true slime molds are known, and many are universally distributed. A few are predominantly tropical, whereas others are temperate only. Myxomycetes live in or on moist soil, wood, dung, or decaying vegetation, where they ingest bacteria, mold spores, and other particulate material. Most species have either "secretive" or minute assimilative bodies that are not commonly observed in nature. The spore-containing reproductive structures are formed in exposed positions and, in some species, are conspicuous.

Life History Outline

Myxomycetes produce moldlike, resistant spores within their fruiting bodies. Under suitable conditions, spores germinate by producing flagellated motile cells, called **swarm cells,** or amoeboid **myxamoebae.** Myxamoebae and swarm cells can ingest food particles—i.e., feed **phagotrophically**—and divide by fission. They can form resistant cysts under unfavorable conditions, but eventually function as gametes. The zygote is amoeboid, and its nuclei divide

Table 4-1 Summary of Main Features Characterizing the Classes of Slime Molds and Related Organisms

Class	Assimilative Structure	Feeding Method	Flagellated Cells
Myxomycetes	2n, amoeboid plasmodium	phagotrophic	anteriorly biflagellate, whip-lash
Dictyosteliomycetes	1n, uninucleate myxamoebae	phagotrophic	none
Plasmodiophoromycetes	n primary, n + n secondary plasmodia	absorptive, phagotrophic	anteriorly biflagellate, whip-lash
Labyrinthulomycetes	1n, uninucleate spindle cells in tubes, or chytridlike	absorptive	anteriorly biflagellate, 1 whip-lash 1 tinsel

mitotically to form a multinucleate, naked, amoeboid **plasmodium.** The plasmodium also ingests food; it can form a hard, resistant structure, or **sclerotium,** under adverse conditions. Plasmodia eventually move to exposed positions and form **sporocarps.** Meiosis occurs after cleavage of uninucleate spores. An outline of the life history is illustrated in Figure 4-1.

Spores Myxomycete spores function in dispersal and are a resistant phase in the life history. The spores, like those of some true fungi, have a wall that is usually pigmented and often marked by a pattern of ridges, warts, or spines (Fig. 4-2A, B). Masses of spores are black, brown, rusty, violet, pink, or yellow.

The spore wall comprises two layers (Fig. 4-2C), the thick inner layer of which is reported to contain cellulose. The spore protoplast is provided with a plasma membrane, nucleus, mitochondria, and other organelles characteristic of eukaryotic cells.

Myxomycete spores are long-lived, some maintaining their viability after storage of more than seventy-five years. Spores of many species germinate readily in water, and aqueous extracts of decayed wood or leaves enhance germination in some. Germination requires from as little as 15 minutes to as much as 18 days. The amount of time required varies with the species, with the collection and its age, and with environmental conditions.

At germination (Fig. 4-3A), each spore liberates one or more protoplasts through either a dissolved pore or a wedge-shaped split. Some protoplasts are flagellated when released; others

are amoeboid and develop flagella after a short time. Flagellated cells are called *swarm cells;* non-flagellated cells are called *myxamoebae.*

Swarm Cells and Myxamoebae Swarm cells have either one or two anterior whiplash flagella (Fig. 4-3B). Some swarm cells are uniflagellate when first formed and become biflagellate after several hours. Typically, one long flagellum is directed anteriorly and a shorter one is recurved, but two long flagella are sometimes present. Electron microscope preparations show two **basal bodies,** so all swarm cells would appear to be potentially biflagellate. Swarm cells are pear-shaped or cylindric when swimming, but they are capable of amoeboid movement, during which their form is variable.

Myxamoebae lack flagella (Fig. 4-4A) and are similar in motility and appearance to true amoebae. They are potentially flagellated and can become swarm cells; conversely, swarm cells can become myxamoebae. Flagella are often lacking in the absence of free water—e.g., on agar media. However, in some species, only swarm cells are formed if such water is present.

Both swarm cells and myxamoebae ingest food particles and absorb dissolved nutrients. Swarm cells have an adhesive posterior to which particles such as bacterial cells adhere. Engulfment can then occur; the myxamoeba cell "surrounds" the particle (Fig. 4-4B). Engulfed particles are enclosed within digestive vacuoles, and digestion appears to be similar to that in many protozoa.

Myxamoebae can undergo fission directly, but swarm cells retract their flagella before dividing. During mitotic division, the nuclear membrane disintegrates—i.e., division is **extranuclear.** Under unfavorable conditions, such as drying, myxamoebae and swarm cells often encyst. If conditions remain favorable for their growth and development, they eventually function as gametes. Syngamy can be by fusion of two myxamoebae or between two swarm cells. Fusing gametes of some species are of distinct mating classes, and fusion requires bringing together unlike types—e.g., a_1 and a_2.

Zygote The zygote is at first amoeboid or flagellated, depending on the gametes. If flagellated, its flagella are withdrawn and the zygote becomes amoeboid. Fusion of gamete nuclei occurs soon after the gametes have fused. Within a few hours, the zygote nucleus divides mitotically, and subsequent nuclear divisions are synchronous. If a zygote is spatially isolated, it can develop into a plasmodium by itself; if other zygotes are nearby, coalescence often takes place. In such instances, there is fusion of the cytoplasm but not of nuclei.

Plasmodium The main assimilative phase in the myxomycete life history is that of the plasmodium (Fig. 4-5). These assimilative structures are multinucleate, acellular, and lack rigid walls. Plasmodia are of several types, the simplest being microscopic and sluggishly amoeboid, with undifferentiated, granular protoplasm (Fig. 4-12). When sporulation occurs, such a plasmodium produces a single, microscopic fructification. Other plasmodia are larger, but they have transparent, nongranular protoplasm and are inconspicuous. The cytoplasm in these plasmodia exhibits streaming in a veinlike network of strands or thin sheets.

Conspicuous plasmodia, of the type shown in Figure 4-5, consist of a changing pattern of veinlike, often opaque strands. Migrating plasmodia of this type are roughly fan-shaped; they have conspicuous, infrequently branched strands posteriorly and abundant branching near the broad anterior. Toward the anterior, subdivisions are progressively smaller and, at the advancing margin, the protoplasm forms a continuous

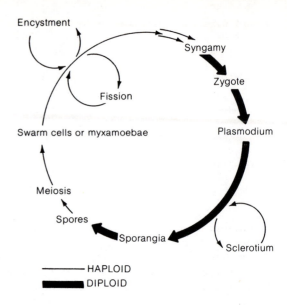

Figure 4-1 Outline of a myxomycete life history.

sheet. The protoplasm exhibits remarkably rapid streaming within the network of strands.

Movements of the assimilative plasmodium are oriented responses to various external stimuli. These responses generally are negative to light and positive to increasing moisture and dissolved nutrients.

Plasmodia (Fig. 4-6) are bounded by a plasma membrane and by a surface layer of slime. If a strand of a living plasmodium is examined microscopically, two cytoplasmic zones can be seen. The peripheral layer is transparent and appears to form a gel-like wall around the granular internal cytoplasm of the strand. The granular cytoplasm undergoes rapid, reversible streaming that is commonly rhythmic. Flow occurs in one direction for a few seconds, slows, stops, and then reverses direction.

An extensive series of invaginations, or microchannels, is visible in thin sections of some plasmodia (Fig. 4-6). The outer layer of cytoplasm contains few organelles, but the granular appearance of the inner cytoplasm is caused by nuclei, mitochondria, other organelles, pigment

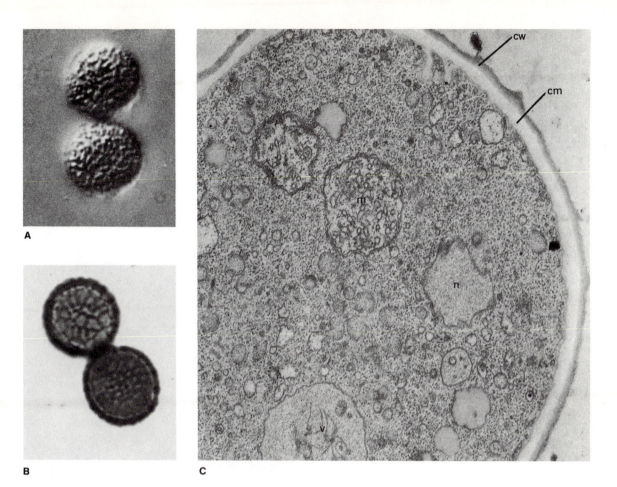

Figure 4-2 A, B, myxomycete spores showing characteristic surface patterns, ×2000. **C,** *Fuligo.* Thin section through portion of spore, ×25,000: *cm,* cytoplasmic membrane; *cw,* cell wall consisting of dark outer layer and light inner layer; *m,* mitochondrion; *n,* nucleus; *v,* vacuole. Note: only a small lobe of the nucleus is visible in this photograph. (Photograph courtesy of Nancy Ricker.)

granules, and so forth. Minute fibrils are present near the microchannels, in a loose sheath outside the plasma membrane, and in the slime track. Some of these fibrils have properties similar to those of the contractile protein of muscle, actomyosin. As with actomyosin, the contractile proteins of plasmodia undergo reversible changes in viscosity when treated with ATP (adenosine triphosphate). At the present time, contractile proteins are thought to be responsible for both the cytoplasmic streaming and the motility of the plasmodium.

Plasmodia, like myxamoebae, ingest food phagotrophically. When grown in pure culture, plasmodia of some species eventually decline and die. These species appear to have a strong dependence upon the bacteria typically associated with plasmodia. However, other species grow well on chemically defined media.

Mitosis occurs at regular intervals in growing plasmodia; almost all nuclei divide simultaneously. In contrast to mitotic divisions in the haploid cells, those of the plasmodium are within a persistent nuclear membrane.

In culture, plasmodia often divide into two or more smaller plasmodia, which then go their separate ways. The smaller plasmodia can fuse again to form a single plasmodium. However, if plasmodia of the same species but of different genetic strains meet, they may or may not fuse. Fusion is genetically controlled, and several genetic loci can regulate the process.

Sclerotium If conditions remain favorable, a plasmodium continues to take in food and grow. Under unfavorable conditions (e.g., slow drying or low temperature) the plasmodium converts to a hardened, resistant structure called a *sclerotium*. Sclerotia are irregular masses of numerous small, cell-like compartments. Sclerotia generally remain viable for no more than three years, although germination has been reported after longer periods. When conditions are again favorable for growth, a plasmodium develops from the sclerotium.

Sporulation The factors governing change from the assimilative to the reproductive phase are not completely known. In culture, starved plasmodia fruit more rapidly than those with abundant food, but starvation alone will not induce fruiting. Drying, changes in pH, and temperature can also stimulate this change. In species with pigmented plasmodia, exposure to light is necessary before fruiting will occur.

Plasmodia move to an exposed position before sporulation occurs. Those living inside decaying wood or compost may move only as far as the surface of the substrate; others travel up the stems or leaves of living or dead plants to positions favorable to spore dispersal. There they form one or more fructifications or sporocarps.

The commonest type of sporocarp is called a *sporangium* and, with the exception of microscopic plasmodia, a single plasmodium produces many sporangia. Sporangial development varies greatly among different genera of slime molds. That of *Didymium*, shown in Figure 4-7A–C, is characteristic of many species. The plasmodial protoplasm accumulates at points in the branched network. These accumulation points eventually become isolated from one another,

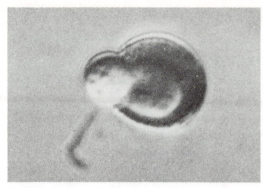

A

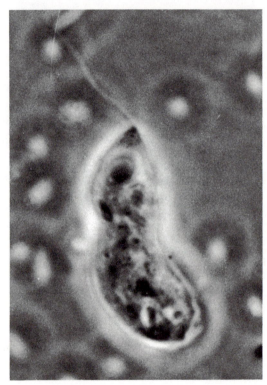

B

Figure 4-3   A, B, *Fuligo.* **A,** germinating spore, ×2700; **B,** swarm cell, ×3400.

and each develops into a sporangium. The isolated portions of protoplasm first assume a columnar form, then differentiate into an apical

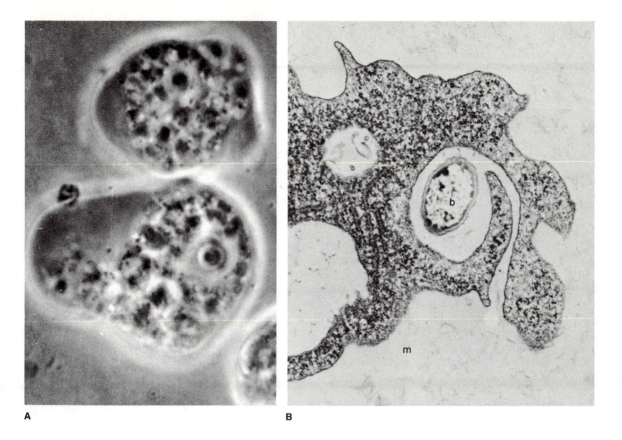

A   B

Figure 4-4 **A,** *Didymium myxamoeba,* ×3550. **B,** electron micrograph of *Physarum* myxamoeba engulfing a bacterial cell, ×27,650: *m,* myxamoeba; *b,* bacterium. (Photograph courtesy of F. Y. Kazama and H. C. Aldrich, with permission of *Mycologia.*)

sporangial region and a stalk. At this stage in development, the soft, slimy fructifications of many species are bright pink, yellow, or white.

Mitosis occurs within the sporangial region of the sporocarp, and the cytoplasm becomes cleaved into uninucleate portions. After cleavage, the thin outer wall layer and associated spines, warts, or other decoration, form outside the spore membrane (Fig. 4-8A, B). The thick inner wall layer is then deposited immediately inside the cleavage membrane.

Recent studies indicate that meiosis takes place after cleavage of the spores. Structures called **synaptonemal complexes** (Fig. 4-8C), considered valid indicators of the onset of meiosis, generally appear at that time. In some species,

nuclear degeneration occurs after meiosis, and the mature spore is then uninucleate and haploid.

As spores are developing, threadlike structures, collectively known as the **capillitium,** are produced in the sporangium. Capillitial strands are interspersed with, but not attached to, the spores. The surface layer of the sporangium develops into a wall, or **peridium,** in many myxomycetes. In others, only the membrane is present at maturity.

The relationship of various parts of the sporangium to one another is shown in Figure 4-9A. A stalk is lacking in many species; in some, the stalk apex continues into the spore case and is called a **columella.** The capillitium may be a branched system attached to the columella; it may also consist of free threads. The peridium consists of a thin, membranous layer or of such a layer plus limy material. A shiny, membranous, plasmodial remnant, the **hypothallus,** is often visible at the base of the sporocarp, or stalk.

In species not forming sporangia, each plasmodium can produce one to many sporocarps of other types. In the development of **plasmodiocarps** (Fig. 4-9B), the major veins, or parts thereof, appear to be transformed directly into fructifications. Plasmodiocarps are sessile and branched, doughnut-shaped, or simply elongate.

In a third type of sporocarp, the **aethalium** (Fig. 4-9C), an entire plasmodium heaps up into one or a few pillow-shaped or rounded masses. As with sporangia, plasmodiocarps and aethalia have an outer peridium and can contain a capillitium in addition to spores.

Some intergradation is found between different types of sporocarps. Both plasmodiocarps and sporangia can be produced by a single plasmodium. Sporangia of other species are produced in crowded masses in which there is partial or complete fusion. Such masses closely resemble or are indistinguishable from aethalia.

When sporocarps are mature and dry, the spores are unattached but enmeshed in the capillitial network. When the peridium ruptures, the spores are exposed to varying degrees and are distributed by air currents.

Myxomycete Classification

Two subclasses, Ceratiomyxomycetidae and Myxogastromycetidae, are recognized. The subclass Ceratiomyxomycetidae contains a single genus, *Ceratiomyxa*, with few species. *C. fruticulosa* (Fig. 4-10A–D) is of widespread occurrence, but the species is poorly known and of uncertain relationship to other slime molds. Plasmodia of this species inhabit decayed wood, surfacing only to fruit. Fructifications vary in form, but generally consist of columnar or threadlike strands (Fig. 4-10B, C). Many stalked spores (Fig. 4-10D) cover the surfaces of the strands. The mature spore is surrounded by a thin wall and is enclosed in a delicate sheath that is continuous with the stalk. When a *Ceratiomyxa* spore germinates, it

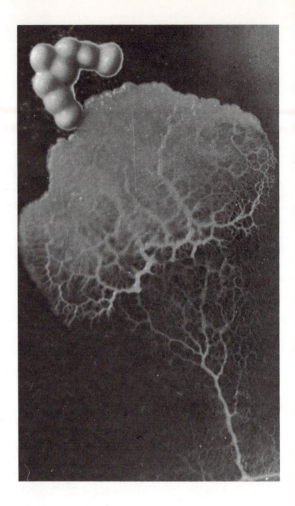

Figure 4-5 Small plasmodium of *Physarum*, shown moving toward mass of yeast cells (*upper left*), ×3.

gives rise to a single, spherical protoplast that elongates, undergoes mitosis, and produces eight swarm cells. Spores of *Ceratiomyxa* are considered by some to be homologs of sporangia in the myxogastromycetes. The production of eight rather than the usual one to four swarm cells may support this view.

Five orders of Myxogastromycetidae are recognized on the basis of spore color, capillitium characteristics, and the presence or absence of lime in the fructifications.

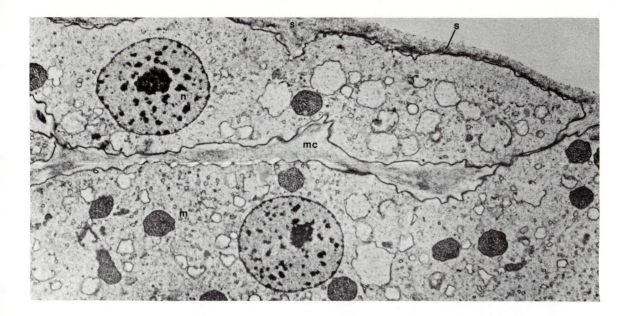

Figure 4-6 Electron micrograph of thin section of *Physarum* plasmodium, ×7300: *s*, surface slime; *mc*, microchannel; *n*, nucleus; *m*, mitochondrion. (Photograph courtesy of J. W. Daniel and U. Järlfors, with permission of *Tissue and Cell*.)

Order Liceales Sporocarps in the order Liceales lack limy materials, have pallid spores, and lack capillitia. In some aethalial forms (e.g., *Lycogala*, Fig. 4-11A), a false capillitium is present. It is composed of remnants of the plasmodial strands or of vestiges of fused sporangial walls.

Lycogala epidendrum is a very common species of cosmopolitan distribution. Aethalia of this species can reach 15 mm in diameter and resemble small puffballs. When mature, the tough, flexible peridium ruptures irregularly, and the sporocarp functions as a small bellows. Raindrops striking the flexible peridium result in air and spores being expelled.

Sporangia of *Dictydium* (Fig. 4-11C) are stalked, and the peridium is intricate. It has a thickened network, the interstices of which are very thin. At maturity, only the thickened, peridial portions remain, and they form a cagelike case for the spores (Fig. 4-11D).

Order Echinosteliales *Echinostelium*, the only genus in the Echinosteliales, has minute pinkish or yellowish sporangia and spores. These often develop on bark, wood, litter, or mosses placed in moist chambers. All species of *Echinostelium* have microscopic plasmodia (Fig. 4-12A, B), each of which produces a single sporangium. The sporangia (Fig. 4-12C) usually have a columella and a capillitium.

Order Trichiales Among the more brightly colored myxomycete fructifications are those of the Trichiales. Their spores typically are white to yellow, orange, or reddish. All species lack limy materials and have capillitia, the strands of which are often beautifully sculptured. In *Hemitrichia* (Fig. 4-13A, B), the strands are marked by regular spiral patterns; those of *Arcyria* (Fig. 4-13D) have prominent cogs, or spines.

Peridia of the Trichiales dehisce irregularly or along preformed lines. In many species, the upper portion of the peridium falls away, and a cuplike base remains. Capillitia can expand permanently, or, in some species, hygroscopic expansion and contraction occur. The movements

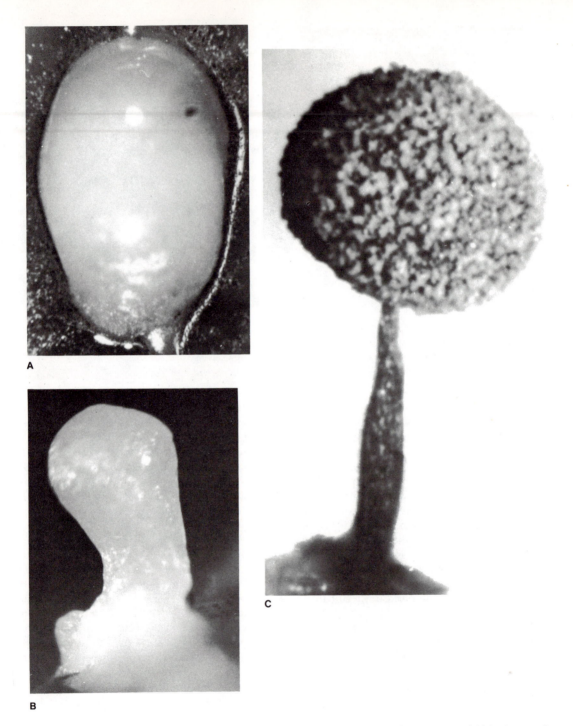

Figure 4-7 Sporangial development in *Didymium*. **A,** pillow-shaped thickening, or primordium, still attached to plasmodial strand, ×50. **B,** the primordium, now isolated from the plasmodium, has become columnar and erect, ×66. **C,** mature sporangium, ×115.

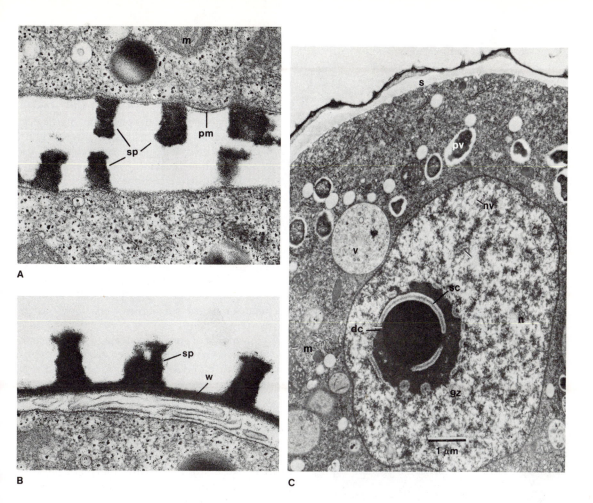

Figure 4-8 A, B, developing spores in *Didymium*. **A,** early stage showing the presence of spines, but no wall layer, ×33,785: *pm,* plasma membrane; *sp,* spines; *m,* mitochondrion. **B,** early development of spore wall (*w*), ×30,635. **C,** *Echinostelium minutum* spore. Thin section showing synaptonemal complexes in spore nucleus (*scale line* shown is 1 μm): *dc,* dense core of nucleolus; *gz,* granular zone of nucleolus; *nv,* nuclear vesicle; *pv,* phagocytotic vacuole; *s,* spore wall; *sc,* synaptonemal complex. (**A, B** courtesy of F. Schuster; **C,** courtesy of E. F. Haskins, A. A. Hinchee, and R. A. Cloney, from "The occurrence of synaptonemal complexes in the slime mold *Echinostelium minutum* de Bary," *The Journal of Cell Biology* 51:898–903, 1971, with permission of Rockefeller University Press.)

of such capillitia assist in loosening and exposing spores to air currents.

Order Stemonitales The spores of Stemonitales are mostly rusty to brown or blackish. Spo-

rangiate genera, such as *Stemonitis, Diachea* and *Lamproderma* (Fig. 4-14A–D) generally are columellate, and the capillitial threads arise as branches of the columella (Fig. 4-14B). The elongate sporangia of *Stemonitis* typically are produced in dense clusters and reach 20 mm in

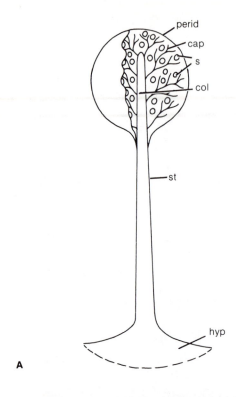

A

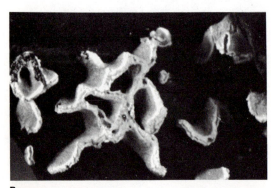

B

C

height in some species. The peridium of *Stemonitis* is membranous and ephemeral; only a coarse network of capillitial threads encloses the spores at maturity. Sporangia of *Lamproderma* (Fig. 4-14D) have a tough, metallic, iridescent peridium at maturity, but are structurally similar to those of *Stemonitis*. The peridium of *Diachea* (Fig. 4-14C) is also more or less persistent. However, sporangia in this genus have conspicuous limy stalks, columellae, and hypothalli. The lime is amorphous calcium carbonate.

Order Physarales The largest order of true slime molds is Physarales. Most species have brown to violaceous or black spores and limy fructifications. Lime is present in the peridium, stalk, or capillitium, or often in all three. Fructifications may be sporangia, plasmodiocarps, or aethalia—the last sometimes very large, as in *Fuligo* (Fig. 4-9C).

Plasmodia of some Physarales (e.g., *Physarum polycephalum*) have been studied relatively intensively. Both plasmodia and myxamoebae of this species have been grown in pure culture; myxamoebae require a complex medium, but plasmodia can be grown on a chemically defined medium.

Physarum, the largest genus of Myxomycetes, contains species with sporangia or plasmodiocarps (Figs. 4-9B). The capillitium consists of fine hyaline tubules connecting knots of lime. The sporangia vary greatly in form, color, presence or absence of a stipe and columella, and other characteristics.

Fuligo septica, also a common species, produces aethalia (Fig. 4-9C) that can reach 20 cm in diameter and 3 cm in thickness. The thick outer layer consists of lime, the color ranging from white to tan, yellowish, greenish, or dull reddish. Within the aethalium, interwoven and poorly defined tubular zones are visible. Within these zones, which may represent vestiges of sporangia, capillitia and spores similar to those of *Physarum* are present.

Figure 4-9. Myxomycete fructifications, **A,** diagram of sporangium, showing types of structures present: *cap,* capillitium; *col,* columella; *perid,* peridium; *s,* spore; *st,* stipe or stalk; *hyp,* hypothallus, **B,** plasmodiocarps of *Physarum bivalve,* ×5; **C,** aethalium of *Fuligo septica,* ×0.5.

Sporangia of *Didymium* (Fig. 4-7) and closely allied forms resemble those of *Physarum*. However, the lime present on the peridium is crystalline rather than amorphous.

Class Dictyosteliomycetes (Dictyostelids)

Dictyosteliomycetes, or *dictyostelids,* are predominantly inhabitants of soil, dung, or litter, where they "prey" upon bacteria. Dictyostelids are designated as cellular slime molds because they have unicellular, amoeboid assimilative phases. Like the slime bacteria (p. 63), the assimilative cells aggregate to form communal fruiting bodies. Most species can be cultured readily if grown with bacteria; some have also been grown on synthetic media in pure culture. In recent years, these economically unimportant but biologically interesting organisms have become subjects of intensive laboratory study.

The life history of *Dictyostelium discoideum,* the best-known species, is outlined in Figure 4-15. The following is based primarily upon this species.

Life-History Outline

Spores In *D. discoideum* (Fig. 4-16A), the spores have thick walls (Fig. 4-16B) composed primarily of cellulose. The outermost layer consists of mucopolysaccharide. Within the plasma membrane, a nucleus, mitochondria, and other organelles are present (Fig. 4-16C).

When spores of *D. discoideum* germinate, the three outer layers rupture; the innermost layer is apparently digested before protoplast release occurs.

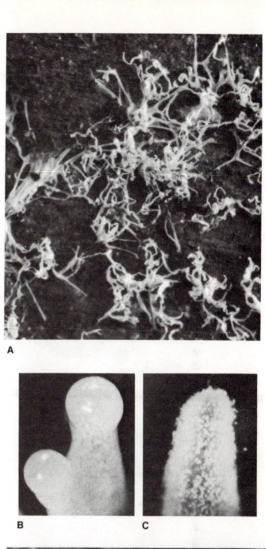

A

B   C

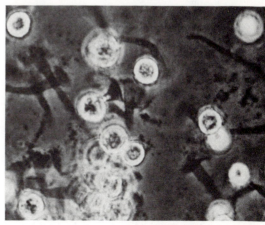

D

Figure 4-10 A–D, *Ceratiomyxa fruticulosa.* **A,** habit of mature fructification, × 12.5; **B,** single strand or pillar during early development, × 150; **C,** pillar nearing maturity, × 160; **D,** spores and individual stalks, × 1000.

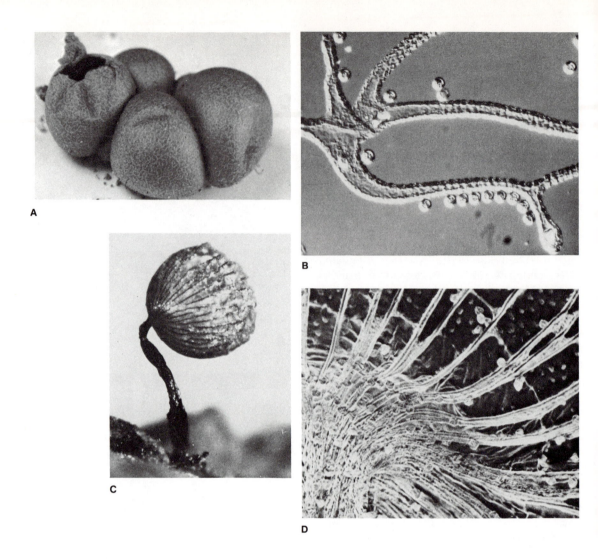

Figure 4-11 **A, B,** *Lycogala epidendrum.* **A,** aethalium, ×6; **B,** false capillitium and spores, ×565. **C, D,** *Dictydium.* **C,** sporangium, ×31; **D,** scanning electron micrograph of peridium at base of sporangium, ×1630.

Myxamoebae A single amoeboid cell, or myxamoeba, emerges from each spore. Myxamoebae, morphologically like soil amoebae, move over the substrate and engulf bacterial cells (Fig. 4-17A). They enlarge and divide by fission. During this assimilative period, they act independently of one another; individual myxamoebae tend to move away from one another, probably in response to a chemical stimulus. They respond positively to (i.e., are attracted by) cyclic AMP (3′,5′-adenosine monophosphate) and the vitamin cholic acid produced by bacteria.

Myxamoebae of *D. discoideum* and related forms have organelles similar to those of other amoebae. These include food vacuoles, mitochondria, and a single nucleus (Fig. 4-17B). Filaments of a musclelike protein are present in the amoebae. Evidence is accumulating that such ac-

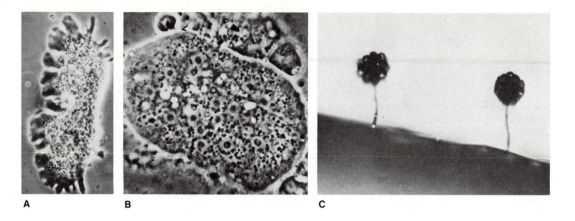

A   B   C

Figure 4-12 **A–C,** *Echinostelium minutum.* **A, B,** plasmodia showing amoeboid form. **A,** ×1200; **B,** ×930. Note the frilled advancing margin in **A** and the numerous nuclei in **B. C,** two sporangia, ×135. (Photographs courtesy of E. F. Haskins, reproduced by permission of the National Research Council of Canada from the *Canadian Journal of Microbiology* 14:1309–15, 1968.)

tinlike proteins are involved in cytoplasmic streaming and changes in cell morphology in many different organisms.

The assimilative phase can go on indefinitely if environmental conditions are favorable and if there is an adequate food supply. Depletion of food supply, changes in humidity, crowding of myxamoebae, and other factors affect the duration of the assimilative phase. At the end of this phase, myxamoebae enter an interphase in which they stop feeding. Subsequent development is at the expense of food reserves; there is a decrease in the size of amoebae, granules appear in the cytoplasm, and digestive vacuoles disappear. Changes in the cell membrane also occur, including increased adhesiveness. Myxamoebae still behave as a population of independent individuals.

Aggregation   The next phase in the life history, **aggregation,** involves convergence or movement of the cells toward a common point, or *aggregation center* (Fig. 4-18A). The myxamoebae move together, forming branched streams that lead to an aggregation center. An amoeba that contacts a moving stream adheres to and becomes oriented with other amoebae in the stream. Within the streams, the cells are elongate and have definite polar organization.

It is not known what causes a given point or group of cells to function as an aggregation center. A single cell, called a *founder cell,* is involved in the aggregation of some species, but this does not appear to be so in *D. discoideum.* Movement toward the aggregation center is thought to be controlled through the interaction of a hormone, an enzyme that inactivates the hormone, and an inhibitor of the enzyme. The hormone, called **acrasin,** is probably cyclic AMP.

Aggregation is a response of myxamoebae to a chemical signal (cyclic AMP) diffusing out from the aggregation center. Neither the signal nor the response to it are continuous, the hormone being released intermittently. Movement of responding cells is also intermittent, and bands or spiral patterns of moving and stationary cells are visible in early aggregation (Figure 4-18B). As myxamoebae respond to the signal by moving into the streams, they also produce cyclic AMP. Somehow, individual amoebae can distinguish between signals from the center and those from cells peripheral to themselves.

Migration   A rounded, papillate mass of cells

Figure 4-13 **A, B,** *Hemitrichia*. **A,** sporangia, ×45; **B,** capillitium and spore, ×550. **C, D,** *Arcyria*. **C,** sporangia, ×12; **D,** capillitium and spores, ×1000.

develops as the myxamoebae heap up at the aggregation center (Fig. 4-19A). The mass tilts, elongates, and forms a sluglike structure, the **grex** (Fig. 4-19B) or *pseudoplasmodium*. The grex, although composed of many myxamoebae, has the appearance of a single multicellular organism. Electron micrographs show that some of the cells of the grex are sometimes connected cytoplasmically (Fig. 4-19C).

The grex migrates away from the point of aggregation at a speed of 0.25 to 2.0 mm per hour, and migration can last for several hours. When gliding, the grex secretes a slimy sheath that is left behind as a slime trail.

Cells of the grex, numbering from fewer than 100 to 2000 or more, all appear morphologically alike. However, they have differentiated into two types: those that will become stalk cells and those destined to become spores. Regardless of grex size, prestalk cells and prespore cells will be present in proportions of approximately 1:2. Prestalk cells mainly form the anterior third of the grex; the remaining two-thirds are primarily prespore cells.

Differentiation into prespore and prestalk cells occurs before aggregation commences. However, redifferentiation can occur during migration of the grex. That is, if the grex is cut into two or more portions, and these are induced to continue migration, the 1:2 ratio will be restored.

The mechanism involved in grex movement is not known. The slime sheath remains stationary with respect to the substratum during migration and it could control polarity of grex movement. The sheath is very thin at the anterior and progressively thicker toward the grex posterior. Thus the sheath might make it physically easier for the cells to move forward rather than in any other direction.

Migrating grexes are sensitive to and will move toward faint light, such as that given off by bioluminescent bacterial cultures. They are also sensitive to extremely small differences in temperature and will move toward warmer areas. However, bright light or marked temperature increase will cause the grex to cease migration and form a fructification. Sensitivity to temperature and light reside in the anterior tip of the grex and ensure movement toward areas favorable to spore dispersal.

The grex and its component cells are not sensitive to cyclic AMP. Cells removed from the grex and placed adjacent to aggregating streams of amoebae show no response to the hormone present. However, they regain sensitivity to the hormone after about an hour.

The length of the migration period varies with environmental conditions. If cultures are kept in dark humid conditions, migration can continue indefinitely. Migration can be eliminated entirely if the grex is exposed to drying conditions, bright light, and high pH.

Culmination Eventually, migration ends, and the grex assumes a form similar to that at the end of aggregation. During **sorocarp** formation (Fig. 4-20), prestalk cells become vacuolate, enlarge, and become surrounded by cellulose walls. This process takes place within a noncellular stalk sheath that develops beneath the central papilla. As the stalk initial develops, more prestalk cells migrate to the apex and secrete cell walls. The prespore cells migrate up the developing stalk, or **sorophore;** when the latter is completed, they develop into spores. The spores develop by wall formation around individual amoebae; there is no increase in numbers prior to spore formation. The mature spores are embedded in a droplet of slime, and the mass of spores is designated a **sorus.** A few of the cells in the grex remain at the bottom of the sorophore and are transformed into a basal disc.

Macrocysts and Sexual Reproduction In *D. discoideum* and related forms, sexual reproduction is by formation of large, spherical-to-oval structures called **macrocysts.** The brief account presented here is based upon observations of the development in species of *Dictyostelium* and *Polysphondylium.*

Macrocyst development can be induced in culture by growing myxamoebae under a thin layer of water, maintaining cultures in the dark, and incubating at around 25°C. Media lacking phosphates also stimulate macrocyst formation. Two compatible mating types are required for macrocyst production in *D. discoideum*, but production will occur with but a single strain in some species.

Macrocyst development, like that of the so-

A

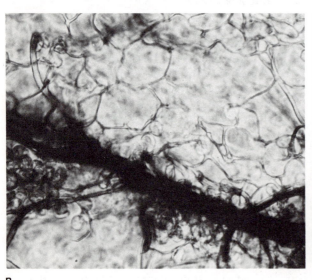

B

C

D

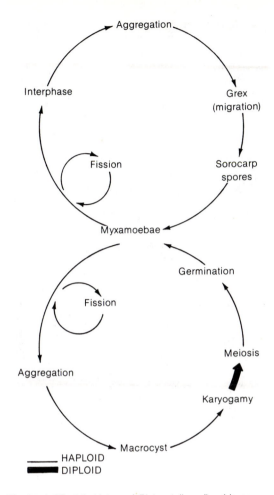

Figure 4-15 Life history of *Dictyostelium discoideum*.

Figure 4-14 Sporangia of *Stemonitis* (**A,** × 16), *Diachea* (**C,** ×60) and *Lamproderma* (**D,** × 15). **B,** photomicrograph of *Stemonitis* sporangium, showing origin of capillitial threads from columella, ×400.

rocarp, commences with aggregation of myxamoebae. Soon after cells have come together, the clumps become surrounded by a thin primary wall layer (Fig. 4-21A, B). A cell at the center of the developing macrocyst enlarges, ingesting the other myxamoebae phagotrophically (Fig. 4-21B, C). Ingestion continues until all myxamoebae have been taken in by the giant cell. Secondary wall layers are then deposited within the primary wall (Fig. 4-21D). The giant protoplast is at first binucleate, then uninucleate. Meiosis appears to occur at the time when most of the ingested myxamoebae have been digested.

The completely developed and apparently mature macrocyst requires an aging period before germination will occur readily. Exposure to light and reduced temperature (15 °C) stimulates germination. During the initial stages of germination, individual cells again become visible within the cyst wall (Fig. 4-22A). The wall layers then break down, and numerous myxamoebae are released (Fig. 4-22B, C). These myxamoebae reinitiate the assimilative phase of the life history or produce a sorocarp at the site of germination.

Although two mating types are needed for macrocyst production in *D. discoideum*, all progeny from one macrocyst are of the same mating type. If progeny of several macrocysts are examined, both mating types can be recovered. This suggests that three of the four meiotic nuclei in a macrocyst degenerate prior to germination.

Other Dictyostelids Free migration of the grex does not occur in most dictyostelid life histories. Instead, sorocarp development commences at the place where the grex is formed. This type of development occurs in species of *Polysphondylium*, as shown diagrammatically in Figure 4-23A. Grex migration is limited to that occurring during stalk formation, the base of the stalk lying flat upon the substratum. The upper portion of the stalk is erect, and as it is added to apically, portions of the grex remain behind at a number of points. These portions produce whorls of branches at right angles to the main axis, and each branch is tipped by a sorus (Fig. 4-23B). A slightly larger sorus is produced at the tip of the main axis.

In *D. polycephalum*, a single aggregation of myxamoebae divides into a variable number of long grexes. Each grex, up to a centimeter in length, migrates away from the aggregation site and eventually subdivides into several parts at culmination. Each portion produces a sorocarp, the stalks of the cluster adhering basally but not apically (Fig. 4-23C).

Class Plasmodiophoromycetes (Endoparasitic Slime Molds)

A single order, the Plasmodiophorales, includes several genera, all species of which are endoparasites of vascular plants or fungi. The Plasmodiophoromycetes are variously classified as true fungi or slime molds. Since they do produce multinucleate plasmodia and anteriorly biflagellate swarm cells, they are treated here as related to true slime molds. They are distinct from that group in nutrition, spore formation, and in the nature of their plasmodia. Two of the Plasmodiophoromycetes cause economically important diseases: *Plasmodiophora brassicae*, causing club root of cabbage and other crucifers, and *Spongospora subterranea*, the cause of powdery scab of potato.

P. brassicae is widely distributed and often severely damages cabbage crops. Although it is the most-studied member of the order, it is still incompletely known, and many reports on this

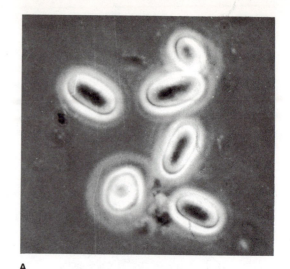

A

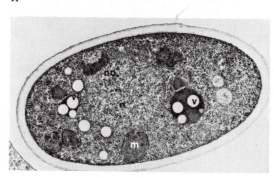

B

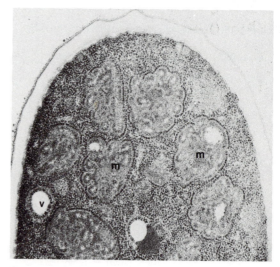

Figure 4-16 A–C, *Dictyostelium discoideum.* **A,** photomicrograph of spores, ×2000; **B,** thin section of spore, ×11,000: *m,* mitochondrion; *n,* nucleus; *no,* nucleolus; *v,* vacuole; **C,** section of spore in which the lobed nature of the mitochondria is plainly visible, ×20,740. (Photographs **B, C,** courtesy of H. R. Hohl and S. T. Hamamoto, with permission of *Journal of Ultrastructure Research,* from H. R. Hohl and S. T. Hamamoto, "Ultrastructure of spore differentiation in *Dictyostelium:* the prespore vacuole," *J. Ultrastruct. Res.* 26:442–53, 1969.)

C

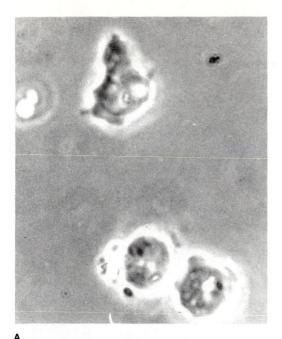

A

B

Figure 4-17 **A,** *Dictyostelium discoideum* myxamoebae during assimilative phase, ×2000. **B,** electron micrograph of *Dictyostelium discoideum* amoeba, ×8640: *fv*, food vacuole; *m*, mitochondrion; *n*, nucleus; *er*, endoplasmic reticulum. (Photograph **B** courtesy of J. M. Ashworth, D. Duncan, and A. J. Rowe, from "Changes in fine structure during cell differentiation of the cellular slime mold *Dictyostelium discoideum*," *Experimental Cell Research* 58:73–78, 1969, with permission of Academic Press.)

species are contradictory. The life-history diagram shown in Figure 4-24 is derived mainly from recent studies of *Plasmodiophora* in cabbage root tissue culture.

Spores (Fig. 4-25A) of *P. brassicae* are formed within infected root cortex cells and are released into the soil upon decay of host tissue. They remain viable in soil for as much as six years, each releasing a biflagellate swarm cell upon germinating (Fig. 4-25B, C). The swarm cell encysts upon contact with the host root hair, and the cyst protoplast then enters the root hair through a minute pore. A **primary plasmodium** develops (Fig. 4-26A), and the root hair becomes somewhat swollen and contorted in the region of the plasmodium.

After six to seven days, the primary plasmodium contains 50 to 100 nuclei and cleaves into gametangia (Fig. 4-26B). Mitotic divisions occur in the gametangia, and up to 16 gametes are produced in each. Gametes, similar to swarm cells but smaller, are mostly released to the outside of the root hair. Gametes fuse in pairs prior to reinfection, but this fusion is thought to involve plasmogamy only; the nuclei do not fuse.

After reinfection, migration to the cortical cells occurs. In cortical cells, the **secondary plasmodia** enlarge, becoming multinucleate through mitoses. The plasmodia can divide, and several plasmodia can thus occupy a single host cell. If such a host cell divides, both daughter cells may be infected. Plasmodia also can migrate from one cell to another. The presence of plasmodia results in abnormal host cell enlargement and cell division. The pattern of host tissue development is greatly altered, and the root becomes swollen and "clubbed" (Fig. 4-27).

The secondary plasmodium (Fig. 4-28A) exists within the host cytoplasm and is bounded by a membranous envelope. The host cell remains alive during development of the plasmodium and degenerates only when sporulation occurs. Host plastids and mitochondria are sometimes present in vacuoles within plasmodia; thus, some phagotrophic activity occurs.

Late in the development of secondary plasmodia, the haploid nuclei fuse in pairs. **Karyogamy** is followed soon by meiosis, and cleavage into uninucleate segments occurs. Spiny resting spores (Fig. 4-28B) develop from the segments.

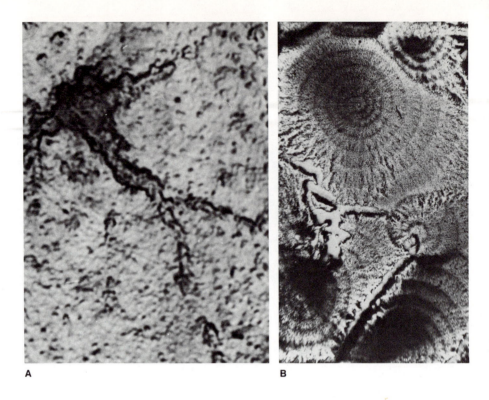

A
B

Figure 4-18 A, late aggregation phase in *Dictyostelium discoideum,* × 125. **B,** aggregation phase in populations of *Dictyostelium discoideum,* showing pattern created by intermittent release of hormone and response. (Photograph **B,** courtesy G. Gerisch, from "Periodische signale steuern die Musterbildung in Zellverbänden," from *Naturwissenschaften* 58:430–38, 1971, with permission of Springer-Verlag.)

Class Labyrinthulomycetes (Labyrinthulids and Thraustochytrids)

The labyrinthulids and thraustochytrids are mostly parasites of algae, higher plants, or animals, but some occur on dead plant materials. Several species can be maintained in dual culture with yeasts or on complex media in pure culture. They are predominantly marine or freshwater organisms.

A serious disease of eel grass (*Zostera marina*), a marine angiosperm, has been attributed to parasitism by *Labyrinthula macrocystis*. Eel grass is extremely important; the plants are eaten by some marine animals, and extensive beds provide protective cover for the larval stages of many animals. A second labyrinthulid, *L. maxima*, causes a serious disease of a commercial oyster (*Crassostrea virginica*) in North America.

The assimilative phase in *Labyrinthula* consists of uninucleate, spindle-shaped cells (Fig. 4-29A). These cells are "typical" eukaryotic cells but, though motile, are neither flagellated nor amoeboid. *Labyrinthula* cells appear to be bounded only by a membrane in some species, but a thin wall is present in others. The cells do not move freely, but glide within a network of "slime ways," the **ectoplasmic net** (Fig. 4-29B). The net is composed of tubular elements, the outer and inner surfaces of which are membrane limited (Fig. 4-30A, B). No organelles are visible in the matrix of the ectoplasmic net. Each assim-

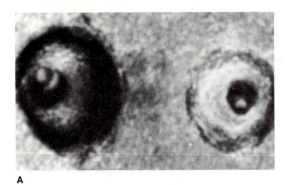

A

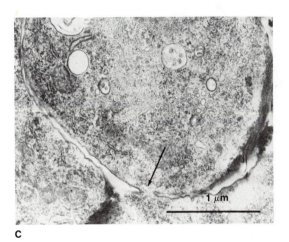

C

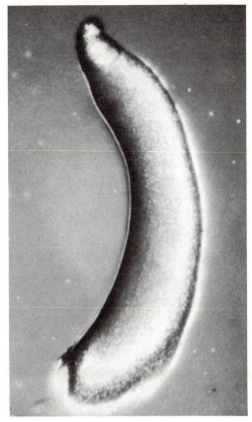

B

Figure 4-19 *Dictyostelium discoideum.* **A,** aggregates of myxamoebae at the end of the aggregation phase, ×58. **B,** migrating grex, ×80. **C,** thin section of *Dictyostelium discoideum* grex, showing cytoplasmic continuity (*arrow*) between adjacent amoebae (*scale line* shown is 1 μm). (Photograph **C,** courtesy of D. Kirk, W. E. McKeen, and R. Smith, reproduced by permission of the National Research Council of Canada from the *Canadian Journal of Botany* 49:19–20, 1971.)

ilative cell has an organelle, the **bothrosome** (Fig. 4-30B), thought to function in the production of the membranes and matrix of the ectoplasmic net.

The ectoplasmic net is produced as a network ahead of the advancing cells of a colony. The cells put forth filaments up to ten times their length; filaments of adjacent cells can fuse to pro-

duce a network. Although the network matrix lacks organelles, digestion of food organisms occurs only when the net is in contact with such organisms.

Within the network, assimilative cells reproduce by mitotic nuclear division followed by cell division. Numerous vesicles appear between the daughter nuclei, fuse, and elongate to form a continuous cleavage vesicle (Fig. 4-30C–E) across the cell. Two or four daughter cells result from the division.

Cells of *Labyrinthula* can heap together within the ectoplasmic tubules and form a membrane-bound sorus of sporelike bodies. When released, each spore gives rise to an assimilative cell.

Biflagellated zoospores occur in several Labyrinthulomycetes. In *Labyrinthuloides yorken-*

Figure 4-20 Culmination of migration and development of the sorocarp in *Dictyostelium discoideum*, ×35.6. (Photograph by K. B. Raper, from John Tyler Bonner, *The Cellular Slime Molds*, second revised and augmented edition, Princeton University Press, 1967. Reprinted by permission of Princeton University Press.)

sis, the flagella are subapically inserted: a tinsel flagellum anteriorly directed and a short, posteriorly directed whiplash flagellum (Fig. 4-31). Amoebic and plasmodial stages have also been reported for some species of the group.

Recently, several organisms previously included among the true fungi have been found to be structurally similar to the Labyrinthulomycetes and accordingly have been transferred to this class. In *Thraustochytrium*, the assimilative structure (Fig. 4-32A) resembles that of many chytrids (see *Rhizophydium*, p. 148). However, the rhizoidlike structures of *Thraustochytrium* are cytologically similar to the ectoplasmic net in *Labyrinthula*. Bothrosomes are present within these structures (Fig. 4-32B), and this organelle is unique to Labyrinthulomycetes.

Thraustochytrium produces zoospores similar to those of *Labyrinthuloides yorkensis*. Following

their release, the zoospores encyst upon dead algae or similar substrates, and the "rhizoidal" system develops. As food is assimilated, the cyst enlarges and is later transformed into a zoosporangium. At maturity of the zoospores, or just prior to it, the sporangium bursts, and the upper half of it is torn away. Zoospores repeat the cycle.

Relationships of the Myxomycota

Interrelationships of the four classes of slime molds, as well as their relationships to other groups of organisms, are unclear. Since no fossil records exist, all ideas on relationships are based

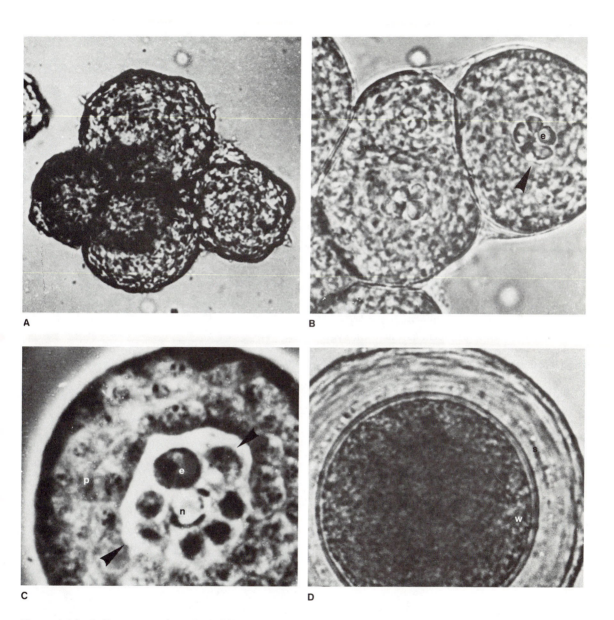

Figure 4-21   A–D, macrocyst formation in *Dictyostelium mucoroides.* **A,** precyst masses, each consisting of numerous amoebae, ×375. **B, C,** the primary cyst wall has developed and engulfment of amoebae by the giant cell (*arrows*) has commenced. **B,** ×1040; **C,** ×2190: *e,* engulfed cells; *n,* nucleus; *p,* peripheral cells. **D,** mature macrocyst, ×1370: *s,* fibrillar sheath; *w,* macrocyst wall. (Photographs from M. F. Filosa and R. E. Dengler, "Ultra-structure of macrocyst formation in the cellular slime mold, *Dictyostelium mucoroides:* Extensive phagocytosis of amoebae by a specialized cell," *Developmental Biology* 29:1–16, 1972, with permission of Academic Press.)

upon similarities or differences among extant forms.

The class Myxomycetes has generally been classified with either the Protozoa (as Mycetozoa) or with true fungi (as Myxomycetes). It is often classified as a subdivision, Myxomycotina, of the fungal division Mycota or as the phylum Gymnomyxa of the kingdom Protista. Several important characteristics of true slime molds do appear to indicate a relationship to protozoa rather than to true fungi. These features include phagotrophy, amoeboid movement, and lack of a cell wall around the assimilative structure, as well as similarities of myxamoebae to true amoebae. Presence of actinlike proteins might be considered an animallike characteristic, but similar proteins are produced by bacteria and plants.

Classification of Myxomycetes together with fungi is based upon resemblances of reproductive structures in the two groups. However, myxomycete sporangial development and the finished structure do not closely resemble those of true fungi. Spores of Myxomycetes do resemble those of some true fungi, but their germination and germination products do not. Many fungal zoospores are capable of amoeboid movement, but none feed phagotrophically or reproduce by fission.

The class Dictyosteliomycetes appears to be related to the Myxomycetes. Until recently, the two groups seemed clearly demarcated by the absence of both flagellated cells and plasmodia in dictyostelids. However, a recently discovered group, the Protostelida, share some characteristics with both dictyostelids and true slime molds. Some protostelids produce anteriorly biflagellate, motile cells, and although most have uninucleate, amoeboid assimilative structures, a few have

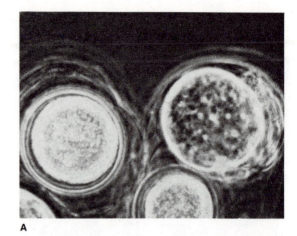

A

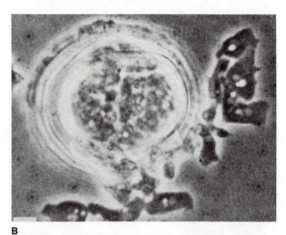

B

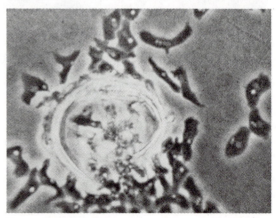

C

Figure 4-22 A–C, *Dictyostelium mucoroides.* **A,** dormant (*left*) and germinating macrocyst (*upper right*), ×752. The germinating macrocyst contains numerous myxamoebae. **B, C,** rupture of the outer cyst wall and release of myxamoebae. **B,** ×800; **C,** ×575. (Photographs courtesy of A. W. Nickerson and K. B. Raper, with permission of *American Journal of Botany.*)

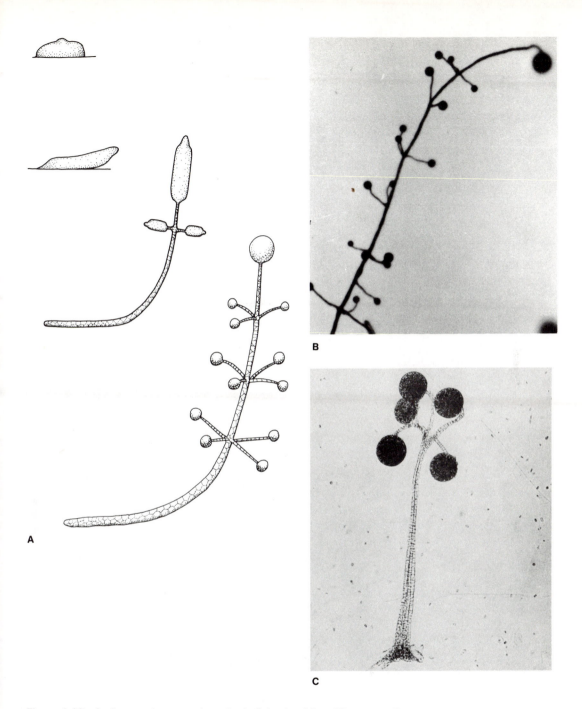

Figure 4-23 **A,** diagram of sorocarp formation in *Polysphondylium*. After aggregation, grex migration and stalk formation commence; portions of grex separate and produce lateral branches. **B,** mature sorocarp of *Polysphondylium,* ×50. **C,** sorocarp of *Dictyostelium polycephalum,* ×115. (**C,** photograph courtesy K. B. Raper, from "*Dictyostelium polycephalum* n. sp.: a new cellular slime mold with coremiform fructifications," *Journal of General Microbiology* 14:716–32, 1956, with permission of Cambridge University Press.)

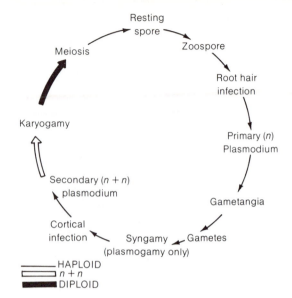

Figure 4-24 Diagram of suggested life history of *Plasmodiophora brassicae.*

HAPLOID
n + n
DIPLOID

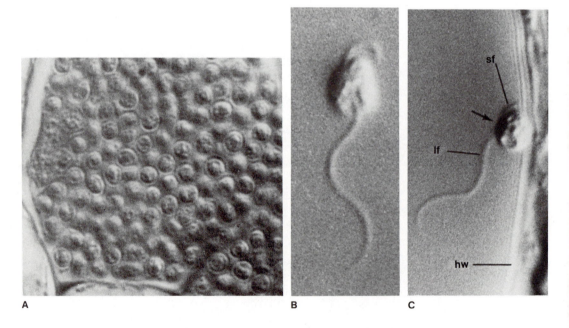

A   B   C

Figure 4-25 *Plasmodiophora brassicae.* **A,** spores in cabbage root cell, ✕1500. **B, C,** living zoospores, one of which (**C**) has just attached to a cabbage root hair. **B,** ✕4400; **C,** ✕3375: *sf,* short flagellum; *lf,* long flagellum; *hw,* host wall; *arrow* indicates point of flagellar insertion. (Photographs **B, C,** courtesy of J. R. Aist and P. H. Williams, reproduced by permission of the National Research Council of Canada from the *Canadian Journal of Botany* 49:2023–34, 1971.)

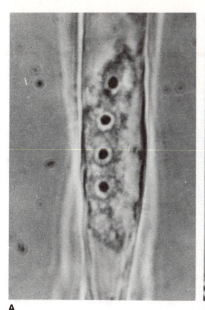
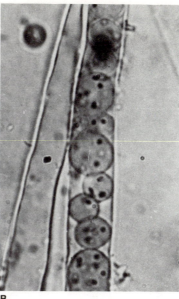

A   **B**

Figure 4-26 *Plasmodiophora brassicae.* **A,** primary plasmodium in cabbage root hair, × 1500. **B,** sporangia in root hair, × 1000. (Photograph **A,** courtesy of P. H. Williams, S. J. Aist, and J. R. Aist, reproduced by permission of the National Research Council of Canada, from the *Canadian Journal of Botany* 49:41–47, 1971; **B,** courtesy of D. S. Ingram and I. C. Tommerup, with permission of the Royal Society, London.)

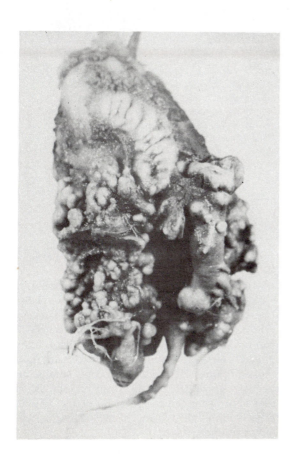

Figure 4-27 *Plasmodiophora brassicae.* Infected roots of cabbage plant, showing characteristic growth form, × 1.

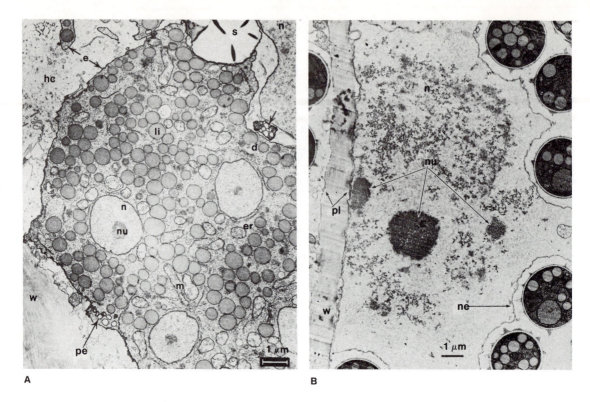

Figure 4-28 **A, B,** *Plasmodiophora brassicae.* **A,** secondary plasmodium in root cell (*scale line* shown is 1 μm): *m,* mitochondrion; *li,* lipid; *n,* nucleus; *nu,* nucleolus; *er,* endoplasmic reticulum; *hc,* host cytoplasm; *w,* host cell wall; *s,* host starch grain; *d,* dictyosome; *pe,* plasmodial envelope. **B,** spiny resting spores in dead host cell (*scale line* shown is 1 m): *n,* disintegrating host nucleus; *ne,* host nuclear envelope; *w,* host wall; *pl,* host plasma membrane. (Photographs courtesy of P. H. Williams and S. S. McNabola, reproduced by permission of the National Research Council of Canada from the *Canadian Journal of Botany* 45:1665–69, 1967.)

small, multinucleate plasmodia. Most protostelids produce a single spore per stalk, and both the spore wall and the stalk are thought to consist of cellulose.

Plasmodiophoromycetes have often been classified with true fungi rather than with slime molds. Because of the anteriorly biflagellate reproductive cells and naked plasmodial stages, the group is included with slime molds in this treatment. The occasional presence of host cell organelles within the plasmodia of *Plasmodiophora* suggests that phagotrophy sometimes occurs. However, the plasmodia do not appear to

be diploid as in myxomycetes; they are either *n* or *n* + *n*.

Finally, labyrinthulids. and thraustochytrids seem at this time to be unrelated to other classes included in this chapter. The ectoplasmic net and gliding cells of labyrinthulids differ markedly from structures found in other groups of slime molds. There are resemblances to certain algae, but additional electron microscope studies are needed before definite links can be established. The bothrosome and the nature of their motile cells definitely link the thraustochytrids and labyrinthulids.

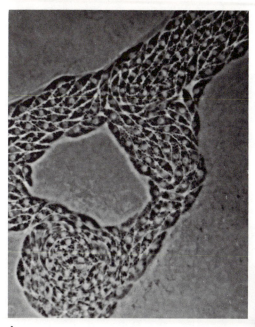

A

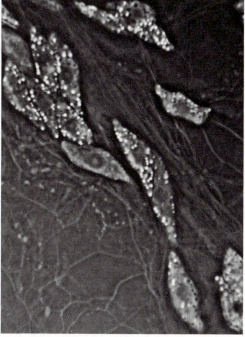

B

Figure 4-29 **A,** *Labyrinthula vitellina.* Plasmodiumlike mass of spindle cells, the ectoplasmic nets of which are not visible, ×1300. **B,** spindle cells and ectoplasmic net of
Labyrinthula minuta, ×1100. (Photographs courtesy of S. W. Watson, from ''*Labyrinthula
minuta* sp. nov.,'' by S. W. Watson and K. B. Raper, *Journal of General Microbiology*
17:368–77, 1957, with permission of Cambridge University Press.)

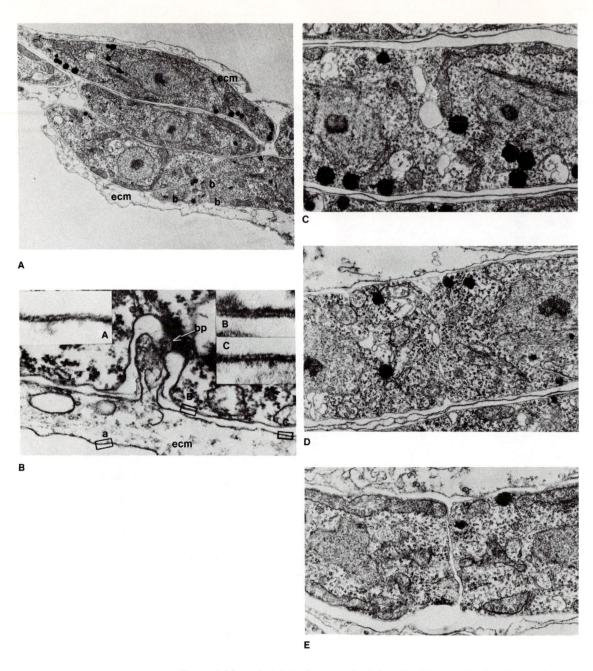

Figure 4-30 *Labyrinthula.* **A,** group of spindle cells within ectoplasmic net element, ×5640: *ecm,* ectoplasmic matrix; *b,* bothrosomes. **B,** enlarged bothrosome, ×65,600: *bp,* basal portion of bothrosome; *ecm,* ectoplasmic matrix; rectangles *A* and *C* indicate outer and inner membranes of ectoplasmic net element, ×46,850; rectangle *B* encloses a portion of the plasma membrane of the spindle cell. **C–E,** cell division in *Labyrinthula,* ×14,700. **C,** series of large vesicles is visible in the area between the two nuclei; **D,** vesicles have formed a narrow but interrupted line; **E,** division is complete. (Photographs courtesy of D. Porter, with permission of *Protoplasma.*)

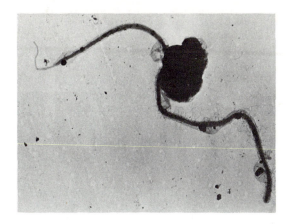

Figure 4-31 Zoospore of *Labyrinthula vitellina,* ×5115. Note that the anteriorly directed flagellum is of the tinsel type. (Photograph courtesy of D. Porter, with permission of *Veröffentlichungen des Institut für Meeresforschung Bremerhaven.*)

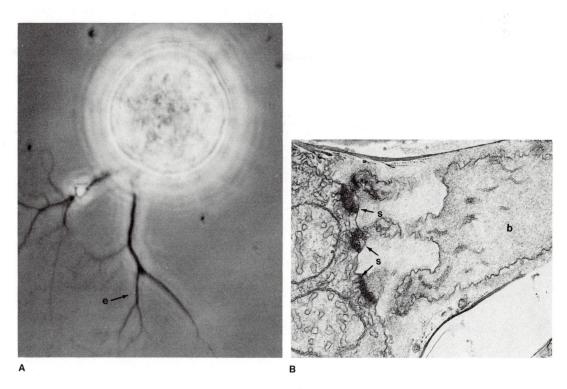

A   B

Figure 4-32 *Thraustochytrium.* **A,** assimilative body consisting of spheroidal portion and rhizoidlike ectoplasmic net (*e*), ×960. **B,** thin section showing battery of bothrosomes at proximal end (*arrows,* near main part of body) of ectoplasmic net, ×29,350. (Photographs courtesy of F. O. Perkins, with permission of *Archiv für Mikrobiologie.*)

References

Alexopoulos, C. J. 1966. Morphogenesis in the Myxomycetes. In Ainsworth, G. C., and Sussman, A. S., eds., *The fungi*, vol. 2, pp. 211–34. Academic Press, New York.

———. 1973. The Mxyomycetes. In Ainsworth, G. C.; Sparrow, F. K.; and Sussman, A. S., eds., *The fungi*, vol. 4B, pp. 39–60. Academic Press, New York.

Alexopoulos, C. J., and Mimms, C. W. 1979. *Introductory mycology*. 2nd ed. John Wiley & Sons, New York.

Ashworth, J. M., and Cappuccinelli, P., eds. 1977. *Development and differentiation in the cellular slime molds*. Elsevier/North-Holland Biomedical Press, Amsterdam.

Ashworth, J. M., and Dee, J. 1975. *The biology of slime molds*. Edward Arnold, London.

Bonner, J. T. 1967. *The cellular slime molds*. 2nd ed. Princeton Univ. Press, Princeton, N. J.

———. 1971. Aggregation and differentiation in the cellular slime molds. *Ann. Rev. Microbiol.* 25:75–92.

Erdos, G. W.; Nickerson, A.; and Raper, K. B. 1972. Fine structure of macrocysts in *Polysphondylium violaceum. Cytobiologie* 6:351–66.

———. 1973. The fine structure of macrocyst germination in *Dictyostelium mucoroides. Develop. Biol.* 70:1823–30.

Erdos, G. W.; Raper, K. B.; and Vogen, L. K. 1975. Sexuality in the cellular slime mold *Dictyostelium giganteum. Proc. Nat. Acad. Sci.* 72:970–73.

Garrod, D., and Ashworth, J. M. 1973. Development of the cellular slime mold *Dictyostelium discoideum*. In Ashworth, J. M., and Smith, J. E., eds., *Microbial differentiation*, pp. 407–35. Cambridge Univ. Press, London.

Gray, W. D., and Alexopoulos, C. J. 1968. *Biology of the Myxomycetes*. Ronald Press Co., New York.

Ingram, D. S., and Tommerup, I. C. 1972. The life history of *Plasmodiophora brassicae* Woron. *Proc. Roy. Soc. London*, B, 180:103–12.

Karling, J. S. 1968. *The Plasmodiophorales*. 2nd ed. Hafner Press, New York.

Loomis, W. F. 1975. Dictyostelium discoideum: *a developmental system*. Academic Press, New York.

Martin, G. W., and Alexopoulos, C. J. 1969. *The Myxomycetes*. Univ. of Iowa Press, Iowa City.

Olive, L. S. 1975. *The mycetozoans*. Academic Press, New York.

Perkins, F. O. 1972. The ultrastructure of holdfasts, "rhizoids", and "slime tracks" in thraustochytriaceous fungi and *Labyrinthula* spp. *Arch. Mikrobiol.* 84:95–118.

———. 1974. Phylogenic considerations of the problematic thraustochytriaceous-*Dermatocystidium* complex based on observations of fine structure. In *Marine Mycology*. 2nd Internat. Symposium (Bremerhaven, Sept. 1972). *Veröff. Inst. Meeresforsch. Bremerh.* (Suppl.) 5:45–63.

Porter, D. 1972. Cell division in the marine slime mold, *Labyrinthula* sp., and the role of the bothrosome in extracellular membrane production. *Protoplasma* 74:427–48.

———. 1974. Phylogenetic considerations of the Thraustochytriaceae and Labyrinthulaceae. In *Marine Mycology*. 2nd Internat. Symposium (Bremerhaven, Sept. 1972). *Veröff. Inst. Meeresforsch. Bremerh.* (Suppl.) 5:19–44.

Raper, K. B. 1973. Class Acrasiomycetes. In Ainsworth, G. C.; Sparrow, F. K.; and Sussman, A. S., eds., *The Fungi*, vol. 4B, pp. 9–36. Academic Press, New York.

Waterhouse, G. 1973. *Plasmodiophoromycetes*. In Ainsworth, G. C.; Sparrow, F. K.; and Sussman, A. S., eds., *The Fungi*, vol. 4B., pp. 75–82. Academic Press, New York.

This chapter includes a general introduction to all fungi (including those covered in Chapter 6) and gives a detailed treatment of only lower groups (divisions Oomycota, Hyphochytridiomycota, Chytridiomycota, and Zygomycota). Higher fungi and lichens are treated in detail in Chapter 6.

Since the early history of man, two kingdoms of living organisms, plants and animals, have been recognized. Fungi were considered a part of the plant kingdom because large conspicuous types are anchored to the soil like plants. Additionally, these fungi are often equipped with rootlike structures extending into the soil and have rigid, plantlike cell walls. However, many biologists now feel that three basic nutritional types—autotrophy, **lysotrophy,** and phagotrophy—have been characteristic of plants, fungi, and animals since early time. Accordingly, they would divide living organisms among three kingdoms: plants, animals, and fungi (in some modern classification systems, five or more kingdoms are recognized).

In this text, fungi have been retained with plants more for historical than for phylogenetic reasons. However, some aspects of traditional classification systems have been dropped. Slime molds have funguslike sporangia, but are unlike fungi in many important features. Accordingly, they have been treated as a division separate from the other fungi.

About 100,000 species of fungi have been described, and estimates of the total number range to 250,000 or more. Although some species produce conspicuous reproductive structures (e.g., mushrooms, bracket fungi, and puffballs), most fungi are inconspicuous or microscopic in all phases of their life histories. Fungi are eukaryotic heterotrophs that are dependent upon other organisms for certain organic nutrients. Some are parasites of plants or animals, others gain nutrients from mutualistic symbionts with which they are associated, and dead plant or animal remains are nutrient sources for most. Nutritionally, they also resemble bacteria in the production of extracellular enzymes that digest complex substances (e.g., cellulose and starch) to produce smaller molecules that can be absorbed.

Both sexual and asexual reproduction occur in many species of fungi; in others, only asexual reproduction is known. Numerous types of sex-

5

LOWER FUNGI

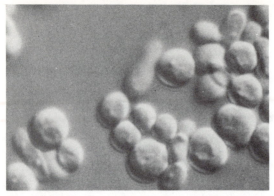

A

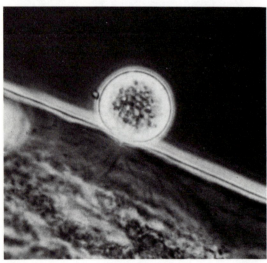

B

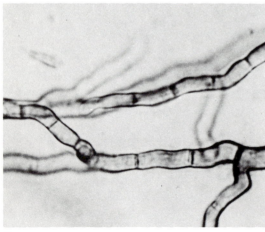

C

ual and asexual **propagules** generally are designated by the term *spores*. Fungal spores differ from one another in method of formation, general structure, and method of germination. Classification of fungi is based largely upon the kinds of spores that are produced by each major group.

General Characteristics of Fungi

The Thallus

The assimilative structure, or thallus, of a fungus can be uninucleate or multinucleate. Uninucleate thalli are characteristic of yeasts (Fig. 5-1A) and some Chytridiomycota. Many of these fungi live in habitats where they are more or less bathed in nutrient solutions, e.g. host tissues in the case of parasites, water, or sugary solutions in nectaries of flowers. In the Chytridiomycota, uninucleate thalli are often equipped with a system of **rhizoids** (Fig. 5-1B) that functions in anchorage and in absorption of nutrients. The growth of uninucleate thalli is **determinate;** it is limited to within a certain range genetically determined for the species.

Multinucleate fungal thalli are characteristic of most fungi, and these typically consist of **hyphae** (Fig. 5-1C), most of which are indeterminate in their growth pattern. Hyphae of a single thallus are usually branched and form a network called a **mycelium.** Hyphae of many primitive fungi are coenocytic; they do not have transverse walls that divide the structure into compartments. However, transverse walls, or **septa,** are characteristic of most fungal hyphae. Fungal septa often have a central pore (see Fig. 6-1)

Figure 5-1 Fungal thalli. **A,** cells of baker's yeast, ×1450. **B,** thallus of *Rhizophydium* (Chytridiomycota) with nucleated spheroidal portion and rhizoids penetrating algal cell, ×1280. **C,** hyphae, ×650. (**B,** photograph courtesy of D. J. S. Barr; **C,** courtesy of M. Dunn.)

through which the cytoplasm is continuous from cell to cell.

Although either hyphae or single cells are characteristic of most fungal species, **dimorphic** fungi produce both kinds of thalli. This feature is most common among primitive ascomycetes and basidiomycetes.

In culture, mycelial thalli often consist of both aerial and submerged portions. However, in nature, mycelial thalli are mostly embedded in decaying plants, wood, soil, or other substrates; typically, only the reproductive structures are exposed. Hyphae embedded in host tissue are often equipped with special absorptive branches that penetrate the host cell (see Fig. 5-23).

Increase in length of hyphae is generally restricted to a minute area at the hyphal tip (Fig. 5-2). As the apex grows through plant or animal material, it digests a pathway through the material. Digestion occurs through the activity of exoenzymes; the digestion products and other small molecules present are absorbed and used as nutrients.

Cell Walls

Cell walls of fungi are thicker and more complex than those of bacteria. The initial wall layer is deposited at the hyphal apex, where extension is occurring. In the cytoplasm of the tip region (Fig. 5-2), numerous small vesicles are present. These membrane-bound vesicles transport enzymes and wall materials required for extension. The vesicle membranes are thought to fuse with, and thus permit extension of, the plasma membrane.

Several zones, or lamellae, are visible in thin sections of fungal walls (Fig. 5-3A). The innermost layer consists or interwoven microfibrils (Fig. 5-3B) of chitin, cellulose, noncellulosic glucans, or other polysaccharides. This fibrillar "fabric" is embedded in a matrix of amorphous proteins and polysaccharides. Lipids and proteins are firmly bound in the structure of the wall and may contribute to its strength. Additional protein is loosely attached and probably consists of enzymes. The outermost wall layer is often capsulelike and consists of slimy or mucilaginous polysaccharides.

About 80–90% of the wall is composed of

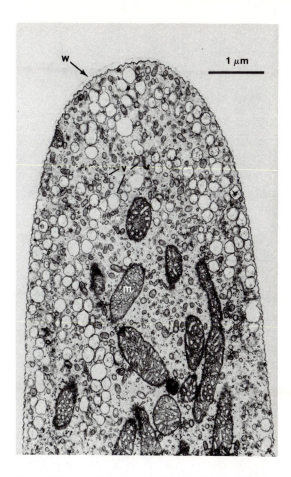

Figure 5-2 Growing apex of hypha, showing characteristic vesicles (*scale line* shown is 1 μm): *w*, wall; *v*, vesicles; *m*, mitochondria. (Photograph courtesy of S. N. Grove and C. E. Bracker, with permission of *Journal of Bacteriology*.)

polysaccharides, and the remainder is protein, lipid, and sometimes pigments. Chitin and cellulose constitute only a small percentage of the total wall material. However, these and other wall substances correlate well with morphological and reproductive features of major fungal groups. Cellulose occurs in the walls of two groups: the Hyphochytridiomycota and the Oomycota. Chitin is also present in walls of the Hyphochytridiomycota, as well as in the Chytridiomycota, Zygomycota, and Eumycota.

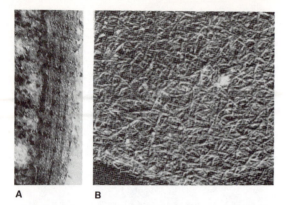

A   **B**

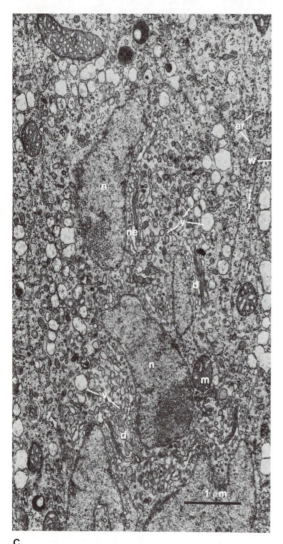

C

The Protoplast

Within the cell wall, a plasma membrane surrounds the fungal protoplast (Fig. 5-3C). The cells contain typical eukaryotic organelles, including nuclei, mitochondria, vacuoles, and others. Vacuoles typically remain small and do not constitute a large percentage of the cell volume in actively growing cells. Ribosomes are abundant in the cytoplasm and often attached to the endoplasmic reticulum. Granular reserve materials, such as **glycogen,** are also present in the cytoplasm. Lipids, common reserve materials in most fungi, are visible in the cells as conspicuous globules, called *guttules.*

Nuclei and Nuclear Division

Fungal nuclei, typically small, have an envelope consisting of a double membrane with pores. A nucleolus is present. The nuclear membrane commonly remains intact during mitotic divisions (Fig. 5-4); it may also do so during meiosis. Because of the small size, and the persistent nuclear membrane, chromosomes are difficult to discern during mitosis. Mitotic chromosomes rarely form a typical metaphase plate, and their separation at anaphase is often asynchronous. At telophase, the nuclear membrane constricts be-

Figure 5-3 Hyphal walls. **A,** thin section of hyphal wall, showing layered structure, ×22,930. **B,** inner wall layer consisting of interwoven microfibrils, ×27,000. **C,** thin section of fungal hypha, showing organelles (*scale line* shown is 2 μm): *n,* nucleus; *d,* Golgi dictyosome; *er,* endoplasmic reticulum; *ne,* nuclear envelope; *v,* wall vesicles. (Photographs **A, B,** courtesy D. Hunsley and J. H. Burnett, from "The ultrastructural architecture of the walls of some hyphal fungi," *Journal of General Microbiology* 62:203–18, 1970, with permission of Cambridge University Press; **C,** courtesy of S. N. Grove and C. E. Bracker, with permission of *Journal of Bacteriology.*)

tween the two poles, and two nuclei are formed. Centrioles are present in nuclei of fungi with motile cells (Fig. 5-5A, B); a structure possibly homologous to the centriole, the centriolar plaque (Fig. 5-5C; 5-6C), is present in higher fungi. Chromosomes are generally more conspicuous during meiosis (Fig. 5-6), and these divisions differ only in detail from meiotic divisions in other organisms.

Nuclear division occurs together with cell division in hyphal filaments with uni- or binucleate cells. However, cell division and nuclear division are independent of one another in many hyphae, and the numbers of nuclei per compartment thus vary.

As with bacteria, numerous analyses have been made of fungal DNA composition. Percentages of guanine + cytosine for all fungal taxa range from 27.0–70.0% G + C. Since ranges for major categories overlap, this feature has limited value in characterizing such groups. However, G + C ratios seem to have predictive value at lower levels of classification (e.g., in determining affinities of fungi lacking sexual stages).

Nutritional and Environmental Requirements

Nutritional requirements for most fungi are relatively simple and include a carbon-energy source, water, and mineral nutrients. Many species also require an external supply of one or more vitamins. In nature, fungi obtain these nutrients from living or dead organisms. **Saprobic** species are extremely versatile scavengers that utilize simple or complex organic substances of many types. Proteins, cellulose, and other complex polymers are digested externally; the digestion products and other small molecules present are absorbed and utilized. Parasitic fungi obtain their nutrients from living hosts and may be either **facultative parasites** or **obligate parasites.** The former can exist either as saprobes or as parasites; the latter must depend upon living hosts for nutrients.

Water requirements vary among fungi. Many species live only in aquatic or semiaquatic environments; numerous terrestrial species grow only when moisture is abundant and become

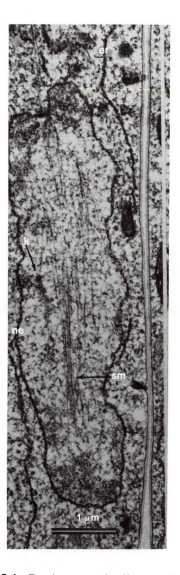

Figure 5-4 *Fusarium oxysporium* (Ascomycotina). Late anaphase of mitotic division (*scale line* shown is 1 μm): *sm*, spindle microtubules; *ne*, nuclear envelope; *k*, kinetochore; *er*, endoplasmic reticulum. Note that the nuclear envelope is still intact. (Photograph courtesy of J. R. Aist and P. H. Williams, from "Ultrastructure and time course of mitosis in the fungus *Fusarium oxysporum*," *Journal of Cell Biology* 55:368–89, 1972, with permission of Rockefeller University Press.)

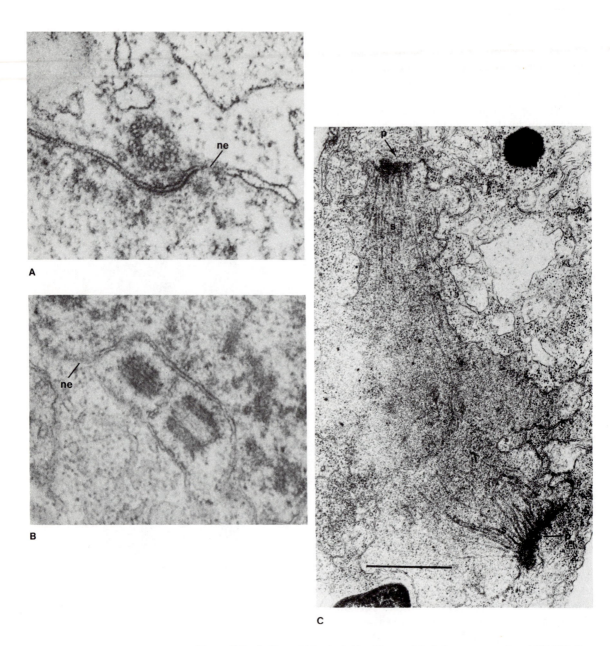

Figure 5-5 A, B, centrioles in *Achlya* (Oomycota). **A,** transverse section, ×61,100; **B,** longitudinal section through paired centrioles, ×46,800: *ne,* nuclear envelope. **C,** centriolar plaques in *Erysiphe* (Ascomycotina): *p,* plaques; *s,* spindle fibers (*scale line* shown is 1 μm). (Photographs **A** and **B,** courtesy of N. Ricker; photograph **C,** courtesy of W. E. McKeen, reproduced by permission of the National Research Council of Canada from the *Canadian Journal of Microbiology* 18:1915–22, 1972.)

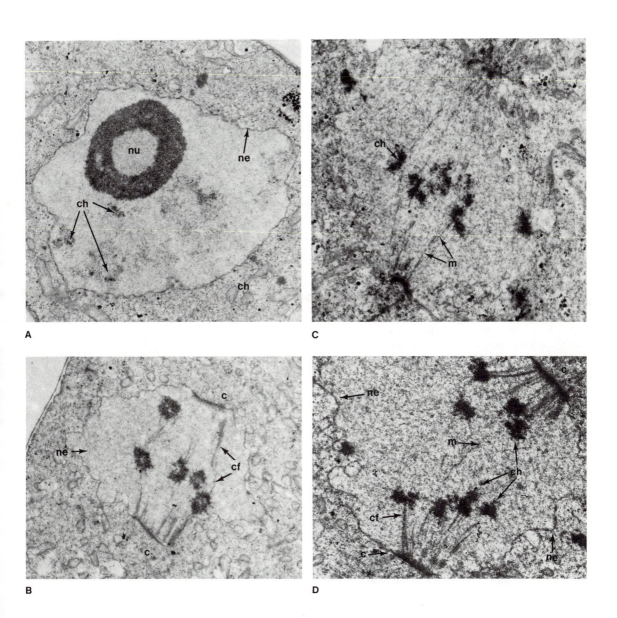

Figure 5-6 Meiotic division (Ascomycotina). **A,** 2n nucleus, prophase, ×9470. **B,** metaphase I, ×8820. **C,** anaphase I, ×11,485. **D,** metaphase II, ×10,820: *c,* centriolar plaque; *ch,* chromosomes; *cf,* chromosomal fiber; *m,* microtubules; *ne,* nuclear envelope; *nu,* nucleolus. (Photographs courtesy of K. Wells, with permission of *Mycologia.*)

dormant during dry periods. However, some common fungi are capable of carrying on slow growth in stored grains, other foods, and building timbers, where little moisture is present.

Fungi are predominantly aerobic, but some noteworthy exceptions exist among aquatic fungi (see under *Aqualinderella fermentans,* p. 141, and *Blastocladia,* p. 152) which are obligately fermentative. Additionally, some filamentous fungi and yeasts are facultatively fermentative.

Most fungi grow best at temperatures in the range 20–30 °C. However, some fungi are adapted to higher temperatures (40–60 °C); these occur in rotting compost, bird nests, soil, and other warm habitats. Some fungi are also adapted to low temperatures (−10 °C). These cold-loving fungi can grow under snow and are responsible for *snow-mold* diseases of wheat and other plants. As with other factors affecting fungal growth, most species are tolerant of wide extremes of temperature.

Two additional factors commonly affecting growth are pH and light. As a rule, fungi grow best in acid media (pH 4–6), but many species grow in more acid conditions or in alkaline habitats. Exposure to light is required for induction of the reproductive phase in numerous fungi. Light may stimulate or inhibit assimilative growth; light-tropic responses (**phototropism**) occur in both assimilative and reproductive structures.

Reproduction

In morphologically simple fungi, the entire assimilative protoplast can be converted to reproductive units when reproduction occurs. In such fungi, the differentiation associated with reproduction is largely internal. However, most fungi produce special reproductive structures that are distinct from the assimilative body. Here, reproduction does not use the entire mass of protoplasm, but reproductive and assimilative growth can occur concurrently. Asexual reproduction typically occurs over a wide range of conditions that permit assimilative growth. On the other hand, sexual reproduction may require special environmental conditions as well as the chance encounter of gametes or their equivalent.

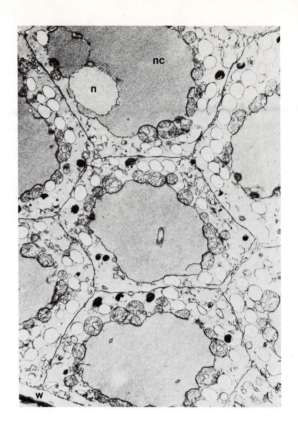

Figure 5-7 *Allomyces.* Portion of a sporangium immediately after cleavage of the protoplast into uninucleate segments, ×2475. In this example, the segments mature into zoospores: *n,* nucleus; *nc,* nuclear cap. (Photograph courtesy of R. T. Moore, with permission of the *Journal of the Elisha Mitchell Society.*)

Asexual Reproduction Most primitive fungi reproduce asexually by forming sporangia and sporangiospores. Sporangiospores develop through cleavage of the sporangial protoplast into segments containing one or more nuclei (Fig. 5-7). Each segment can then develop into either a naked, motile cell, a zoospore (**planospore**), or a walled, nonmotile spore (**aplanospore**).

Two types of flagella, called *tinsel* and *whiplash* types, occur in lower fungi. The tinsel flagellum (Fig. 5-8A) possesses two rows of lateral processes called flimmers or **mastigonemes.** The whiplash flagellum has a rigid basal portion and a thinner, flexible tip; it resembles an old-

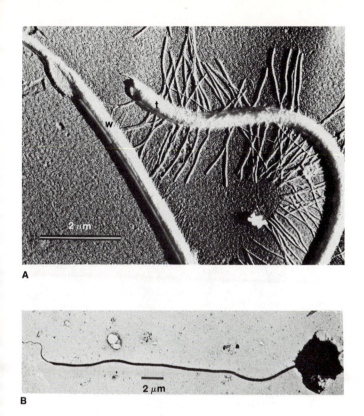

A

B

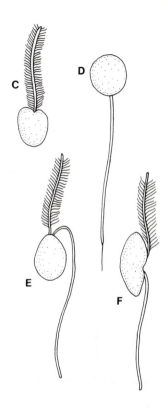

Figure 5-8 **A,** tinsel (*t*) and whiplash (*w*) flagella (Oomycota); *scale line* shown is 2 μm. Note that whiplash of this species also has minute lateral hairs. **B,** whiplash flagellum (Chytridiomycota; *scale line* shown is 2 μm): *L,* lipid body. **C–F,** diagrams showing type and arrangement of flagella of fungal motile cells. **A,** single anterior tinsel flagellum (Hyphochytridiomycota); **B,** single posterior whiplash flagellum (Chytridiomycota); **C, D,** anterior and lateral whiplash and tinsel flagella (Oomycota), the tinsel flagellum anteriorly directed in both types. (Photograph **A,** courtesy of P. R. Desjardins, G. A. Zentmyer, and D. A. Reynolds, reproduced by permission of the National Research Council of Canada from the *Canadian Journal of Botany* 47:1077–79, 1969; **B,** courtesy of F. Y. Kazama, from "Ultra-structure and phototaxis of the zoospores of *Phlyctochytrium* sp., an estuarine chytrid," *Journal of General Microbiology* 71:555–66, 1972, with permission of Cambridge University Press.)

fashioned buggy whip. The zoospores (Fig. 5-8C–F) can have one anterior tinsel flagellum (Hyphochytridiomycota), one posterior whiplash flagellum (Chytridiomycota), or one of each type (Oomycota). Zoospores in the Oomycota have lateral or anterior insertion of the two flagella. The tinsel flagellum is anteriorly directed and the whiplash flagellum is posteriorly directed in both types of insertion.

Fungal zoospores typically are capable of both swimming and amoeboid movement. However, unlike myxomycete swarm cells, they neither feed phagotrophically nor do they undergo fission. Dispersal is by water currents and by swimming movement.

Walled sporangiospores (Zygomycota) presumably evolved from motile spores and are an adaptation to terrestrial existence. When released, these spores are dispersed by air currents, by insects, or by splashing rain.

All higher fungi (Eumycota) lack sporangia. Where present, asexual reproduction is by **budding** or by formation of **conidia.** In budding, the main method of reproduction in yeasts, an outgrowth develops from the parent cell (Fig. 5-9A). The outgrowth enlarges, receives a nucleus and other organelles, and finally separates from the parent cell. Conidia are essentially separable portions of hyphae or of specialized hyphal branches which function as spores (Fig. 5-9B). A single species of Eumycota can produce one or more types of conidia—or none. In the Fungi Imperfecti, conidia are the only types of spores produced. Conidia are dispersed by air currents, insects, or water.

Sexual Reproduction   Fungi are as varied with respect to sexual reproduction as they are in their assimilative structures and asexual reproduction. Motile gametes are found only in the Chytridiomycota, where **syngamy** can occur through the fusion of like gametes (**isogamy**), of unequal motile gametes (**anisogamy,** Fig. 5-10), or of fusion of a motile sperm with a large, nonmotile egg (**oogamy**). In isogamy and anisogamy, gametes are released from the gametangia and syngamy occurs in the surrounding water. In oogamy (Monoblepharidales), the egg remains in its gametangium and is fertilized there by the sperm.

In other lower fungi, sexual reproduction can occur through fusion of whole assimilative

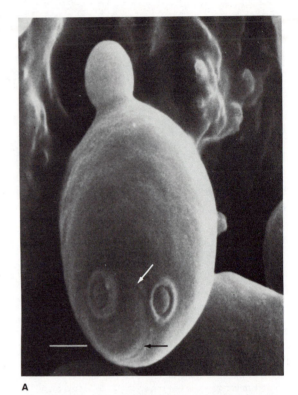

A

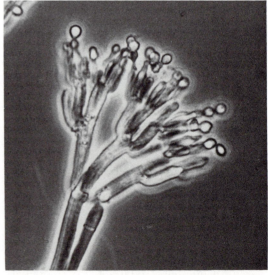

B

Figure 5-9 Budding in *Saccharomyces* (baker's yeast, Ascomycotina). **A,** early stage in bud development (*scale line* shown is 1 µm): *arrows* indicate margin of birth scar formed when mother cell was produced; the two craterlike structures are bud scars marking sites of earlier reproduction. **B,** phase-contrast photograph of specialized hyphal branch bearing conida in *Penicillium,* ×1145 (Ascomycotina). (Photograph **A,** courtesy of L. T. Talens, M. Miranda, and M. W. Miller, with permission of *Journal of Bacteriology.*)

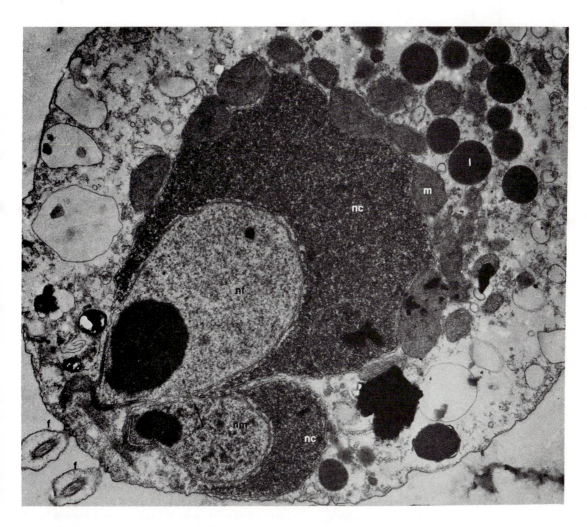

Figure 5-10 Zygote resulting from anisogamous fusion in *Allomyces* (Chytridiomycota), ×19,160: *nc,* nuclear cap; *nf,* female nucleus; *nm,* male nucleus; *m,* mitochondria; *l,* lipid droplets; *f,* flagella. Note that the nucleus and nuclear cap of the female gamete (*left*) are larger than those of the male; most of the mitochondria and lipid droplets are from the female. The *arrow* indicates the point at which fusion of the nuclear envelopes has occurred. (Photograph courtesy of M. S. Fuller, with permission of Longman Group Ltd.)

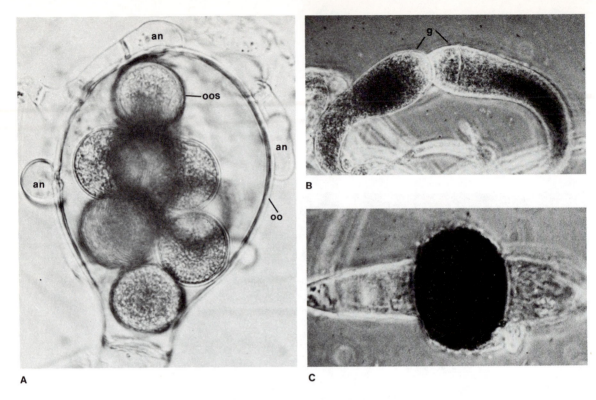

Figure 5-11 **A,** sexual reproductive structures in the Oomycota, ×460: *an,* antheridium; *oo,* oogonium; *oos,* oospore. **B, C,** sexual reproduction in the Zygomycota, showing essentially equal gametangia and mature zygospore (*bottom*), ×335.

thalli (some Chytridiomycota, Hyphochytridiomycota), or by production of specialized gametangia. In the Oomycetes, antheridia and oogonia develop (Fig. 5-11A) and meiosis occurs in the formation of gamete nuclei. The antheridium contacts the oogonium, and nuclei are transferred to the oogonium where they fuse with oogonial nuclei.

In Zygomycetes the gametangia (Fig. 5-11B) fuse. These gametangia are alike in most instances and are not designated as male and female. Following gametangial fusion, a resistant zygospore (Fig. 5-11C) develops within which nuclear fusion and meiosis occur.

In all of the above fungi, syngamy or its equivalent typically leads to the formation of a single diploid resistant structure or a diploid thallus. In most Eumycota, the two phases of syngamy or its equivalent, **plasmogamy** and **karyogamy,** are separated from one another in time and in space. Plasmogamy, the coalescence of protoplasts, occurs through fusion of either hyphae or specialized reproductive cells. Pairs of compatible nuclei brought together through plasmogamy are called **dikaryons,** and cells or hyphae containing such pairs are said to be dikaryotic. Within dikaryotic hyphae (Fig. 5-12E), synchronous mitotic divisions lead to the production of many pairs of compatible nuclei. Finally, numerous **asci** or **basidia** (Fig. 5-12B, D) are produced within which karyogamy and meiosis take place. Asci and basidia commonly are produced in large numbers within complex fruiting bodies, the **sporocarps.**

In Ascomycotina, the dikaryotic phase is usually initiated on a female gametangium and is

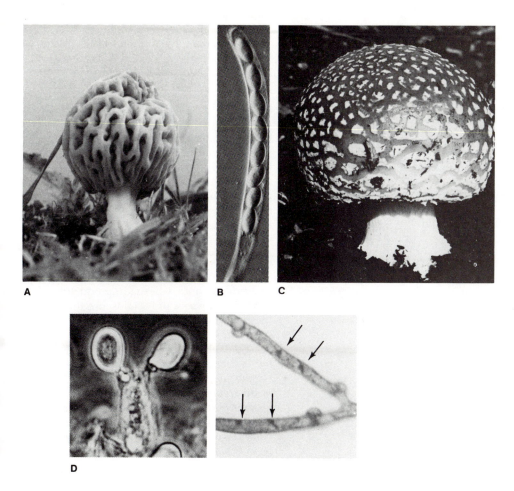

Figure 5-12 **A,** sporocarp of *Morchella* (Ascomycotina), ×0.5. **B,** ascus and mature ascospores, ×560. **C,** sporocarp of *Amanita* (Basidiomycotina), ×0.5. **D,** mature basidium and basidiospores, ×1000. **E,** dikaryotic hypha stained to show nuclei (Basidiomycotina), ×925.

of limited extent. A single sporocarp, containing numerous asci and ascospores, typically develops after plasmogamy. The dikaryotic phase of most Basidiomycotina is initiated through fusion of assimilative hyphae, and the dikaryotic mycelium is of potentially unlimited duration. Dikaryotic mycelia of many Basidiomycotina are perennial, producing sporocarps whenever environmental conditions are satisfactory. Many sporocarps are usually produced by a single dikaryotic thallus.

Types of reproduction and other major fea-

tures of the main fungal groups are summarized below.

1. *Division Oomycota.* Zoospores with anterior or lateral flagella, a tinsel flagellum directed anteriorly, and a whiplash directed posteriorly. Sexual reproduction by formation of antheridia and oogonia, contact of these, and transfer of gamete nuclei from the antheridium to the oogonium. Meiosis occurs in antheridia and oogonia prior to nuclear

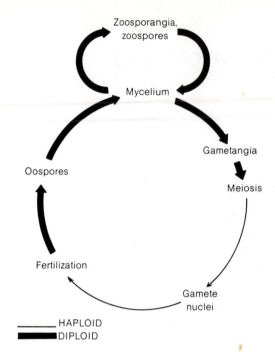

Zoosporangia,
zoospores

Mycelium

Gametangia

Oospores

Meiosis

Fertilization

Gamete
nuclei

—— HAPLOID
▬▬ DIPLOID

Figure 5-13 Life history characteristic of filamentous Oomycota.

transfer and fertilization; the assimilative structure is diploid. Unicellular to mycelial, coenocytic. Cell walls contain cellulose and other glucans; reserve carbohydrate is mycolaminaran. Percentage G + C, 40.5–62.0.

2. *Division Hyphochytridiomycota.* Zoospores with single anterior tinsel flagellum. Sexual reproduction essentially unknown; by fusion of whole protoplasts in one species. Assimilative structure probably haploid throughout division, predominantly unicellular. Cell walls containing chitin, cellulose. Percentage G + C not known.

3. *Division Chytridiomycota.* Zoospores with one posterior whiplash flagellum. Sexual reproduction isogamous, anisogamous, or oogamous; motile gametes structurally similar to zoospores. In some, sexual reproduc-

tion by fusion of thalli. Assimilative structure haploid or with alternation of haploid and diploid thalli, unicellular to mycelial. Cell walls contain chitin and glucans; reserve materials are glycogen and lipids. Percentage G + C not known.

4. *Division Zygomycota.* Sporangiospores nonmotile, walled; asexual reproduction by conidia in some. Sexual reproduction by fusion of gametangia (conjugation) and formation of multinucleate or uninucleate zygospore. Assimilative structure haploid, predominantly mycelial. Cell walls contain chitin, chitosan, glucans; reserve materials are glycogen and lipids. Percentage G + C range: 27.5–59.0.

5. *Division Eumycota.* Neither motile cells nor sporangia produced. Asexual reproduction by budding or conidia. Sexual reproduction through ascospore or basidiospore formation; these spores commonly produced in complex sporocarps. The thallus can be a single-celled yeast (lower ascomycetes and basidiomycetes) but is commonly composed of hyphae. Hyphae are septate and can be either haploid or dikaryotic. Cell walls contain chitin and glucans and, especially in yeasts, mannans; reserve materials are glycogen and lipids. Percentage G + C: 29–60 in ascomycetes, 44–70 in basidiomycetes, and 35.5–64.5 in Fungi Imperfecti.

The group Fungi Imperfecti is a nonsexual category composed mainly of higher fungi. Species of this group reproduce only by conidia or exist as sterile hyphae or modified hyphal structures.

Fungi are treated in two chapters in this text primarily as a matter of convenience. That is, although Ascomycotina and Basidiomycotina appear to be closely related, the four divisions placed in Chapter 5 are not considered to be closely linked. These four divisions, once collectively designated *Phycomycetes*, probably had independent origins. Ascomycotina and Basidiomycotina, treated as a single division here (Eumycota), are sometimes classified as separate divisions (Ascomycota and Basidiomycota).

Lower Fungi

Division Oomycota

The division Oomycota includes a single class, Oomycetes, and four orders of fungi.

The Saprolegniales, Leptomitales, and Peronosporales contain mostly well-developed mycelial fungi. Their hyphal walls contain cellulose and β-glucan, the latter being the most abundant. In at least a part of the group, the reserve carbohydrate is **mycolaminaran.** The hyphae are multinucleate and (apart from septa that separate off the reproductive structures) coenocytic. Most recent studies indicate that the assimilative thallus is diploid. Asexual reproduction is by production of zoospores or, in some terrestrial species, by sporangia that function as conidia. In the latter instance, the sporangia are deciduous and have the ability to germinate by direct formation of a hypha. Oomycete zoospores can have either anterior or lateral insertion of flagella, and both types appear in the life histories of some forms.

Sexual reproduction in mycelial species is by production and contact of male and female gametangia, called respectively *antheridia* and *oogonia*. Following meiosis in the gametangia, nuclei are transferred from antheridium to oogonium. One to many nuclei in the oogonium can function as gametes, and after fertilization, thick-walled oospores are produced.

The generalized life-history pattern of filamentous Oomycetes is diagrammed in Figure 5-13.

Order Saprolegniales The Saprolegniales are freshwater or soil-inhabiting fungi, most of which grow on dead plant or animal materials. A few species parasitize algae, higher plants, or animals. Filamentous species, especially of *Saprolegnia* and related genera, are among the best-known aquatic fungi. These species grow and reproduce readily in culture, and they have been the object of extensive laboratory studies.

Species of Saprolegniaceae, the best-known and most frequently encountered Saprolegniales, have coenocytic hyphae (Fig. 5-14) and sporangia that are only slightly modified hyphal tips. The

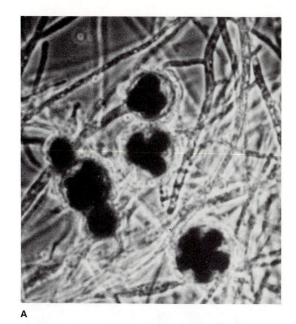

A

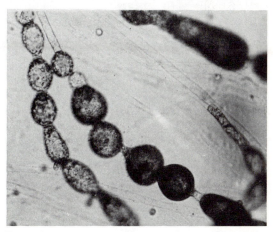

B

Figure 5-14 *Saprolegnia.* **A,** mycelium, $\times$200. **B,** hyphae with gemmae, $\times$160. (Photograph **B** courtesy of G. Neish.)

zoospores are of two types (Fig. 5-8E, F): primary zoospores that are pear-shaped and anteriorly biflagellated, and secondary zoospores that are bean-shaped and have laterally inserted flagella. The whiplash flagellum of some species is provided with lateral flimmers, or hairs, that are much smaller than those on the tinsel flagellum. Similar hairs have been observed on whiplash

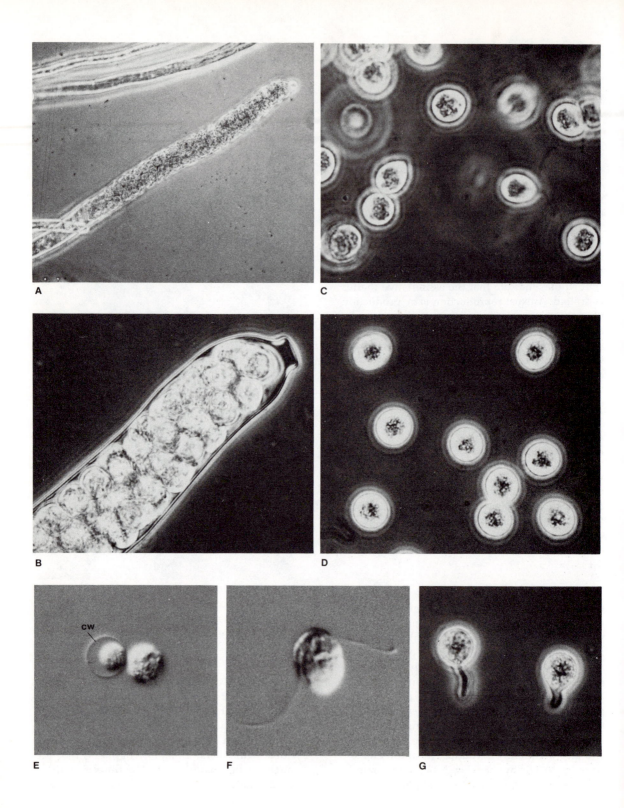

A

B

C

D

E

cw

F

G

flagella of some Peronosporales, but the frequency and importance of their occurrence are not yet known.

Oogonia usually develop as terminal cells on short lateral branches, but intercalary oogonia are produced by some species. Each oogonium contains from one to twenty or more **oospores** at maturity. The antheridium can arise from the same hypha as the oogonium, from a different hypha, or from a different thallus.

Saprolegnia and allied genera once were referred to as *water molds*. However, we now know that there are many other groups of aquatic fungi, and that saprolegniaceous fungi are common in soil as well as in water. Extensive mycelium can often be seen growing on seeds, dead insects, and other organic material in fresh water.

In some species, walls can delimit short intercalary cells that develop into structures called **gemmae** (Fig. 5-14B). Gemmae are thick-walled structures that function as vegetative propagules.

Sporangia of *Saprolegnia* develop from only slightly enlarged hyphal tips (Fig. 5-15A, B). The sporangial protoplast is multinucleate and, following cleavage, uninucleate primary zoospores are produced. Zoospores are released through a pore that dissolves in the wall at the tip of the sporangium. Zoospore release is not passive but is brought about by an internal pressure that rapidly forces all zoospores from the sporangium. Once released, zoospores swim away from the sporangium.

Primary zoospores of *Saprolegnia* are motile for up to an hour, then withdraw their flagella, and encyst (Fig. 5-15C, D). Primary cysts germinate (Fig. 5-15E, F) after a short period, giving rise to laterally flagellated secondary zoospores. The secondary zoospore also is motile for a time and then encysts. Depending upon conditions, this secondary cyst (Fig. 5-15G) can germinate by producing either another secondary zoospore or

a **germ tube.** The latter structure is the start of a new mycelial thallus.

Genera of saprolegniaceous fungi are distinguished mainly by differences that occur in asexual reproduction. *Achlya,* morphologically similar to *Saprolegnia,* differs in the behavior of its primary zoospores. These typically encyst at the mouth of the sporangium (Fig. 5-16A–C), the secondary zoospores thus being the only motile stage. In a third genus, *Dictyuchus,* encystment of primary spores occurs within the sporangium. The name *Dictyuchus* is derived from the netted appearance produced by cyst walls within the sporangium (Fig. 5-16D). Secondary zoospores of this genus are released through individual pores; each cyst produces a tube (Fig. 5-16E) of variable length, through which the zoospore escapes.

In all the above genera, the release of zoospores is usually followed by renewed growth of the hypha below the sporangium. In *Saprolegnia,* this growth occurs through the base of the spent sporangium and is called *internal proliferation* (Fig. 5-16F). Successive sporangia can be produced in this way, and several old sporangial walls can ensheath a newly formed sporangium. In *Achlya* and *Dictyuchus,* renewed growth of the hypha below the sporangium occurs laterally.

The sexual reproductive structures of *Saprolegnia* are shown in Figure 5-17A. Oogonia typically develop on short, lateral branches of main hyphae. Antheridial branches can develop from the same hypha, from a different hypha, or from a different mycelium. One to several antheridia are attached to each oogonium; they and the oogonium are separated from their parent hyphae by septa. Meiosis occurs in the gametangia, and a number of haploid gametic nuclei are produced. Only a few nuclei in the oogonium are functional; the others degenerate. Functional nuclei are not delimited by walls, but each such nucleus and a zone of cytoplasm around it constitutes an egg or **oosphere.** The antheridia produce **fertilization tubes** that penetrate the oogonial wall; nuclei are transferred through these tubes to the oogonium. After fertilization, each zygote develops into a thick-walled, resistant structure: the oospore.

Oospores are known to retain their viability for several years, but the factors that initiate germination are not known. When germination

---

Figure 5-15 *Saprolegnia.* **A,** mature sporangium with basal septum and apical exit papilla, ×300. **B,** sporangial apex just prior to zoospore discharge, ×900. **C,** primary zoospores, ×900. **D,** primary cysts, ×900. **E,** germinating primary cyst, ×900. **F,** secondary zoospore, ×1400. **G,** germinating secondary cysts producing germ tubes, ×900.

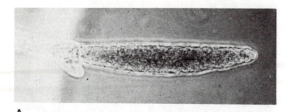

A

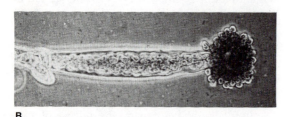

B

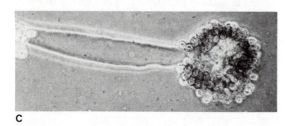

C

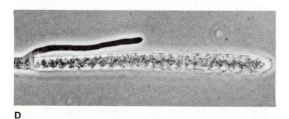

D

E

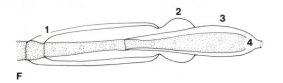

F

does occur, it is by the production of a short hypha tipped by a sporangium.

The antheridia and oogonia of some saprolegniaceous species develop on separate thalli, and sexual reproduction occurs only if two such thalli exist side by side. For example, in *Achyla bisexualis* and *A. ambisexualis,* single thalli reproduce only asexually. When compatible thalli are brought together, antheridia develop on one thallus and oogonia are produced by the other. Initiation and development of sexual structures occurs in a series of clearly defined steps in these species:

1. Production of antheridial filaments on the male thallus.

2. Production of oogonial initials on the female thallus.

3. Growth of antheridial filaments toward oogonial initials and delimitation of antheridia after contact.

4. Delimitation of the oogonium and differentiation of oospheres.

5. Formation of fertilization tubes and nuclear transfer.

Experiments have shown that the developmental stages are controlled by hormones (Fig. 5-18). Only one of the hormones, hormone *A,* has been isolated and identified chemically. It is a sterol and has been given the name *antheridiol.* Antheridiol functions in the initiation of antheridial branches and controls their directional

Figure 5-16 *Achylya.* **A–C,** sporangium before, during, and after discharge of primary zoospores, ×235 (note encysted primary zoospore at exit pore of **C**). **D, E,** *Dictyuchus.* **D,** young sporangium, ×205; **E,** mature sporangium containing encysted primary spores, ×455. Note germ pore (*arrow*), through which individual secondary zoospore escaped. **F,** sporangial proliferation in *Saprolegnia,* ×375 (approx.). Renewed hyphal growth is through the base of the spent sporangium; walls of previously formed sporangia (*numbered*) enclose the base of the new sporangium.

growth and the delimitation of the antheridium. After antheridial branches are initiated, they start to produce hormone *B*, which initiates the development of oogonial initials on the female thallus. Hormone *D* is produced by the antheridium; it initiates septum formation at the base of the oogonium and delimitation of the oospheres. The formation of fertilization tubes and nuclear transfer are also thought to be controlled by hormones, but experimental evidence for these is lacking.

Although saprolegniaceous fungi are predominantly saprobes, some species of *Achlya* and *Saprolegnia* infect living fish, fish eggs, and amphibians. These fungi occur as wound parasites on adult fish, but once established, the mycelium rapidly invades healthy tissues. Especially if crowded, fish in aquaria and in hatcheries often become infected. Salmon, moving into the fresh water on their spawning journey, are also frequently infected.

Order Leptomitales The Leptomitales are freshwater saprobes that grow on submerged plant materials. Species of this order are found in both clean, aerobic water and in essentially anaerobic waters rich in organic matter. These fungi are most easily obtained, like many of the Blastocladiales, by "baiting" water bodies with fruits such as apples, pears, or rose hips. The fruits are kept in water for several weeks and, when retrieved, are colonized by a variety of zoosporic fungi.

The thallus of these fungi can be a mycelium, but it can also be a club-shaped or irregularly branched structure of limited growth. In *Aqualinderella fermentans* (Fig. 5-19A), thalli are of the latter type, the large trunklike portion being equipped with an extensive rhizoidal system at its base. Large numbers of thin-walled sporangia are borne directly on the massive "trunk" in *Aqualinderella*; in some related forms, the sporangia are on hyphae arising from this structure. Sexual reproduction is not known in *Aqualinderella*, but thick-walled, resistant structures are produced (Fig. 5-19B). These are produced in oogoniumlike cells, and it is possible that they are *apogamously* developed (without fertilization) oospores.

Although almost all fungi are aerobic, *A. fer-*

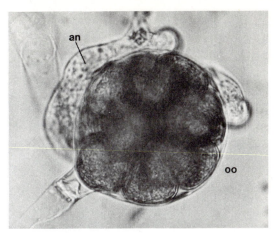

A

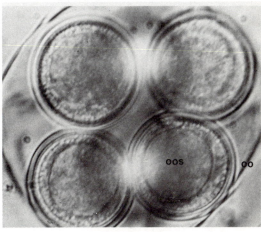

B

Figure 5-17 *Saprolegnia.* **A,** young oogonium with appressed antheridia and developing oospores, ×430. **B,** mature oospores, ×1180. *oo,* oogonium; *an,* antheridia; *oos,* oospore. (Photographs courtesy of G. Neish.)

mentans does not require oxygen. This species occurs in warm waters where oxygen is meager or lacking and where carbon dioxide concentrations are very high. *A. fermentans* is obligately fermentative and breaks down sugars anaerobically. The fungus obtains only a portion of the energy present in sugar molecules, lactic acid being produced in the process. The fungus lacks the cristate mitochondria and cytochromes characteristic of aerobic organisms. It grows best in an atmosphere of 20% carbon dioxide, but it can tol-

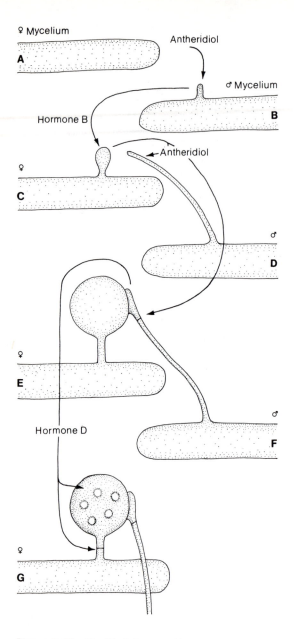

♀ Mycelium

Antheridiol

A

♂ Mycelium

B

Hormone B

♀

Antheridiol

C

♂

D

♀

E

♂

Hormone D

F

♀

G

Figure 5-18 Simplified diagram of hormonal control of sexual reproduction in *Achlya bisexualis*. **A, B,** paired ♀ and ♂ thalli; the hormone antheridiol produced by the ♀ thallus has stimulated formation of an antheridial branch on the ♂ thalla. Stimulated ♂ thallus (**B**) produces Hormone *B*, which causes oogonial initial to develop on thallus (**C**). Antheridiol causes directional growth of antheridial filament (**D–F**) to oogonium and antheridium delimitation. Hormone *D*, produced by antheridial filament, stimulates oogonial delimitation and oosphere formation (**G**).

erate much higher concentrations of this gas. Oxygen is neither required, nor does its presence or absence appear to have any effect on the fungus.

Order Peronosporales Sexual reproduction similar to that of Saprolegniales and Leptomitales is characteristic of Peronosporales. Asexual reproduction of soil- and water-inhabiting species also differs only in detail from that of Saprolegniales. However, more advanced members of Peronosporales are all obligate parasites of higher plants. Their sporangia are deciduous and function in dispersal; they can release zoospores or, alternatively, can produce a hypha directly. If zoospores are produced, they are of the secondary (laterally biflagellated) type (Fig. 5-20D). The hyphae of most obligately parasitic Peronosporales grow between cells of the host and are equipped with **haustoria.**

Species of *Pythium* and *Phytophthora* occur in freshwater habitats and in soil; a few species of the latter genus are also common in marine waters. Certain soil-inhabiting species of both genera are responsible for numerous diseases of plants—e.g., *"damping off," root rots,* and *fruit rots. Pythium* species, predominantly saprobic, can rapidly invade the tissues of seedlings in crowded seedbeds where moisture levels are high to cause damping off. This invasion occurs at the soil line, and rapid disintegration of tissue occurs there. Infected seedlings collapse, and the disease can spread to uninfected individuals from these fallen plants.

Pythium debaryanum (Fig. 5-20A, B) is one species commonly associated with damping off of seedlings. This species has narrow hyphae that bear numerous, roughly globose sporangia. At maturity, a short tube develops through which the sporangial protoplast is extruded. The protoplast, surrounded only by a membrane, cleaves into a number of zoospores. The zoospores can be seen to squirm within the restraining membrane as they complete their development. Finally, the membrane ruptures and zoospores quickly swim away.

The oogonia and antheridia of *Pythium* are similar to those of saprolegniaceous fungi, but the oogonia contain a single egg.

Many species of *Phytophthora* are important

plant parasites. Most are inhabitants of terrestrial or freshwater habitats, but some marine species are known.

Phytophthora infestans, cause of *late blight* of potato, is the most notorious parasitic species. The mycelium in this species rapidly invades the tissues of infected plants, causing death of the infected parts. Sporangia are produced on special aerial hyphae (Fig. 5-20C) called **sporangiophores,** which grow out through stomata. The lemon-shaped sporangia are deciduous, and the sporangiophore branches proliferate and produce successive sporangia. The sporangia are airborne and, depending upon environmental conditions, can germinate by releasing zoospores or producing a hypha. At temperatures of about 10–15 °C, germination is by zoospore release; at higher temperatures, germ tubes develop. From electron miscroscope studies, it has been found that flagella always form in sporangia, but if germination occurs by germ tube, the flagella disappear.

During periods of high humidity and relatively cool temperatures, sporangia of *P. infestans* cause the rapid spread of blight in potato fields. The productivity of potato plants is greatly reduced because the death of leaf tissues reduces photosynthesis. Infection and destruction of tubers can also occur. *P. infestans* caused complete destruction of the potato crops of Ireland and parts of Europe during the years 1845 to 1847. At that time, the diet of many poorer people consisted largely of potatoes, and the destruction of the crop resulted in the Irish famine. Thus, this plant disease was the cause of much human suffering and, indirectly, was responsible for the emigration of many Irish to North America.

In *P. infestans*, sexual reproduction requires the presence of two compatible strains. The sexual stage is common in Mexico, but only one of the mating types is present in most other areas. In the absence of the oospore stage, the mycelium of *P. infestans* overwinters in decayed potato tubers.

Zoospore encystment, the stage preceding germ tube production, can occur in a remarkably short time in some *Phytophthora* species. Cyst formation can be stimulated by agitating a zoospore suspension; vigorous agitation of such a suspension will cause most zoospores to encyst within 60–120 seconds. During this period, flagella are

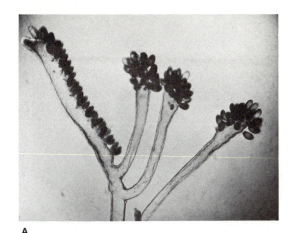

A

B

Figure 5-19 *Aqualinderella fermentans.* **A,** thallus with numerous sporangia, ×30. **B,** thick-walled resistant structures, possibly representing apogamously developed oospores, ×410. (Photographs courtesy of R. Emerson, with permission of *American Journal of Botany.*)

retracted, and a thin but rigid glucan wall develops around the protoplast.

The ability to capture small animals and use them as a source of nutrients is characteristic of several unrelated groups of fungi. *Zoophagus insidians,* a fungus related to *Pythium* and *Phytophthora,* captures rotifers and is nutritionally dependent upon these animals. *Z. insidians* hyphae are provided with numerous, short, ad-

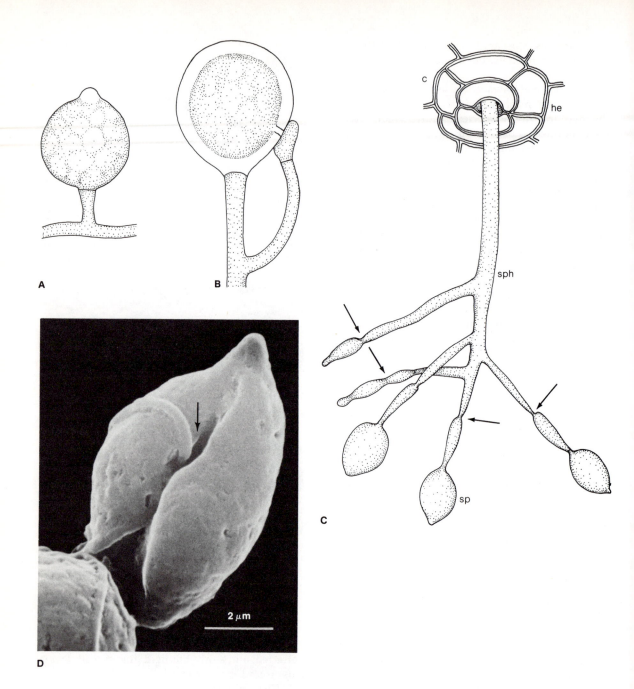

Figure 5-20 **A, B,** *Pythium debaryanum.* **A,** hypha and zoosporangium, ×1000; **B,** oogonium, antheridium, and oospore, ×1000. **C,** *Phytophthora infestans* sporangiophore and sporangia (*arrows* indicate points of previous sporangium formation), ×470. **D,** scanning electron micrograph of *Phytophthora palmivora* zoospore (*scale line* shown is 2 μm; *arrow* indicates groove in which flagella are located.) (**D,** photograph courtesy of P. R. Desjardins, M. C. Wang, and S. Bartnicki-Garcia, with permission of *Archiv für Mikrobiologie.*)

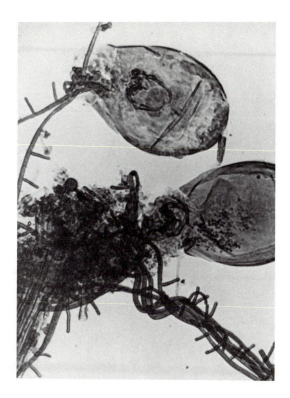

Figure 5-21 *Zoophagus insidians* hyphae with numerous adhesive branches and two captured rotifers, ×225.

hesive branches (Fig. 5-21). When the mouth parts of a rotifer contact such a branch, adhesion occurs; the animal body is then invaded by absorptive hyphae. This activity has given *Z. insidians* the descriptive nickname *lethal lollipop.*

The remaining Peronosporales are obligate parasites of terrestrial plants. These fungi cause two types of diseases, *downy mildews* and *white rusts,* which affect many cultivated and wild plants.

Downy mildew fungi cause important diseases of grape, lettuce, onions, sugar beets, corn, cabbage, and turnips; white rusts of cabbage and brussels sprouts are also economically important diseases. Hyphae of these fungi grow between cells of the host and produce haustoria (Fig. 5-23B) that penetrate the host cell walls. As in *Phytophthora,* the sporangia of downy mildews are formed on aerial sporangiophores. Sporangia of white rusts (*Albugo* spp.) develop subepidermally on the host.

Genera of downy mildews are differentiated from one another primarily on the basis of sporangiophore branching patterns. Those of *Peronospora* and *Plasmopara* are shown in Figure 5-22. The sporangia are deciduous, and the growth of the sporangiophore is determinate: renewed growth does not occur after sporangia are released. In *Plasmopara* and other genera, the lemon-shaped sporangia can germinate either by germ tube or by releasing zoospores. However, in *Peronospora* (Fig. 5-22A) sporangia can only germinate by germ tube. Because of the tendency to germinate by germ tube, the sporangia of downy mildews are often called conidia, and the specialized hyphae that bear them are then called **conidiophores.**

Plasmopara viticola (Fig. 5-22B), the cause of downy mildew of grape, is native to the North American continent. In the latter half of the last century, this fungus was accidentally introduced into Europe on plants imported to combat another North American pest, the *root louse.* Within the space of a few years, *P. viticola* very nearly destroyed the grape and wine industry in France. The disease was finally brought under control through the discovery of "Bordeaux mixture," a concoction of lime and copper sulfate. This mixture was later found to be effective against a number of fungal pathogens of plants.

Phytophthora infestans also made a contribution of sorts to the science of plant pathology. At the time of the Irish famine, the causes of plant diseases were still unknown. The concept of spontaneous generation of life was generally held, and microorganisms were thought to arise from decaying substances, rather than causing decay. The destruction of the potato crop led to arguments over the cause and, eventually, to more careful study of the disease. Although French and Italian botanists had ascribed diseases of plants to fungus infection many years previous to 1850, their ideas had not generally been accepted. The study of plant pathology can be considered as having begun with the appearance of potato blight in Europe and England.

White rusts, caused by species of *Albugo* (Fig. 5-23A), occur on members of the mustard family and on plants of a few other groups. Like downy mildews, species of *Albugo* produce intracellular hyphae with haustoria (Fig. 5-23B). Sporangia are produced in chains on compactly arranged, subepidermal sporangiophores (Fig. 5-24A). The presence of asexual structures produces a blisterlike swelling of the host epidermis, which eventually ruptures. Powdery white masses of sporangia, responsible for the common name of the disease, are splashed away by raindrops or dispersed by wind. In the presence of moisture, the sporangia germinate by producing zoospores (Fig. 5-24B–D).

Sexual structures of *Albugo* (Fig. 5-25A–B) are produced in intercellular spaces within stems or leaves of the host. Within the oogonium, a single large oospore develops; the oospore wall is distinctively sculptured in each species. After a dormant period of several months, the oospore germinates by releasing zoospores. Under suitable conditions, the outer wall cracks, and a vesicle containing zoospores is extruded.

Change from an aquatic to a terrestrial existence has occurred in the evolution of several groups of organisms. However, the relatively small order Peronosporales is unusual in this respect. Species of this order exhibit a complete range from strictly aquatic to well-adapted terrestrial types. In primitive species, sporangia remain attached to their hyphae and can produce only zoospores. In more specialized species, the sporangia are deciduous, airborne, and can produce either germ tubes or zoospores. Finally, in *Peronospora*, sporangia germinate only by producing a germ tube.

Division Hyphochytridiomycota

The small division Hyphochytridiomycota contains a single class and order (Hyphochytridiales). The species of this group are morphologically similar to the simplest Oomycetes and chytrids in Chytridiomycota. However, the zoospore differs from that of other zoosporic fungi, and similarities in thallus structure are attributed to parallel evolution. The zoospores (Fig. 5-26A) have a single anterior flagellum of the tinsel type. Hyphochytridiomycota are unusual in their wall chemistry, in that both chitin and cellulose have been detected. Sexual reproduction is almost unknown in the division. In one parasitic species, sexual reproduction occurs by fusion of whole protoplasts within the host cell. A thick-

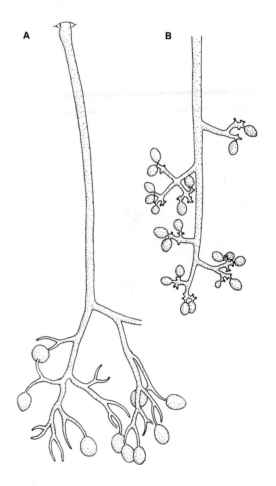

Figure 5-22 **A,** sporangiophore of *Peronospora,* ✕255. **B,** sporangiophore and sporangia of *Plasmopara viticola,* ✕255.

walled, resistant phase is known in other species, but its development has not been observed.

Species of *Rhizidiomyces* are mainly saprobes that live in soil or water and have thalli similar to those of some chytrids (e.g., *Rhizophydium,* p. 155). Zoospores encyst on the substrate surface at germination, the flagellar tinsels appearing as hairs on the cyst (Fig. 5-26B). A bulbous swelling and a rhizoidal system are produced within the substrate, and the cyst proper enlarges (Fig. 5-26C–E) to become the main part of the fungus body. At maturity, the enlarged cyst functions as a zoosporangium; a discharge tube and pore develop, and the zoosporangial protoplast is ex-

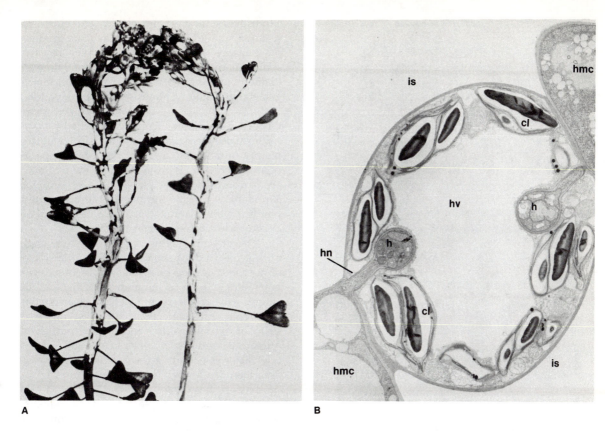

A

B

Figure 5-23 **A,** *Albugo cruciferarum*-induced "white rust" of *Capsella*, × 1.5. **B,** haustorium of *Albugo*, × 4240: *hmc*, haustorial mother cell; *cl*, host chloroplast; *is*, intercellular space; *hv*, host vacuole; *h*, haustorium; *hn*, haustorial neck. (Photograph courtesy of M. D. Coffey, reproduced by permission of the National Research Council of Canada from the *Canadian Journal of Botany* 53:1285–99, 1975.)

truded. The zoospores are not discernible when discharged, but they develop rapidly, mature, and are released from the enclosing membrane. Sexual reproduction is unknown in *Rhizidiomyces*.

There is some disagreement concerning the classification of Hyphochytridiomycota. In some treatments, the group has been classified with protozoa rather than with other fungi.

Division Chytridiomycota

The thallus in Chytridiomycota ranges from unicellular through intermediate to mycelial types; their cell walls contain chitin and glucans. Both sporangiospores and gametes can be motile in this division, motile cells having a single whiplash flagellum. Sexual reproduction ranges from isogamy through anisogamy to oogamy in those species with motile gametes. In some species, sexual reproduction involves fusion of assimilative cells rather than motile gametes. Chytridiomycota are predominantly saprobic aquatic fungi, but many exist as parasites of other fungi, algae, aquatic animals, or terrestrial plants.

Three orders, Chytridiales, Blastocladiales, and Monoblepharidales, are recognized. Chytridiales are mainly simple forms, in which sexual reproduction is isogamous, and the zygote develops into a **resistant sporangium.** Blastocladiales generally are more complex structurally than are

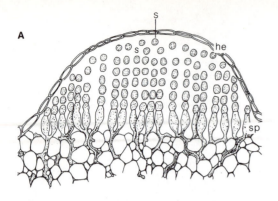

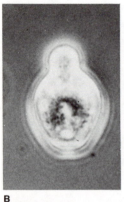

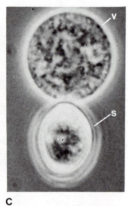

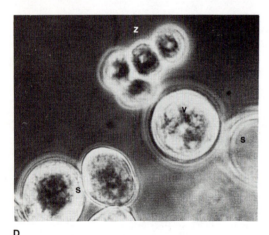

Figure 5-24 *Albugo cruciferarum.* **A,** section through host tissue, showing sporangia and sporangiophores, ×80: *he,* host epidermis; *sp,* sporangiophore; *s,* sporangium. **B,** sporangium just prior to germination by vesicle formation, ×735. **C,** germinated sporangium (*s*) and vesicle (*v*), ×650. **D,** zoospores prior to release from vesicle (an undifferentiated vesicle is also present), ×650.

Chytridiales, and their usually brownish resistant sporangia are not the immediate products of syngamy. Monoblepharidales have resistant sporangia that develop from the zygote; they are filamentous fungi with distinct male and female gametangia and oogamous sexual reproduction.

Order Chytridiales Commonly called *chytrids,* Chytridiales are structurally simple fungi and are mainly inhabitants of fresh water. Some species occur in soil or in marine habitats, and a number are parasites of higher plants. Chytrids are most abundant as parasites on planktonic animals and plants, or on materials such as pollen or insect exoskeletons in fresh water.

The chytrid thallus can consist of a single uninucleate cell, of such a cell with rhizoids, or of a primitive type of mycelium. In those with filamentous thalli, a network of rhizoidal filaments interconnects broader, nucleated compartments. Although the thallus is simple in structure, variations in it and in life histories are numerous, and the number of species is large. Life histories of only a few of the many species described are known; in most, only asexual reproduction has been observed. In most, the thallus produces a single type of reproductive structure—gametangium or sporangium—and much or all of the protoplast is utilized in the process.

Chytrid zoospores have a posterior whiplash flagellum and a large, refractile lipid body (Fig. 5-27). Their motion is often jerky, "hopping," or sometimes amoeboid.

The first step in the germination of a zoospore involves encystment, the flagellum being either withdrawn or cast off. Encystment occurs on the surface of the host or substrate and is usually followed by the development of a tube of varied length. Two common developmental patterns, commencing with the encysted zoospore, are shown in Figure 5-28. In the *Olpidium* type (Fig. 5-28A), the cyst tube penetrates only the host cell wall; the cyst protoplast is then discharged into the host cell and develops there. In the *Rhizophydium* type (Fig. 5-28B), the minute tube that penetrates the host wall or other substratum continues to grow and develops into a rhizoidal system. In this thallus type, the cyst itself enlarges to become the main part of the structure. The rhizoidal system is enucleate, the single nucleus residing in the cyst. This nucleus eventually divides when the thallus is mature, and gametes or zoospores are produced.

Motile cells identical to zoospores function as gametes in some genera (e.g., *Olpidium* and *Synchytrium*); in others (e.g., *Rhizophydium*), two adjacent thalli fuse. The zygote develops into a diploid thallus, and a thick-walled resistant sporangium is later formed.

The life histories of simple chytrids such as *Olpidium* and *Synchytrium* follow the general pattern diagrammed in Figure 5-29. The motile cells produced by a single thallus can all function either as zoospores or as gametes. In some instances, the function is determined by the amount of water present during development. Gametes are often formed more abundantly as the season progresses and the weather becomes drier.

Species of *Olpidium* parasitize plant or animal hosts, and several occur on economically important plants. *O. brassicae* (Fig. 5-30A, B) parasitizes the root cells of many plants, but does little visible harm to the host. During infection, the zoospore encysts on the root surface and produces a minute tube that penetrates the host cell wall. The cyst protoplast is then discharged into the host cell, where it soon develops a cell wall. The cell absorbs nutrient and enlarges. There are no rhizoids or other specialized absorptive structures; at maturity, the entire thallus is converted into a zoosporangium. One or more discharge tubes develop, the tips of which extend through the host cell wall. Zoospores are released through a pore that dissolves in the apex of the discharge tube.

In species of *Olpidium* in which sexual reproduction has been observed, cells indistinguishable from zoospores function as gametes. Following syngamy, the biflagellated zygote swims to the host surface, encysts, and infects a cell in the same manner as a zoospore. The diploid thallus enlarges and becomes surrounded by a thick wall that is smooth or convoluted. This resistant sporangium germinates by releasing zoospores that are thought to be haploid.

Although *O. brassicae* itself causes little apparent harm to its hosts, the zoospores of this species can function as vectors in the transmission of viral diseases.

Synchytrium, which differs from *Olpidium*

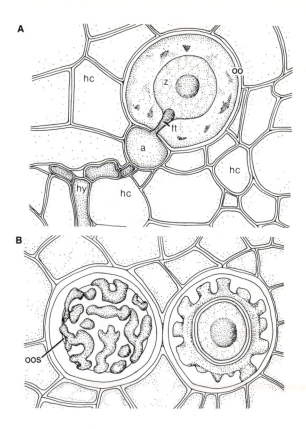

Figure 5-25 Sexual reproduction in *Albugo cruciferarum.* **A,** oogonium with attached antheridium, ×665. **B,** mature oospores (that on the *right* shown in sectional view), ×665: *oo*, oogonium; *an*, antheridium; *ft*, fertilization tube; *oos*, oospore; *hc*, host cell.

only in minor morphological details, infects stems and leaves of vascular plants. The presence of these chytrids induces both cell enlargement and cell division in the host cells around the infection site. Wartlike protuberances of host tissue result. *S. endobioticum*, which causes *black wart* of potato, occurs in many regions where potatoes are grown. As in *Olpidium*, both thin-walled and thick-walled resistant sporangia (Fig. 5-31) are produced in the host. These sporangia extrude their protoplasts, and zoosporangia or gametangia develop externally. The black wart disease can cause extensive damage to potato crops; control of the disease is mainly through use of resistant potato varieties.

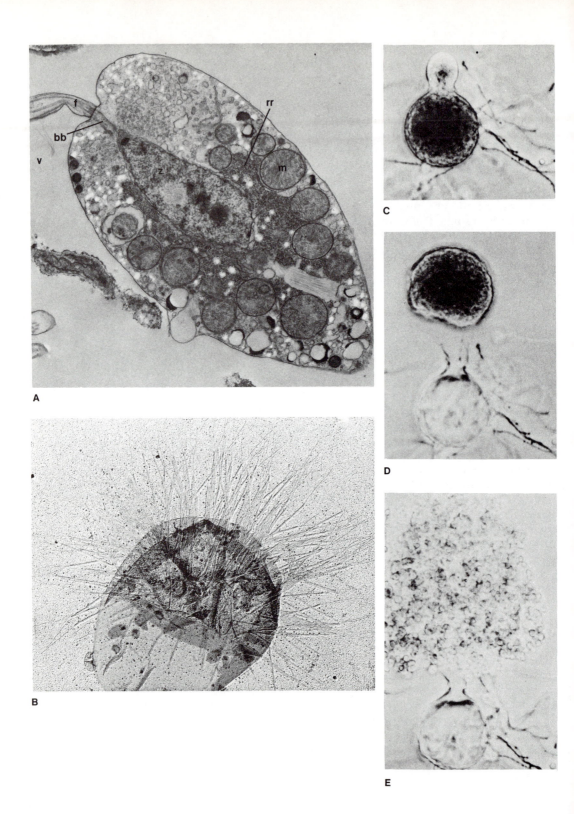

In *Synchytrium* and *Olpidium*, the thallus is not divided into assimilative and reproductive parts but consists of a more or less spherical cell. In a slightly more complex thallus type, illustrated by *Rhizophydium* (Figs. 5-32, 5-33), the structure consists of a cyst plus an absorptive system of rhizoids. Here the zoospore encysts on the surface of dead organic matter or a host, depending upon the species. The germ tube that subsequently develops from the cyst elongates (Fig. 5-32A, B), branches, and forms the system of rhizoids. The cyst enlarges and, at maturity, is transformed into a zoosporangium (Fig. 5-32C–E). The zoospores are released through a pore that develops at the apex of a papilla on the sporangium (Fig. 5-33A).

Sexual reproduction in *Rhizophydium* is preceded by the clumping of zoospores on the host or substratum. After encystment and formation of rhizoids, a tube grows from one cyst to another (Fig. 5-33B–D). The entire protoplast of one thallus then passes through the tube and fuses with the protoplast in the second thallus. The emptied cyst shows no further signs of development, but the other enlarges and develops into a thick-walled resistant sporangium (Fig. 5-33E).

The chytrids just discussed have a single center of development; one zoospore eventually results in the formation of a single zoosporangium. Some chytrids, such as *Nowakowskiella* (Fig. 5-34), have more complex thalli that produce numerous sporangia. The *Nowakowskiella* thallus comprises a network of narrow filaments interconnecting swellings. Nuclei are present in swollen zones, and constrictions or septa are usually present in narrow portions. Sporangia can develop as either terminal or intercalary structures, and after zoospore discharge,

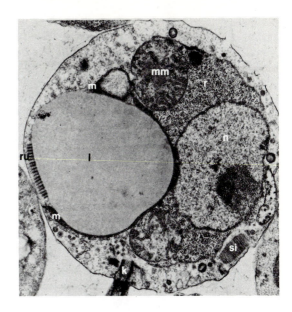

Figure 5-27 Longitudinal section of a chytrid zoospore, × 14,690: *mm,* mitochondrion; *m,* microbody; *l,* lipid body; *n,* nucleus; *r,* ribosomes; *si,* striated inclusion; *k,* kinestosome. (Photograph courtesy of D. J. S. Barr and V. E. Hartmann, reproduced by permission of the National Research Council of Canada from the *Canadian Journal of Botany* 54:2000–2013, 1976.)

Figure 5-26 *Rhizidiomyces.* **A,** thin section of zoospore, showing anterior tinsel flagellum, × 14,515: *bb,* basal body; *f,* flagellum; *rr,* ribosome-containing region; *m,* mitochondrion. **B,** encysted zoospore, × 16,520 (the fibers radiating from the cyst are the remains of the lateral hairs of the flagellum). **C–E,** single thallus showing stages in the development and zoospore release, × 335. **C,** growth of discharge tube; **D,** extrusion of protoplast and zoospore release. (Courtesy of M. S. Fuller; **A, B,** with permission of *Mycologia;* **C–E,** with permission of *American Journal of Botany.*)

a second sporangium often develops within the empty wall. Thick-walled resistant structures are also produced, but the details of their development and germination are not known.

When zoospore release occurs in *Nowakowskiella,* a preformed lid opens at the apex of the discharge tube (Fig. 5-34C). The lid, called an **operculum,** is also present in many simple chytrids.

Order Blastocladiales The zoospores and gametes of Blastocladiales resemble those of the chytrids. The thallus is also chytridlike in some blastocladiaceous fungi, but a well-developed mycelium is present in others. These fungi produce thick-walled resistant sporangia, the wall of which is often pitted and contains a brownish melanin pigment. Blastocladiales are mainly saprobes that inhabit fresh water or soil. However, some species parasitize aquatic fungi or the larvae of aquatic insects, nematodes, or liver flukes.

Sexual reproduction in Blastocladiales is by fusion of motile isogametes or anisogametes. In some species of *Allomyces* (Fig. 5-35), the zygote encysts and germinates to produce a spore-bearing thallus. This diploid thallus is filamentous and has a branched rhizoidal system embedded within the substratum. The young thallus is dichotomously branched, and false septa, consisting of thickened rings, are present at the branching points. Both thin-walled sporangia (Fig. 5-35A; see also Fig. 5-7) and thick-walled resistant sporangia are borne on the diploid thallus. Thin-walled sporangia release zoospores that germinate to produce more diploid thalli.

Resistant sporangia are less susceptible to being damaged or killed by heat or desiccation than are assimilative structures and thin-walled sporangia. Resistant sporangia usually undergo a short dormancy period, and meiosis occurs before germination. At germination, the thick, outer wall layer ruptures, and a thinner, inner wall layer protrudes through the crack. Haploid zoospores are released through exit pores that develop. After a short motile period, the zoospores encyst and germinate to give rise to gamete-bearing thalli (Fig. 5-35C–G). The haploid thalli resemble diploid thalli in all gross characteristics, but they produce male and female gametangia.

In *A. arbuscula*, the male gametangia are orange and are usually borne below slightly larger, colorless, female gametangia (Fig. 5–35G). The orange color results from the presence of carotenoid pigments in male gametes. The gametes are anisogamous, the female being somewhat larger than the male, but otherwise morphologically similar; they are released through pores that dissolve in gametangial walls. Female gametes secrete minute quantities of a hormone, called **sirenin,** that attracts male gametes. Sirenin has been isolated and identified chemically as bicyclic sesquiterpene.

Following syngamy, the biflagellated zygote (Fig. 5-10) is motile for a period, then encysts and produces a new spore-bearing thallus. *A. arbuscula* and related species thus have an alternation of essentially isomorphic generations—that is, an alternation of morphologically similar haploid and diploid phases. *Allomyces* species are common in soils, especially of warm regions.

Species of *Blastocladia* produce thin-walled

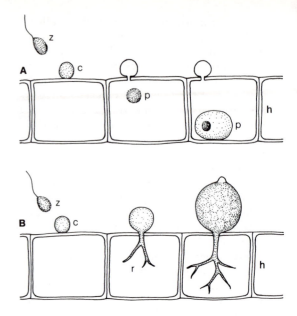

Figure 5-28 Diagrams showing common zoospore encystment and developmental patterns in the simple chytrids. **A,** *Olpidium* type. **B,** *Rhizophydium* type.

and resistant sporangia on thalli that resemble those of some Leptomitales. Several *Blastocladia* species grow in the same habitats as *Aqualinderella fermentans* and, like that species, require little or no oxygen. They are also similar to *A. fermentans* in having only vestigial mitochondria and in lacking functional cytochrome systems.

Blastocladiella has chytridlike thalli (Fig. 5-36A–D), each of which produces a single, thin-walled, colorless sporangium; a thin-walled, orange sporangium; or a thick-walled resistant sporangium. In culture, the type of sporangium produced can be influenced by manipulating the culture medium. Adding small quantities of carbonate to the culture medium causes almost all germlings, excluding those destined to produce orange sporangia, to develop resistant sporangia. The effect of bicarbonate can be duplicated by increasing the carbon dioxide concentration in the culture atmosphere. In light, the fixation of carbon dioxide is greatly enhanced in *Blastocladiella*. This increase can lead to greatly increased

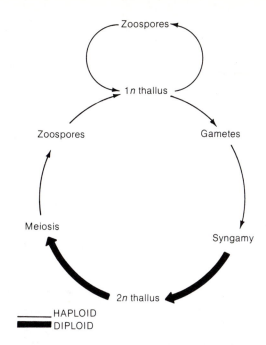

Figure 5-29 Diagram of a common life-history pattern among simple chytrids.

HAPLOID
DIPLOID

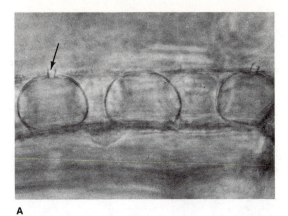

A

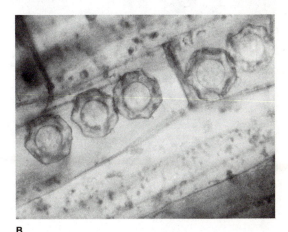

B

Figure 5-30 *Olpidium* in root cells of cereal grain plants. **A,** thalli, ×785 (*arrow* indicates exit tube). **B,** resistant sporangia, ×800. (Photographs courtesy of D. J. S. Barr.)

growth and, in one species, induces the formation of resistant sporangia.

Zoospores of *Blastocladiella* have been the subject of a number of studies. Each zoospore has a single nucleus and nucleolus, a relatively massive nuclear cap, a single mitochondrion, and a typical whiplash flagellum. The nuclear cap consists of an aggregate of ribosomal particles and accounts for most of the RNA of the spore. In the absence of nutrients, the *Blastocladiella* zoospore can swim for many hours. The energy required is obtained from reserve substances in the spore. Prior to germination, the zoospore stops swimming and retracts its flagellum. Following retraction, the nuclear cap is dispersed, and the cyst puts forth a germ tube.

Order Monoblepharidales The Monoblepharidales occur in clean, fresh water. They grow on a variety of substrates but are most common on waterlogged twigs. In the classical sense, these are the only fungi in which oogamy occurs.

That is, the female gamete is large and generally nonmotile, and male gametes are small and motile. Limited amoeboid movement occurs in the female gametes of some species.

Monoblepharis (Fig. 5-37) produces narrow hyphae, the cytoplasm of which often appears frothy or vacuolate. Thin-walled zoosporangia are produced, and the zoospores give rise to more haploid thalli. Gametangia are distinguishable by their form as well as by the nature of their gametes. Gametangia occur in pairs, with the small antheridium laterally attached to the oogonium or basal to it. At maturity, the anther-

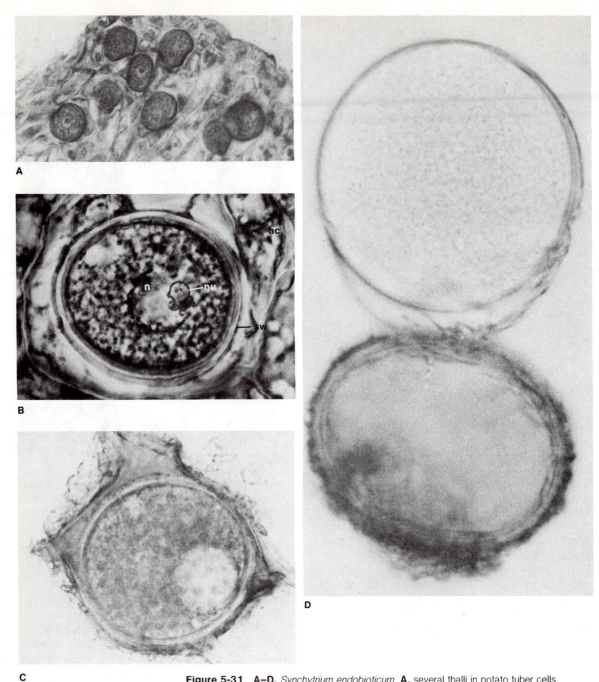

Figure 5-31 A–D, *Synchytrium endobioticum.* **A,** several thalli in potato tuber cells, ×200. **B,** summer spore, ×790; *n,* nucleus; *nu,* nucleolus; *hc,* host cell; *sw,* spore wall; **C,** thick-walled resistant spore, ×1100; **D,** germinated resistant spore and extruded protoplast, ×3115. (Photographs **C, D,** courtesy R. Sharma, from R. Sharma and R. H. Cammack, "Spore germination and taxonomy of *Synchytrium endobioticum* and *S. succisae,*" *Transactions of the British Mycological Society* 66:137–147, 1976, with permission of Cambridge University Press.)

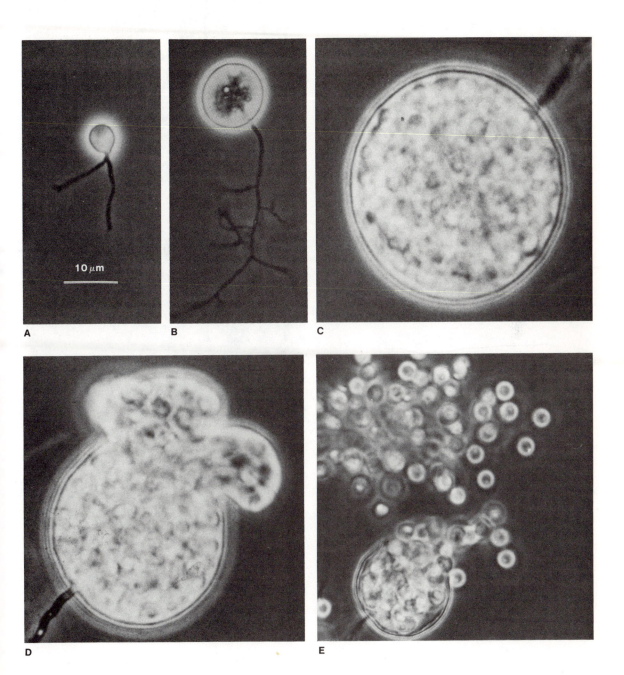

Figure 5-32 *Rhizophidium sphaerocarpum,* asexual reproduction. **A, B,** development of thallus from encysted zoospore (*scale line* shown is 10 μm). **C,** mature thallus, ×2135. **D, E,** release of zoospores. **D,** ×1800; **E,** ×1410. (Photographs courtesy of D. J. S. Barr, reproduced by permission of the National Research Council of Canada from the *Canadian Journal of Botany* 48:1067–71, 1970; and 53:164–78, 1975. **A, B,** 1970; **C–E,** 1975.)

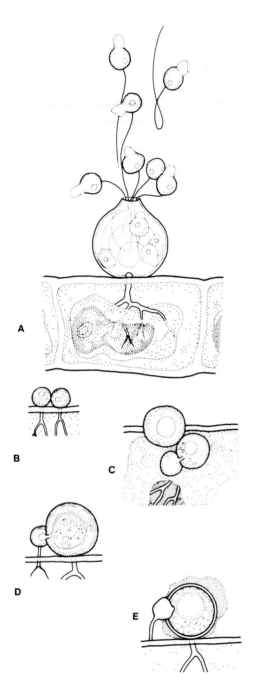

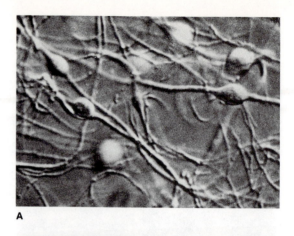

A

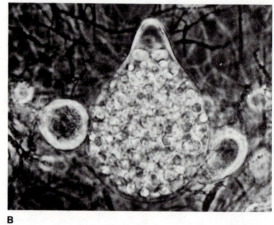

B

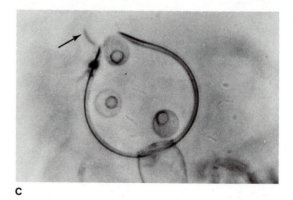

C

Figure 5-33 Sexual reproduction in *Rhizophydium*. **A,** release of zoospores, ✕ 1035. **B,** young developing thalli, ✕ 1035. **C, D,** conjugation of adjacent thalli, ✕ 1035. **E,** thick-walled resistant structure developed from zygote, ✕ 1035. (After Sparrow, with permission of *Mycologia*.)

Figure 5-34 *Nowakowskiella*. **A,** photomicrograph of portion of thallus, showing narrow filaments and swollen nucleate portions, ✕200. **B,** mature sporangium, ✕1050. **C,** sporangium after release of zoospores, ✕1000 (note operculum, *arrow*). (Photograph **C,** courtesy of C. J. Anastasiou.)

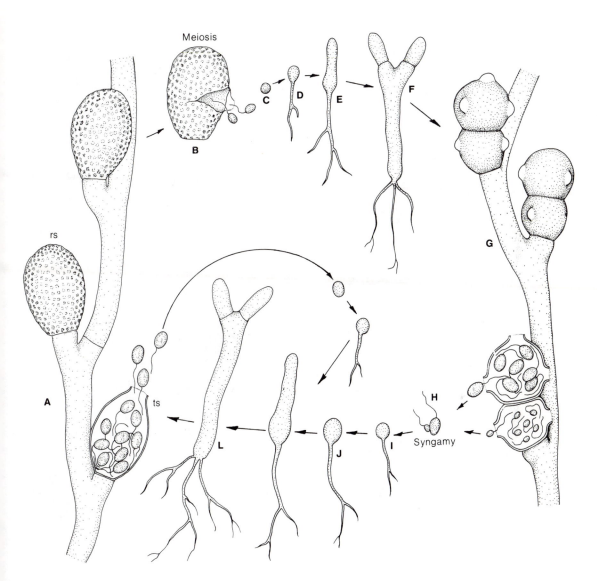

Figure 5-35 Life history and development of *Allomyces arbuscula*. **A,** diploid thallus bearing one thin-walled sporangium and two resistant sporangia; zoospores produced in thin-walled resistant sporangia give rise to new diploid thalli. Zoospores released from resistant sporangia are meiospores and develop into haploid thalli (**B–G**). The haploid thallus (**G**) bears paired gametangia; when released, gametes fuse and the zygote produces a new diploid thallus (**H–K**). **A, G,** ×200 (approx); other figures not drawn to scale.

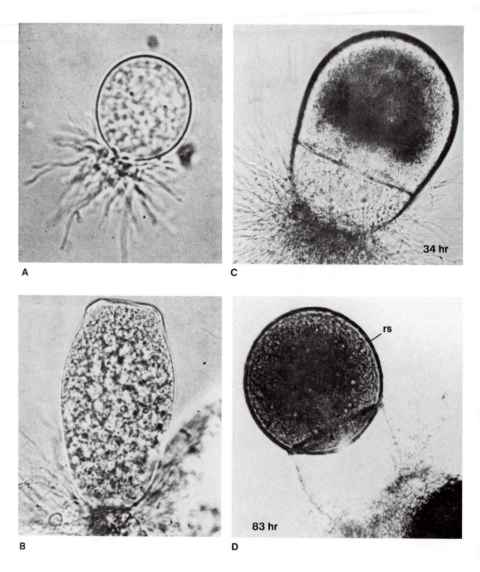

Figure 5-36 *Blastocladiella emersonii.* **A,** immature colorless thallus, ×1100, and **B,** mature colorless thallus, ×500, the latter with developing exit papillae. **C,** immature and **D,** mature resistant sporangium-producing cells; **C,** ×330; **D,** ×295: *rs,* resistant sporangium. (**A, B,** courtesy of A. Goldstein and E. C. Cantino, from "Light stimulated polysaccharide and protein synthesis by synchronized, single generations of *Blastocladiella emersonii,*" *Journal of General Microbiology* 28:689–99, 1962, with permission of Cambridge University Press; **C, D,** courtesy of J. S. Lovett and E. C. Cantino, with permission of *American Journal of Botany.*)

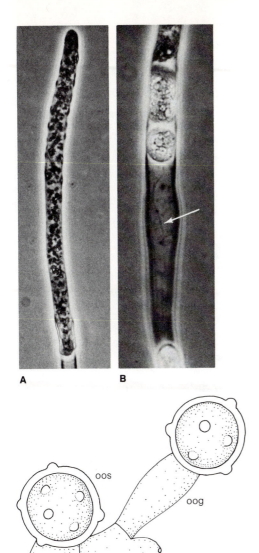

Figure 5-37 *Monoblepharis*. **A,** immature sporangium, ×700, and **B,** mature sporangium, ×465 (*arrow* indicates flagellum of lowermost zoospore). **C,** semidiagramatic illustration of antheridia (*an*), oogonia (*oo*), and oospore (*oos*) that develop at mouth of oogonium.

idium releases a small number of flagellated sperms through an apical pore. The oogonium contains a single, large egg and, at the time of fertilization, has an apical pore. Fertilization takes place in the oogonium; the zygote then either emerges and forms a thick-walled resistant spore at the apex of the oogonium or develops into a resistant spore within the oogonium. The resistant spore germinates by forming a germ tube, and meiosis is thought to occur at this time.

Division Zygomycota

The division Zygomycota includes two classes of fungi, Zygomycetes and Trichomycetes. Zygomycetes are saprobic or parasitic fungi, and most are characterized by a well-developed mycelium. Trichomycetes all occur in association with arthropods. Most have a thallus consisting of a short, branched or unbranched hypha attached to the arthropod digestive tract by a holdfast.

Asexual reproduction in Zygomycota is by the formation of nonmotile sporangiospores or conidia. Sexual reproduction typically is by the fusion of whole gametangia, called *conjugation*, and the subsequent development of a thick-walled zygospore. The life histories of many species follow the pattern shown in Figure 5-38.

Class Zygomycetes

There are several orders of the class Zygomycetes; two of them, Mucorales and Entomophthorales, are discussed here. The Mucorales are mainly saprobes that reproduce asexually by means of sporangiospores. The sporangia can be multispored, or they can be reduced structures with few spores, or one. Parasitic Mucorales occur on other fungi, plants, and animals. The Entomophthorales are mainly insect parasites, but some species exist as soil saprobes. For the most part, these fungi reproduce asexually by forming modified sporangia or conidia.

Order Mucorales Among the Mucorales are some of the commonest soil- and dung-inhabiting fungi; they are also abundant in decaying vegetation. Although few of these fungi parasitize higher plants, they often cause decay of stored fruits and vegetables. Certain species are also capable of growth at reduced temperatures (2–6 °C) and can cause damage to stored meat. Many parasitic species attack other Mucorales, but parasites of plants and animals also occur in this group. Among these, species of *Mucor* and *Rhizopus* sometimes cause severe infections of man and domesticated animals. These infections, called **mycoses,** often occur in individuals suffering from other afflictions, such as cancer and diabetes. Some Mucorales are utilized in the preparation of fermented soybean foods and in other industrial processes, one of which is the transformation of sterols in the manufacture of steroid hormonal preparations.

In most Mucorales, the mycelium consists of both aerial and submerged hyphae. The submerged hyphae function in obtaining nutrients; rhizoids, if present, serve only to anchor reproductive structures. The hyphae may be either coenocytic or septate, the latter occurring in those forms considered to be most advanced.

In asexual reproduction, aerial sporangiophores bear sporangia with one to many spores. In general, those with reduced numbers of spores are considered advanced, and it is these species that commonly have septate hyphae. In those with a single spore per sporangium, the sporangial wall and that of the spore may be fused.

In sexual reproduction, two gametangia fuse and the resultant cell develops into a thick-walled **zygospore.** The terms **heterothallism** and **homothallism** were first applied to this group of organisms. In heterothallic species, fusing gametangia arise on separate mycelia. These two types of mycelia are morphologically indistinguishable but are genetically distinct and are designed as + and −. Each can reproduce only by asexual means when grown alone. In homothallic species, fusing gametangia arise as branches from a single thallus. Within the fusion cell of both homothallic and heterothallic species, compatible nuclei pair up and later fuse. In most species, numerous nuclear pairs are present in the fusion cell, which is commonly called a **coenozygote.** The coenozygote becomes enclosed in a thick wall and is then called a *zygospore.* Meiosis can occur soon after the

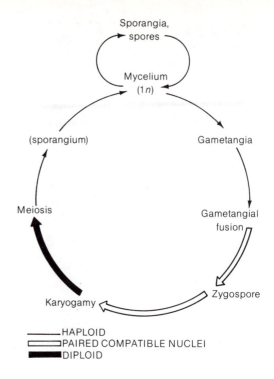

Figure 5-38 Diagram of life cycle common to many zygomycetes.

zygospore has formed, or it occurs after a long resting period and prior to germination.

Rhizopus stolonifer, often called *black bread mold,* occurs on stored fruits and vegetables and in soil; it is also a common laboratory contaminant. *Rhizopus* produces an extensive, submerged mycelium, and rapid cytoplasmic streaming is often visible in the hyphae. Aerial hyphae, called **stolons,** also develop and grow from place to place on the substratum. At points of contact, stolons become attached to the substrate by means of coarse rhizoids. Above each system of rhizoids, a cluster of sporangiophores and sporangia develops from the stolon (Fig. 5-39A).

As the sporangium develops, a bulbous septum, or **columella,** is deposited between the sporangium proper and its sporangiophore. The sporangial protoplast cleaves into segments containing one or a few nuclei. Each segment devel-

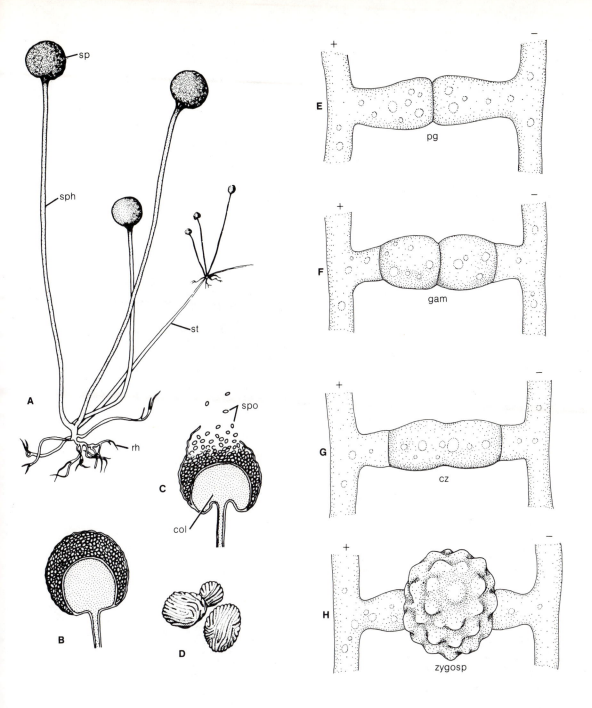

Figure 5-39 *Rhizopus stolonifer.* **A–D,** asexual reproduction. **A,** habit sketch, ×50: *sp,* sporangium; *rh,* rhizoids; *st,* stolon; *sph,* sporangiophore. **B,** sectional view of mature sporangium. **C,** ruptured sporangium, ×110: *col,* columella; *spo,* sporangiospores. **D,** sporangiospores. **E–H,** sexual reproduction. **E,** progametangia developing on + and − hyphae, **F,** gametangia; **G,** coenozygote; **H,** mature zygospore; figures **E–H,** ×136.

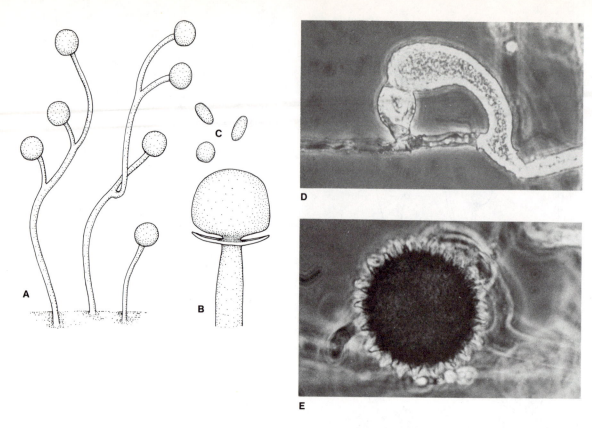

Figure 5-40 *Zygorrhynchus moelleri.* **A,** sporangia and sporangiophores, ×320. **B,** columella after spore release, ×390. **C,** spores, ×390. **D,** gametangia, showing unequal size and origin from a single hypha, ×750. **D,** mature zygospore, ×925.

ops into a walled spore (Fig. 5-39B–D). The mature sporangial wall is dry and brittle; it fragments to release spores.

R. stolonifer is a heterothallic species, in which sexual reproduction requires the coming together of compatible + and − strains. In culture, the mycelia of opposite mating strains grow toward one another, then produce special aerial reproductive branches. The reproductive branches of opposite strains grow toward one another; those of the same strain repel one another. The tips of these branches, called **progametangia,** meet end to end; a septum develops in each, and gametangia are thus delimited (Fig. 5-39E, F). The gametangial end walls dissolve at their point of contact, and a single, multinucleate cell (Fig.

5-39G) results. Within this cell, + and − nuclei pair and then undergo karyogamy and, eventually, meiosis. Since many zygote nuclei occupy the same cytoplasm, the cell is called a **coenozygote.** The coenozygote is soon transformed into a thick-walled resistant spore, the zygospore (Fig. 5-39H), by deposition of a new wall. The zygospore typically germinates by forming a short sporangiophore and sporangium containing haploid spores.

The formation of specialized reproductive branches in sexual reproduction is known to be hormonally controlled in several Mucorales. The production of these branches is stimulated by substances called *trisporic acids* that are produced by both + and − mycelia. A second type of

Figure 5-41 *Pilobolus sporangiophores*, ×43. (Photograph courtesy of M. Higham.)

Although referred to as black bread mold, *R. stolonifer* is notorious for the damage it does to stored fruits and vegetables. Succulent fruits, such as peaches and strawberries, can be quickly destroyed. As with many Mucorales, the growth of the mycelium is rapid, and abundant, airborne sporangiospores are soon formed.

Species of *Rhizopus* and related forms are used in preparing *tempeh* in Indonesia and *sufu* in China. Hulled soybeans are boiled, then inoculated with appropriate mucoraceous fungi. Inoculated beans are incubated for about 24 hours before being consumed. During incubation, growth of the fungi in the soybeans renders the beans more digestible and may add to their nutrient content.

The + and − gametangia of *Rhizopus* are alike in both size and morphology. In *Zygorrhynchus moelleri* (Fig. 5-40A–D), a common soil fungus, fusing gametangia differ in size and form (Fig. 5-40C, D). *Z. moelleri* is homothallic, and progametangia arise as branches of a single, aerial hypha. In homothallic species such as this, sexual reproduction regularly accompanies growth and asexual reproduction. In contrast, sexual reproduction of heterothallic species can only occur if + and − thalli grow together—something which appears to happen infrequently in nature.

Species of *Pilobolus* and *Phycomyces* are common dung-inhabiting fungi. *Pilobolus*, which means *hat-thrower*, gets its name from the fact that the mature sporangium (Fig. 5-41) is fired away from its sporangiophore. As the sporangium matures, pressure is built up in the swollen sporangiophore until a weakened zone at the base of the sporangium gives way. When this happens, a jet of cell sap is emitted from the sporangiophore tip; the jet carries the sporangium a distance of up to 200 cm. The sporangium has a cutinized, waterproof layer on its upper surface and a mucilaginous, hydrophilic substance at its base. If it and the accompanying cell sap strike a nearby object, such as a plant, the sporangium bobs to the surface of the liquid with its cutinized surface outermost. Drying of the liquid brings the mucilaginous substance into contact with the plant, and the sporangium adheres. If an herbivore eats the plant, the sporangiospores pass through the digestive tract unharmed and will germinate in the excrement.

growth substance, as yet unknown chemically, is responsible for stimulating trisporic acid production in associated + and − thalli.

Much variation occurs among species of Mucorales with respect to karyogamy and meiosis in the coenozygote and zygospore stages. The paired nuclei can fuse almost immediately, or they can remain separate during a long resting period of the zygospore. Meiosis can occur soon after karyogamy, or it can be delayed until immediately before germination of the zygospore. In most species, meiosis is followed by degeneration of one of the nuclear types, and only + or − spores are produced in the germination sporangium.

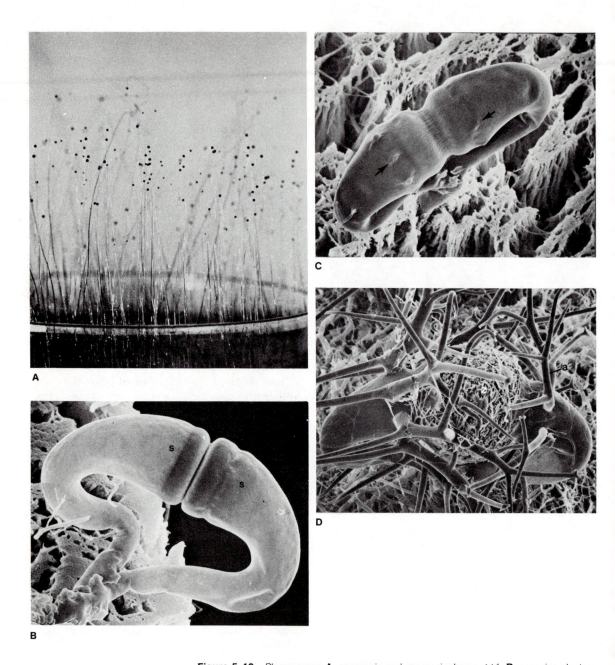

Figure 5-42 *Phycomyces.* **A,** sporangia and sporangiophores, ✕1. **B,** scanning electron micrograph of conjugating hyphae, ✕180. **C,** the gametangial walls and appendage initials are visible, ✕147. **D,** mature zygospore, ✕90: *s,* suspensor; *a,* sterile appendage; *z,* zygospore. (Photographs **B–D,** courtesy of K. L. O'Donnell and G. R. Hooper, reproduced by permission of the National Research Council of Canada from the *Candian Journal of Botany* 54:2573–86, 1976.)

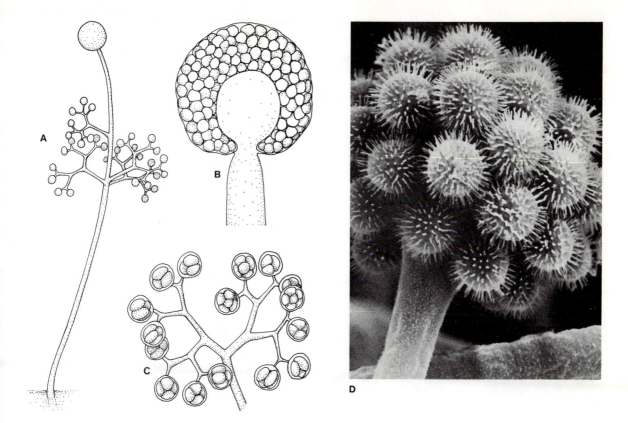

Figure 5-43 **A–C,** *Thamnidium elegans.* **A,** sporangiophore with columellate and reduced sporangia, ×80; **B,** sporangium, ×250; **C,** reduced sporangia, ×340. **D,** scanning electron micrograph of sporangiophore apex and single-spored sporangia of *Cunninghamella echinulata,* ×1300. (**D,** courtesy of M. Higham.)

The sporangiophore of *Pilobolus* is specialized not only for producing the pressure and jet of cell sap, but for "aiming" the sporangium. The sporangiophore is sensitive to light and is positively phototropic.

In *Pilobolus,* sporangia develop over a 24-hour period, development being regulated by an internal "clock." After releasing their sporangia, the sporangiophores collapse, and a new crop starts to develop.

Phycomyces sporangiophores (Fig. 5-42) can reach a height of 20 cm, and the large sporangia contain up to 100,000 spores. Because of their size, and the ease and speed with which they can be grown, these sporangiophores have become important tools in studies of fungal growth and development. The sporangiophore is positively phototropic; it also shows rapid changes in growth rate with exposure to pulses of light.

Gametangia of *Pilobolus* and *Phycomyces* do not develop at the initial point of contact of conjugating hyphae. At contact, these hyphae adhere, but growing tips curve outward away from one another and then recurve to make contact again. The structure produced, as shown in *Phycomyces* (Fig. 5-42B, C), resembles a pair of tongs. Its subsequent development is like that in other Mucorales. The mature zygospore of *Phycomyces* (Fig. 5-42D) is surrounded by sterile appendages that arise from conjugating hyphal branches.

A

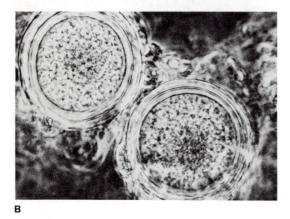

B

Figure 5-44 *Endogone pisiformis*. **A,** habit of sporocarp, ×9. **B,** section of sporocarp showing thick-walled spores, ×735.

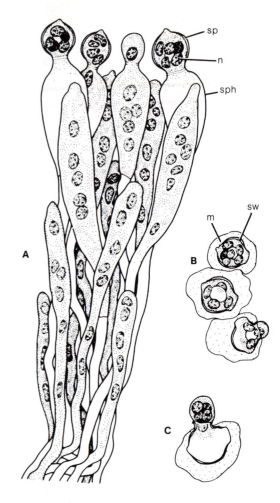

Figure 5-45 A–C, *Entomophthora muscae*. **A,** cluster of sporangiophores with developing sporangia, ×700; **B,** three mature sporangia that have been fired from sporangiophores, ×700; **C,** germinating sporangium, ×700.

All Mucorales described above have sporangia with many spores and with a columella. In some genera (e.g., *Thamnidium*, Fig. 5-43A–C), such sporangia occur together with small, few-spored sporangia that lack columellae. Still others produce only reduced sporangia with single spores that are often referred to as *conidia* (e.g., *Cunninghamella*, Fig. 5-43D). Here, the spore wall and that of the sporangium are fused.

Zygospores of some Mucorales—e.g., *Endogone* (Fig. 5-44)—are produced in complex fruiting bodies, or **sporocarps.** These sporocarps consist of sterile, interwoven hyphae, within which spores or sporangia are embedded. Pea-sized sporocarps are formed at the soil surface or just below it and often constitute a substantial portion of the diet of small rodents.

Forms closely related to *Endogone* exist in mutualistic associations with higher plant roots called **mycorrhizae.** Hyphae of the fungus penetrate between cells of the root, some also penetrating the cell walls. Within the cells, hyphae often branch repeatedly or form vesicles. Zygomycetous mycorrhizal fungi are associated with numerous important crop plants, such as corn and soybeans. The fungus has been shown to make the host plant more resistant to wilting,

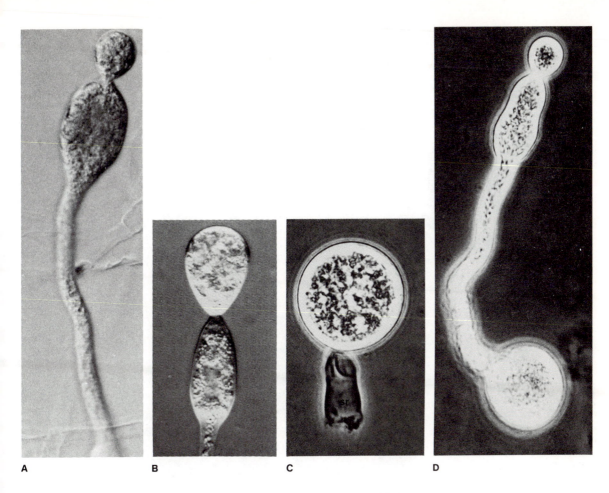

Figure 5-46 *Basidiobolus ranarum.* **A, B,** developing sporangiophores and modified sporangia (**A,** ×825; **B,** ×660). **C,** mature sporangium that has been fired from the sporangiophore, ×1000. **D,** sporangiophore and sporangium production by germinating sporangium, ×825.

and it also assists in the uptake of mineral nutrients, especially in mineral-poor soils. The higher plant provides nutrients and probably growth factors required by the fungus.

Order Entomophthorales Many Entomophthorales are parasites of insects; others parasitize lower plants and humans or exist as saprobes. Species of *Entomophthora*, a commonly encountered group, all parasitize and kill various types of insects. *E. muscae* (Fig. 5-45) infects the com-

mon housefly, the mycelium growing within the host's body. Infected flies are often found attached to windows in late summer or fall, the body surrounded by a halo of whitish spores. The mycelium of *E. muscae* grows into all parts of the insect body, utilizing protein there. Hyphae then grow out between segments of the host exoskeleton, and sporulation occurs (Fig. 5-45A). The spores, considered to be reduced sporangia, are borne singly at the apices of broad sporangiophores. The sporangiophore is somewhat swollen, as in *Pilobolus;* it also functions in a man-

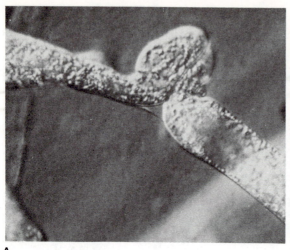

A

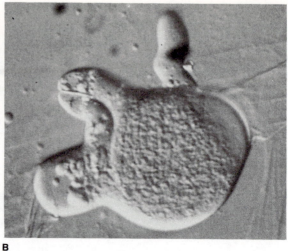

B

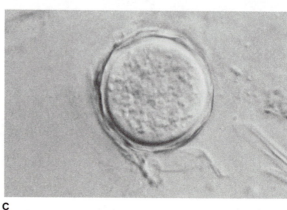

C

Figure 5-47 Sexual reproduction in *Basidiobolus ranarum*. **A,** early stage in conjugation of adjacent hyphal cells, ×875. **B,** nuclear transfer and karyogamy have occurred, ×1000. **C,** mature zygospore, ×800.

ner similar to that of *Pilobolus* and projects the spore for a distance of up to 15 mm. The spore (Fig. 5-45B) is surrounded by a layer of adhesive material, which causes it to adhere to flies or other objects. Germination can be through the production of a hypha that infects the host. Alternatively, the structure can germinate by producing a new spore identical to the first (Fig. 5-45C). Sexual reproduction in *Entomophthora* is by conjugation similar to that of Mucorales. However, in *E. muscae* conjugation is unknown, and the zygosporelike structures that are produced are considered to be asexual spores.

Entomophthora and related genera of insect parasites are thought to be potentially useful in biological control of insect pests. Unfortunately,

most attempts to control insects by use of parasitic fungi have been unsuccessful.

Basidiobolus species can be isolated readily from the excrement of insectivorous reptiles and amphibians, or from soil. These fungi produce septate hyphae, the cells of which are uninucleate. Small sporangiophores are produced, each bearing a terminal modified sporangium (Fig. 5-46A, B). The sporangium is shot away from its sporangiophore in a manner similar to that of *Pilobolus*. However, in *Basidiobolus*, a portion of the sporangiophore accompanies the sporangium (Fig. 5-46C), and it is the jet of liquid from this structure that propels the sporangium. During flight, the sporangiophore segment and its sporangium usually separate. The sporangium can

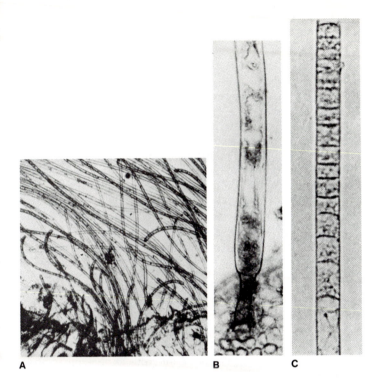

A   B   C

Figure 5-48 *Enterobryus.* **A,** thalli attached to segment of millipede gut lining, ×66. **B,** holdfast at base of hypha, ×480. **C,** one of several types of spores the thallus is capable of producing, ×480. (Photographs courtesy of R. W. Lichtwardt, with permission of *Mycologia.*)

later become divided internally into sporangiospores, but this does not usually happen in culture. Either direct germination (i.e., by germ tube) or **repetitive germination** can occur. In the latter, the sporangium produces a new sporangiophore and a sporangium that is shot away (Fig. 5-46D).

Sexual reproduction in *Basidiobolus* (Fig. 5-47A–C) is unusual in several respects and bears at least a superficial resemblance to that of some Hemiascomycetes. Conjugation occurs between adjacent cells of a hypha after the nuclei in each cell divide. Short, tubular projections are formed at adjacent ends of the two cells, and a daughter nucleus migrates into each projection. Septa form in the projections, and the two nuclei degenerate there. One of the two remaining nuclei then migrates into the adjacent compartment, fuses with the second nucleus, and a zygospore is formed (Fig. 5-47C).

Species of *Basidiobolus* have been isolated from infections in man and other mammals. However, these infections appear to be rare.

Class Trichomycetes

Trichomycetes are as yet poorly known. These organisms all live either within the digestive tracts or attached externally to the exoskeletons of millipedes, insects, or other arthropods. Most species have short, nonseptate hyphae of limited development; the hyphae are attached to the gut or exoskeleton by a holdfast. *Enterobryus* (Fig. 5-48) attaches to a thin, chitinous band lining part of the gut in millipedes. Although structurally simple, thalli can produce either sporangiospores or one of several different types of conidia. Sexual reproduction is unknown in *Enterobryus*, but conjugation resembling that of Zygomycetes occurs in some Trichomycetes.

Relationships of Lower Fungi

Studies of fossil fungi have not been numerous, and the few that have been made tell us little about the evolution of fungi. Too often, fungus fossils are either fragmentary and insufficient for characterizing species or well-preserved but of relatively recent origin.

Fossil fungi have been reported from Precambrian sediments, but there is doubt concerning these reports. Rhynie cherts, of Devonian age (ca. 374–390 million years ago), contain numerous fossilized hyphae. Both septate and nonseptate hyphae are present in the cherts, and some of the latter apparently belonged to mycorrhizal fungi associated with primitive land plants. This suggests that fungi colonized terrestrial habitats at about the same time as green land plants or earlier. The nonseptate hyphae probably belonged to oomycetous or zygomycetous fungi, certain extant species of which form mycorrhizae.

Most theories on the origin of fungi and the classification systems based upon them are derived from the study of extant forms. Several theories date from the latter half of the last century; all have been modified somewhat, but none currently has sufficient factual evidence supporting it to exclude all others. Basically, concepts of fungal origin assume either algal ancestry or protozoan ancestry.

Schemes in which fungi are derived from algae are based upon morphological similarities of certain representatives of the two groups. Probably one of the earliest comparisons made was that of oomycetous fungi, such as *Saprolegnia*, with coenocytic filamentous algae, such as *Vaucheria* (see p. 315). Vegetative thalli of these two genera of extant organisms are morphologically similar, and their sexual reproductive structures bear a superficial resemblance to one another. Although their zoospores are not similar, those of oomycetous fungi are similar to *Vaucheria* sperms in their flagellar type and insertion.

Other alga-fungus similarities include those found in the sexual reproduction (conjugation) of Zygomycota and the green alga *Spirogyra* and allied forms (see p. 279). Both these groups lack motile cells of any kind, but the structure and chemistry of their assimilative bodies differ markedly. Their types of conjugation also differ, and it is extremely doubtful that the two groups have more than a remote relationship.

Red algae have also been suggested as ancestors of ascomycetes and rusts in different phylogenetic schemes (see under relationships of higher fungi, p. 249). Since it would appear unlikely that these higher fungi gave rise to zoosporic types, acceptance of red algal ancestry theories precludes acceptance of monophyletic theories.

Those holding the view that fungi arose from simple protozoan ancestors consider alga-fungus similarities to be the result of parallel evolution. They point out the ancient age of fungus groups, differences in nutritional types, and differences in cell wall chemistry and zoospore types. If slime molds are classified with true fungi, a morphological link to the Protozoa is apparent. It should be noted here that Coelomomycetaceae (Blastocladiales), with wall-less thalli, have been mentioned in discussions of possible relationships between slime molds and true fungi. The differences between certain primitive fungi, such as *Synchytrium* and *Olpidium*, and some protozoa are not extremely great.

Cell wall chemistry and flagellar characteristics suggest that Oomycota and Hyphochytridiomycota might also have had an origin separate from that of other fungus groups. Oomycetous fungi have cell walls containing some cellulose, chitin being absent. Their motile cells and the reserve material mycolaminaran suggest a relationship to certain algae—e.g., the brown or yellow-green algae. The diploid thalli also tend to set Oomycetes apart from most other fungal groups.

The small group Hyphochytridiomycota is of uncertain relationships. These fungi bear a striking resemblance to chytrids and to simple Oomycetes, but the similarities are generally attributed to parallel evolution.

One theory that should be mentioned concerns a possible relationship of three groups of fungi having motile cells. According to this theory, two types of uniflagellated cells (Hyphochytridiomycota and Chytridiomycota) could have been derived from the biflagellated type (Oomycota). In Oomycota, the two types of flagella can be either anteriorly or laterally inserted; in both types of insertion, however, the tinsel flagellum is anteriorly directed, and the whiplash flagellum is posteriorly directed. Through the loss of either flagellum, a biflagellated ancestral form could conceivably have given rise to either the chytrid type or hyphochytrid type of cell. Additionally, some uniflagellated motile cells contain a structure that has been interpreted as a vestigial blepharoplast. However, differences in cell wall chemistry, chemistry of storage products, and other features do not support this concept. At the present time, the three groups with motile cells are best considered as having evolved from three different ancestral groups.

Zygomycota, like other groups of lower fungi, may represent an isolated division of unknown ancestry. Amoeboid cells occur in one group of trichomycetous fungi, but flagellated cells are lacking in the division. The presence of chitin in the cell walls and the production of sporangiospores could indicate a distant relationship to chytridiaceous fungi. Structurally and reproductively, more advanced zygomycetes do resemble certain simple ascomycetes. In some monophyletic schemes, ascomycetes are considered to have evolved from such an ancestral group. It is also worth noting that biochemical similarities have given rise to the suggestion that Eumycota descended directly from chytridlike ancestors.

References

All Fungal Groups

Ainsworth, G. C. 1971. *Ainsworth & Bisby's dictionary of the fungi.* 6th ed. (including Lichens, by P. W. James and D. L. Hawksworth). Commonwealth Mycological Institute, Kew, Surrey.

Ainsworth, G. C., and Sussman, A. S., eds. 1965–73. *The fungi: an advanced treatise,* vol. 1, 1965; vol. 2, 1966; vol. 3, 1968; vol. 4A, 1973; vol. 4B, 1973 (with F. K. Sparrow). Academic Press, New York. (These volumes contain chapters by numerous specialists on various aspects of the biology of fungi and include keys to all groups of fungi.)

Alexopoulos, C. J., and Mimms, C. W. 1979. *Introductory mycology.* 3rd ed. John Wiley & Sons, New York.

Barron, G. L. 1977. *The nematode-destroying fungi.* Topics in Mycology, no. 1. Can. Biol. Publ., Guelph, Ontario.

Bartnicki-Garcia, S. 1970. Cell wall composition and other biochemical markers in fungal phylogeny. In Harborn, J. B., ed. *Phytochemical phylogeny,* p. 81–103. Academic Press, New York.

Beckett, A., and Heath, I. B. 1974. *An atlas of fungal ultrastructure.* Longman, London.

Bessey, E. A. 1950. *Morphology and taxonomy of the fungi.* The Blakiston Co., Toronto.

Burnett, J. H. 1976. *Fundamentals of mycology.* 2nd ed. Edward Arnold, London.

Burnett, J. H., and Trinci, A. P. J. 1979. *Fungal walls and hyphal growth.* Cambridge Univ. Press, Cambridge.

Christensen, C. M. 1975. *Molds, mushrooms, and mycotoxins.* Univ. of Minnesota Press, Minneapolis.

Cooke, R. C. 1978. *Fungi, man and his environment.* Longman, London.

Deverall, B. J. 1969. *Fungal parasitism.* Studies in Biology, no. 17. Edward Arnold, London.

Duddington, C. L. 1956. *The friendly fungi: a new approach to the eelworm problem.* Faber & Faber, London.

Fincham, J. R. S.; Day, P. R.; and Radford, A. 1979. *Fungal genetics.* 4th ed. Univ. of California Press, Berkeley and Los Angeles.

Gray, W. D. 1959. *The relation of fungi to human affairs.* Henry Holt and Co., New York.

Hanlin, R. T., and Ulloa, M. 1979. *Atlas of introductory mycology.* Hunter Publishing Co., Winston-Salem.

Hawker, L. E., and Linton, A. H. 1979. *Micro-organisms: function, form and environment.* 2nd ed. Edward Arnold, London.

Heath, I. B., ed. 1978. *Nuclear division in the fungi.* Academic Press, New York.

Hudson, H. J. 1972. *Fungal saprophytism.* Institute of Biology's Studies in Biology, no. 32. Edward Arnold, London.

Ingold, C. T. 1971. *Fungal spores, their liberation and dispersal.* Clarendon Press, Oxford.

Müller, E., and Loeffler, W. 1976. *Mycology, an outline for science and medical students.* Translated by B. Kendrick and F. Barlocher. George Thieme Publishers, Stuttgart.

Peberdy, J. F. 1980. *Developmental microbiology.* John Wiley & Sons, New York.

Ross, I. 1979. *Biology of the fungi.* McGraw-Hill Book Co., New York.

Smith, J. E., and Berry, D. R., eds. 1978. *The filamentous fungi,* vol. 3, *Developmental Mycology.* John Wiley & Sons, New York.

Webster, J. 1980. *Introduction to fungi.* 2nd ed. Cambridge Univ. Press, London.

Lower Fungi

Ainsworth, G. C.; Sparrow, F. K.; and Sussman, A. S., eds. 1973. *The fungi,* vol. 4B. Academic Press, New York. (Keys to the genera of slime molds, phycomycetes, and basidiomycetes. Chapters by F. K. Sparrow, G. M. Waterhouse, M. W. Dick, and C. W. Hesseltine and J. J. Ellis cover the lower fungi.)

Benjamin, R. K. 1959. The merosporangiferous Mucorales. *Aliso* 4:321–43.

Fitzpatrick, H. M. 1930. *The lower fungi: Phycomycetes.* McGraw-Hill Book Co., New York.

Fuller, M. S., ed. 1978. *Lower fungi in the laboratory.* Palfrey Contributions in Botany, no. 1. Univ. of Georgia Press, Athens.

Jones, E. B. G., ed. 1976. *Recent advances in aquatic mycology.* Paul Elek, London (Although this book is concerned with all groups of fungi, a substantial portion is devoted to ecology, physiology, fine structure, etc., of lower fungi.)

Sparrow, F. K. 1960. *Aquatic Phycomycetes.* 2nd ed. Univ. of Michigan Press, Ann Arbor.

Zycha, H.; Siepmann, R.; and Linnemann, G. 1969. *Mucorales.* J. Cramer, Lehre, Germany.

A general introduction to all fungi is given in Chapter 5. In Chapter 6, a detailed treatment is presented of the division Eumycota, including the subdivisions Ascomycotina and Basidiomycotina, and the form class Fungi Imperfecti.

Examples of the higher fungi, such as mushrooms, puffballs, and morels, are familiar to all. These types illustrate the complex sporocarps that characterize most Eumycota. Less conspicuous higher fungi are yeasts used in baking and brewing, fungi causing athlete's foot and ringworm, and common food molds, such as *Penicillium* species.

Most Eumycota are saprobes that decay plant and animal materials of many types. Although the process of decay is absolutely essential in nature, valuable commodities are also destroyed by decay fungi. Many higher fungi are parasites of plants; these species cause losses in crop yields amounting to billions of dollars annually. A few species of higher fungi also parasitize animals, including humans.

Many Eumycota exist as mutualistic symbionts with plants or animals. The largest group of mutualistic fungi are those in lichens, a major part of the Ascomycotina. A second group, mycorrhizal fungi, exist in mutualistic symbioses with the roots of many higher plants.

Higher fungi are most abundant in warm, moist regions, such as the tropics and wet temperate areas. However, some species are adapted to arid habitats, and others occur in cold regions. The Eumycota are predominantly terrestrial organisms; the number found in marine or freshwater habitats is relatively small.

Structure and Reproduction

In some ascomycetous yeasts, the growth form is unicellular in all phases of the life history. Unicellular, yeast-type haploid phases are also characteristic of dimorphic ascomycetes and basidiomycetes. However, most higher fungi produce hyphae throughout their life histories.

Hyphae in Eumycota are septate, the compartments containing from one to many nuclei. In most Ascomycotina and in some Basidiomycotina, hyphal septa have a simple central pore

6

HIGHER FUNGI AND LICHENS

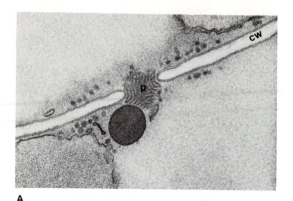

A

(Fig. 6-1A). In such hyphae, cytoplasm is continuous from compartment to compartment, and organelles can be transported readily from one to another. Septa of most Basidiomycotina are more complex, as shown in Figure 6-1B, C. Here, the presence of membranous structures probably prevents the movement of organelles between compartments.

Cell walls in hyphal Eumycota contain chitin and glucans; in budding forms, mannans are a major constituent.

Eumycota reproduce asexually by budding or by conidial production. Conidia are often produced under most conditions that allow growth, but sexual reproduction can require special nutritional or environmental conditions. Sexual and asexual stages of higher fungi can thus be separated in time and space. The complete absence of a sexual stage in many higher fungi, and the infrequent appearance of a sexual stage in others, presents problems of identification in systems based upon sexual structures. This problem has been solved by establishing the form class Fungi Imperfecti. This form class, a "catch-all" for asexually reproducing higher fungi, permits us to identify both species that lack known sexual stages and those with only asexual structures present in the material on hand.

Sexual reproduction in Eumycota involves the formation of **asci** and **ascospores** (Ascomycotina) or **basidia** and **basidiospores** (Basidiomycotina). Asci and basidia are specialized cells within which karyogamy and meiosis occur; ascospores develop within asci (Fig. 6-2A), and basidiospores are produced exter-

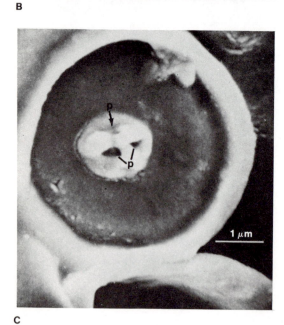

Figure 6-1 A, septum and pore characteristic of many Ascomycotina, ×38,310: *p*, pore; *cw*, crosswall. **B,** complex septal pore found in many Basidiomycotina, ×62,660. Note membranous structures around pore (*arrows*): *p*, pore; *cw*, crosswall. **C,** scanning electron micrograph of basidiomycete septum viewed through cut end of hypha (*scale line shown is 1 μm*): *p*, pore. (**A,** courtesy of C. E. Bracker; **B,** courtesy K. Wells; **C,** courtesy of N. Lisker, J. Katan, and Y. Henis, reproduced by permission of the National Research Council of Canada from the *Canadian Journal of Botany* 53:1801–4, 1975.)

nally on basidia (Fig. 6-2B). In some yeasts, plasmogamy occurs through fusion of whole cells, and nuclei can pair and fuse immediately. However, in most sexually reproducing higher fungi, plasmogamy and karyogamy are separated in time and space. In Ascomycotina, plasmogamy commonly occurs through gametangial contact; in most Basidiomycotina, plasmogamy occurs through fusion of compatible hyphae. Following plasmogamy, compatible nuclei pair, but do not fuse. The paired nuclei divide conjugately, side by side, in subsequent hyphal growth. This **dikaryotic** phase is usually closely associated with the production of sexual spores in Ascomycotina. In Basidiomycotina, the dikaryotic phase can be the main assimilative phase.

Asci and basidia develop in complex sporocarps in most higher fungi. The sporocarps are composed of interwoven hyphae, the nature of which can differ from those constituting the assimilative phase. Factors causing the change from assimilative to reproductive growth have been studied and are known in many species. These factors can include reduced nutrient levels, exposure to light, increasing moisture, temperature changes, and others. However, little is known of the specific control mechanisms required for the change from a diffuse mycelial growth in the assimilative structure to the compact, patterned growth that characterizes sporocarps.

The hyphae composing sporocarps can be thick-walled or thin-walled. The cells can become inflated to produce a "tissue" resembling parenchyma of higher plants. This parenchyma-like growth is called **pseudoparenchyma;** it often constitutes a large part of the sporocarp. Hyphal end cells also become highly modified to produce asci, basidia, and sterile hairs on sporocarp surfaces or interspersed among asci and basidia.

In addition to sporocarps, many higher fungi produce complex hyphal strands that function in conduction. Other higher fungi produce compact resistant **sclerotia** (see *Claviceps*, p. 202) that function as storage and resistant bodies. Sclerotia, which can be as much as 20 cm in diameter, often give rise to sporocarps after a dormant period. Another type of complex structure, the **stroma,** is produced by many ascomycetes, e.g., *Xylaria* and *Daldinia* (p. 202). Like sclerotia, stromata typically have a rindlike outer layer;

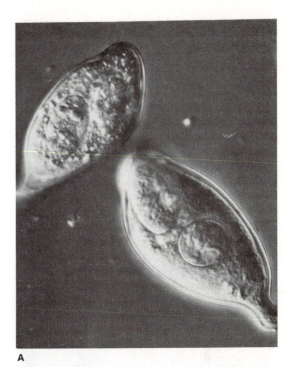

A

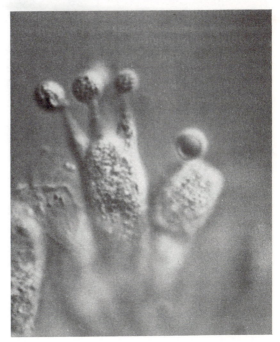

B

Figure 6-2 A, ascus and developing ascospores, ×795. **B,** basidium and developing basidiospores, ×1690.

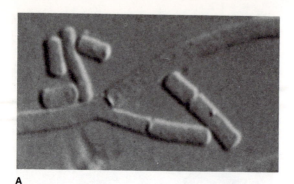

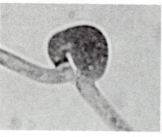

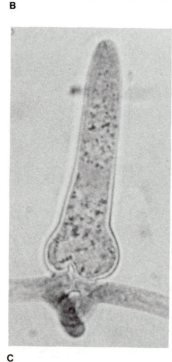

Figure 6-3 *Dipodascus aggregatus.* **A,** conidia, × 1665.
B, C, conjugation and early stage of ascus formation: **B,**
× 1290; **C,** × 1465. **D,** mature ascus, × 1215.

they are of various forms and generally bear sporocarps on or within their structure.

Subdivision Ascomycotina

"Sac" fungi, or ascomycetes, is the largest of all
fungal groups. In addition to numerous saprobic
and parasitic species, the subdivision includes almost all of the 20,000 or more lichen fungi. Ascomycetes are mainly terrestrial, but some are
found in freshwater and marine habitats. Many
ascomycetes are saprobes that occur on plant or
animal remains, on dung, or in soil; others are
parasites of plants or animals, or they live as mutualistic symbionts with other organisms.

Although all ascomycetes produce asci and
ascospores, species vary widely in life history,
morphology, and other features. In this chapter,
the group is divided into two classes, Hemiascomycetes and Euascomycetes. The main distinguishing features of each group are described
below.

1. *Hemiascomycetes.* Thallus unicellular or
   hyphal. Asexual reproduction by budding,
   fission, or conidia. Distinct male and female
   gametangia are lacking, the ascus developing
   through direct transformation of an assimilative cell or of a specialized hyphal branch.
   Sporocarps are lacking and, with a few exceptions, no dikaryotic phase is present.

2. *Euascomycetes.* Thallus mycelial. Asexual reproduction by one or more types of conidia.
   Sexual reproduction initiated by gametangial
   contact and nuclear transfer or equivalent; a
   dikaryotic phase follows. Asci develop
   within ascocarps from terminal cells of dikaryotic hyphae.

Class Hemiascomycetes

Hemiascomycetes are all simple, microscopic
forms, and they are generally considered to be
the most primitive ascomycetes. These fungi are
common in soil, water, and on surfaces of plants

and animals. Exudates of injured plants and the surfaces of fruits are habitats of some; others, such as Taphrinales, are plant parasites. Many form mutualistic associations with beetles or other insects.

The Hemiascomycetes is not a large class, but the group is an important one to humans. It includes yeasts used in baking, in manufacturing all types of alcoholic beverages, and in producing some vitamins and other substances of value. Two orders, Endomycetales and Taphrinales, are discussed here.

Order Endomycetales The Endomycetales are mycelial or budding fungi. In *Dipodascus* species, which occur in association with beetles or on plants attacked by these insects, the hyphae are septate and the compartments are uninucleate. Some hyphal branches segment, and individual cells separate and function as conidia (Fig. 6-3A).

Sexual reproduction in *Dipodascus* occurs through conjugation between adjacent hyphal cells (Fig. 6-3B, C), as in some zygomycetes. Outgrowths develop on either side of the septum separating two cells; the nuclei divide, and one nucleus migrates into each of the two outgrowths. These outgrowths then become separated from their parent cells by transverse walls and can be considered as uninucleate gametangia. They fuse to form a single cell, and the two nuclei undergo karyogamy. The fusion cell (or young ascus) enlarges, and the diploid nucleus undergoes meiosis and several mitotic divisions. Each haploid nucleus and a portion of the cytoplasm become surrounded by a cell wall and is transformed into an ascospore (Fig. 6-3D). At maturity, ascospores are discharged through a pore at the ascus tip; each ascospore germinates to produce a new fertile mycelium.

Dipodascus and some related fungi are carried by beetles that have special pouches in which the fungi are maintained and grow. When such an insect bores into wood, it deposits spores or cells of the fungus; these then grow in the insect galleries. Larvae of the insect feed upon the fungus hyphae.

The most important yeast, and one of the oldest domesticated organisms, is *Saccharomyces cerevisiae*, commonly called baker's yeast, or brewer's yeast. Perhaps because of its usefulness,

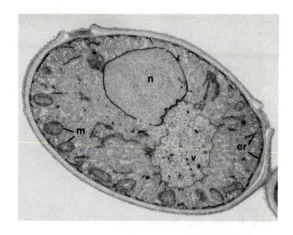

Figure 6-4 Electron micrograph of thin section through *Saccharomyces* cell, ×7200: *n*, nucleus; *v*, vacuole; *m*, mitochondria; *er*, endoplasmic reticulum. (Photograph courtesy of R. F. Illingworth, A. H. Rose, and A. Beckett, with permission of *Journal of Bacteriology*.)

S. cerevisiae is among the most intensively studied of all fungi. Still, much remains unknown concerning this structurally simple organism.

Cells of *S. cerevisiae* (Fig. 6-4) are very small, but they contain vacuoles, a membrane-enclosed nucleus, and other structures. Several mitochondria are often visible, but examination of serial sections reveals a single large mitochondrion in some instances.

The life history of *S. cerevisiae* includes both budding diploid and budding haploid phases. During budding (Figs. 6-5 and 5-9A) a small protuberance develops, gradually expands, and eventually reaches a size and form comparable to that of the parent cell. During enlargement of the bud, the nucleus divides, and one daughter nucleus is then present in the bud. At maturity of the bud, a plug develops between the parent and daughter cells and they separate. A "bud scar" remains on the parent cell at the site of attachment, and a "birth scar" is present on the bud (see Fig. 5-9A).

Saccharomyces cell walls consist mainly of glucan and a protein-mannan complex; chitin is mostly present in the bud scars. Enzymes are secreted into the wall during budding and soften the rigid glucan. As the bud expands, more

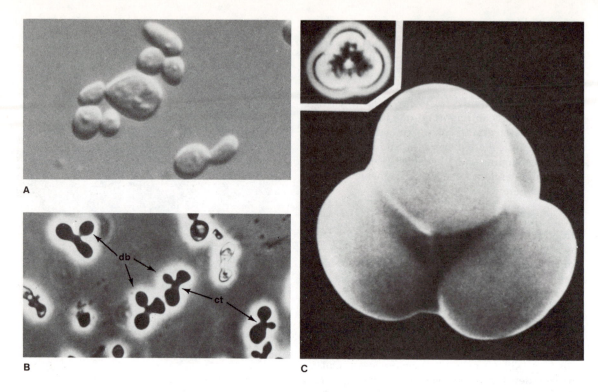

Figure 6-5 *Saccharomyces.* **A,** budding haploid cells, × 1850. **B,** conjugation, 2½ hours after mixing of compatible strains, × 1320; *db,* diploid buds; *ct,* conjugation tubes. **C,** scanning electron and phase-micrograph (*inset*) of mature ascus and ascospores, × 9260. (**B,** courtesy of Eng-Hong Lee, C. V. Lusena, and B. F. Johnson, reproduced by permission of the National Research Council of Canada from the *Canadian Journal of Microbiology* 21:802–6, 1975; **C,** courtesy of P. Rousseau, H. O. Halvorson, L. A. Bulla, Jr., and G. St. Julian, with permission of *Journal of Bacteriology.*)

glucan and other wall substances are inserted into the expanding wall. Each bud develops at a new site in *S. cerevisiae,* and the number of buds from a single cell is thus limited.

S. cerevisiae is heterothallic; the two mating strains are referred to as α and *a* (or + and −). Fusion of two haploid cells is by short, broad tubes (Fig. 6-5B), and karyogamy soon follows plasmogamy. The zygote produces buds, and a budding diploid phase is initiated. If diploid cells are transferred to a sporulation medium, ascospore formation occurs. The nucleus undergoes meiosis, and four ascospores are produced (Fig. 6-5C). No dehiscence mechanism is present; the spores are eventually released by the break-

down of the ascus wall. The ascospores germinate by bud formation (Fig. 6-6).

Several hormones regulate the initial phases of conjugation in *S. cerevisiae.* One of these, a peptide, is produced by α mating type cells, and it inhibits cell division and causes enlargement and elongation of *a*-type cells. Similar substances are probably produced by *a* cells and affect α cells.

Electron microscopy has helped to clarify some aspects of nuclear structure and division in *S. cerevisiae;* other aspects remain obscure. In both mitotic and meiotic divisions, the nuclear membrane remains intact. In meiosis, centriolar plaques and a spindle apparatus are present (Fig. 6-7A). The nucleus becomes four-lobed, and the

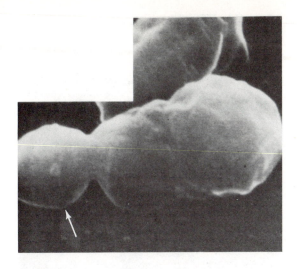

Figure 6-6 Scanning electron micrograph of germinating ascospore of *Saccharomyces*, ×8485 (*arrow* indicates bud). (Courtesy of P. Rousseau, H. O. Halvorson, L. A. Bulla, Jr., and G. St. Julian, with permission of *Journal of Bacteriology*.)

lobes are partially enclosed in spore-delimiting membranes before division is completed (Fig. 6-7B). Spore delimiting membranes are double, the ascospore wall material being deposited between the two membranes (Fig. 6-7C). In this respect, *S. cerevisiae* resembles other ascomycetes. Mature ascospores contain a nucleus, mitochondria, and other organelles.

The genus of fission yeasts, *Schizosaccharomyces*, differs from *Saccharomyces* in both method of reproduction and life history. Assimilative cells of *Schizosaccharomyces* (Fig. 6-8A) undergo fission rather than budding. After transverse walls are deposited, the daughter cells eventually separate. Conjugation of vegetative cells (Fig. 6-8B, C) is followed by ascus development. In *Schizosaccharomyces*, only the zygote nucleus is diploid; the remainder of the life history is haploid.

Saccharomyces cerevisiae is used in the baking, brewing, and distilling industries. It is a source of ergosterol, ephedrine, enzymes, and other substances. Ergosterol is a sterol that can be converted to vitamin D or to other steroid substances. Yeast cells from brewing and distilling industries, a good source of B vitamins, are used as dietary supplements for both humans and livestock. Other species of *Saccharomyces* are used in producing wine and ale. Yeasts can quickly convert a variety of cheap substances into protein; the use of this protein as human food is a future possibility.

Order Taphrinales *Taphrina*, the principal genus of the Taphrinales, is a small genus of obligate parasites. *T. deformans*, causing peach leaf curl, is a species of widespread occurrence. Hyphae of *T. deformans* penetrate between the cells of host leaves, twigs, or other parts. Their presence there induces abnormal growth of host tissues, causing thickening and curling of infected leaves (Fig. 6-9). Some hyphae grow out between the host epidermal cells and produce asci on the surfaces of infected parts.

In ascus formation (Fig. 6-10A–D), karyogamy occurs in a terminal hyphal cell. A mitotic division follows, and the cell is divided into two by a transverse wall. The outermost cell then functions as an ascus, its nucleus undergoing meiosis and then mitosis. Eight ascospores are formed, but because of their budding (Fig. 6-10E), the ascus often contains many cells.

Budding is the typical method of ascospore germination in *Taphrina*, and the budding phase can be maintained indefinitely in culture. Conjugation of the yeastlike cells occurs on the host, and a dikaryotic, septate mycelium develops. In *T. deformans*, the dikaryotic phase has been reported to develop from a single budding cell.

Class Euascomycetes

Euascomycetes are all mycelial fungi that produce their asci in sporocarps. In this class, asci develop on dikaryotic hyphae that arise from the female gametangium following gametangial contact. One or more types of conidia can be produced in asexual reproduction.

The mycelium is extensive in most euascomycetes. With the exception of some parasites and most lichen fungi, this mycelium is usually embedded in wood or other substrates. Hyphae of these fungi have one or more nuclei in each compartment, the number often varying with time.

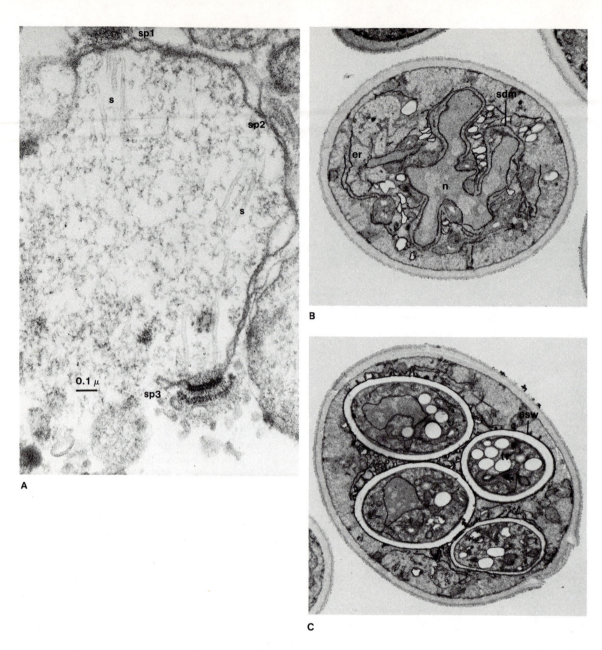

Figure 6-7 A, B, meiosis in *Saccharomyces*. **A,** Electron micrograph of nucleus during meiosis II (*scale line* = 0.1 μm): *sp1, sp2, sp3,* spindle plaques; *s,* spindle microtubules; **B,** thin section showing lobed nucleus (*n*) and spore-delimiting membranes (*sdm*) enveloping lobes: *er,* endoplasmic reticulum, ×10,115. **C,** electron micrograph of ascus and ascospores, ×9160: *asw,* ascospore wall; *tv,* electron transparent vesicle. (**A,** from J. B. Peterson, R. H. Gray, and H. Ris, "Meiotic spindle plaques in *Saccharomyces cerevisiae*," *Journal of Cell Biology* 53:837–41, 1972, with permission of Rockefeller University Press; **B, C,** courtesy of R. F. Illingworth, A. H. Rose, and A. Beckett, with permission of *Journal of Bacteriology*.)

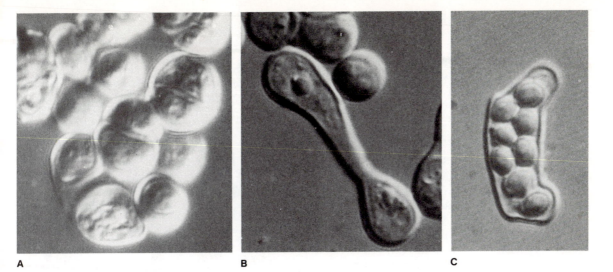

A   B   C

Figure 6-8 *Schizosaccharomyces.* **A,** haploid cells with transverse walls produced during fission, ×2000. **B,** conjugation of two haploid cells, ×1500. **C,** mature ascus and ascospores, ×1500.

Heterokaryosis and Parasexuality Hyphae developing from a uninucleate spore often contain nuclei of a single genetic type, but genetically different nuclei of several types can occur in a single thallus. This condition, called **heterokaryosis,** is common in hyphae of many higher fungi. Heterokaryotic mycelia develop as a result of mutation, germination of heterokaryotic spores, or by fusion of hyphae from two separate mycelia. The different nuclei within a heterokaryotic mycelium exist side by side in a common cytoplasm; each nuclear type increases by mitosis, and each type exerts its influence on the growing mycelium.

Within a heterokaryotic mycelium, haploid nuclei sometimes fuse. The diploid nuclei so formed can also divide mitotically and increase in number together with haploid nuclei. Infrequently, crossing over occurs during mitoses of such diploid nuclei. Haploidization, probably through loss of chromosomes, can also take place. These phenomena lead to genetic variability, just as in sexual reproduction. However, the events involved neither take place in a regular, orderly sequence, nor do they occur in specialized cells or structures. The phenomena are referred to collectively as the **parasexual cycle,** or parasexuality.

Parasexuality and heterokaryosis are two factors responsible for the variability found in certain ascomycetes, Fungi Imperfecti, and basidiomycetes. In species producing conidia, any of the nuclear types can be incorporated into conidia to be dispersed.

Asexual Reproduction Developmental details of conidial formation and conidia themselves differ greatly among euascomycetes. Conidiophores, specialized branches bearing conidia, are often present. Conidia and conidiophores of two ascomycetes are shown in Figure 6-11A, B. Conidiophores are produced singly on the hyphae of some euascomycetes; in others they are produced within special sporocarps. Two types of the latter are **pycnidia** and **acervuli** (Fig. 6-11C, D). A pycnidium is like a **perithecium** (see p. 185), but it contains only conidia and conidiophores. The acervulus consists of a saucer-shaped aggregate of conidiophores and conidia produced subepidermally on plants.

Conidial stages constitute a repeating or reinfecting stage in many parasitic ascomycetes. That is, conidia are formed abundantly during the growing season of the host, and they infect new plants. The sexual stages often commence development late in the growing season and mature the following spring. Conidial stages are also the stages most commonly encountered in many saprobic euascomycetes.

Sexual Reproduction Sexual reproduction commences with contact and plasmogamy of appropriate cells, followed by nuclear transfer. Both male and female gametangia can be present on a single thallus. Such a thallus is hermaphroditic and self-fertile if it can reproduce sexually by itself. In hermaphroditic, self-sterile species, cross-fertilization is required by a genetically distinct but morphologically similar thallus. The terms heterothallism and homothallism, though not entirely appropriate, often are applied to self-sterile and self-fertile forms respectively.

Gametangia and the mechanisms for accomplishing plasmogamy are varied; a common form is illustrated in diagrams in Figure 6-12A–C. Where present and distinct from ordinary hyphae, the female gametangium, or **ascogonium,** can be equipped with a tubular appendage, the **trichogyne.** Contact occurs between **antheridium** and trichogyne if the latter is present, or between antheridium and ascogonium in the absence of a trichogyne. Following contact, plasmogamy occurs, and one or more nuclei migrate from the antheridium and pair with the nucleus or nuclei in the ascogonium. Dikaryotic hyphae then develop from the ascogonium. Nuclei within the hyphae undergo mitosis, the members of each pair dividing simultaneously and side by side.

Two types of ascogonia are illustrated in Figure 6-12D, E. Antheridia are absent in many euascomycetes, and fertilization is brought about by conidia or by special cells called **spermatia.** Spermatia are small, uninucleate, sporelike cells produced on special hyphal branches (Fig. 6-12F). Spermatia or conidia (Fig. 6-12E) can be carried by wind, water, or insects; upon contact with the ascogonium, they function in nuclear transfer and dikaryotization.

Figure 6-9 Peach leaf curl, caused by *Taphrina deformans,* ×0.66.

Ascogenous Hyphae. Dikarotic hyphae are called **ascogenous hyphae,** and they eventually give rise to asci. Each compartment in an ascogenous hypha has a pair of nuclei, the septa developing when the nuclei divide. In many euascomycetes, cells of ascogenous hyphae develop through formation of **croziers.** Crozier formation (Fig. 6-13) commences by the hyphal tip growing back upon itself to form a broad crook. The two nuclei present divide (Fig. 6-13B), with a daughter of one nucleus going into the hyphal tip and a daughter of the second into the base of the crook cell. Transverse walls then develop between sister nuclei and at right angles to the division plane. The tip cell continues to grow, contacts the basal cell, and becomes confluent with it by dissolution of the intervening wall. Renewed growth can then occur from what is now the terminal (penultimate) cell.

The development of ascogenous hyphae re-

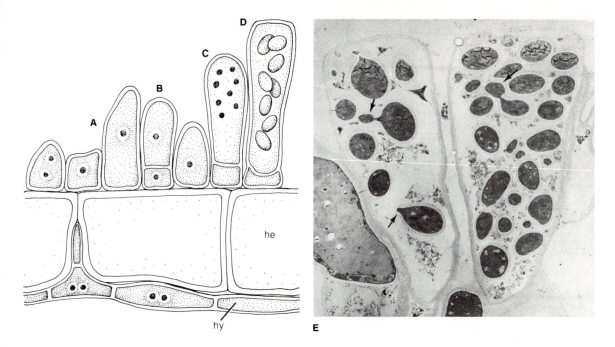

Figure 6-10 Ascus development in *Taphrina deformans*. **A,** diploid cells after karyogamy. **B,** mitosis has occurred, resulting in ascus mother cell (*above*) and basal cell. **C,** ascus after meiosis. **D,** mature ascus and ascospores; all ×895. **E,** electron micrograph of two mature asci, ×3020 (budding indicated by *arrows*). (Courtesy of M. Syrop and A. Beckett, reproduced by permission of the National Research Council of Canada from the *Canadian Journal of Botany* 54:293–305, 1976.)

sults in an increase in the number of compatible nuclear pairs. These hyphae branch repeatedly in most euascomycetes, and tip cells of the final branches develop into asci.

The Ascus. Asci typically develop from terminal cells of ascogenous hyphae (Fig. 6-14A–E). Within the terminal cells, called **ascus mother cells,** nuclei unite. The ascus mother cell enlarges, the nucleus undergoes meiosis, and ascospores are produced. In most euascomycetes, a mitotic division follows meiosis, and the ascus has eight nuclei. Walls develop around each nucleus, together with cytoplasm and organelles, to form ascospores. Cytoplasm and organelles that are not enclosed degenerate as the ascus matures.

The type of ascus and certain biochemical and structural details of it are considered basic in the classification of euascomycetous fungi. In most euascomycetes, the ascus wall appears as a single layer by light microscopy (Fig. 6-14F–I). Asci of some species (Fig. 6-14F, G) open by a minute lid, the **operculum,** when ascospores are released. In others, evagination of an apical pore occurs (Fig. 6-14H, I) when ascospores are blown from the ascus. The asci of some species simply dissolve at maturity, and ascospore release is passive.

Ascocarp Development. The development of the sporocarp, specifically an **ascocarp,** commences soon after plasmogamy. Haploid hyphae composing the ascocarp arise from around gametangia, their growth producing a structure characteristic for each group. Stages in this process are shown in development of *Neurospora* ascocarps in Figure 6-15A–D. Although individual hyphae are discernible as such in early stages of

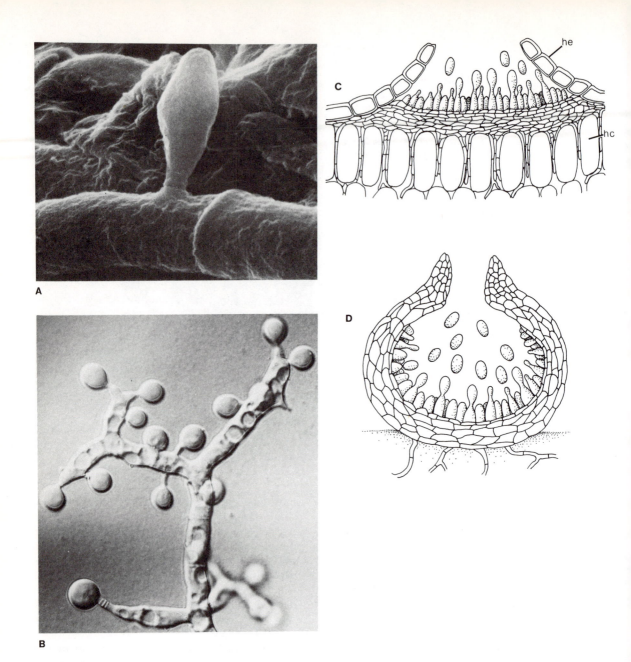

Figure 6-11 Conidial stages of ascomycetes. **A,** scanning electron micrograph of single conidium arising from hypha, ×9100. **B,** branched conidiophore with numerous conidia, ×550. **C,** semidiagrammatic illustration of a subepidermal acervulus with compactly arranged conidiophores. **D,** semidiagrammatic illustration of section through pycnidium, showing arrangement of conidiophores. (**A,** courtesy of R. D. Koehn and G. T. Cole, reproduced by permission of the National Research Council of Canada from the *Canadian Journal of Botany* 53:2251–59, 1975; **B,** courtesy of K. L. O'Donnell and G. R. Hooper, reproduced by permission of the National Research Council of Canada from the *Canadian Journal of Botany* 54:1084–93, 1976.)

development, they can become pseudoparenchymatous. Commonly encountered types of ascocarps (Fig. 6-16A–D) are (1) **cleistothecia,** (2) **perithecia,** (3) **apothecia,** and (4) variously shaped **ascostromata.** Relative positions of gametangia in ascocarps are indicated in Figure 6-16.

Cleistothecia typically are completely closed ascocarps in which asci are scattered irregularly. Most cleistothecia are small, spherical structures; the wall can be a loose network or a highly developed pseudoparenchymatous structure. Perithecia are small, flask-shaped structures, the walls of which are composed of pseudoparenchyma. Asci occur in a basal tuft or fertile layer, the **hymenium,** in the perithecium; they are often accompanied by sterile filaments called **paraphyses.** An opening, the **ostiole,** is present in the apex of the perithecium, and it is lined with sterile hairs. The apothecium, often discoid or bowl-shaped, has an exposed hymenium consisting of asci and paraphyses. Other portions of the ascocarp can be either obviously hyphal or pseudoparenchymatous.

Many euascomycetes produce perithecia on or in stromata; however, in Loculoascomycetes, ascus-containing chambers develop within the stroma itself. These chambers are called **locules** and are produced by digestion of stromatal tissue; they have no wall of their own, unlike embedded perithecia. Stromata of this type, called **ascostromata,** mostly contain asci that are conspicuously double-walled (see *Pleospora,* p. 209).

Life Histories Although euascomycetes are a structurally diverse group, the life histories of many follow a similar pattern (Fig. 6-17). Asexual and sexual reproduction can occur concomitantly on a single thallus. More commonly, conidia are produced on a young thallus that later produces ascocarps. Since many species are self-sterile, ascocarps cannot develop until fertilization has occurred. Until then, only asexual reproduction can occur. Several types of conidia are produced by some species; others produce none.

Lichens More than 20,000 species of euascomycetes exist in complex algal-fungal associations called *lichens.* Such associations also are characteristic of a few basidiomycetes and Fungi Imperfecti. Although lichens generally are treated as a separate group, the classification of lichen fungi as a category separate from other fungi is unnatural. Lichens are polyphyletic, the symbiotic association having originated separately among a number of different groups of fungi and with different algae. The name (binomial) of a lichen applies specifically to its fungal component, or **mycobiont;** the algal component, called the **phycobiont,** has a different binomial. Most mycobionts are euascomycetes that produce apothecia, but many produce ascocarps of other types.

One of the most distinctive features of many mycobionts is the fact that their mycelium is largely exposed and can produce a mass of relatively regular form. By contrast, most nonlichenized ascomycetes have a submerged mycelium, and only reproductive structures are exposed. Lichen fungi also produce some submerged mycelium; in fact, some have most of the thallus embedded in bark.

Lichen thalli can mostly be categorized as **crustose, foliose,** or **fruticose,** on the basis of thallus morphology. Crustose lichens consist of a thin layer closely adherent to the underlying substrate or sometimes submerged in it (Fig. 6-18A). As seen in transverse section (Fig. 6-18B), such a thallus commonly consists of an outer layer, the **cortex;** an underlying algal layer; and a **medulla** of varied thickness. The cortex consists of tightly packed, mucilaginous hyphae. The algal layer contains cells of the phycobiont plus a loose network of hyphae, some of which are in intimate association with algal cells. The medulla is composed of loosely arranged hyphae. Some of the lowermost hyphae penetrate into the substratum.

The foliose thallus (Fig. 6-18C) is leaflike, and the margins are free from the substratum. The thallus is either broadly attached or attached by one or more small bundles of hyphae. In transverse section, the foliose thallus (Fig. 6-18D) is similar to the crustose type, but a lower cortex layer is often present.

The fruticose lichen thallus is branched, shrublike, and can be erect or pendent (Fig. 6-18E). In transverse section, the branches of this thallus also have an outer cortex, an underlying

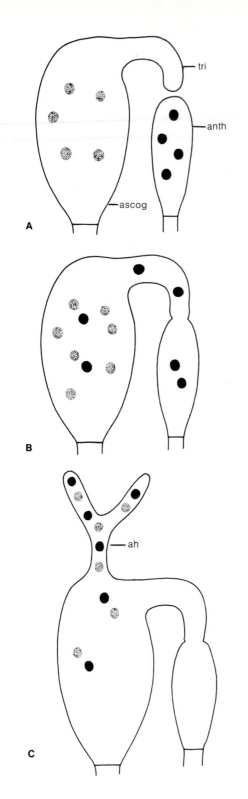

A

B

C

tri

anth

ascog

ah

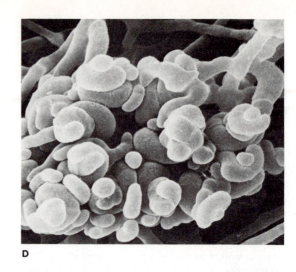

D

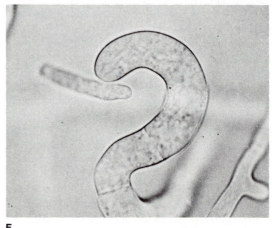

E

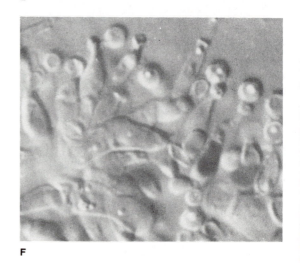

F

algal layer, and a medulla. The central area is either hollow or occupied by a central strand of tough hyphae.

All lichen types described above are stratified (i.e., the algal cells occupy a narrow zone). In a few lichen thalli, algal cells constitute the bulk of the thallus; in others, the algal cells are scattered (i.e., unstratified). In nature, a given lichen fungus regularly occurs with the same species, or, in a few instances, two algal species are in a single thallus. The phycobiont generally is a green (Chlorophyceae) or bluegreen (Cyanophyceae) alga; the most common are species of *Trebouxia* (Chlorophyceae), which are rarely found outside of lichen thalli. *Trebouxia, Trentepohlia* (Chlorophyceae), and *Nostoc* (Cyanophyceae) species are the phycobionts of approximately 90% of all lichens.

Some mycobionts produce haustoria that penetrate the cell walls of the phycobiont. More commonly, mycobiont and phycobiont cells have very thin walls at points of contact, and haustoria are absent.

Vegetative Reproduction. In many lichens, the cortex ruptures (Fig. 6-19A, B) to expose small structures called **soredia.** A soredium consists of a few algal cells held together by mycobiont hyphae. Soredia are often produced in abundant, dusty masses that can be carried away by air currents. More highly organized vegetative structures, called **isidia** (Fig. 6-19C), are produced by some lichens. Isidia also consist of hyphae and

phycobiont cells; they readily break free of the thallus and probably function like soredia. Both isidia and soredia can establish new lichen thalli.

Sexual Reproduction. The phycobiont reproduces only asexually within the lichen; in culture, however, it will reproduce sexually. The only sexual structures of the lichen, those of the fungus, are predominantly ascocarps. Lichen ascocarps are structurally similar to those of other euascomycetes, but they grow more slowly and produce and release their spores over a longer time period.

Growth and Physiology. The growth of lichens is slow; in some crustose species, the annual increase in radius is only 0.1 to 2.0 mm. That of foliose and fruticose thalli ranges from 0.5 to 10 mm per year. Crustose and foliose lichens grow at the margin, the center often dying and eroding away with age. Measurements of the total radius and annual increments indicate ages up to 4500 years for some thalli.

Lichens generally tolerate extremes in temperature, illumination, and desiccation. During dry periods, the thallus is inactive. Water is quickly absorbed by the blotterlike thallus, but water loss also occurs rapidly. Photosynthesis is dependent upon the presence of water and, with drying, declines rapidly. Minerals, also in short supply in some habitats, are obtained from water, from airborne particles, or from the underlying substrate.

During periods of photosynthesis, the phycobiont produces the sugar alcohol ribitol or glucose, and much of these substances is "leaked" to the mycobiont. The mycobiont rapidly converts them to others, such as mannitol, which cannot be utilized by the alga.

Much interest in lichens, and a portion of their taxonomy, is based upon unusual chemicals that accumulate outside hyphal walls as water-insoluble crystals. Many of these chemicals are brightly colored, and they are often responsible for the color of the thallus. Usnic acid, for example, is responsible for the bright yellow color of many lichens. Although these lichen substances once were thought to be produced only by the symbiotic association, cultural studies indicate that some can be produced by the fungus alone.

Both mycobiont and phycobiont can be

Figure 6-12   A–C, diagrammatic representation of ascogonia and antheridia. (**A**), plasmogamy (**B**), and the development of ascogenous hyphae (**C**): *tri,* trichogyne; *anth,* antheridium; *ascog,* ascogonium; *ah* ascogenous hyphae. **D,** coiled gametangia, the ascogonia and antheridia indistinguishable and arising from the same or different hyphae, ×1715. **E,** ascogonium growing toward conidium that will function in spermatization, ×1300. **F,** cluster of spermatium-producing cells and spermatia, ×3000. (**D,** courtesy of K. L. O'Donnell and G. R. Hooper, reproduced by permission of the National Research Council of Canada from the *Canadian Journal of Botany* 54:572–77, 1976; **E,** Courtesy of J. W. Paden and M. A. S. Linton, reproduced by permission of the National Research Council of Canada from the *Canadian Journal of Botany* 54:1784–92, 1976.)

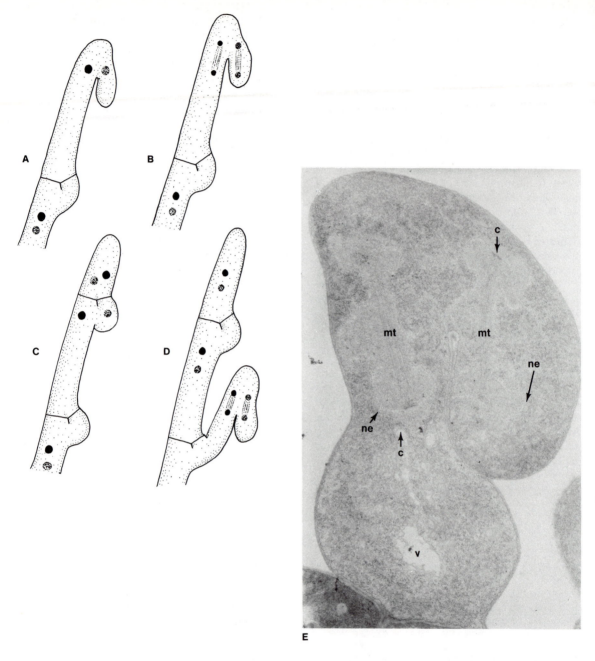

Figure 6-13 **A–D,** diagram showing development of a crozier. **A,** hook with dikaryon; **B,** nuclear division; **C,** fusion of tip with hypha and septum formation; **D,** two dikaryotic cells with renewed growth of new tip cell. **E,** electron micrograph of thin section of crozier during nuclear division, ×9795: *c,* centriolar plaque; *mt,* microtubules; *ne,* nuclear envelope; *v* vacuole. (**E,** courtesy of C.-Y. Hung and K. Wells, from "Light and electron microscopic studies of crozier development in *Pyronema domesticum*," *Journal of General Microbiology* 66:15–27, 1971, with permission of Cambridge University Press.)

grown separately in culture; some mycobionts also grow alone in nature. Some phycobiont species do occur as free-living forms. Lichens are abundant in tropical, temperate, and polar regions. In the harsh, antarctic environment, they are the dominant forms of "plants," and they occur abundantly in desert and alpine habitats. Whatever the nature of their relationship, it seems that the combination can exist where neither symbiont could live alone.

Classification of Euascomycetes

The class Euascomycetes is often subdivided into several series of subclasses. We divide the group in this text into five series: Plectomycetes, Pyrenomycetes, Loculoascomycetes, Laboulbeniomycetes, and Discomycetes. The names have no official taxonomic status as used here, but they are useful in designating general groups of ascomycetes. The main features of each series are as follows:

1. *Plectomycetes.* Ascocarps mostly cleistothecia, the asci scattered and dissolving at maturity.

2. *Pyrenomycetes.* Ascocarps perithecia, the asci in a hymenium; ascospores mostly forcefully discharged.

3. *Discomycetes.* Ascocarp an apothecium, the asci arranged in an exposed hymenium; ascospores forcefully discharged.

4. *Loculoascomycetes.* Asci produced in locules, the ascus wall distinctly double; ascospores forcefully discharged after rupture of the outer wall layer.

5. *Laboulbeniomycetes.* Ascocarp reduced, peritheciumlike; mycelium lacking, thallus consisting of a haustorium, foot cell, and ascocarp. Obligate parasites of arthropods.

Series Plectomycetes

Order Eurotiales The Eurotiales include many economically important species, some of which are among the most intensively studied fungi. Some species are used in industrial fermentation processes; others are responsible for the destruction of stored foods, leather goods, and even fine lenses. Some Eurotiales also cause diseases of humans, other mammals, and birds.

Eurotiales are characterized by cleistothecia within which asci dissolve at maturity. Cleistothecia of *Shanorella* (Fig. 6-20A) and closely related genera have a wall that is rudimentary; it is essentially an open network or mesh of hyphae. Sterile appendages extend outward from the walls of many of the cleistothecia. Ascospores are released into the ascocarp when the asci (Fig. 6-20B, C) dissolve, and they probably sift out through large openings in the cleistothecial wall.

Many soil fungi related to *Shanorella* have the ability to break down keratin. These species mainly grow saprobically on feathers, hair, horn, or other keratinous substrates in soil. However, some species cause athlete's foot, ringworm, and similar disorders in humans and domesticated animals.

The common green or blue molds growing on citrus fruit, jams, jellies, bread, and other foods are usually species of *Aspergillus* and *Penicillium*. These names were applied to the conidial stages of certain Eurotiales long before the sexual stages were known; the names are still used for the asexual stages. The name *Penicillium* (Figs. 5-9B, 6-21F) is derived from the Latin term *penicillium*, meaning a small brush; *Aspergillus* (Fig. 6-22A, B) is so named because of the resemblance of the conidiophore to a holy water sprinkler, or *aspergillum*. In these genera, conidia are produced in long chains on special conidiophore cells called **phialides** (Fig. 6-22B). Hyphae of these fungi are essentially colorless, but because of masses of bluish or greenish conidia, the colonies can be brightly colored.

A number of genera of Eurotiales have asexual stages of the *Aspergillus* and *Penicillium* type; only two of these are described here. *Talaromyces*, with conidia of the *Penicillium* type, has ascogonia and antheridia (Fig. 6-21A) that develop on separate hyphae of a single mycelium. Contact of antheridium and oogonium occurs, and a small pore develops between the two structures. Following plasmogamy, hyphae grow up around the gametangia and produce a cleistothecium

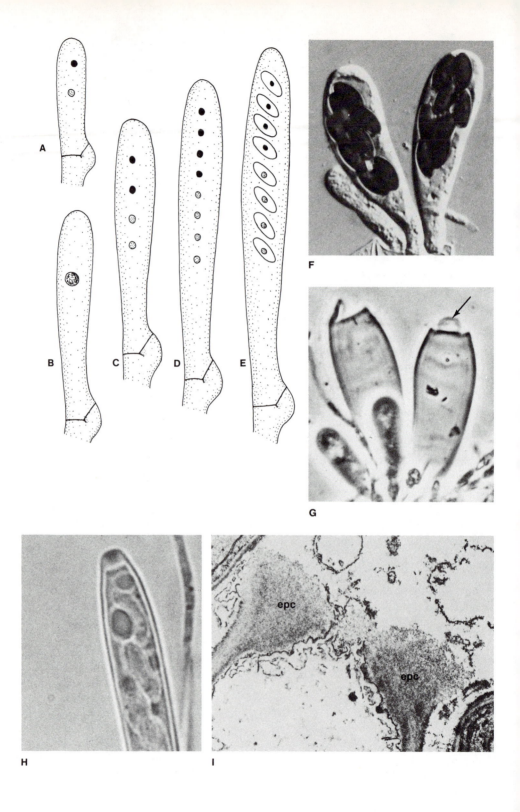

(Fig. 6-21B) with walls of closely interwoven hyphae. Asci (Fig. 6-21C, D) are borne on ascogenous hyphae within the ascocarp, and ascospores are eventually released through irregular dehiscence of the ascocarp wall.

Eurotium has conidiophores of the *Aspergillus* type. The cleistothecium develops in much the same way as in *Talaromyces*, but the ascocarp wall is pseudoparenchymatous (Fig. 6-22C, D).

Species of *Aspergillus* and *Penicillium* are used in producing antibiotics, organic acids, and blue cheeses. Penicillin, the first useful antibiotic and still the most used, is obtained from *Penicillium chrysogenum* and others. Citric acid, gluconic acid, itaconic acid, and gallic acid are all produced commercially using species of *Aspergillus* or *Penicillium*. These acids are used in the preparation of foods, beverages, plastics, and medicines; they are also used in industrial processes of other types. Blue cheeses, such as Roquefort, Gorgonzola, and Stilton, all owe their characteristic flavors to the action of *Penicillium* species. The blue coloration in these cheeses is caused by masses of conidia. Penicillia are also used in the manufacture of certain uncolored cheeses, notably Camembert and Brie.

In recent years, toxic substances called *mycotoxins* have been the subject of numerous studies. Some species of *Aspergillus* and *Penicillium* produce especially potent mycotoxins that, in addition to being directly toxic, are highly active carcinogenic substances. The growth of mycotoxin-producing species on peanuts or other foods renders them unfit for human or animal consumption.

Another important plectomycete, *Ceratocystis ulmi*, causes Dutch elm disease—the most serious disease of shade trees in North America. The disease was introduced into North America around 1930 and has since destroyed countless elms.

Series Pyrenomycetes

Order Erysiphales Members of the order Erysiphales are obligate parasites of vascular plants and cause a type of disease called *powdery mildew*. Many mildews cause little visible damage to their hosts, but serious losses of gooseberries, grapes, hops, apples, and other crops can occur. Erysiphales are unusual among nonlichenized ascomycetes, in that their mycelia are almost entirely external to the host (Fig. 6-23A, B). Their hyphae grow on the surfaces of leaves, twigs, or fruits and produce haustoria (Fig. 6-23C, D) that penetrate the host epidermal cells or those immediately below the epidermis.

Masses of conidia impart a powdery white appearance to infected host surfaces. Borne in chains on short, erect conidiophores (Fig. 6-23B), the conidia are wind-dispersed and bring about further infection of the host species. Although fungal spores generally have a low moisture content, conidia of the Erysiphales typically contain 50–75% water. This water permits germination under very dry conditions, such as would prevail at times on host surfaces.

Ascocarps also develop on the infected host surface, appearing somewhat later than the conidial stage. These ascocarps are peritheciumlike in their development, but as is shown in *Sphaerotheca* (Fig. 6-24), they lack an ostiole. The mature ascocarps have a series of radiating appendages. The appendages are hyphalike and unmodified in *Erysiphe* (Fig. 6-25A); those of *Uncinula* (Fig. 6-25B) are hooked or coiled at the tip, and in *Podosphaera*, the appendage tips are branched (Fig. 6-25C). The appendage characteristics and numbers of asci per ascocarp are features that distinguish various genera. Ascocarps of *Erysiphe* contain several asci; those of *Sphaerotheca* (Fig. 6-24) have a single ascus.

Figure 6-14 **A–E,** diagrams of asci development, **A,** dikaryotic ascus mother cell; **B,** karyogamy; **C,** meiosis; **D,** mitosis; **E,** the eight nuclei, each with a small amount of cytoplasm, become surrounded by ascospore walls. **F, G,** operculate asci before and after ascospore release, ×1200 (*arrow* indicates operculum). **H, I,** inoperculate ascus with ascospores, ×1500, and similar ascus showing everted ascus apex after ascospore discharge, ×15,300: *epc,* everted pore cylinder. (**I,** courtesy of M. Corlett, reproduced by permission of the National Research Council of Canada from the *Canadian Journal of Botany* 52:1459–63, 1974.)

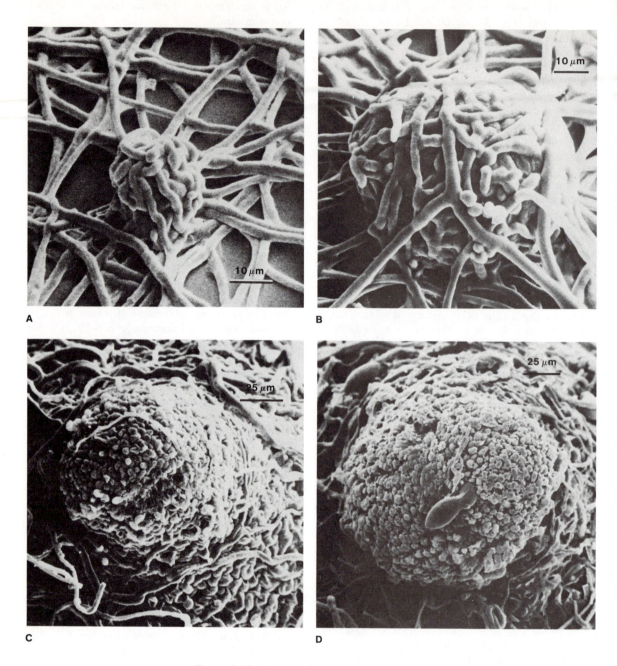

Figure 6-15 Ascocarp (perithecium) development in *Neurospora crassa*. **A,** possible ascocarp initial prior to crossing (*scale line* = 10 μm). **B,** early development of ascocarp, (*scale line* = 10 μm). **C,** ascocarp nearing mature form. **D,** mature ascocarp with ascus tip and two ascospores protruding from opening (*scale line* = 25 μm). Note that in **C** and **D,** hyphal cells of ascocarp wall have lost their form and become parenchymalike. (Courtesy of J. L. Harris, H. B. Howe, Jr., and I. L. Roth, with permission of *Journal of Bacteriology*.)

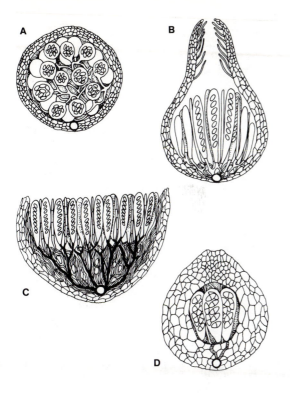

Figure 6-16 Diagrams of ascocarp type (approximate positions of gametangia indicated by circles). **A,** cleistothecium; **B,** perithecium; **C,** apothecium; **D,** ascostroma.

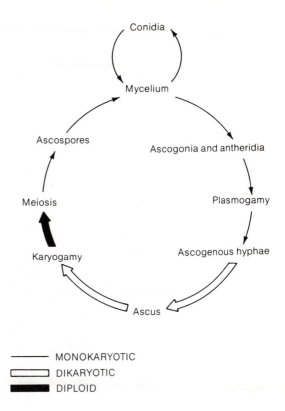

MONOKARYOTIC
DIKARYOTIC
DIPLOID

Figure 6-17 Diagram of common ascomycete life histories.

In most genera of powdery mildews, the ascocarps remain on leaves or other dead host parts until spring. At that time, the ascocarp of *Erysiphe* absorbs water, swells, and bursts, scattering asci; individual asci next absorb water, burst, and scatter ascospores.

Fungi that are obligate parasites often show a high degree of host specificity; that is, one species of parasite can attack only one species or a few varieties of hosts. However, most species of powdery mildews parasitize numerous hosts. *Erysiphe graminis* occurs on about 100 species of grasses, and the species can be divided into subspecific categories on the basis of the hosts that it can infect. That is, the strain of *E. graminis* from barley can infect other species of *Hordeum,* but not wheat; the strain from wheat can infect other species of *Triticum,* but not *Hordeum.*

Order Chaetomiales The perithecia of the small order Chaetomiales are covered by characteristic hairlike appendages. The appendages are variously branched or coiled and are often relatively bright bluish, greenish, or yellow-green. *Chaetomium* (Fig. 6-26A), the largest genus, contains species that are common in soil, on dung, or on decaying plant materials. The perithecium contains a basal cluster of asci (Fig. 6-26B) of different ages; as ascospores approach maturity, the ascus dissolves. Spores and a mucilaginous substance are extruded through the ostiole and accumulate at the top of the perithecium.

Many species of *Chaetomium* break down cellulose rapidly and can destroy cotton textiles, paper, and other cellulose materials. Some species that grow on dead forage plants also produce mycotoxins that are toxic to livestock.

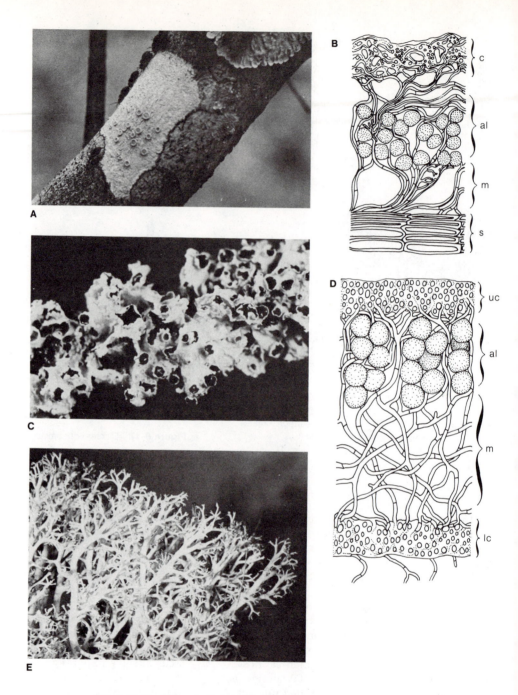

Figure 6-18 Lichen thalli. **A,** habit of crustose thallus on tree branch, ×1. **B,** diagram of section through crustose thallus: *c,* cortex; *al,* algal layer; *m,* medulla; *s,* substrate. **C,** habit of foliose lichen thallus, ×1.5. **D,** diagram of section through foliose thallus: *uc,* upper cortex; *al,* algal layer; *m,* medulla; *lc,* lower cortex. **E,** habit of fruticose thallus of *Cladonia,* ×0.75

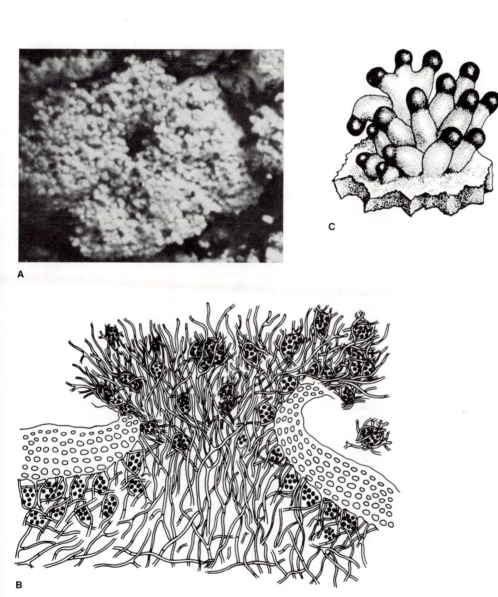

Figure 6-19 Lichen reproduction. **A,** soredium-producing area of thallus, ×25. **B,** section showing soredial production, ×140. **C,** habit sketch of isidia, ×200.

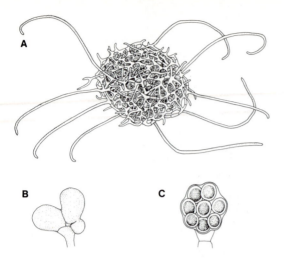

Figure 6-20 *Shanorella.* **A,** cleistothecium with coiled appendages, ×100. **B,** ascogenous hypha and young asci, ×700. **C,** ascus and mature ascospores, ×1330.

Order Xylariales Members of the Xylariales produce perithecia that have dark-colored, pseudoparenchymatous walls. The perithecia can be superficial on the substrate, or they can be submerged within it or within a stroma. One of the largest in the Euascomycetes, the order includes both saprobes and parasites.

Sordaria fimicola (Fig. 6-27A) occurs on the dung of herbivorous animals and in soil; it reproduces only by means of ascospores. Within the perithecium, there is a basal cluster of asci (Fig. 6-27B) in different stages of development. The perithecium has a short, beaklike extension, or neck, which grows toward light. At maturity, an ascus elongates until its apex protrudes through the ostiole. Ascospores are then shot from the ascus; the empty ascus collapses, and another repeats the process. Like the perithecial neck, the ascus tip shows a positive phototropic response; these responses help to ensure that ascospores are fired away from the dung on which they were produced. The spores have a mucilaginous surface layer that causes them to adhere to any plants that they contact when fired from the ascus. If eaten by an herbivorous animal, the ascospores pass unharmed through the digestive tract; in fact, their germination is enhanced.

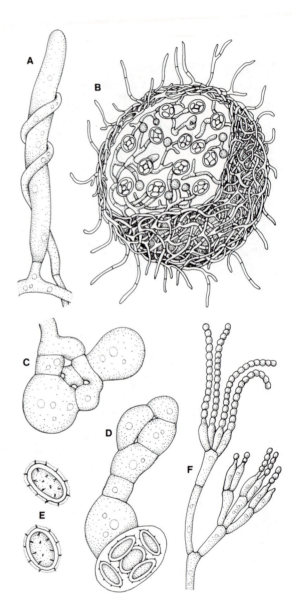

Figure 6-21 *Talaromyces.* **A,** ascogonium and coiled antheridium, ×1860. **B,** habit of ascocarp with section showing wall structure and scattered asci, ×115. **C, D,** asci and ascogenous hyphae, ×1485. **E,** two ascospores, ×1930. **F,** conidiophore and conidia of *Penicillium* type, ×1210.

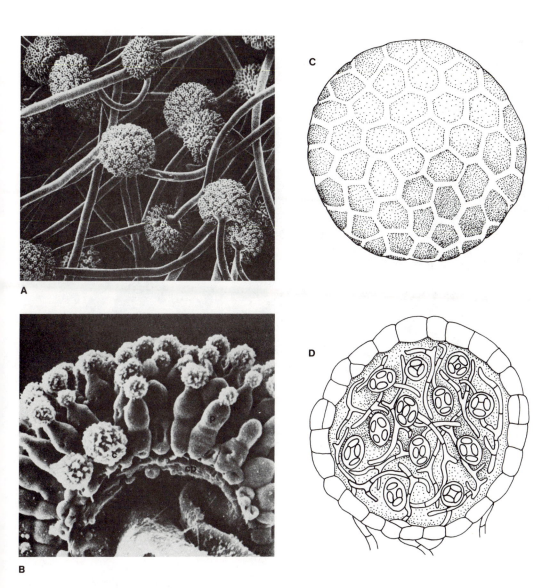

Figure 6-22   A, scanning micrograph of *Aspergillus* conidiophores, ×455. **B,** sporogenous apex of conidiophores showing phialides and conidia, ×2190: *p,* phialide; *c,* conidium; *cp,* conidiophore. **C, D,** *Eurotium.* **C,** habit of ascocarp showing surface structure of wall, ×350; **D,** section showing wall structure and asci and ascospores, ×350. (**A,** courtesy of J. Tokunaga, M. Tokunaga, H. Yotsumoto, and Y. Hataba, with permission of the *Japanese Journal of Medical Mycology;* **B,** courtesy of M. Tokunaga, J. Tokunaga, and K. Harada, with permission of the *Journal of Electron Microscopy.*)

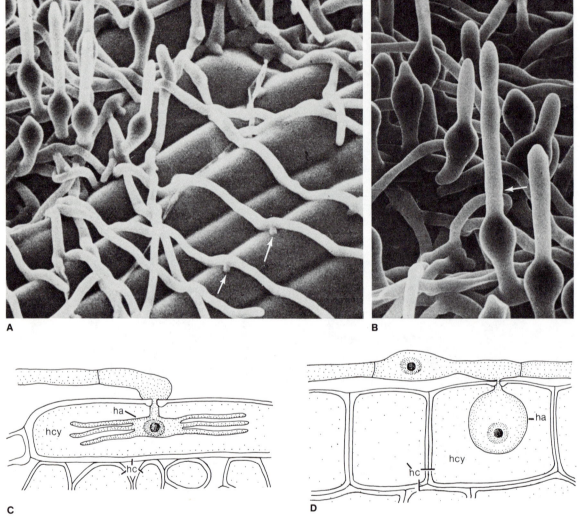

Figure 6-23 A, B, scanning electron micrographs of *Erysiphe graminis*. **A,** view showing hyphae and points where haustoria penetrate epidermal cells (*arrows*), ×615; **B,** enlarged view of conidiophores and developing conidia, ×910 (*arrow* indicates top of conidiophore). **C, D,** haustoria of powdery mildews. **C,** *Erysiphe graminis,* ×1150: *hc,* host cell; *hcy,* host cytoplasm; *ha,* haustorium; **D,** haustorium of *Unicinula salicis,* ×1260. (Photographs **A, B,** from P. R. Day and K. J. Scott, "Scanning electron microscopy of fresh material of *Erysiphe graminis* f. sp. *hordei,*" *Physiological Plant Pathology* 3:433–35, 1973, with permission of Academic Press.)

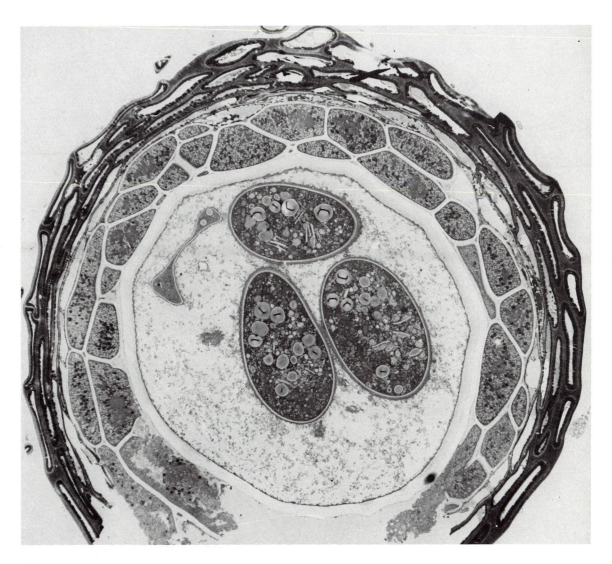

Figure 6-24 *Sphaerotheca.* Electron micrograph of section through ascocarp, ×2442. The ascocarp contains a single ascus and three matured ascospores; an aborted ascospore is also present. (Courtesy of M. Martin, F. L. Gay, and G. V. H. Jackson, from "Electron microscopic study of developing and mature cleistothecia of *Sphaerotheca mors-uvae*," *Transactions of the British Mycological Society* 66:473–87, 1976, with permission of Cambridge University Press.)

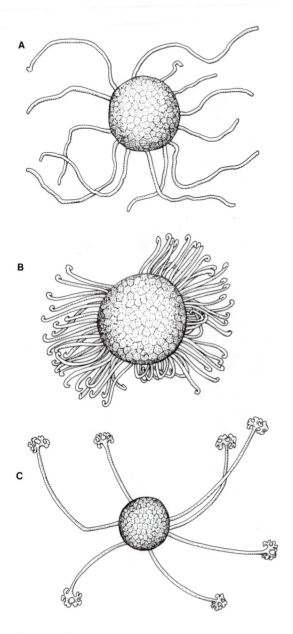

A

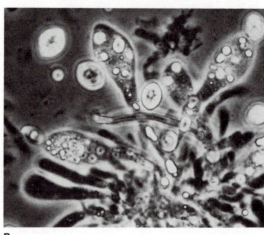

B

Figure 6-25 Ascocarps of powdery mildews, with attached appendages. **A,** *Erysiphe*, ×96. **B,** *Uncinula,* ×113. **C,** *Podosphaera,* ×110.

Figure 6-26 *Chaetomium.* **A,** ascocarp, ×80. **B,** cluster of asci, ×650.

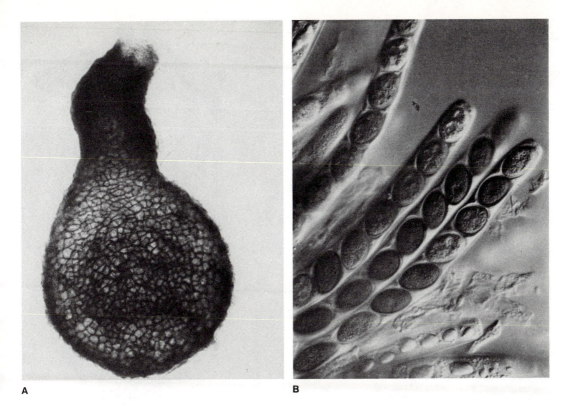

A   B

Figure 6-27 *Sordaria fimicola.* **A,** perithecium, × 160. **B,** mature asci and ascospores, ×490.

Thalli of *S. fimicola* are self-fertile, and ascocarps will develop on a mycelium from a single ascospore. However, hyphal fusions and nuclear transfer can occur between two adjacent thalli of different strains.

Neurospora (Fig. 6-15), a genus closely related to *Sordaria*, contains species that are genetically among the best-known of all organisms. Much of our knowledge of the biochemical aspects of genetics was originally derived from studies of *N. crassa* and *N. sitophila*. In these species, thalli are self-sterile, each producing ascogonia with trichogynes in perithecial initials. Two types of conidia are produced on each thallus, and these can germinate to form new mycelia or function as spermatia. The perithecial initial develops after spermatization.

Neurospora species can be grown on simple

media, and they rapidly complete their life histories. The products of meiosis, the ascospores, form a linear and ordered series, the individual spores of which are readily isolated. Since the spores give rise to haploid mycelia, mutations are easily detected. These characteristics are responsible for the value of *Neurospora* species in genetic studies.

Conidial stages of *Neurospora* species have often been referred to as *red bread molds*. These fungi grow rapidly in the warm, humid environment of a bakery, and the masses of conidia they produce make them difficult to eradicate. Conidia become established on the cooled, sliced loaves and cause rapid deterioration.

In many Xylariales, distinctive stromata are produced, and perithecia are embedded within these structures. During early development of

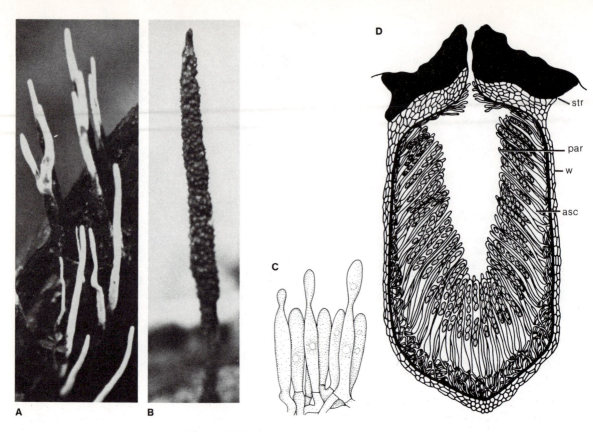

Figure 6-28 A–D, *Xylaria.* **A,** young stroma, the white portions of which are covered by a layer of conidiophores and conidia, ×1; **B,** mature stroma, the uneven surfaces caused by embedded perithecia, ×1.4; **C,** conidiophores and conidia, ×1190; **D,** section through portion of stroma and one perithecium, ×150: *str,* stroma; *par,* paraphyses; *w,* perithecial wall; *asc,* ascus.

the stroma, a layer of conidiophores and conidia (Fig. 6-28A, C) is present on the stromatic surface. Stromata of *Xylaria* (Fig. 6-28A, B) are stalked and relatively regular in form; those of *Daldinia* (Fig. 6-29) are sessile. These stromata all have a dark outer rind beneath which perithecia are embedded (Fig. 6-28D). A number of stromatic Xylariales are facultative parasites of woody plants and, in some instances, cause economically important diseases.

Although superficially resembling those of other Xylariales, the stromata of *Daldinia* (Fig. 6-29A, B) are highly modified structures that function in water storage. If *Daldinia* stromata are removed from their substrates and placed in a dry room, sporulation can continue for several weeks through use of this reserve water.

Order Clavicipitales Members of the Clavicipitales differ from the Xylariales in details of ascus and ascocarp structure.

Claviceps purpurea, the cause of ergot of rye and other grasses, is a common temperate fungus. Spores of this species infect the ovaries of grasses, and the mycelium produces numerous minute conidia (Fig. 6-30A). Conidia are exuded from infected flowers in a sticky fluid, called **honeydew,** which may attract insects to the spore mass. The conidia are then dispersed by the insects, or by splashing rain, and these propagules infect additional flowers.

The hyphae of *Claviceps* continue to grow in the ovary and produce a hard, purplish sclerotium (Fig. 6-30B). At maturity, the sclerotium has the form of a grain produced in a healthy

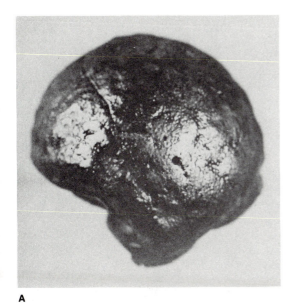

A

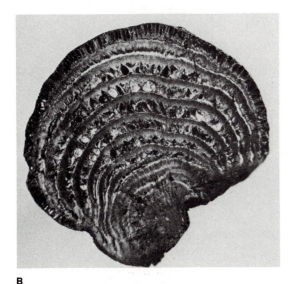

B

Figure 6-29 *Daldinia*. **A, B,** intact and sectioned stromata, respectively, the latter showing characteristic concentric pattern and single row of embedded perithecia (**A, B,** × 1.3).

ovary, but it is larger than such grains. Sclerotia, or *ergots,* fall to the ground and function there as resistant, overwintering structures. They require exposure to low temperatures for further development and retain their viability in soil for several years. Upon germinating, each sclerotium produces one or more stalked stromata (Fig. 6-30C), within which are embedded numerous perithecia (Fig. 6-30D). Asci of *Claviceps* are extremely long and narrow and have a thickened apex perforated by a canal (Fig. 6-30E). The ascospores are threadlike and multiseptate; they fragment readily upon release from the ascus.

Sclerotia of *Claviceps* contain a number of potent alkaloids; others are produced by cultures of the species. "St. Anthony's Fire," a serious type of intoxication that results from eating ergots, once was relatively common among humans. Poisoning usually resulted from eating bread or other foods made with contaminated rye flour, and it was often fatal. Livestock grazing on infected grasses are also poisoned by sclerotia. Several *Claviceps* alkaloids are used medicinally at the present time.

Cordyceps, a genus closely related to *Claviceps,* contains numerous species that parasitize insects. Adult insects or larvae that are attacked are mummified, the body essentially being transformed into a sclerotium. Later, stalked stromata arise from the sclerotized insect body (Fig. 6-31).

Series Discomycetes

Discomycetes produce apothecia with asci arranged in an exposed, palisadelike layer, the hymenium. Although many apothecia are disc- or cup-shaped, other forms exist; they range in size from minute, almost microscopic structures to large forms such as morels. Three orders are discussed here; two of them (Helotiales and Lecanorales) have inoperculate asci, and the third (Pezizales) has operculate asci. The largest group, the Lecanorales, includes all of the lichen-forming discomycetes.

Order Helotiales The Helotiales are apothecial fungi with inoperculate asci; the ascus opens by a pore or irregular tear. The generally small apothecia are common on wood, dead leaves, or other plant materials; some species are destructive plant parasites.

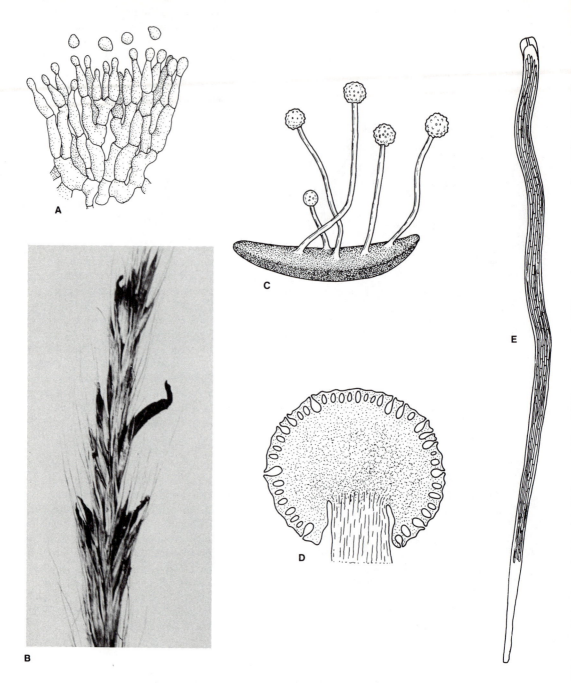

Figure 6-30 *Claviceps purpurea.* **A,** pallisadelike layer of conidiophores and conidia in infected grass flowers, ×805. **B,** sclerotia or "ergots" in grass inflorescence, ×1.5. **C,** germinated sclerotium with stalked stromata, ×3. **D,** section through stroma and embedded perithecia, ×36. **E,** ascus and ascospores, ×1500.

Figure 6-31 Stalked stromata of *Cordyceps*, ✕2.

A   B

Figure 6-32 *Bisporella*. **A,** habit of apothecia on decaying wood, ✕1.7. **B,** mature ascus, ✕760.

Apothecia of *Bisporella* (Fig. 6-32A), common on decaying wood, have a short stalk, or stipe. The fertile zone, or hymenium, consists of asci and paraphyses (Fig. 6-32B). The asci have a minute, plugged pore, the contents of which stain bright blue in iodine solution.

Species of *Monilinia* and related genera infect fruits or other parts of many plants. *M. fructicola* (Fig. 6-33A–D) causes *brown rot* of various stone fruits. Infected fruits rot and eventually are transformed into shriveled *mummies*. Abundant conidia are produced on the fruit surface while active growth of the fungus occurs (Fig. 6-33C, D).

The mummies contain an extensive mycelium and, after overwintering, function as sclerotia. One or more stalked apothecia develop (Fig. 6-33A) from each mummy. The apothecia are similar to those of *Bisporella* in structure.

Botrytis cinerea (Fig. 6-33E), a widespread conidial fungus, has an infrequently seen apothecial stage of the type found in *Monilia*. This species is found on moribund plant materials of many types and is responsible for diseases of strawberry, grape, lettuce, and other plants. Infected grapes, called "botryotized grapes" are used to produce a highly prized wine.

Although apothecia of the type found in *Bisporella* and *Monilinia* are commonest among Helotiales, other forms do occur. For example, *Geoglossum* species produce spatula-shaped apothecia (Fig. 6-34), the hymenium of which covers the flattened upper portion. The black apothecia, often called *earth tongues,* occur on soil in moist habitats.

Order Lecanorales The largest order of lichens, Lecanorales, includes 8000 or more species, all of which produce apothecia. Species of this group occur on soil, rocks, or trees; the thalli are crustose, foliose or fruticose.

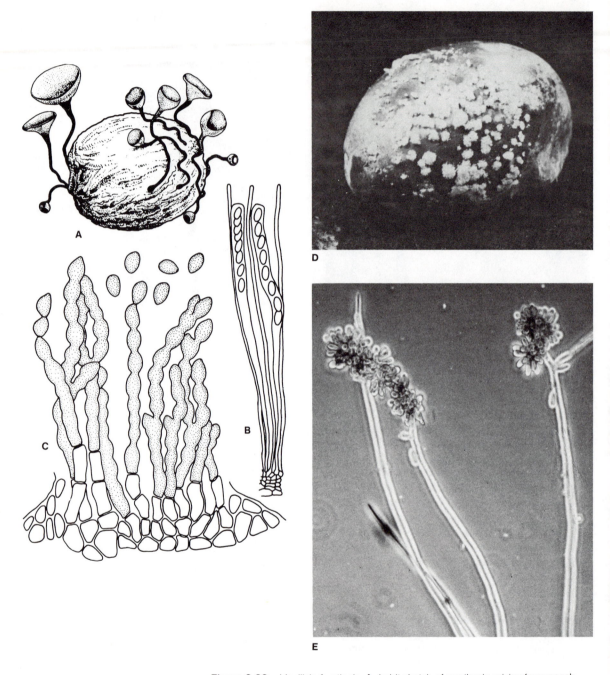

Figure 6-33 *Monilinia fructicola.* **A,** habit sketch of apothecia arising from peach mummy, ×1. **B,** asci and paraphyses, ×750. **C,** conidiophores and conidia, ×415. **D,** infected plum with conidiophores and conidia developing at surface, ×1.

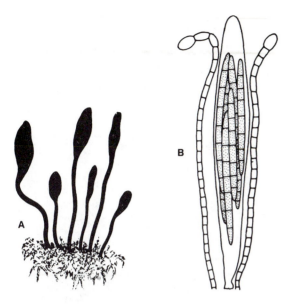

Figure 6-34 *Geoglossum.* **A,** ascocarp, ×0.5. **B,** asci and paraphyses, ×460.

The apothecia of lichens develop slowly and, unlike most nonlichenized apotheciate species, continue to produce asci and spores over a period of years. The hymenium consists of asci and paraphyses, the latter often having branched apices that form a tough covering over the asci.

Foliose thalli of *Physcia* (Fig. 6-35A–C) occur on trees or rocks. The greyish or green thalli bear small, sessile apothecia with dark-colored hymenia. These apothecia resemble those of Helotiales, but cortical and algal layers extend upward around the ascocarp margin.

Cladonia thalli (Fig. 6-36A, B) often combine characteristics of crustose, foliose, and fruticose types. Small, foliose scales or a crustose structure can be present at first, and the typical fruticose thallus then develops. The primary growth of scales or crustose structure often dies, leaving only the fruticose portion. Apothecia are borne singly on branch tips in some species; in others, a series of apothecia develop around the periphery of an erect, trumpet-shaped structure. The trumpet-shaped structure may proliferate, and apothecia are then present at several levels.

Letharia and *Usnea* (Fig. 6-36C, D) also have fruticose thalli, those of *Usnea* being pendent. *Letharia vulpinia* thalli contain large amounts of the

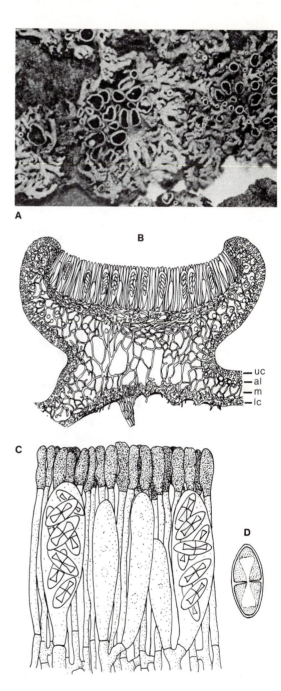

Figure 6-35 *Physcia.* **A,** thallus with apothecia, ×1. **B,** section through apothecium and portion of thallus (semi-diagrammatic), ×82, showing hymenium, subhymenium and a portion of the foliose thallus: *uc,* upper cortex; *al,* algal layer; *m,* medulla; *lc,* lower cortex. **C,** portion of hymenium, ×825. **D,** ascospore, ×1775.

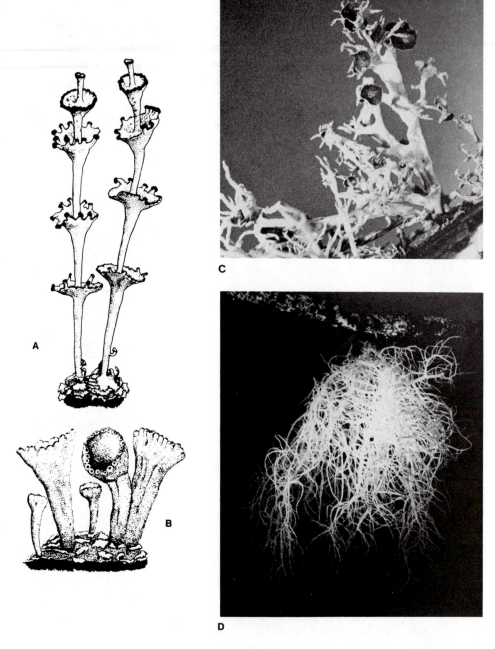

Figure 6-36 **A, B,** *Cladonia.* **A,** *C. verticellata;* **B,** *C. deformis,* both ×2. The small scalelike lobes characteristic of young thalli are visible on the substrate; apothecia are borne on the erect portion. **C,** *Letharia,* fruticose thallus on tree branch, ×1.2; note apothecia. **D,** *Usnea,* a pendant lichen, ×0.6.

toxin vulpinic acid, and are said to have been used to poison foxes in Scandinavia. *Usnea* thalli contain the more common lichenic acid, usnic acid.

Cladonia rangiferina, called *reindeer moss*, forms extensive stands on Arctic soils. This species and species of *Usnea* and *Alectoria* constitute an important part of the diet of some herbivores, especially during winter months.

Order Pezizales Many Pezizales produce apothecia similar in form to those of Helotiales; however, those of Pezizales are often larger; their asci open by an operculum or, less commonly, by an irregular tear. Most members of the order occur on dung or soil.

Apothecia of *Peziza* (Fig. 6-37E) on dung or soil often exceed 10 cm in diameter. These apothecia are usually bowl-shaped when young but tend to become flattened and distorted with age. The initiation of apothecium development around the ascogonium is shown in Figure 6-37A–C. The mature apothecium consists of a thick supporting layer (Fig. 6-37D) of pseudoparenchyma and the hymenial layer. At maturity, ascus tips of *Peziza* and related fungi are positively phototropic, bending toward light. Many asci in an ascocarp can mature but not immediately release their ascospores. If an ascocarp containing such asci is touched, a visible cloud of ascospores can be discharged; the release is often accompanied by an audible hiss.

If an apothecium has a stalk, or stipe, the fertile portion is called the **pileus.** In *Helvella* (Fig. 6-38A) the pileus is essentially discoid, but it is folded over the stipe apex and is often somewhat contorted. *Gyromitra esculenta* (Fig. 6-38B), the *brain mushroom*, has a hollow pileus with numerous grooves and ridges. These distortions greatly increase the surface area of the pileus. The hymenium of *Morchella* (Fig. 5-12) lines numerous large pits that are separated by sterile ridges.

Morchella species, called *sponge mushrooms* or *morels*, are among the most highly prized edible fungi. Although considered a choice edible mushroom in some parts of the world, *Gyromitra esculenta* can be extremely toxic. The toxins can be removed from pilei by boiling before use, or by cooking thoroughly.

In one order of ascomycetes (Tuberales), ascocarps are produced beneath the soil surface. Transitional forms suggest a close relationship between these subterranean forms and Pezizales. The best-known subterranean ascomycetes are fungi called *truffles;* their tuberlike ascocarps are hunted commercially with the assistance of trained dogs or pigs. Truffles have distinctive odors that attract animals, and the animals disperse the spores. In recent seasons, the best truffles have sold for up to $400 per pound in Europe.

Series Loculoascomycetes

Many Loculoascomycetes superficially resemble perithecial fungi, with which they were once included. However, all differ in having asci with a conspicuously two-layered ascus wall. They also differ in the manner in which the ascocarp develops, since their asci develop within chambers dissolved in a stroma. Lichens once considered as perithecial (Pyrenulales) are almost all loculoascomycetes.

Order Pleosporales The asci of the loculate fungi Pleosporales are produced in stromata that are small and often peritheciumlike in form. In most instances, the double nature of the ascus wall is apparent at the time of ascospore release. The inner wall layer of the ascus extends, eventually rupturing and protruding through the outer wall layer (Fig. 6-39C, D). Spores are then released through a pore at the tip of the inner wall layer.

Pleospora species are common parasites or saprobes on many herbaceous plants. Conidial stages, often of the type classified in *Alternaria* (Fungi Imperfecti, p. 241), can precede development of stromata.

Stromata of *Pleospora* (Fig. 6-39A, B) develop beneath the epidermis of herbaceous plants, maturation often occurring between fall and spring.

Venturia inaequalis, the cause of *apple scab* (Fig. 6-40A), is similar to *Pleospora*; it occurs wherever apples are grown and causes one of the most destructive diseases of apples. Initially, the hyphae penetrate only into the cuticle of leaves, fruits, and young twigs; a thin stroma is produced there. Elongate conidia (Fig. 6-40B) develop on

conidiophores arising from the stroma, and these conidia spread the infection. The presence of conidal stromata is responsible for the scablike lesions that give the disease its name. Hyphae later penetrate into deeper tissues to produce **ascostromata** (Fig. 6-40C). Ascostromata mature in the spring and release ascospores (Fig. 6-40D) that initiate the infection.

In *Apiosporina morbosa* (Dothidiales, Fig. 6-41) the extensive stroma develops on the branches of cherry, producing a disease called *black knot*. Much of the "knot" results from abnormal host cell division and enlargement in the infected tissue. Locules of *Apiosporina* develop in numerous, minute protrusions of the stromatic surface, and these contain asci and ascospores (Fig. 6-41B).

Order Pyrenulales Ascostromatic lichen fungi of the order Pyrenulales have crustose thalli. For example, species of *Verrucaria* form extensive crustose thalli on rocks; these are conspicuous just above the tide line along rocky sea coasts of the Northern Hemisphere. Hyphae of the thin, blackish thalli often penetrate the underlying stone to some depth. *Pyrenula* species grow mostly on bark, with the thallus and ascocarps embedded. The ascostromata are relatively large, and their presence causes the surface of the young bark to protrude (Fig. 6-42A). The ascospores are septate (Fig. 6-42C). Like many other lichens, *Pyrenula* species produce pycnidia. However, it is not known whether the spores produced in these can function in establishing the lichen elsewhere or are spermatizing agents.

Series Laboulbeniomycetes

Laboulbeniomycetes, with the single order Laboulbeniales, are obligate parasites of insects or, in a few instances, other arthropods. About 5000 species of these fungi have been described, but the total number is thought to be much larger. The laboulbeniomycete thallus is attached to the external surface of its host. It can reach a maximum of 3 mm in height but does not exceed 1 mm in most species.

Laboulbenia thalli (Fig. 6-43A, B) illustrate a number of features common to most species. Attachment to the host is by means of a **foot cell** that is embedded in the chitinous exoskeleton; a haustorium is often present at the base of the foot cell. The cells composing the thallus have relatively thick walls, and adjacent cells are interconnected by large pores. Male and female gametangia develop on a single thallus or on separate thalli, depending upon the species. The ascogonium is equipped with a trichogyne, and fertilization occurs through transfer of spermatia to this structure. After nuclear transfer, asci develop at the base of the ascocarp, and two-celled ascospores are produced. Upon contacting an insect and germinating, one cell of the spore develops into the foot and the other produces the remainder of the thallus.

The laboulbeniomycetes are unique among parasitic fungi in the high degree of specificity shown. Species are generally able to parasitize only a single host species; in many instances, species can attack only male or female insects. Finally, species can also be position-specific and then are always found attached to the host in the same position. The host does not appear to be greatly harmed or hampered by these parasites.

Subdivision Basidiomycotina

Most large, conspicuous fungi encountered in woods and fields—the mushrooms, puffballs, and bracket fungi—are representatives of the subdivision Basidiomycotina. In addition to conspicuous wood- and soil-inhabiting species, the subdivision includes numerous inconspicuous

Figure 6-37 *Peziza.* **A,** scanning electron micrograph of ascogonium near conidiophore, ×2150. **B, C,** early stages in development of monokaryotic hyphae around ascogonium to produce ascocarp. **B,** ×1110; **C,** ×1370. **D,** scanning electron micrograph of young, apothecial, freeze-fractured preparation, ×54: *h,* hymenium. **E,** habit of ascothecium of *Peziza,* ×1. (**A–D,** courtesy of K. L. O'Donnell, G. R. Hooper, and A. O. Ackerson, reproduced by permission of the National Research Council of Canada from the *Canadian Journal of Botany* 54:2254–67, 1976.)

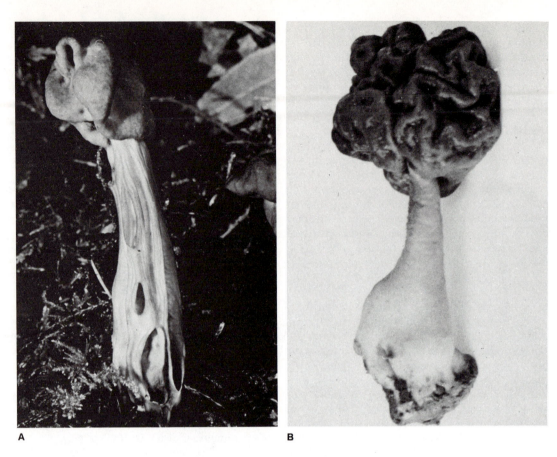

A **B**

Figure 6-38 **A,** ascocarps of *Helvella lacunosa*, ×0.8. **B,** ascocarp of *Gyromitra esculenta*, ×0.075.

forms, many of which are plant parasites. Basidiomycetes are primarily terrestrial fungi, but some aquatic species are known.

Assimilative Stage In basidiomycetes, germinating basidiospores can give rise to a yeast phase, a monokaryotic mycelium, or a dikaryotic mycelium. Yeast phases occur in some primitive genera of Heterobasidiomycetes. In these, conjugation of compatible yeast cells (see *Tremella*, p. 218) leads to the establishment of a dikaryotic mycelium similar to that of higher basidiomycetes.

In homothallic mycelial basidiomycetes, di-

karyotic hyphae arise directly from basidiospores. However, most basidiomycetes are heterothallic, and their basidiospores germinate by producing a monokaryotic mycelium (Fig. 6-44A). This **primary mycelium** can grow indefinitely in many basidiomycetes. Dikaryotization typically occurs through the fusion of compatible hyphae of primary mycelia; spermatia or conidia that function as spermatia bring about dikaryotization in some basidiomycetes.

In primitive basidiomycetes, such as Uredinales and Ustilaginales, thalli may all be + and − in heterothallic species. However, in other groups, compatibility is determined by multiple alleles at one or more loci.

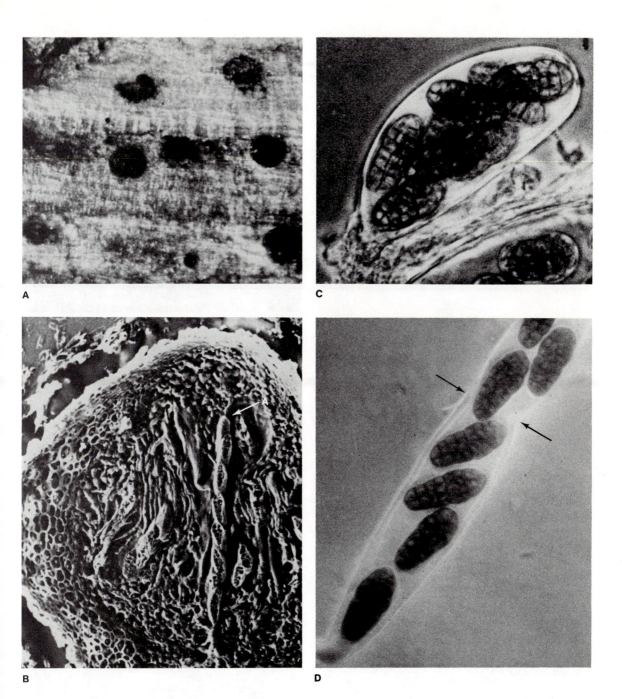

Figure 6-39 *Pleospora.* **A,** habit of ascostroma on dead herbaceous stem, ×22. **B,** scanning electron micrograph of section through stroma, ×350. **C, D,** asci before and after rupture of outer wall layer (*arrow* indicates edge of broken outer layer). **C,** ×585; **D,** ×665. (**B,** courtesy of M. Corlett, with permission of *Nova Hedwigia.*)

A

B

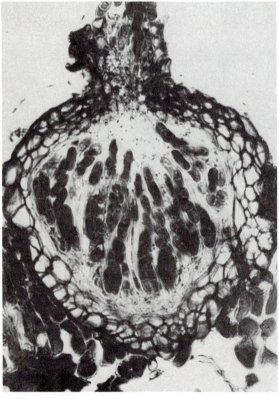

C

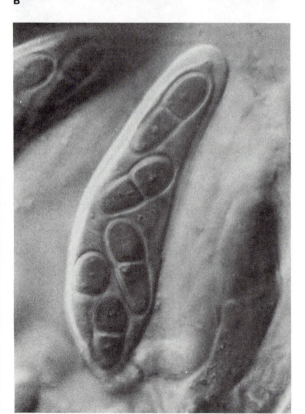

D

Following spermatization or hyphal fusions, nuclear migration takes place, and a dikaryotic, **secondary mycelium** is produced. Cells of the secondary mycelium (Fig. 6-44B) are binucleate. As with ascogenous hyphae, nuclear divisions of paired nuclei occur simultaneously, and septa typically form at this time. In many basidiomycetes, crozierlike structures, called **clamp connections** are produced during divisions (Fig. 6-44C–F). The clamp connection starts as a subterminal branch, rather than by the bending back of the hyphal apex. After division of the two nuclei, septa develop as in croziers.

As in the Euascomycetes, most basidiomycetous hyphal walls contain glucans and chitin. As in that group also, the walls of budding forms contain mannans. Septal pores of many primitive basidiomycetes are simple, but most have the complex type shown in Figure 6-1B, C.

The secondary mycelium, like other mycelia, grows out radially from the initiation point. Mycelia growing in wood, dung, or similar substrates are limited in their spread by the size and shape of the substrate. However, in perennial, soil-inhabiting species, the buried mycelial mass increases in diameter each year; older parts at the center eventually die. Crops of basidiocarps on the surface, called *fairy rings* (Fig. 6-45), mark the form and extent of the underground mycelium. By measuring the annual increase in diameter, and the total diameter of the ring, one can accurately estimate how long the mycelium has been growing. In undisturbed soils, this often exceeds 500 years.

Asexual Reproduction In its details, budding in asexual reproduction of basidiomycetous yeasts differs from budding in Hemiascomycetes, but the end result is the same. Conidia can be produced on the primary mycelium, the secondary mycelium, or both. Some primitive basidiomycetes produce several types of asexual spores (e.g., Uredinales and Tremellales). In general, however, asexual reproduction plays a less prominent role in the life histories of basidiomycetous fungi than in those of Ascomycotina.

Sexual Reproduction The initial phase of sexual reproduction occurs when the dikaryotic mycelium is initiated. Production of sexual spores can commence soon after initiation of the dikaryotic phase in simple basidiomycetes. However, in most basidiomycetous fungi, the secondary mycelium constitutes the main assimilative phase, and a long period of growth occurs before sexual spores are formed. After such growth has occurred, environmental factors such as temperature, moisture, and light are important in bringing about the formation of sporocarps.

Basidiocarp Formation. The simple basidiomycetes produce no sporocarps; their basidia are borne directly on dikaryotic hyphae, or they develop from thick-walled spores produced by the hyphae. However, most basidiomycetes produce sporocarps, specifically **basidiocarps.** These are produced in annual crops by many perennial mycelia.

In contrast to ascocarps, basidiocarps are entirely composed of dikaryotic hyphae. Basidia are borne on or in the basidiocarp, either irregularly or in a definite hymenial layer. Sterile parts of the basidiocarp are generally composed of recognizable hyphae, but pseudoparenchyma can be present. In addition, numerous modifications of the hyphae occur in the basidiocarp.

Basidia. The cytological events that occur in the developing basidium (Fig. 6-46A–F) are similar to those in the ascus. However, meiosis is generally not followed by mitosis, and (typically) four basidiospores are produced. Basidiospores develop externally (Figs. 6-46F and 6-47), usually on narrow, tubular structures called **sterigmata.** One haploid nucleus and a portion of the

Figure 6-40 *Venturia inaequalis.* **A,** scablike, conidial lesions on apple, ×0.9. **B,** scanning electron micrograph of conidiophores and conidia, ×1575. **C,** transverse section through host leaf and ascostroma of *Venturia,* ×540. **D,** ascus and ascospores, ×1450. (**B,** courtesy of M. Corlett, from *Canada Agriculture,* Spring 1970.)

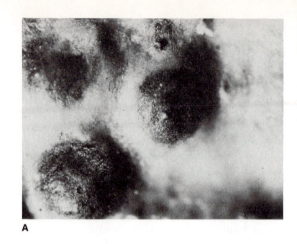

A

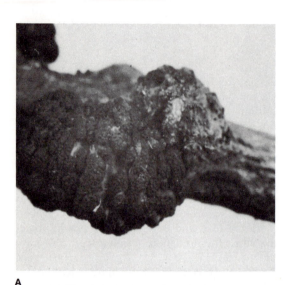

A

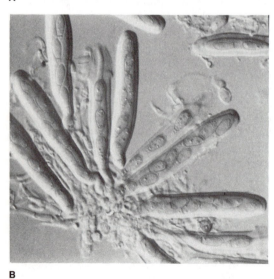

B

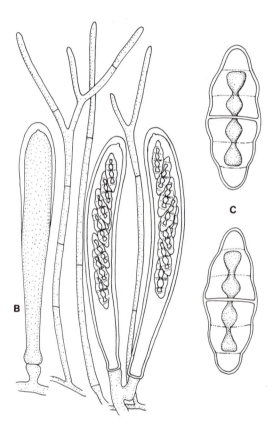

Figure 6-41 *Apiosporina morbosa.* **A,** photograph of stem showing characteristic "black knot" appearance, ×3. **B,** cluster of asci from single locule, ×695.

Figure 6-42 *Pyrenula.* **A,** surface view of thallus embedded in tree bark; the dark swollen areas are the sites of embedded ascocarps, ×29. **B,** asci, ×550. **C,** two ascospores, ×1590.

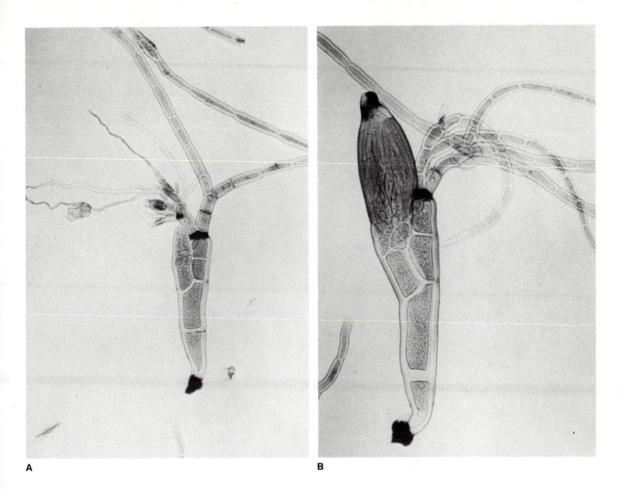

A

B

Figure 6-43 *Laboulbenia.* **A,** immature thallus, ×196 approx. **B,** mature thallus, ×250 approx. Note that the trichogyne has degenerated at this stage of development. (Courtesy of R. K. Benjamin.)

basidial cytoplasm pass through the sterigmata into each developing basidiospore.

The mature basidiospores of many basidiomycetes are shot from the basidium. Immediately before discharge, a bubblelike protrusion develops on the spore near its point of attachment. The appearance of this bubble or droplet has given rise to the phrase *water-drop abstriction mechanism,* but the actual mechanism has not yet been satisfactorily explained. Spores are shot away with a force sufficient to carry them free of the hymenium; gravity and air currents then carry out dispersal.

Classification The subdivision Basidiomycotina is divided into two classes, Heterobasidiomycetes and Homobasidiomycetes. The basidia of most heterobasidiomycetes are either septate or deeply divided; they arise from thick-walled resistant spores in many species. The basidia of homobasidiomycetes are typically club-shaped and nonseptate; they do not arise from resistant structures. The basidiospores of many heterobasidiomycetes germinate by budding, by forming conidia, or by forming secondary basidiospores. The latter are produced on sterigmata and are fired away in the same manner as

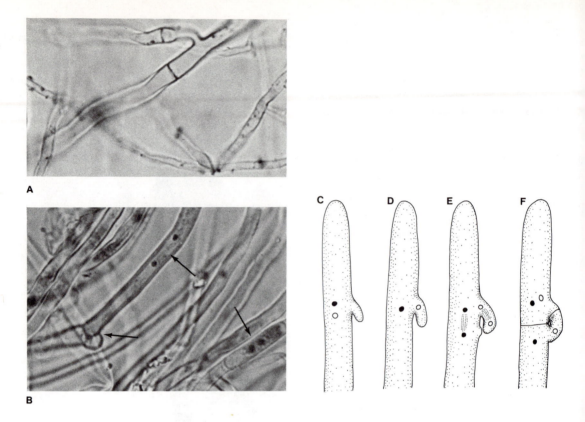

Figure 6-44 **A,** monokaryotic primary mycelium with simple septa, × 1185. **B,** secondary mycelium of same species, showing paired nuclei and clamp connection (*arrows*), × 1185. **C–F,** diagrams of clamp connection formation. **C, D,** hyphal apex with loop of clamp starting to develop; **E,** nuclear division; **F,** completed clamp connection after septum formation.

basidiospores. Basidiospores of homobasidiomycetes typically germinate by germ tubes.

Class Heterobasidiomycetes

Heterobasidiomycetes include Tremellales, or *jelly fungi*, and two of the most important groups of plant parasites: the *rusts* (Uredinales) and *smuts* (Ustilaginales). Tremellales are mainly wood-inhabiting decay fungi, but some species are parasites of other fungi, plants, and insects.

Order Tremellales The name *jelly fungi* is derived from the gelatinous consistency of the basidiocarps of many species. In gelatinous basidiocarps, hyphae are embedded in a tough, gelatinlike matrix.

In *Tremella* (Fig. 6-48A–G), the gelatinous basidiocarps are cushion-shaped to irregularly folded or lobed. Most of the exposed surface is covered by a hymenium consisting only of basidia or of these structures and conidiophores. Within the roughly globose basidium, karyogamy and meiosis occur; the basidium becomes divided into four compartments during meiotic divisions (Fig. 6-48B–D). After division is complete, a tubular extension and sterigma develop from each of the compartments. The extensions protrude beyond the gelatinous matrix, and the basidiospores are shot away at maturity.

Species of *Tremella* are dimorphic, and iso-

Figure 6-45 Fairy ring of *Marasmuis oreades*, ×0.09. The ring of basidiocarps marks the presence of a buried dikaryotic thallus.

lated basidiospores germinate by budding (Fig. 6-48E, F) to produce yeastlike colonies. If compatible strains are mixed, conjugation (Fig. 6-48G) occurs, and a dikaryotic mycelium develops.

Dacrymyces basidiocarps (Fig. 6-49A) resemble those of *Tremella*, but their basidia are nonseptate and resemble tuning forks (Fig. 6-49B). Each basidium bears two curved, septate basidiospores (Fig. 6-49C). Conidia (Fig. 6-49D) can also be produced in the basidiocarp or in an asexual sporocarp of similar appearance.

More highly developed basidiocarps of some Tremellales, such as *Auricularia* (Fig. 6-50A), are divided into sterile and fertile regions. The earlike basidiocarps of *Auricularia* grow out laterally from trees or logs. The upper surface is covered by sterile hairs, and the hymenium is restricted to the lower surface. Such an arrangement protects the hymenium from wetting and therefore helps to prevent entrapment of basidiospores in water films on the basidiocarp. Basidia of *Auricularia* (Fig. 6-50B, C) are cylindrical and transversely septate, as in the Uredinales and some Ustilaginales.

Order Uredinales Commonly known as *rusts*, Uredinales are obligate parasites of numerous vascular plants. Hyphae of rusts are intercellular and have haustoria that penetrate host cell walls (Fig. 6-51). As many as five types of spores are produced during the life histories of some rusts, and many species require two different host species in order to complete their development. Where two hosts are required, they are always from different plant groups: a gymnosperm and an angiosperm, a fern and a gymnosperm, or a monocot and a dicot.

One of the most destructive species of fungi, and one that has been the subject of numerous studies, is the wheat rust fungus, *Puccinia graminis* The life history of *P. graminis*, cause of *stem rust* of wheat and other small grasses (Fig. 6-52), is typical of more complex rusts. Basidiospores are released in early spring and infect leaves of the common barberry, *Berberis vulgaris*. Small, flask-shaped structures, called **pycnia** (Fig. 6-53A), then develop mainly on the upper surface of the leaves. Pycnia produce numerous minute **pycniospores** (also called *spermatia*) as well as special **receptive hyphae.** Pycniospores are exuded in a sugary liquid, called honeydew, and are dispersed by insects.

Each pycnial thallus is one of two mating types, + or −. In order for further development to occur on barberry leaves, + pycniospores must be transferred to − receptive hyphae or vice versa (spermatization). Transferred pycniospores adhere to receptive hyphae, pores develop where contact occurs, and the pycniospore nucleus migrates into the receptive hypha to initiate the dikaryotic stage.

The second structure produced on the barberry leaves, called an **aecium** (Fig. 6-53B), is dikaryotic. Aecial initials are present before spermatization, but they do not complete their development until dikaryotization has occurred. Formed mainly on the underside of the leaf, aecia produce dikaryotic spores, called **aeciospores.** Aeciospores develop in chains; their continued production eventually causes the aecial wall and host epidermis to rupture. These dikaryotic spores are airborne and only infect wheat or related grasses.

Soon after infection of the wheat plant, subepidermal structures called **uredia** develop (Fig. 6-53C). These are blisterlike and produce rust-colored **uredospores,** from which the common name is derived. The host epidermis ruptures

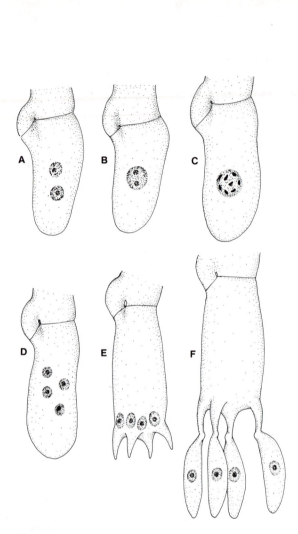

Figure 6-46 Diagram of basidial development (Homobasidiomycetes). **A,** young binucleate basidium. **B,** diploid basidium. **C, D,** meiosis. **E,** development of sterigmata. **F,** basidium and developing spores.

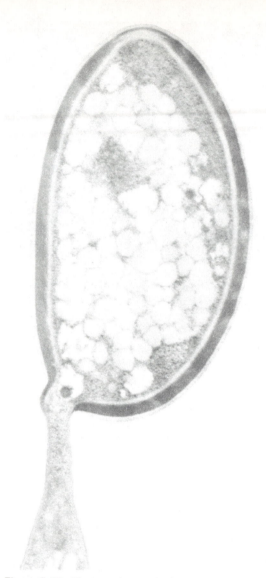

Figure 6-47 Electron micrograph of section through basidiospore and sterigma, × 14,730. (Photograph courtesy of K. Wells.)

and exposes the uredospores, which are then dispersed by air currents. Uredospores are repeating spores; that is, they spread the disease to other wheat plants during the growing season. Like aeciospores, uredospores are dikaryotic.

As the host plant approaches maturity, **teliospores** are produced. These can occur in uredia, but they commonly develop in separate structures called **telia** (Fig. 6-53D). Teliospores remain attached to wheat culms and are the resistant, overwintering stage of the fungus. Each teliospore of *P. graminis* has two binucleate compartments; karyogamy occurs within each compartment soon after the spore has formed. In spring, teliospores germinate *in situ,* each cell producing a basidium similar to that of *P. mal-*

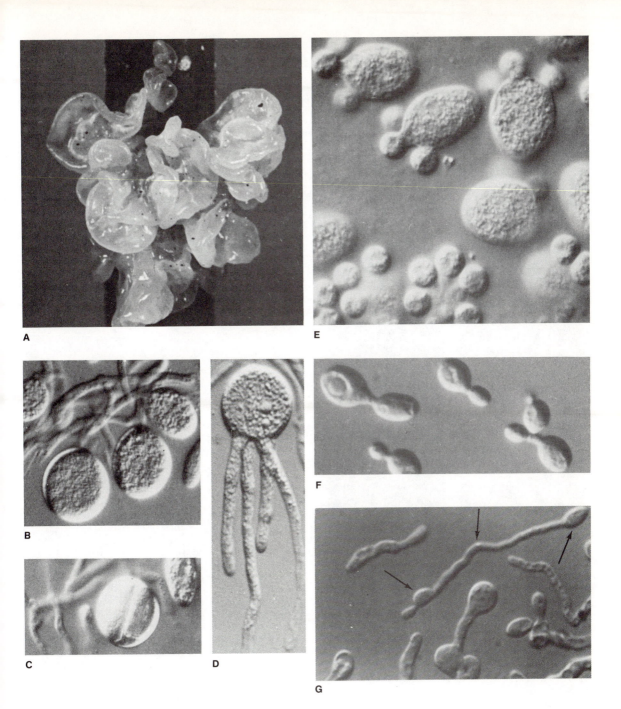

Figure 6-48 *Tremella.* **A,** basidiocarp, ×1. **B, C,** stage in the development of basidia, ×945. **D,** basidium just before basidiospore production, ×1025. **E,** basidiospores germinating by budding, ×1475. **F,** budding haploid cells, ×1750. **G,** conjugating cells, 12 hours after mixing compatible isolates (*arrows* indicate original cells and point of fusion), ×1070.

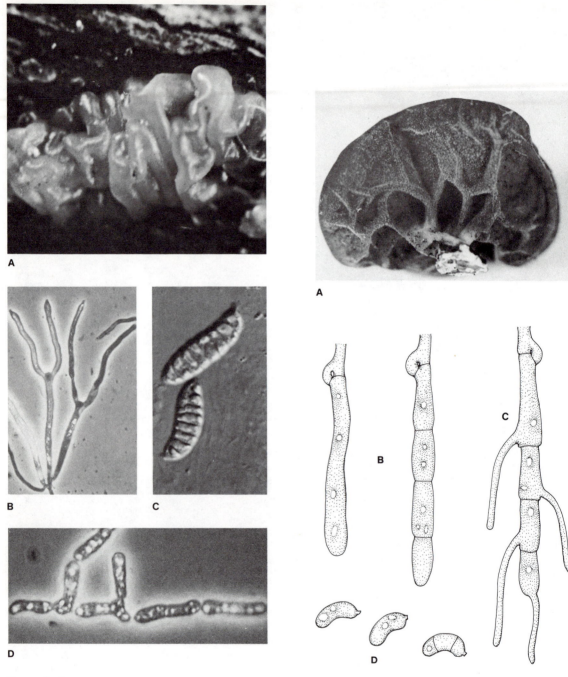

Figure 6-49 *Dacrymyces.* **A,** basidiocarp, ×3.5. **B,** basidia, ×515. **C,** basidiospores, ×1120. **D,** conidia, ×920.

Figure 6-50 *Auricularia.* **A,** basidiocarp, ×2. **B, C,** young and mature basidia, ×895. **D,** basidiospores, ×895.

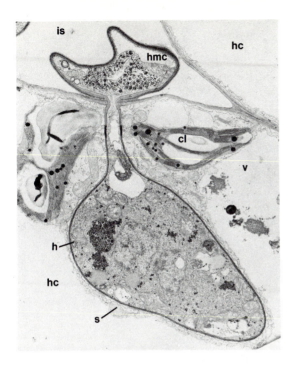

Figure 6-51 Electron micrograph of a rust haustorium, ×7670: *hmc*, haustorial mother cell; *hc*, host cytoplasm; *cl*, chloroplast; *h*, haustorium; *is*, intercellular space; *s*, electron transparent sheath. (Courtesy of M. D. Coffey, B. A. Palevitz, and P. J. Allen, reproduced by permission of the National Research Council of Canada from the *Canadian Journal of Botany* 50:23–40, 1972.)

vacearum (Fig. 6-54B). The basidium becomes four-celled during meiosis, each cell producing a sterigma and spore.

Basidiospores of rusts are shot from their basidia and are airborne. Many are capable of germinating by forming secondary basidiospores if conditions or substrates are unfavorable for further development.

Because of its function as an alternate host for *P. graminis*, the common barberry has been eradicated in much of North America. However, eliminating the alternate host has not prevented serious losses of grains through stem rust infection. Uredospores are sufficiently resistant to survive mild winters in the southernmost wheat-growing regions of North America. Infected plants occur progressively northward during the growing season, the infections originating from windborne uredospores. The elimination of the alternate host has prevented the genetic variability associated with sexual reproduction, but variation still occurs through a type of parasexual activity.

As in many other highly evolved host-parasite systems, different resistant varieties of hosts and virulent strains of rusts occur. Genes determining resistance in the host are matched by a series of virulence genes in the parasite. Infection can occur only if the rust strain has the correct virulence genes to overcome resistance genes in the particular wheat variety.

Puccinia malvacearum, which infects mallows (hollyhock, hibiscus, etc.) has a greatly reduced life history. It produces teliospores (Fig. 6-54A, B) similar to those of *P. graminis*, but they soon germinate, and basidiospores infect new mallows of the same species. Only those teliospores formed late in the growing season function as resistant spores; they produce basidia and basidiospores the following spring. The dikaryotic stage is thought to be initiated by hyphal fusions between adjacent, compatible mycelia in the host.

The life history of *Cronartium ribicola*, the cause of *white pine blister rust*, resembles that of *Puccinia graminis* in that two hosts are required, and the same spore stages are produced. *C. ribicola* produces spermatia and aeciospores on white pines, and its uredial and telial stages develop on currant or gooseberry plants. The thin-walled teliospores of this species germinate soon after they are produced; overwintering occurs by way of mycelium in the coniferous host. The name *blister rust* is based upon the large orange, blisterlike aecia (Fig. 6-55).

C. ribicola is native to Europe and was accidentally introduced into North America on diseased nursery stock. Since its introduction around 1900, the fungus has spread over much of the range of eastern and western white pine species. It threatens the existence of all native white pines, although some control could be achieved by eliminating alternate hosts.

Order Ustilaginales Most smuts, like rusts, are parasites of vascular plants. Haploid stages of the parasitic smut fungi can exist as saprobes, and they grow readily in culture. Dikaryotic phases of parasitic species grow less readily in culture and, in nature, are obligate parasites. Saprobic forms are now known that grow and reproduce sexually in culture.

All parasitic smuts complete their development on a single host plant species, and all form only one type of dikaryotic spore. The resistant, dikaryotic spore, called a *teliospore*, is thought to be homologous to that of the rusts. Like that of the rusts, it is at first dikaryotic, then diploid; upon germinating, it produces a cylindrical basidium. In *Ustilago*, the largest genus of smuts, the life cycle generally is as outlined in Figure 6-56. The basidium of *Ustilago* typically is four-celled, and it produces many basidiospores. These develop as successive buds from the basidial cells (Fig. 6-57); upon release, they continue to bud. A dikaryotic mycelium is produced if conjugation of budding cells occurs on the surface of a suitable host. Infection of the host by a monokaryotic hypha either does not occur or is of limited duration in most species.

Teliospores of some *Ustilago* species adhere to host seeds and infect seedlings at germination. In other species, flowers of the host are infected by budding cells, and ovules become smutted during the same season. Alternatively, infection is through the stigma, and an infection hypha grows down the style and into the developing embryo. There, a bit of viable hypha remains and commences growth at the time of seed germination.

Within the developing host plant, the growth of smut hyphae often keeps pace with that of the host. The hyphae are intercellular, with haustoria. In **meristematic** zones, infected flowers, or other portions of the host, masses of teliospores are produced. The hyphae in the mass consist at first of numerous, short, dikaryotic cells (Fig. 6-58C). Each cell becomes swollen, and a new wall is deposited within the old hyphal wall, forming a teliospore (Fig. 6-58D, E). Disintegration of host parts eventually exposes the powdery mass of dark-colored teliospores.

The basidia in a single collection of teliospores usually show a variety of forms. However, some structural features are constant, and basidial form is important in smut classification. For example, basidia of *Tilletia* (Fig. 6-59A) are cylindrical and are either nonseptate or have a single, transverse septum. The narrow, curved

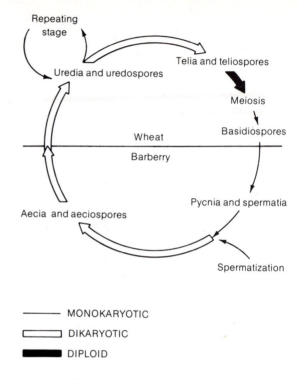

Figure 6-52 Outline of the life history of *Puccinia graminis*.

spores (Fig. 6-59B) are borne terminally and are shot from the basidium; they often conjugate while still attached to the basidium. The mycelium of *Tilletia* often bears conidia that are produced on sterigmata and are fired away in the same manner as the basidiospores of other basidiomycetes (Fig. 6-59C).

As in *Ustilago*, teliospores of *Tilletia* are produced in masses in the host tissue and are powdery when exposed. In some smuts, teliospores are scattered in the host tissue and germinate *in situ*.

Smuts parasitize many important plants, especially cereal grasses. Since these fungi commonly infect inflorescences or fruits, yields are often greatly reduced. Species of *Ustilago* and *Tilletia* parasitize wheat, barley, oats, corn, and other grains; annual losses amount to many millions of bushels. *Ustilago maydis* (Fig. 6-58B), the cause of corn smut, accounts for more than one-fourth of the yield loss from all corn diseases.

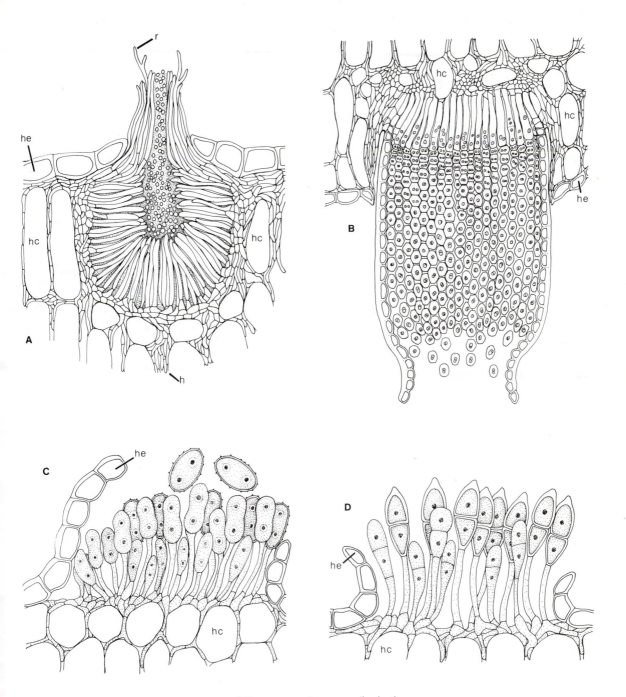

Figure 6-53 *Puccinia graminis.* **A,** pycnium, ×985: *sp,* spermatia; *r,* receptive hyphae; *he* host epidermis; *hc,* host cell; *h,* hyphae. **B,** aecium and aeciospores, ×220. **C,** uredium and uredospores, ×345. **D,** telium and teliospores, ×345.

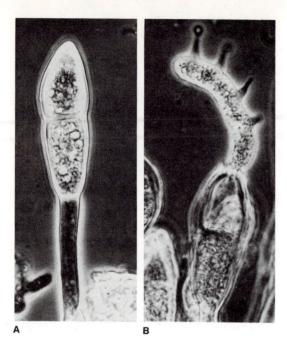

A   B

Figure 6-54 *Puccinia malvacearum.* **A,** mature teliospores, ×560. **B,** germinating teliospore with basidium and developing basidiospores, ×705.

Figure 6-55 Blisterlike aecial stage of *Cronartium ribicola* on *Pinus,* ×0.75.

Several genera of yeastlike saprobic fungi are now known to have life histories and structural features essentially like those of *Ustilago.* These occur in water, soil, and on surfaces of plants and animals.

Class Homobasidiomycetes

Homobasidiomycetes are mostly saprobes that live in soil, dung, litter, and wood. Their decay of litter and wood is beneficial and vital in nature; in the case of wood, it can also lead to great financial loss of standing and sawn timber. Few homobasidiomycetes parasitize green plants, but many form mycorrhizal associations with plant roots. Some basidiomycetes are also associated with algae in lichens, but the number of these is small.

Life Histories The life histories of homobasidiomycetous fungi generally are simpler and more uniform than in the heterobasidiomycetes. The species that have been examined are predominantly heterothallic, and the life history diagrammed in Figure 6-60 is typical of many of these. Dikaryotization is by hyphal fusion and nuclear transfer in most species; the dikaryotic mycelium is the main assimilative phase in this group.

Conidia can develop on the monokaryotic hyphae, on dikaryotic hyphae, or both. However, many homobasidiomycetes produce no conidia.

Mycorrhizae Mycorrhizae, literally "fungus roots," are associations between fungal hyphae and plant roots (Fig. 6-61). Mycorrhizal associa-

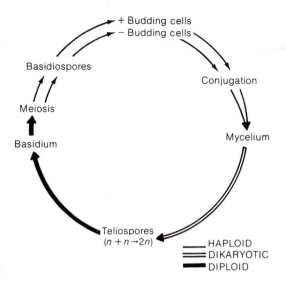

Figure 6-56 Diagram of life history of *Ustilago*.

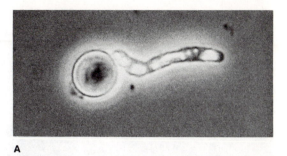

A

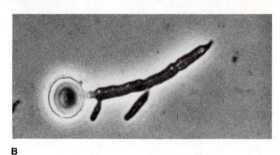

B

Figure 6-57 *Ustilago.* **A, B,** germination of telispore and development of basidiospore, × 1250.

tions can involve species of lower fungi (see *Endogone*, p. 166), ascomycetes, or basidiomycetes. Plants associated with these fungi include species from most vascular plant families; similar associations also occur with some bryophytes. In mycorrhizal associations, the roots of the plant are invaded; other underground structures can also be invaded. Basidiomycetous mycorrhizae are of several types. In the case of forest trees (e.g., most conifers, beeches, oaks), the fungus hyphae form a dense mantle around the root (Fig. 6-61B); hyphae also penetrate between cells of the cortex.

Root morphology is altered by the presence of the mantle; root hairs are absent and the branching pattern is also distinctive. The absence of root hairs is more than compensated for in terms of surface area by hyphae extending out into the surrounding soil. Basidiomycetes in these mycorrhizae include many common mushrooms, puffballs, and related basidiomycetes. Mycorrhizal basidiomycetes obtain sugars and other nutrients from their associated green plants. The fungus absorbs minerals and water from the surrounding soil, and these are supplied to the tree. Mycorrhizae of this type are best developed in nutrient-poor soils where they are necessary to the establishment and normal growth of forest trees.

Orchid seedlings are dependent upon the establishment of suitable mycorrhizal associations in nature. The seed itself contains virtually no food, and the seedling is dependent upon sugars provided by its mycorrhizal associate. Mature orchids have varying amounts of chlorophyll and probably are less dependent upon their mycorrhizal associates. Owing to their lack of chlorophyll, Indian pipe (*Monotropa*) and other "saprophytic" plants are dependent upon their mycorrhizal associates even as mature plants.

Classification Homobasidiomycetes can be divided into two series on the basis of general basidiocarp features: "Hymenomycetes" and "Gasteromycetes." In Hymenomycetes, basidia are arranged in a hymenial layer that is exposed at maturity. These fungi **abstrict** their basidiospores, and spores are then wind-dispersed. Gasteromycetes have basidia that are enclosed at maturity and are often scattered rather than in a hymenial layer. Basidiospores of gasteromycetes are released by enzymes that digest the basidia.

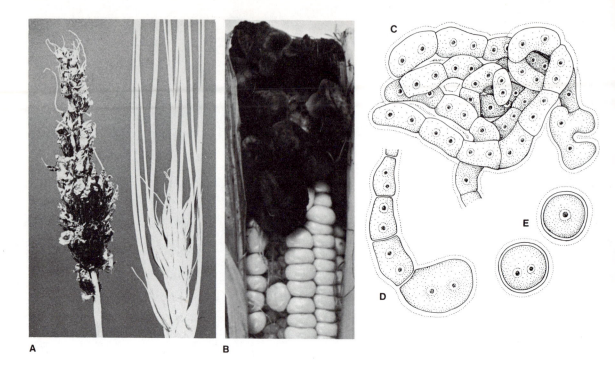

Figure 6-58 A, smutted and healthy barley inflorescences, the latter attacked by *Ustilago hordei,* ×1. **B,** corn smut, caused by *Ustilago maydis,* ×0.75; the infected kernels, top of ear, are greatly enlarged and contain numerous teliospores. **C–E,** development of teliospores in *Ustilago.* **C,** dikaryotic hyphae, ×3000; **D,** early stage in development of teliospore, ×3000; **E,** mature teliospores, ×3000.

Once released, basidiospores are dispersed by a variety of mechanisms and methods. Hymenomycetes include most mushrooms, bracket fungi, coral fungi, and tooth fungi; gasteromycetes are puffballs, stinkhorns, and bird's nest fungi.

Series Hymenomycetes

Order Aphyllophorales The Aphyllophorales are predominantly wood- and soil-inhabiting fungi. A few are important plant parasites, but these fungi are of much greater significance because of their wood-decay activities. They are most abundant in forested areas, and some are important mycorrhizal fungi.

Basidiocarps of aphyllophoraceous fungi can be fleshy, tough, corky, or woody; many can dry, then revive upon wetting. The simplest basidiocarps in the order consist of little more than a hymenium appressed to wood or other substrates. Such basidiocarps are usually of indeterminate size and continue to spread on the substrate if environmental conditions are favorable. This **effused** basidiocarp (Fig 6-62A) is characteristic of many Aphyllophorales, including *Peniophora* and related genera. In *Peniophora,* the structure includes the hymenium and a relatively thick supporting layer (Fig. 6-62A, B). Effused basidiocarps are the ultimate in economy, since there is very little waste in their construction. They usually occur on the undersides of logs or branches, where abstricted

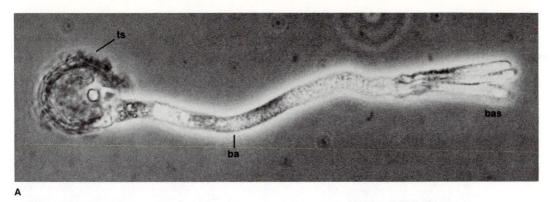

A

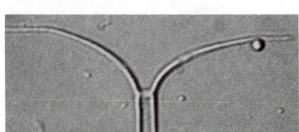

B

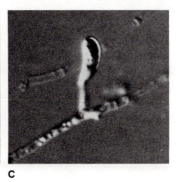

C

Figure 6-59 *Tilletia.* **A,** germinated teliospore, × 1225: *ts,* teliospore; *ba,* basidium; *bas,* basidiospores. **B,** two basidiospores connected by conjugation tube, × 1470. **C,** mycelium with sterigma and developing conidium, × 1470.

basidiospores will fall free of the hymenium. The slightly more complex basidiocarp of *Stereum* (Fig. 6-63A, B) is partially effused, but it has a shelving (reflexed) portion as well. The upper surface of this **effused-reflexed** fructification is sterile; the hymenium occurs on the lower surface and on effused portions. A reflexed part affords some protection against wetting of the hymenium by rain and thus increases the efficiency of spore dispersal.

Fomes (Fig. 6-63C) has sessile, pileate basidiocarps that are laterally attached to the substratum. The pileus of *Fomes* and those of **stipitate** forms, such as *Polyporus* and *Hydnellum* (Fig. 6-64A, B), possess a sterile, upper zone called the **context.** Pileate basidiocarps have their hymenia only on the lower surface.

Basidiocarps of some Aphyllophorales are erect and club-shaped or shrublike, as in *Clavulina* (Fig. 6-65). In these **coralloid** basidiocarps, the hymenium covers most of the exposed surface apart from a sterile, basal portion.

Basidiocarp texture varies, not only with the species, but with age and moisture content. However, some differences in texture result from differences in the kinds of hyphae making up the basidiocarp. Corky or woody basidiocarps are composed largely of thick-walled hyphae; fleshy basidiocarps are composed of thin-walled hyphae that are often inflated.

The hymenium of *Peniophora* and *Stereum* is smooth. In other genera, the hymenium is borne upon toothlike spines (Fig. 6-66A, B), as in *Hydnellum,* or it lines small tubules, as in *Fomes* and *Polyporus* (Fig. 6-66C, D). A few Aphyllophorales also have their hymenia on gills, a more common

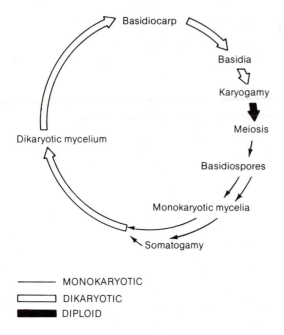

MONOKARYOTIC
DIKARYOTIC
DIPLOID

Figure 6-60 Diagram of characteristic life cycle of Homobasidiomycetes.

A

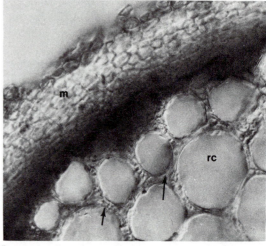

B

Figure 6-61 Mycorrhizae. **A,** ectomycorrhizal root, showing characteristic, stubby, lateral branches, ×1.6. **B,** section of root, showing hyphae (*arrows*) penetrating between root cortical cells, ×190; note thick mantle (*m*) of fungal cells at surface of root: *rc*, root cortex.

feature in Agaricales. Teeth, pores, and gills greatly increase the surface area of the hymenium and, consequently, permit the production of more basidia and basidiospores. The different hymenial types are not restricted to any particular basidiocarp form. That is, effused or effused-reflexed basidiocarps can have smooth, toothed, or poroid (in tubules) hymenia; pileate basidiocarps also can have any of these types of hymenia. Although hymenia themselves are microscopically similar, numerous differences exist in the types of basidia and sterile structures present. The most common sterile structures, called **cystidia,** are mostly of unknown function and occur in both Aphyllophorales and Agaricales (Figs. 6-62, 6-67). Cystidia are of many different types and, like basidia, are of some value in determining relationships.

Although effused basidiocarps require the least expenditure of protoplasm and other materials to construct, complex types are more efficient in spore dispersal. In coralloid and effused forms, the hymenium is subject to wetting by

rain, and water films entrap abstricted basidiospores. Moreover, many basidia are poorly positioned from the standpoint of spore discharge. In pileate species, the hymenium is protected from wetting, and the basidia are all directed laterally or downward. Abstricted basidiospores are fired away from the hymenium, fall clear of the pileus, and are carried away by air currents.

From the standpoint of basidiospore production, the large, perennial basidiocarps of *Ganoderma applanatum* (Fig. 6-68) are noteworthy. A single, large basidiocarp of this species has been estimated to produce about 30 billion spores per day while active. The age of basidiocarps of this

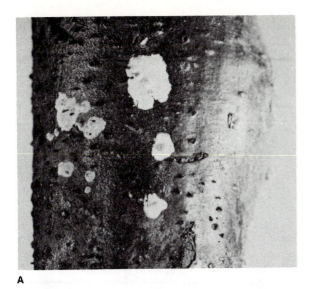

A

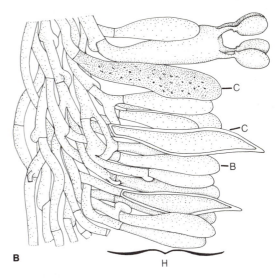

B

H

Figure 6-62 *Peniophora.* **A,** habit of effused basidiocarp on tree branch, × 1.5. **B,** vertical section through basidiocarp, showing hymenium and supporting layer: *h,* hymenium; *b,* basidium; *c,* cystidium; × 1050.

and other perennial species can be determined by examining the tube layers (Fig. 6-68).

Order Agaricales Most Agaricales have soft, fleshy basidiocarps that do not revive upon drying and subsequent rewetting. All have pileate basidiocarps that are either sessile, as in Pleurotus (Fig. 6-69A), or are provided with a stipe. The hymenium is borne on surfaces of **lamellae** (gills) in most species, but *Boletus* (Fig. 6-69B) and allied genera have poroid hymenia.

The development of many agaric and bolete basidiocarps commences with the appearance of a tuft or knot of hyphae near the substrate surface. This **primordium** continues to grow and develops into a miniature mushroom, or *button.* Stages in the development of basidiocarps of *Agaricus bisporus,* the common cultivated mushroom (Fig. 6-70A), are shown in Figure 6-70B–D. If the button is split lengthwise, lamellae and other parts are visible. Lamellae of these buttons are protected by a membranous layer, the **partial veil,** extending from pileus edge to stipe. As the button expands, the veil ruptures; its remnants usually remain attached to the stipe (Fig. 6-70D)

as an **annulus** or ring. Buttons of *Amanita* have another membrane, the **universal veil,** that encloses most of the basidiocarp. Remnants of the universal veil form a cuplike structure, the **volva** (Fig. 6-71C, D), near the base of the stipe and often form wartlike patches on the pileus. Only a universal veil is present in some genera; others have both this and the partial veil, or neither of these.

The development of mushrooms from primordia to mature basidiocarps is often rapid, but this "growth" is unusual in some respects. The primordium is analogous to a fruit, and it develops as a result of the transfer of materials from the buried assimilative structure. Additionally, much of the rapid increase in size from button to mature stage is the result of expansion of existing cells.

As in Aphyllophorales, a sterile context layer is present in the mushroom pileus (Fig. 6-70D). Basidiospores (Fig. 6-70E–G) develop at the gill or tubule surfaces; they are fired from their basidia and fall vertically. During elongation of the stipe, responses to light and gravity help to ensure that lamellae will be vertically oriented at maturity. The control of such responses is known

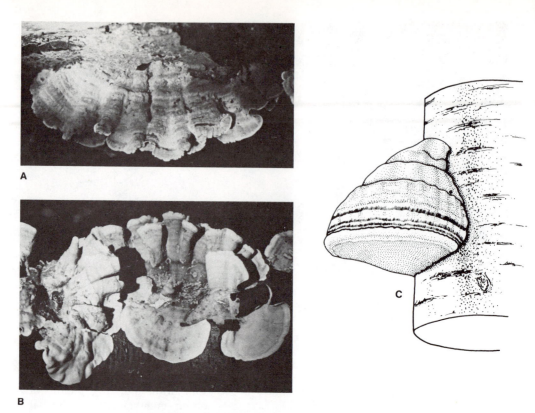

Figure 6-63 A, effused-reflexed basidiocarp of *Stereum,* viewed from above, ✕0.75. **B,** same basidiocarp as seen from the underside (note that only a portion of the basidiocarp is shelving, i.e., extends out from the substrate), ✕0.60. **C,** sessile pileus of *Fomes,* ✕0.5.

to be regulated by substances produced in the pileus or lamellae, but the chemicals involved have not yet been identified. Cystidia (Fig. 6-67) similar to those in Aphyllophorales are frequently present in the hymenium, on the gill edges, or on other surfaces of the basidiocarp.

Perhaps the most complex of hymenomycete basidiocarps are those of some *Coprinus* species (Fig. 6-71A, B). The lamellae are extremely thin, numerous, and closely spaced; their separation is maintained by the presence of large cystidia. Basidia of *Coprinus* mature in groups, starting at the lower edge of the lamella and moving upward. Following the discharge of basidiospores by the first group, autodigestion of the spent portion of the lamella occurs. Another group of spores matures and is released, and autodigestion

again dissolves the spent part of the gill. In this way, spore release and autodigestion occur until all parts of the lamellae have been digested. Autodigestion results from enzyme activity, and the dark, inky fluid produced has given rise to the common name, *inky cap.*

Agaricales are predominantly saprobic fungi that inhabit soil, wood, or dung. A few species are important plant parasites; many more are mycorrhizal fungi. Several mushrooms have been cultivated for food for many years, including the common market mushroom of Europe and North America, *Agaricus bisporus,* and two species in Asia, *Lentinus edodes* and *Volvariella volvacea.* Large quantities of wild mushrooms are also harvested and sold in markets in Europe or are canned and sold. In Russia alone, the annual har-

A　　　　　　　　　　　　　　　　　　　　　　**B**

Figure 6-64　Stipitate basidiocarps of *Polyporus* (**A**), ×0.6, and *Hydnellum* (**B**), ×0.5.
(Courtesy of W. McLennan.)

vest amounts to approximately 30,000 tonnes annually.

Many mushrooms contain toxins that can
cause illness or death. Among the most poisonous are *Amanita phalloides* and the closely related *A. verna* (Fig. 6-71C, D), often called *death
angels* or *destroying angels*. Toxins of many other
mushrooms cause less severe forms of poisoning.

Basidiomycetous Lichens

Basidiomycetous lichens are few in number and
sometimes go unnoticed. For example, *Omphalina
ericetorum* (Fig. 6-72A), a common temperate species, produces basidiocarps similar in all respects
to those of other small mushrooms. Algal cells of
this basidiolichen can be found only if the hyphae leading into the base of the stipe are examined carefully. The hyphae produce a close-fit-
ting mantle around associated algal cells (Fig.
6-72B).

Series Gasteromycetes

Gasteromycetes are thought to have evolved
from several distinct groups of hymenomycetes.

Only three orders are discussed in detail here.
Gasteromycete basidia are enclosed in their
basidiocarps, and all fungi in this group lack
basidiospore abstriction. Basidiospores typically
are either sessile on the basidium, or they develop on extremely long sterigmata. They are released by autodigestion.

Many gasteromycetes are adapted to arid
conditions; desert and semidesert regions have
an abundance of species. Others are common in
wet, tropical regions or in mesic, temperate
zones. Gasteromycetes are mainly soil fungi, and
some species form mycorrhizae with forest trees.
A few species are found on decaying wood or
other vegetation.

Order Lycoperdales　The order Lycoperdales
includes many of the largest and most common
puffballs—e.g., species of *Lycoperdon*, *Bovista*, and
Calvatia. Externally, the young basidiocarps of
Lycoperdon (Fig. 6-73A) resemble button stages of
developing mushrooms. In section, however, little differentiation is visible in the young, pear-
shaped basidiocarp. An outer layer, the peridium, and a fleshy, inner zone, the gleba, are present. The basidia develop within minute cavities
in the gleba and have long, narrow sterigmata
(Fig. 6-73C). As maturity approaches, autodiges-

tion occurs, leaving only basidiospores and a fibrous mass of special hyphae, the **capillitium** (Fig. 6-73B, D). The basidiocarp dries rapidly after autodigestion, and the peridium is then thin and flexible. An ostiole develops at the top of the basidiocarp, and any force applied to the peridium (e.g., by falling raindrops) results in a bellowslike action. Air, together with spores, is expelled through the ostiole, and spores are dispersed by air currents.

Basidiocarps of *Bovista* and *Calvatia* lack ostioles, the peridium cracking irregularly to expose the capillitium and spores at maturity. In *Bovista* (Fig. 6-74A), the mature basidiocarp is weakly attached to soil; it is readily broken free and tumbled about by winds, scattering the basidiospores. The capillitium prevents both rapid loss of spores and collapse of the globose basidiocarp. The largest *Calvatia* basidiocarps (Fig. 6-74B) remain attached to the soil, and the peridium flakes away irregularly, exposing spores to air currents. Basidiocarps of *Calvatia* species are among the largest produced by fungi; some exceed a meter in length and have been estimated to contain 160 billion spores.

Order Phallales In Phallales, autodigestion of the gleba leaves a slimy, foul-smelling spore mass. The odor, and possibly the basidiocarp color, attracts insects that effect spore dispersal. The odor has given rise to the common name *stinkhorns* for some Phallales.

In *Phallus* (Fig. 6-75A) and *Mutinus* (Fig. 6-75B), early stages of basidiocarp development occur at the soil surface or just below the litter layer. *Phallus* buttons are egglike in appearance and have a tough but flexible peridium. Immediately under the outer layer is a thick zone of colloidal material that functions in water storage. The comparatively small gleba develops on a receptacle on the upper surface of the pileus. At maturity of the button, the stipe elongates and ruptures the outer peridial layer; the latter remains as a volvalike cup at the base of the stipe. The glebal mass, having undergone autodigestion, is exposed on the pileus surface. Flies are attracted to the glebal mass, some of which is eaten; spores can also adhere to the legs or other appendages of the insects.

Figure 6-65 Coralloid basidiocarp of *Clavulina*, ✕ 1.6. (Photograph courtesy of W. McLennan.)

Species of *Mutinus* are similar to those of *Phallus*, but their receptacles are less differentiated. The basidiocarps of many Phallales are brightly colored; in some, petallike lobes surround the glebal mass. All species are adapted to insect dispersal; their spores pass unharmed through the digestive tracts of insects.

Order Nidulariales Basidiocarps of Nidulariales, called *bird's nest* fungi, develop on wood, dung, or soil. Basidia occur in hymenial (Fig. 6-76C) layers that line lens-shaped cavities in the gleba; each such cavity and its basidia become surrounded by firm wall layers. The glebal material external to the lens-shaped bodies, or **per-**

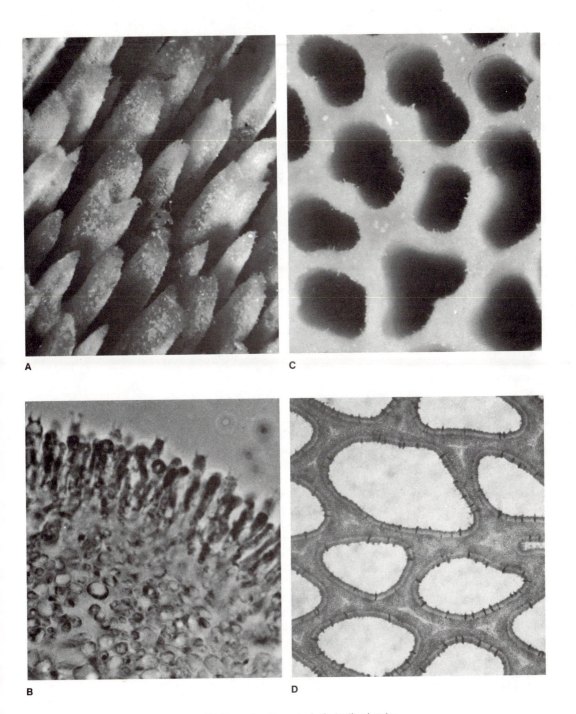

Figure 6-66 A, toothed hymenium, ×50. **B,** section through single tooth, showing hymenial arrangement, ×800. **C, D,** poroid hymenium. **C,** surface view, ×100; **D,** transverse section through pores, ×120.

idioles, undergoes autodigestion. Peridioles are then free from one another in the glebal chamber and are submerged in a liquid resulting from autodigestion of other glebal parts. In *Crucibulum* (Fig. 6-76), peridioles are attached to the peridium by an elastic filament. The mature basidiocarp is an open, cup- or vase-shaped structure that contains exposed peridioles. If a falling raindrop strikes the opening, it tears the peridioles free as it splashes upward and outward. A bit of the peridium remains attached to the end of the peridiole cord (Fig. 6-76B), and the entire structure (peridiole, cord, and peridium remnant) resembles a small bolas. If it strikes a grass blade or similar structure, the cord wraps around the structure. Presumably, further dispersal can take place if an animal then eats the herbage and attached peridiole.

The basidiocarps of *Crucibulum*, specialized for raindrop dispersal, are called *splash cups*. Splash cups also occur in some unrelated fungi and in certain green plants (e.g., gemmae cups of *Marchantia*, see p. 448).

Other Gasteromycetes The Gasteromycetes include many species that differ from mushrooms and boletes only in lacking violent spore abstriction. Other gasteromycetes produce trufflelike subterranean basidiocarps, which are sometimes referred to as *false truffles*. *Hymenogaster* (Fig. 6-77A) basidiocarps develop just below the soil surface. Within the closed basidiocarps, labyrinthiform hymenial cavities are produced. Basidiospores are probably dispersed in the same manner as are true truffles, that is, by rodents that dig and eat basidiocarps. *Hymenogaster* and related genera are mainly mycorrhizal fungi that are associated with forest trees.

Scleroderma (Fig. 6-78B, C) frequently is mistaken for *Lycoperdon*. However, the basidiocarps lack a capillitium and have rigid peridial layers. The peridium ruptures irregularly at maturity, exposing the spore mass.

Form Class Fungi Imperfecti

Only asexual reproduction is known in species classified in the form class Fungi Imperfecti. It

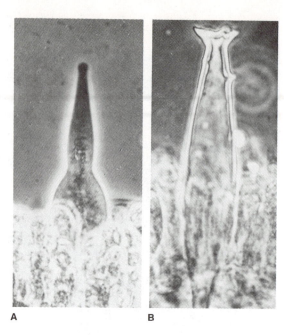

A   B

Figure 6-67 **A, B,** two types of cystidia found in the Hymenomycetes. **A,** × 1070; **B,** ×725.

was initially thought that, as fungi were studied, all species in this group would eventually be placed in one of the natural classes. This has not happened; the numbers of described Fungi Imperfecti increase annually. It is possible that some of these fungi have lost their ability to reproduce sexually, or perhaps they never possessed this ability.

Since asexual and sexual stages of sexually reproducing fungi often occur separately, it is advantageous also to include the conidial stages of these fungi in keys to Fungi Imperfecti. This permits identification in the absence of the sexual material, a common condition in disease fungi and other economically important species. Asexual and sexual stages of a single species can actually be designated by different binomials, a system permitted under the International Code of Botanical Nomenclature.

The form class Fungi Imperfecti is divided into four form orders: Sphaeropsidales, Melanconiales, Moniliales, and Mycelia Sterilia. Conidia of Sphaeropsidales are produced within

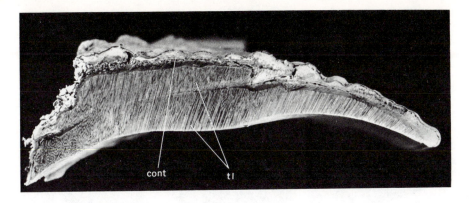

Figure 6-68 *Ganoderma applanatum.* Vertical section through basidiocarp, ×1: *cont,* context; *tl,* tube layers, each representing growth during one year.

pycnidia, and those of Melanconiales develop in acervuli. Moniliales include asexual yeasts as well as filmentous forms with single or clustered conidiophores. In the fourth group, Mycelia Sterilia, no conidia are formed; this group is not discussed here.

Form Order Sphaeropsidales

The Sphaeropsidales include many plant parasites. Species of *Phoma,* which cause diseases of turnips, beets, and other plants, produce pycnidia within the host tissues (Fig. 6-78A, B), but with the ostiole exposed. Numerous conidiophores line the cavity of the pycnidium; conidia are produced in abundance and ooze through the ostiole in a mucilaginous substance. Species of *Septoria* (Fig. 6-78C, D) cause *late blight* of celery, *tomato leaf spot,* and other diseases. Pycnidia of *Septoria* are essentially like those of *Phoma,* but the conidia are extremely long and narrow.

Form Order Melanconiales

Acervuli of Melanconiales are subepidermal on their hosts. These fungi cause plant diseases called *anthracnoses.* As acervuli develop and co-nidia are produced, the host epidermis ruptures and exposes the spore mass. Acervuli of *Gloeosporium* (Fig. 6-79A), common on the leaves of many broadleafed trees, are small but can be present in large numbers. During rainy periods of spring and early summer, infections by *Gloeosporium* can cause massive defoliation of trees. Conidia of *Gloeosporium* develop in a mucilaginous substance and are spread by splashing raindrops or by insects that contact the spore mass. *Pestalotia* (Fig. 6-79B), one of the largest form genera, produces dark, several-celled, appendaged conidia.

Form Order Moniliales

The Moniliales include yeastlike fungi as well as mycelial forms. Some budding fungi placed here possibly are haploid lines of heterothallic hemiascomycetes or heterobasidiomycetes. Asexual yeasts include species pathogenic to humans, the most common being *Candida albicans. C. albicans* (Fig. 6-80A) is often present in the mouths or other body openings of healthy individuals and can exist there as a harmless saprobe. However, under certain conditions, the fungus invades the tissues, and infections sometimes result in fatalities. *C. albicans* has limited hyphal development, and numerous budding cells arise from the hyphae.

A

B

Figure 6-69 A, cluster of sessile basidiocarps of *Pleurotus*, ×0.5; **B,** *Boletus edulis,* basidiocarp, ×0.47. (Photograph **B** courtesy of W. McLennan.)

Aureobasidium pullulans (Fig. 6-80B) is morphologically similar to *C. albicans,* but its colonies show varied degrees of black pigmentation. *A. pullulans* is common on surfaces of healthy leaves, within living stems, in soil, and elsewhere. There is some evidence that its presence on healthy plants may provide the latter with protection against some plant pathogens. *A. pullulans* is also present on painted surfaces, and it is one of several fungi associated with the deterioration of paints.

Fragmentation of hyphae to produce asexual propagules occurs in many Moniliales. *Geotrichum* species (Fig. 6-80C) reproduce by this method, and some species of the genus are important plant and animal pathogens. One species of *Geotrichum* has been found to be capable of infecting both plants and animals.

Most Moniliales produce mycelia with definite conidiophores and specialized conidium-producing cells. The conidia vary greatly in form, but the methods by which they are produced are relatively few. Some conidia—e.g., in *Fusarium* (Fig. 6-81A, B) and *Verticillium* (Fig.

Figure 6-70 *Agaricus bisporus.* **A,** mushrooms in commercial bed, ×0.55. **B–D,** drawings showing development of basidiocarp, ×0.5: *c* cortex; *p,* pileus or cap; *l,* lamellae or gills; *pv,* partial veil; *s,* stipe. **E–G,** scanning electron micrographs. **E,** gill surface (*scale line* = 10 μm); **F, G,** two stages in basidiospore development (*scale line* = 1 μm). (**A,** courtesy of L. Schisler; **E–G,** courtesy of K. M. Saksena, R. Marino, M. N. Haller, and P. A. Lemke, with permission of *Journal of Bacteriology.*)

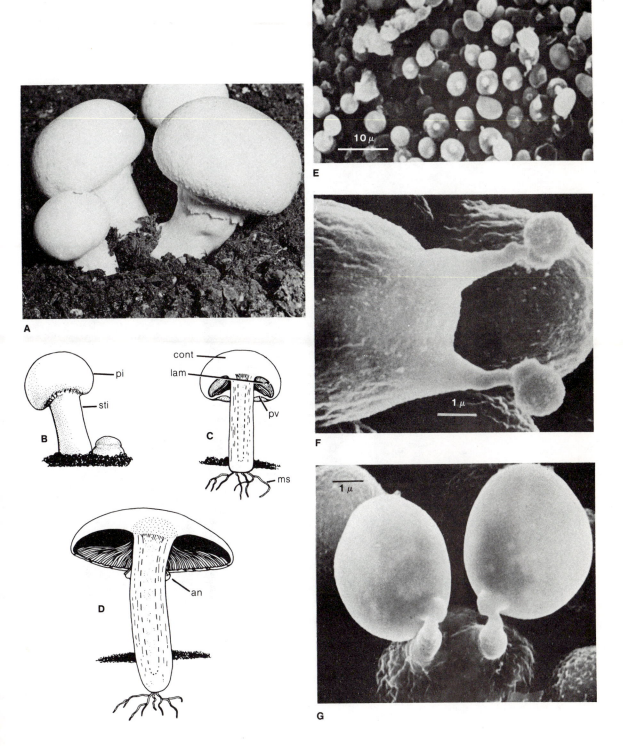

A

cont
lam

pi

sti

pv

C

ms

B

an

D

E

F

G

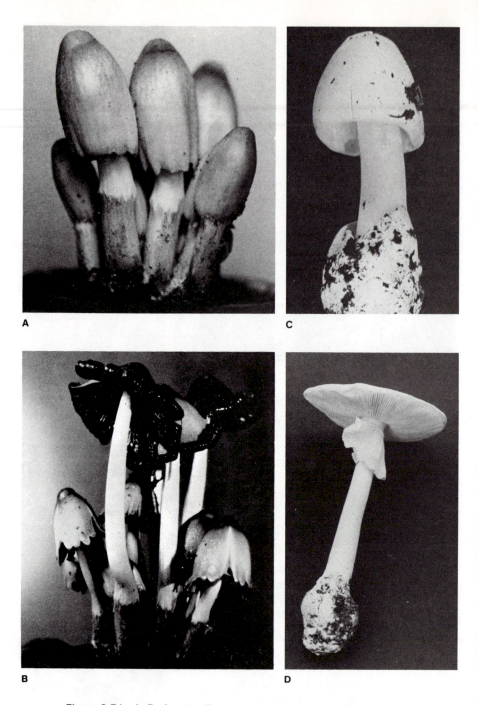

Figure 6-71 **A, B,** *Coprinus.* Two successive photographs of the same cluster of basidiocarps. **A,** just before start of spore release in larger fructifications, ×0.75; **B,** the lamellae of the larger fructifications have undergone autodigestion, ×0.75. **C, D,** *Amanita verna,* one of the *death angels* or *destroying angels*. **C,** young basidiocarp, ×0.7 (note cupulate volva); **D,** mature basidiocarp with skirtlike annulus, ×0.5.

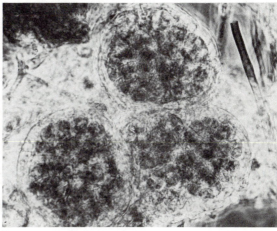

A

B

Figure 6-72 *Omphalina ericetorum,* a temperate lichen species. **A,** basidiocarp, ×0.8.
B, modified hyphal structures enclosing algal cells, ×595.

6-81C)—are produced by specialized cells, called
phialides. Phialides are usually bottle-shaped or
cylindrical cells that produce a succession of conidia at their apices. Both *Fusarium* and *Verticillium* are responsible for *wilt* diseases and
others of vascular plants. In wilt diseases, the
fungus invades vascular tissue of the root or stem
and interferes with water transport. *Aspergillus*
and *Penicillium* species (see p. 189) also have conidia produced by a type of phialide.

Other types of conidial development occur
in *Alternaria* (Fig. 6-82A) and *Cladosporium* (Fig.
6-82B). Conidia of *Alternaria* are extruded, bal-
loonlike, from a pore at the apex of the conidiophore or its branches. In *Cladosporium*, conidia are
essentially budded off the conidiophore. Species
of both *Alternaria* and *Cladosporium* are common
on decaying vegetation and in soil.

In all form orders of Fungi Imperfecti, the
type of conidial development has gained importance in classification. Formerly, highly variable
characteristics of conidia—for example, color,
septation, and form—were used instead of developmental patterns.

Conidia vary greatly in shape. In some instances, form is correlated with habitat type and
is probably related to dispersal within the particular habitat. One of the more striking of these
correlations is found in a group of fungi that
grow on wet, often submerged leaves. In fast-
flowing streams, these decaying leaves support
large numbers of fungi with unusual, branched
conidia. The conidia develop through several
distinctive methods, indicating parallel evolution
and a probable biological advantage of this
shape. However, the exact significance is not yet

Figure 6-73 *Lycoperdon gemmatum.* **A,** cluster of young basidiocarps, ×0.5. **B,** section through immature fructification, ×1.2: *pe,* peridium; *gl,* gleba; *st,* sterile base. **C,** basidia with elongate sterigmata and basidiospores, ×1235. **D,** capillitium and basidiospores, ×1300.

A

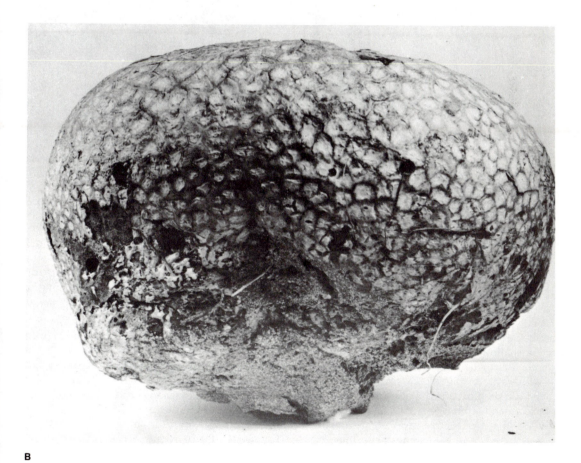

B

Figure 6-74 Mature basidiocarps of *Bovista* (**A,** ×0.75) and *Calvatia* (**B,** ×0.3).

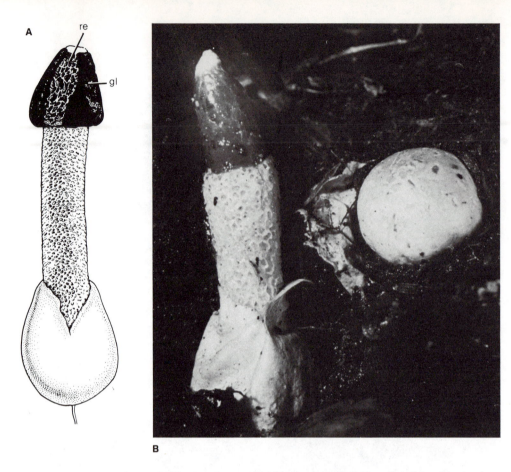

A

B

Figure 6-75 **A,** habit sketch of *Phallus* basidiocarp. A portion of the gleba has been removed to show the pitted nature of the receptacle: *gl*, gleba; *re*, receptacle; ✕1. **B,** button and mature fructification of *Mutinus*, ✕1.4. (**B,** courtesy of W. McLennan.)

known. Foam collected from streams often contains large numbers of these spores. The conidia shown in Figure 6-83A, B were photographed from a foam collection.

A second group of leaf-decay fungi, often abundant in shallow vernal pools, produces conidia that are helicoid. *Helicoma* (Fig. 6-83C) is an example with conidia coiled in a single plane. These coiled conidia float and are also frequently present in foam and scum.

Many Moniliales parasitize invertebrate animals such as insects and nematodes. The most striking of these, called *predaceous* fungi, capture nematodes, which they then digest. Most can live as saprobes; they capture nematodes by

means of constricting rings (snares), or adhesive knobs, rings, or conidia. The snares and rings of two species of *Arthrobotrys* are shown in Figure 6-84. Constricting rings are sensitive to contact on their inner surface. If this portion of the ring is touched by a nematode crawling through the opening, the ring cells suddenly expand to approximately 10 times their normal diameter (Fig. 6-84A, B). The expansion occurs in a fraction of a second and results in capture of the nematode. Adhesive rings develop in complex networks (Fig. 6-84C, D) in some species of *Arthrobotrys*; contact of a nematode with the inner surfaces of such rings results in instantaneous adhesion. Bodies of captured nematodes are invaded by

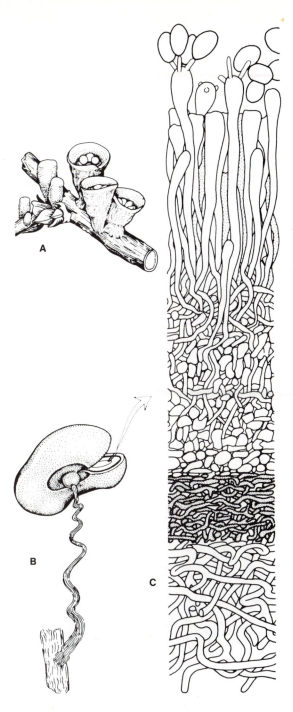

Figure 6-76 *Crucibulum*. **A,** young and mature basidiocarps, ×1.5. **B,** single peridiole and attached cord, ×20. **C,** vertical section through peridiole, showing wall layers and basidia, ×950.

hyphae that digest and absorb most of the body. *Arthrobotrys* species are common in soil and dung.

Relationships of Higher Fungi

As mentioned in the discussion of relationships of lower fungal groups (p. 169), fossil fungi are known, but our knowledge of these is as yet of little assistance in explaining the evolution of fungi. Well-preserved fossils of both basidiomycetous and ascomycetous fungi have been found, but they are of relatively recent origin. For example, some fossil leaf-inhabiting ascomycetes are very well preserved, but they indicate only that certain genera have remained relatively unchanged for the past 50 million years.

Among the evolutionary possibilities that have been considered for higher fungi, two basic theories have received greatest attention. One of these suggests that primitive hemiascomycetes arose from zygomycetous ancestors and that simple ascomycetes gave rise to Basidiomycotina. A second concept involves the origin of ascomycetous fungi and/or rusts (Uredinales) from red algae.

The Zygomycetes-Ascomycotina concept is based on morphological similarities between extant hemiascomycetes and zygomycetes. Both Taphrinales and Endomycetales have been considered as groups of ascomycetes most closely resembling zygomycetes. In the hemiascomycetes and zygomycetes, cell walls contain chitin, but differences in cell wall chemistry do occur. Morphological similarities include lack of motile cells, production of conidia in some zygomycetes, and similarites in sexual reproduction (e.g., in some zygomycetes and Endomycetales).

Suggestions have been made that higher fungi descended from chytridlike ancestors; these suggestions derive from biochemical similarities. Wall chemistry and storage products in the Chytridiomycota are similar to those of higher fungi. Another biochemical similarity concerns the biosynthesis of lysine. In Chytridiomycota, Zygomycota, and Eumycota, lysine is synthesized via aminoadipate; in Oomycota and Hyphochytridiomycota, it is synthesized via diaminopimelic acid.

A

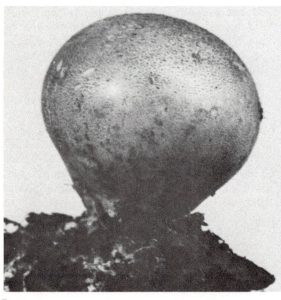

B

C

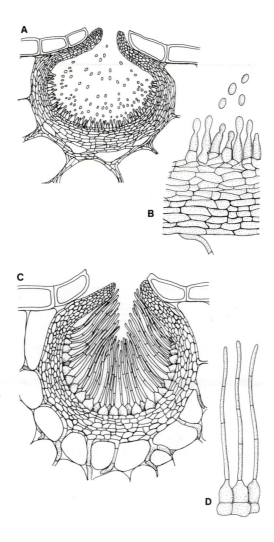

Figure 6-78   A, B, *Phoma.* **A,** vertical section through pycnidium, ✕300; **B,** wall layers and conidiophores and conidia from A above, ✕760. **C, D,** *Septoria.* **C,** section through pycnidium and host leaf, ✕325 approx.; **D,** conidiophores and conidia, ✕575.

Figure 6-77   A, surface (*right*) and sectional (*left*) views of *Hymenogaster* basidiocarps, ✕1.5 approx.; **B, C,** surface and sectional views of *Scleroderma* basidiocarp, showing thick peridial layer, ✕1. (Courtesy of J. M. Trappe.)

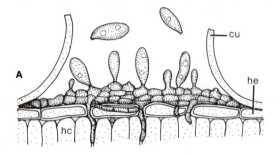

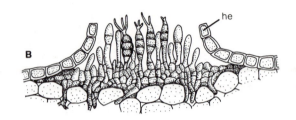

Figure 6-79 A, *Gloeosporium,* section through acervulus, ×700. **B,** *Pestalotia,* section through acervulus. ×315.

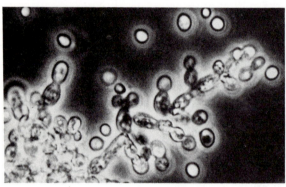

A

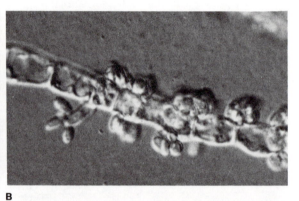

B

Figure 6-80 A, photomicrograph of *Candida albicans* colony, showing limited hyphal development and numerous buds, ×650. **B,** *Aureobasidium pullulans,* hyphae and budding, ×735. **C,** *Geotrichum,* aerial branch fragmenting to form conidia, ×800.

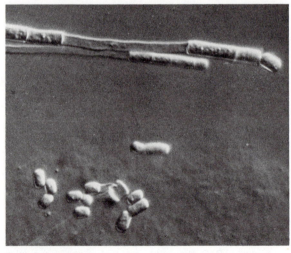

C

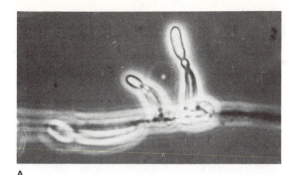

A

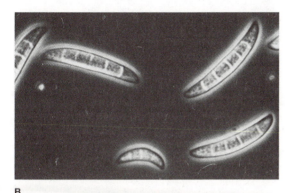

B

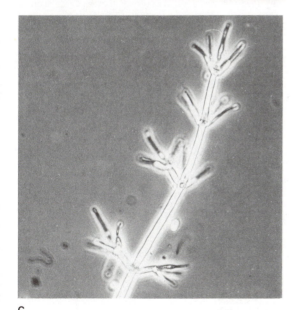

C

Figure 6-81 **A,** conidiophores and **B,** conidia of
Fusarium, both × 1000. **C,** conidiophore of *Verticillium*,
× 525.

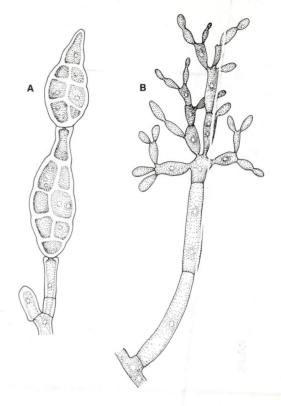

Figure 6-82 **A,** short conidiophore and two conidia of
Alternaria, × 790. **B,** conidiophore and chains of conidia of
Cladosporium, × 1235.

The lack of motile cells and similarities in
fertilization and subsequent development are
cited in aligning higher fungi with red algae
(Florideophyceae). Similarities in wall chemistry
and storage products are cited from time to time,
but our knowledge of these is inadequate.
Chitin, present in all higher fungi, is not known
in red algae. The resemblance in fertilization by
spermatial transfer, receptive hyphae, and subsequent development is striking. In both red algae
and euascomycetes, fertilization is followed by
the development of complex protective structures around the meiospore-producing cells.

Those holding the view that Ascomycotina
arose from zygomycetous ancestors consider the
above similarities to be the result of parallel evolution. It should be noted that there is little evidence of a common origin of spermatial fertiliza-

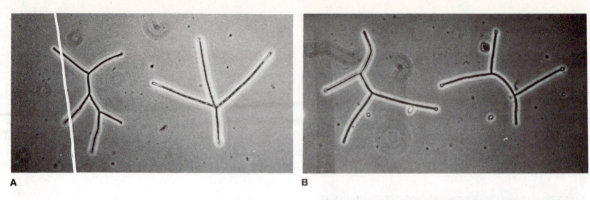

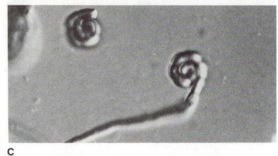

Figure 6-83 A, B, four representative conidia from a single foam sample, **C,** developing and mature conidia of *Helicoma*, ×625.

tion even for the several groups of euascomycetes and heterobasidiomycetes that possess it. The red algal theory presents other problems—e.g. the absence of clamps and croziers, and the dikaryotic stages that they accompany, in red algae. Differences in nutrition (photosynthesis versus heterotrophy) indicate very different biochemical characteristics in these groups. Finally, red algae (especially the Floridophyceae), higher ascomycetes, and rusts are all highly evolved, highly specialized forms. Evolutionary pathways generally run from unspecialized to specialized groups. Ascomycetes most closely resembling red algae are the most complex in Euascomycetes—i.e., perithecial and discomycetous types. If these are derived from red algae, the simple ascomycetes (e.g. yeasts) must be considered as reduced forms.

Whatever their ancestors, Ascomycotina and Basidiomycotina share a number of features: storage of glycogen and lipids; cell walls containing chitin, glucans, or mannans; and other chemical similarities. The similarity in ascus and basidium development is frequently pointed out, the two structures being considered homologous by many mycologists. Clamp connections and croziers are remarkably similar; the presence of the dikaryotic phase in both groups and its absence elsewhere also suggest a relationship. Finally, in simple hemiascomycetes and heterobasidiomycetes, similarities in asexual reproduction also occur.

All current theories concerning the origins of fungal groups are speculative and are based upon inadequate factual information. Fungi are poorly known taxonomically, biochemically, and in every other way. Algal groups for which fungal affinities have been hypothesized are equally poorly known. As information is gained for all these groups, theories concerning their evolution and classification systems based on these concepts must be continuously modified.

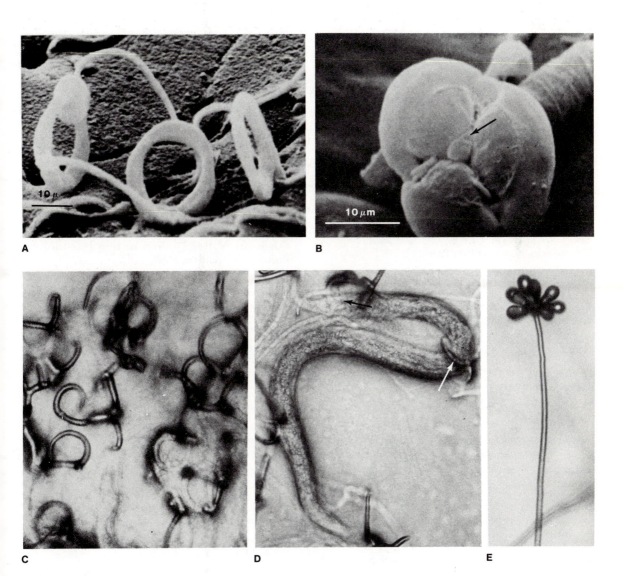

Figure 6-84 *Arthrobotrys.* **A, B,** constricting rings, the latter inflated and with entrapped nematode (*arrow*); *scale line* = 10 μm. **C, D,** networks of rings, the latter with a captured nematode. **C,** ×200; **D,** ×230. **E,** conidiophore with attached conidia, ×250. (**A, B** courtesy of R. H. Estey and S. S. Tzean, from "Scanning electron microscopy of fungal nematode-trapping devices," *Transactions of the British Mycological Society* 66:520–22, with permission of Cambridge University Press.)

References

See also general reference for fungi, p. 171.

Nonlichenized Fungi

Ainsworth, G. C.; Sparrow, F. K.; and Sussman, A. S., eds. 1973. *The fungi: an advanced treatise*, vol. 4A (*Ascomycetes and Fungi Imperfecti*) and vol. 4B (Basidiomycetes and Lower Fungi). Academic Press, New York. (Taxonomic reviews with keys; a number of chapters in volume 2 of this series deal specifically with Ascomycetes, Basidiomycetes or Fungi Imperfecti.)

Arthur, J. C. 1934. *Manual of the rusts in the United States and Canada*. (Illustrated and with a new supplement by G. E. Cummins.) Hafner Press, New York. (Reprint ed, 1962)

Brodie, H. J. 1975. *The bird's nest fungi*. Univ. of Toronto Press, Toronto.

Chadefaud, M., and Emberger, L. 1960. *Traité de botanique systématique*, vol. 1, *Les végétaux nonvasculaires (Cryptogamie)*. Masson et Cie., Paris.

Cole, G. T., and Samson, R. A. 1979. *Patterns of development in conidial fungi*. Pitman Publishing, London.

Corner, E. J. H. 1950. *A monograph of Clavaria and allied genera*. Oxford Univ. Press, London.

Cummins, G. B. 1959. *Illustrated genera of rust fungi*. Burgess Publishing Co., Minneapolis.

Dennis, R. W. G. 1978. *British ascomycetes*. J. Cramer, Lehre, Germany.

Duddington, C. L. 1957. *The friendly fungi*. Faber & Faber, London.

Fischer, G. W., and Holton, C. S. 1957. *Biology and control of the smut fungi*. Ronald Press Co., New York.

Hughes, S. J. 1953. Conidiophores, conidia, and classification. *Can. J. Bot.* 31:577–659.

Ingold, C. T. 1975. *Guide to aquatic Hyphomycetes*. Freshwater Biological Association, Scientific Publication no. 30. The Ferryhouse, Ambleside, England. [An illustrated guide to aquatic and water-borne Hyphomycetes (Fungi Imperfecti), with notes on their biology]

Kendrick, B., ed. 1979. *The whole fungus*, vols. 1, 2. National Museums of Canada and the Kananaskis Foundation, Ottawa.

Kohlmeyer, J., and Kohlmeyer, E. 1979. *Marine mycology: the higher fungi*. Academic Press, New York.

Lodder, J., ed. 1970. *The yeasts: a taxonomic study*. 2nd ed. North-Holland Publishing Co., Amsterdam.

Petersen, R., ed. 1971. *Evolution in the higher basidiomycetes*. Univ. of Tennessee Press, Knoxville.

Phaff, H. J.; Miller, M. W.; and Mrak, E. M. 1978. *The life of yeasts*. 2nd ed. Harvard Univ. Press, Cambridge, Mass.

Raper, J. R. 1966. *Genetics of sexuality in higher fungi*. Ronald Press Co., New York.

Shaffer, R. L. 1968. *Keys to the genera of higher fungi*. 2nd ed. Univ. of Michigan Biological Station, Ann Arbor.

Smith, A. H. 1951. *Puffballs and their allies in Michigan*. Univ. of Michigan Press, Ann Arbor.

Snell, W. H., and Dick, E. A. 1970. *The boleti of northeastern North America*. J. Cramer, Lehre, Germany.

Lichen Fungi

Abbayes, H. des. 1951. "Traité de Lichénologie," *Encycl. Biol.*, vol. 41. Lechevalier, Paris.

Ahmadjian, V. 1967. *The lichen symbiosis*. Blaisdell Publishing Co., Waltham, Mass.

Ahmadjian, V., and Hale, M. E., eds. 1973. *The lichens*. Academic Press, New York. (Includes chapters by numerous authors on all aspects of lichen biology.)

Brown, D. H., and Hawksworth, D. L. 1976. *Lichenology: progress and problems*. Academic Press, London.

Duncan, U. K. 1959. *A guide to the study of lichens*. Arbroath, Buncle.

———. 1963. *Lichen illustrations*. Supplement to *A guide to the study of lichens*. Arbroath, Buncle.

Hale, M. E. 1961. *Lichen handbook: a guide to the lichens of eastern North America*. Smithsonian Institution Press, Washington, D.C.

———. 1967. *The biology of lichens*. Edward Arnold, London.

———. 1979. *How to know the lichens*. Wm. C. Brown Co. Publishers, Dubuque, Iowa.

Kershaw, K. A., and Alvin, K. L. 1963. *The observer's book of lichens*. Warne, London.

Nearing, G. G. 1947. *The lichen book*. Published by the author, Ridgewood, N.J.

Seaward, M. R. D. 1977. *Lichen ecology*. Academic Press, London.

Smith, D. C. 1962. The biology of lichen thalli. *Biol. Rev.*, 37:537–50.

In an evolutionary survey of photosynthetic plants, green algae (Division Chlorophyta) should be considered just prior to any discussion of embryo-producing plants. Logically, algal groups containing pigments that mask chlorophyll *a* might well be studied first, as they have few characteristics in common with land plants (which comprise most plants). However, since Chlorophyta is a large division, with morphological and reproductive diversity, it is the first eukaryotic algal division considered.

Chlorophyta occur throughout the aquatic environment and are known by a variety of names, such as *slime, frog spit, green seaweed,* and other more descriptive (and unpublished) ones. Many occur in fresh water, and a significant number are in the marine habitat. Generally, the marine forms are larger than the freshwater ones. There are over 7000 species in more than 450 genera. Vegetatively, Chlorophyta are relatively simple. The range in morphological diversity includes unicellular, multicellular, motile, nonmotile, filamentous, nonfilamentous, parenchymatous, and siphonous forms, or various combinations of these. They exhibit a greater variety of life histories and methods of reproduction than any other division of plants.

Several classification systems are used for Chlorophyta. In this text, four classes are recognized (Table 7-1). Some workers believe that some of these should be separate divisions; however, this does not necessarily facilitate an overall understanding (especially by the neophyte) of the relationships of these algae. Morphologically, the largest and most diverse class is Chlorophyceae, which in this text is considered in restricted terms. An outstanding feature of Chlorophyceae is that cells are uninucleate. Recognition of a second class, Bryopsidophyceae, is controversial, but its members differ from Chlorophyceae in several ways, including the multinucleate nature of cells. The third class, Charophyceae, is delimited in this text by its very characteristic morphology and reproduction, although in 1975 the class was expanded to include a few genera included in the Chlorophyceae (p. 300). Some botanists consider the Charophyceae unique enough to be a separate division. Finally, recognition of the fourth class, Prasinophyceae, is in a state of flux; however, as defined here, it is

7

CHLOROPHYTA
(Green Algae)

Table 7-1 Characteristics of the Classes of Chlorophyta (items in parentheses refer to some representatives)

Class and Habitat	Size	Nuclei per Cell	Carotenoids*	Cell Wall	Dominant Life History†
Chlorophyceae freshwater marine	microscopic (few μm) to several meters	1	usual	cellulose pectin	haploid haploid/diploid
Bryopsidophyceae marine (freshwater)	macroscopic to several meters	many	usual; additional may be present	xylan, mannan cellulose (CaCO$_3$)	diploid haploid/diploid
Prasinophyceae marine (freshwater)	microscopic (few μm)	1	usual	often lacking (cellulose?, pectin)	?
Charophyceae freshwater (marine)	macroscopic (to 30cm)	1 to many	usual	cellulose (pectin) CaCO$_3$	haploid

*Usual for division includes those occurring in land plants. Additional carotenoid pigments are those present only in Bryopsidophyceae.
†Haploid = zygote only 2n stage (see Fig. 2-8A); diploid = gametes only n stage (see Fig. 2-8B); haploid/diploid = both n and 2n phases multicellular (see Fig. 2-9A).

a small group of green algae with characteristic motile cells.

As Chlorophyceae is the largest class of Chlorophyta and is encountered in most aquatic habitats, it is discussed first and in greatest detail. The features separating the other three classes are considered with regard to those characteristics that set them apart from Chlorophyceae.

Class Chlorophyceae

Cell Structure

Observations of chlorophycean cells confirm that they are eukaryotic. The most obvious part of the cell is the chloroplast, and in larger cells the single nucleus can be readily observed (Figs. 7-1 and 7-3F,G). The overall impression is that there are definite compartments and not the diffuse appearance of the prokaryotic cell.

The protoplast is generally surrounded by a firm, fibrillar wall composed of cellulose with an outer, amorphous, pectic layer. Within the cell

wall and plasma membrane are the usual cell structures, as chloroplast, nucleus, mitochondria, golgi, and endoplasmic reticulum (Fig. 7-2). Also, some cells contain vacuoles (Fig. 7-2B), although other cells seem to lack them.

Chloroplasts contain the photosynthetic and accessory photosynthetic pigments in thylakoids. Chlorophyll a and b are present, plus carotenoids that give the characteristic grass-green color to cells. Sometimes, when there is an abundance of carotenoids (often outside the chloroplast), the cells are orange or brick-red. Thylakoids, arranged in bands with two to six fused thylakoids per band (Fig. 7-2A), can be in stacks (Fig. 7-2C) similar to the grana occurring in the chloroplasts of green land plants. The number of chloroplasts in a cell varies with a given species or genus. When numerous chloroplasts are present, they are small, disc-shaped, and can be interconnected (Fig. 7-3A). More often, only one or two chloroplasts occur, shaped like a half-filled cup, flat band, open ring, net, spiral, or star (Fig. 7-3B–G). The chloroplast position in the cell is peripheral (Fig. 7-3D, F) or central (Fig. 7-3C, G). Often within the chloroplast there is a dense region, the **pyrenoid** (*pyren* = fruit, stone; *oid* = like),

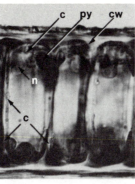

A   **B**

Figure 7-1 A, B Cell structure of Chlorophyceae, light microscope (transverse sections of *Monostroma*); note chloroplast (*c*) with pyrenoid (*py*), nucleus (*n*), and cell wall (*cw*), × 1000. (Courtesy of M. Dube and the *Journal of Phycology.*)

which is proteinaceous. With the light microscope, the pyrenoid appears in the chloroplast as a central, clear area, often hexagonal in form (Figs. 7-1B and 7-3). The electron microscope shows it as an electron-dense area surrounded and traversed by some thylakoids (Fig. 7-2B). The pyrenoid is an area of storage-product formation, although its functions are not completely known. Electron-clear areas surrounding the pyrenoid and throughout the chloroplast comprise starch, the storage product of green algae. In motile cells of some green algae, an eyespot is often present in the chloroplast (Figs. 7-3B and 7-8A). This appears as a bright red or orange dot with the light microscope, but sections show its structure consists of two to several rows of densely packed lipid granules associated with some thylakoids (Fig. 7-20B).

The starch storage product present in the chloroplast is often so abundant that the chloroplast shape is obscured. This gives a rough, granular appearance to the chloroplast, which contrasts with chloroplasts of all other algae (and green plants generally), in which the products of photosynthesis accumulate in the cytoplasm. Thus, chloroplasts of all *but* the green algae are smooth-edged and well-defined in outline when observed with the light microscope.

Most motile cells contain contractile vacuoles, usually at the apical end (Figs. 7-3B and 7-11C); when observed with the light microscope, these are seen to fill and empty alternately. Ultrastructure studies indicate the involvement of endoplasmic reticulum with a contractile vacuole delimited by a single membrane.

In unicellular Chlorophyceae, cell division produces several reproductive, unicellular units (spores) as discussed in Chapter 2. The multicellular Chlorophyceae can also produce spores, but many, particularly the filamentous and parenchymatous ones, undergo cell division that produces larger plants. Thus, in these algae new cells are added that increase the overall size of the plant. In cell division there can be formation of a cell plate or a furrow, which establishes the location of the new plasma membrane and cell wall that separate the two new cells.

The method of cell wall formation varies in Chlorophyceae and can involve the ability of the cell to expand following nuclear division to allow for the process of **cytokinesis** (*cyto* = cell; *kinesi* = movement). In some filamentous forms, cytokinesis is formed by a furrowing inward from the outside of the cell (Fig. 7-4A). In others, cytokinesis and cell wall formation begin from the center of the cell and proceed to form toward the outside (Figs. 7-4B and 7-5B). This latter method superficially resembles that of bryophytes and vascular plants, but ultrastructural details differ. Electron microscope studies show that cell wall formation involves the mitotic spindle, cell microtubules, and the position of the new nuclei at the end of mitosis and during cytokinesis (Fig. 7-5A, C). The position of new nuclei and their distance from one another evidently determine the direction of fibrils forming the cell wall. Nuclei that are close together tend to have a transverse microtubular arrangement (Fig. 7-5A); whereas nuclei that are widely separated have a longitudinal arrangement (Fig. 7-5C), and a true cell plate is formed, as in bryophytes and vascular plants. The method of wall formation is used as an indicator of the evolutionary relationships of the green algae to the embryo-producing plants, as discussed later (see p. 306).

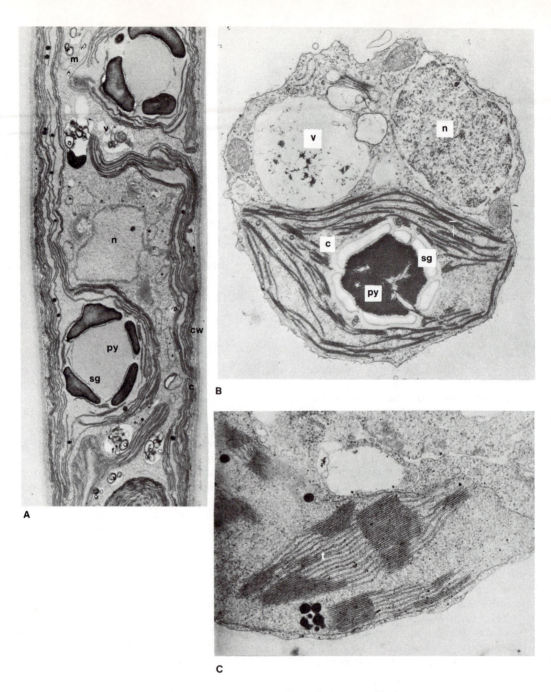

Figure 7-2 Cell and chloroplast structure of Chlorophyceae, electron microscope, cells showing peripheral chloroplasts (*c*) with pyrenoid (*py*), starch (*sg*), and thylakoids (*t*): *n*, central nucleus; *v*, vacuole; *m*, mitochondrion; *cw*, cell wall. **A,** *Stigeoclonium,* ×9200. **B,** *Volvox,* ×12,000. **C,** *Chlamydomonas* with granalike appearance of fused thylakoids, ×28,750. (**A,** courtesy of G. L. Floyd, K. D. Stewart, K. R. Mattox, and *Journal of Phycology;* **B,** courtesy of T. Bisalputra; **C,** courtesy of L. Goff.)

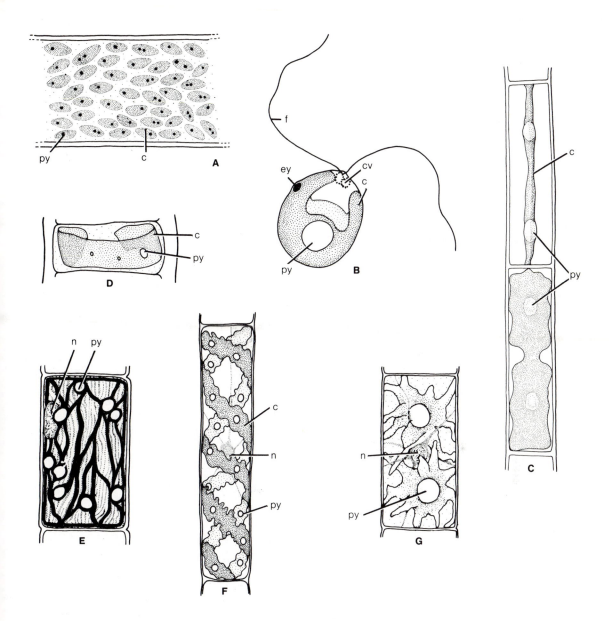

Figure 7-3 Chloroplast arrangement in Chlorophyceae: *c,* chloroplast; *py,* pyrenoid; *cv,* contractile vacuole; *ey,* eyespot; *n,* nucleus; *f,* flagellum **A, B, D–F,** parietal (peripheral) position in cell. **A,** disc-shaped (*Bryopsis*); **B,** cup-shaped (*Chlamydomonas*), with apical eyespot and contractile vacuoles; **D,** ring-shaped (*Ulothrix*); **E,** netlike (*Oedogonium*); **F,** spiral (*Spirogyra*). **C, G,** axile (central) position in cell. **C,** band-shaped (*Mougeotia*); **G,** star-shaped (*Zygnema*).

As noted, some Chlorophyceae have motile vegetative cells. However, for most of the class, only reproductive cells are flagellated. When present, motile cells have two or four anteriorly (or apically) oriented flagella of equal length. However, some Chlorophyceae have subapically oriented flagella and others have flagella of slightly unequal length. In the order Oeodogoniales there is an anterior circle of many short, equal flagella (see Figs. 7-21B and 7-22A,B). The number present is a multiple of two, with as many as 128 flagella reported on some reproductive cells.

The flagella are of the whiplash type (Fig. 7-6A), and some species have submicroscopic, delicate scales (Fig. 7-6E). During normal forward swimming, the flagella beat synchronously in the same plane, much like the arms of a person swimming breast stroke. Occasionally cells swim backwards, possibly under adverse conditions, with flagella undulating and trailing. Investigations of the ultrastructure of motile cells show that a flagellar apparatus consists of the flagellum with its basal body attached to microtubules within the cell. These microtubules form flagellar roots of two types. In the more commonly occurring type, there are four microtubular roots arranged as a cross (Fig. 7-6A, B). In the second type, there is only one broad band or root (Fig. 7-6C, D). Both types of flagellar roots can be striated in appearance. The single, broad band is typical of that in the motile male gametes of bryophytes and vascular plants. Motile cells with the cross arrangement are symmetrical, whereas the other motile cells are asymmetrical (compare Fig. 7-6A and C).

Reproduction

Both vegetative (asexual) and sexual reproduction occur commonly. Many multicellular forms, especially filamentous ones, increase vegetatively by fragmentation. For the filamentous members of the ubiquitous freshwater order Zygnematales (see p. 279), this is the only means of asexual reproduction. Most green algae produce spores that are either motile (zoospore or planospore) or nonmotile (aplanospore). These spores are usually formed by cell divisions not involving the

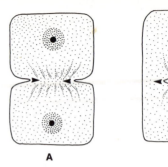

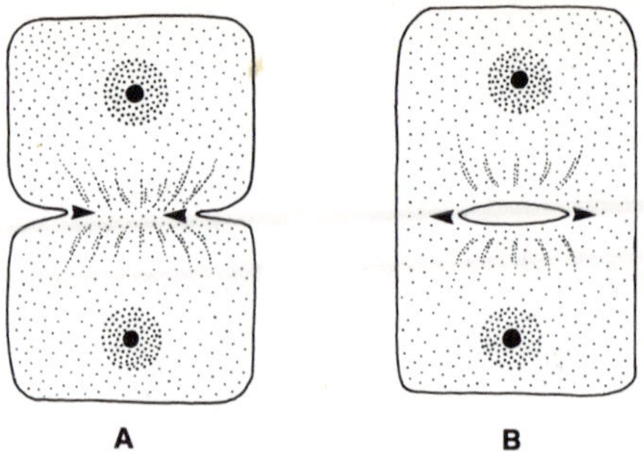

Figure 7-4 Diagrammatic representation of cytokinesis showing cell plate and wall formation; *arrows* indicate direction of formation. **A,** furrowing from outside cell toward center. **B,** from inside cell outward (each cell contains two nuclei).

parental wall in what appears to have been an ordinary vegetative cell. Upon release, spores are dispersed and ultimately settle, producing new plants. Many spores can be released at one time, but only a few grow into mature plants. In a few green algae, the spores are retained within the parental wall until a multicellular colony with a characteristic shape is formed.

Some multicellular Chlorophyceae have a specific number of divisions, and new cells remain united when released from the parental cell. This colony, consisting of a definite number of cells, is known as a **coenobium** (*coen* = common), and the term may be used interchangeably with the word *colony* when referring to such colonies. Most colonial Volvocales (see p. 260 and Figs. 7–8B–F and 7-12A) and some colonial Chlorococcales (see p. 269, Fig. 7-15D–F) are coenobial.

Figure 7-5 Transverse wall formation in Chlorophyceae, electron microscope. **A,** *Ulothrix*, showing transverse microtubules (*tu*) and nuclei (*n*) close together, ×11,500. **B,** *Fritschiella*, showing cell plate (*cp*) formation starting in center, but note microtubules lacking, ×13,500. **C,** *Coleochaete*, showing cell plate with longitudinal microtubules and nuclei well separated, ×8200. Note also *v*, vacuole; *m*, mitochondria; *n*, nucleus; and *c*, chloroplast. (**A,** courtesy of G. L. Floyd, K. D. Stewart, K. R. Mattox, and *Journal of Phycology*; **B,** courtesy of G. E. McBride and *Archiv für Protistenkunde*; **C,** courtesy of H. J. Marchant, and *Cytobios*.)

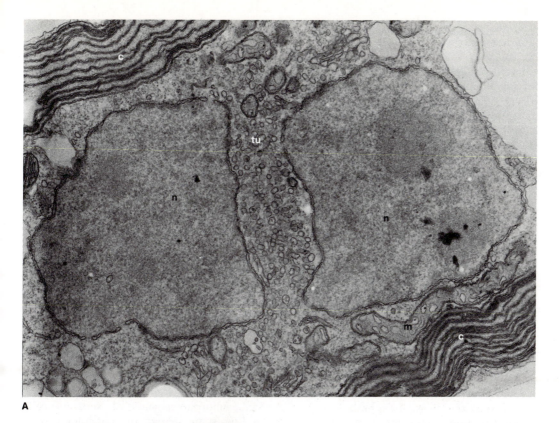

A

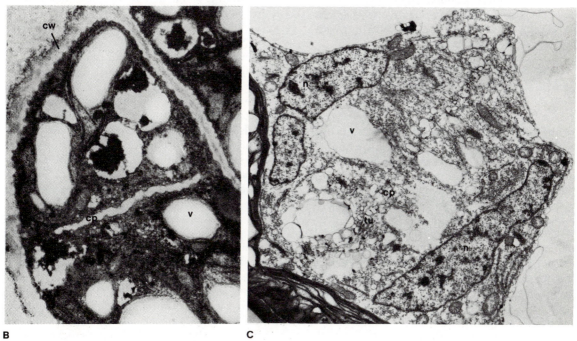

B

C

Many Chlorophyceae, especially in freshwater habitats, undergo extensive and rapid asexual reproduction. In fact, for some this is the only means of reproduction. Asexual reproduction is important in maintaining and increasing algal populations.

In contrast, some Chlorophyceae (especially marine species) have minimal asexual reproduction. Rather, they alternate a haploid, gamete-producing stage with a diploid, meiospore-producing stage. Both types of reproductive cells are motile, and the alternation is often correlated to tides. Massive release of the reproductive cells can be observed in tidepools or similar areas of slack water at low tide. The entire water mass can be green with millions of cells about 10 μm in size. Most of these cells (termed *swarmers,* if it is not known whether they are gametes or meiospores) probably do not produce new plants. Many serve as food for small animals, especially filter feeders, and many die.

The relationship of gamete fusion, or syngamy, and the time of meiosis is more varied in Chlorophyceae than in any other class of photosynthetic plants. Three of the four plant life-history types are present (see Chapter 2, Figs. 2-8 and 2-9) with only the dikaryotic type lacking. The two most common life-history types in Chlorophyceae are the predominantly haploid (see Fig. 2-8A) and the haploid alternating with diploid (see Fig. 2-9A).

Morphological Variation and Classification

In the restricted sense of Chlorophyceae in this text, there are forms that are unicellular (motile or nonmotile) and multicellular (motile or nonmotile), including colonies, simple or branched filaments, and large parenchymatous plants. Representative orders and selected genera, as mentioned in the text, are shown in Table 7-2.

At least three distinct groups (or *lines*) can be delimited in Chlorophyceae. Orders can be arranged in a morphological and possibly evolutionary series within them (Fig. 7-7). These three groups are: (1) the motile, or **volvocine** group (*volv* = roll, turn); (2) the nonmotile, nonfilamentous, or **chlorococcine** (*chloro* = green; *cocc* = berry) group; and (3) the nonmotile, filament-

ous, or **trichocine** group (*trich* = thread). The trichocine group was previously included with the chlorococcine as the **tetrasporine** group in other texts. It has the greatest morphological diversity and has been studied extensively since 1970. There is excellent evidence for accepting at least two main lines in it and possibly two side branches. The siphonous line included by some in the Chlorophyceae constitutes the Bryopsidophyceae in this text; see p. 279.

Volvocine Group The volvocine group (Fig. 7-7A) comprises primarily a single, well-defined order, Volvocales. All representatives are flagellated in the vegetative stage.

Vegetatively, one of the simplest is unicellular *Chlamydomonas* (Fig. 7-8A). It can be considered representative of primitive stock from which most Chlorophyceae may have evolved. *Chlamydomonas* is a biflagellated, motile alga. Under certain environmental conditions, however, it can revert to a nonmotile state, becoming embedded in a gelatinous matrix. In this state, it divides vegetatively and forms a dense, unorganized, amorphous mass of separate cells, resembling some genera in the chlorococcine line (see p. 269).

The volvocine group (Fig. 7-7A) may have evolved from a form similar to *Chlamydomonas* through a series of colonial forms with each cell similar to that of a vegetative *Chlamydomonas* cell (Fig. 7-8A). In cell division, cells remain fastened together by a common matrix, sometimes connected by cytoplasmic strands or attached cell walls. Most colonial Volvocales contain a constant number of cells. This can be shown by a series starting with *Gonium* (Fig. 7-8B, C). *Pandorina* (Fig. 7-8D), and *Eudorina* (Fig. 7-8E), in which all cells are reproductive. In *Pleodorina* (Fig. 7-8F)—sometimes considered as species of *Eudorina*—there are small cells that remain vegetative and large ones that are reproductive. The ultimate is *Volvox*, in which 500 to 50,000 (depending on the species) cells form a large spherical colony with only a few scattered reproductive cells (Fig. 7-12A). Hence, in the volvocine series, there is a trend toward a gradual increase in the number of cells in the colony, with an increase toward specialization of cells. In sexual reproduc-

Table 7-2 Representative Orders of Chlorophyceae, Including Selected Genera

Order and Genera	Cell Arrangement	Dominant Life History*	Habitat	Distinctive Features
Volvocales *Chlamydomonas, Eudorina, Gonium, Haematococcus, Pleodorina, Volvox*	unicellular, colonies	haploid	freshwater	motile, nonfilamentous
Tetrasporales *Tetraspora*	colonies (unicellular)	haploid	freshwater	nonmotile
Chlorococcales *Chlorella, Chlorococcum, Hydrodictyon, Pediastrum, Prototheca, Scenedesmus, Tetraedron, Trebouxia*	unicellular, colonies	haploid	freshwater (marine)	nonmotile, no vegetative divisions
Ulotrichales *Coleochaete, Draparnaldia, Fritschiella, Klebsormidium, Monostroma, Percursaria, Stigeoclonium, Ulothrix, Ulva*	filaments, parenchymatous	haploid haploid, diploid	freshwater marine	heterotrichy can be present; vegetative divisions never unicellular
Oedogoniales† *Oedogonium*	filaments	haploid	freshwater	motile cell with ring of flagella
Zygnematales *Closterium, Cosmarium, Micrasterias, Mougeotia, Spirogyra, Staurastrum, Zygnema*	unicellular, filaments	haploid	freshwater	lack flagellated cells

*See footnote †, Table 7-1.

†Includes Chaetophorales, Ulvales, and some other orders recognized by many phycologists.

tion, there is an apparent evolutionary series: *Gonium* and *Pandorina* are isogamous, and *Eudorina, Pleodorina,* and *Volvox* are oogamous.

Two well-known and ubiquitous genera, *Chlamydomonas* and *Volvox,* both occur in the freshwater habitat and can be easily grown in the laboratory. Cultures are maintained primarily by asexual reproduction.

In *Chlamydomonas,* vegetative (or asexual) reproduction takes place by cell division, usually in the motile condition, forming 4, 8, or 16 biflagellated cells (Fig. 7-9C). These are morphologically similar to the parent cell but are smaller when released and grow to adult size.

A generalized life history for an isogamous species is shown in Figure 7-9. Vegetative cells (Fig. 7-9A, B) function as gametangia by dividing to produce 4–32 biflagellated gametes (Fig. 7-9D, E), termed *plus* and *minus* mating types. After gametes are released (Fig. 7-9F, G), they swim about, then fuse to form a **planzygote** (Fig. 7-9H, I), which rounds up and secretes a heavy, cellulose wall (Fig. 7-9J). This resting zygote (also termed *zygospore*) is dormant in most freshwater species, permitting the species to tolerate extreme changes in moisture and temperature that can occur over a period of time. When appropriate conditions of moisture, light, and temperature are provided, the zygote germinates (Fig. 7-9K). The zygote nucleus divides by meiosis (and sometimes subsequently mitotically) to produce at least four biflagellated meiospores (Fig. 7-9L–O). These zoospores are released, and each enlarges to form a typical vegetative cell. These haploid cells are similar in morphology and behavior to zoospores produced by mitosis (Fig. 7-9C). However, they differ genetically as a result of the recombination of genes during syngamy and segregation during meiosis.

Some species can be manipulated in sexual

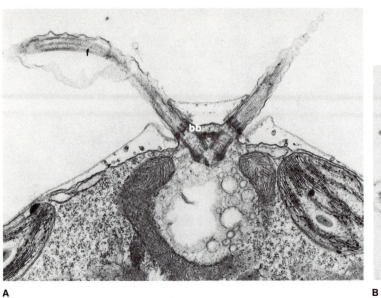

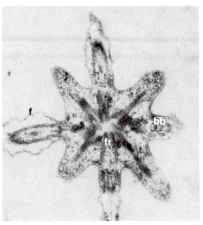

A　　　　　　　　　　　　　　　　　　　　　　**B**

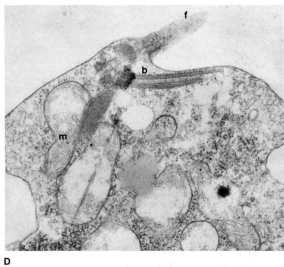

C　　　　　　　　　　　　　　　　　　　　　　**D**

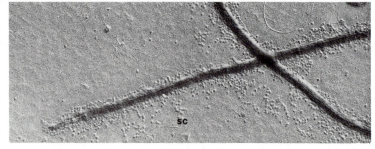

E

reproduction by regulating environmental parameters. The vegetative (haploid) stage is maintained in a mineral medium in liquid or on agar. Cells grown on agar, when suspended in water or dilute mineral solution and illuminated, become flagellated. Within two to four hours, these motile cells behave as gametes; when they are mixed with gametes of the opposite type (plus or minus), mating occurs. In mating, the initial step is a clumping (Fig. 7-10A) of compatible gametes, resulting from a cell surface enzyme produced by the flagellar tips that serves for cellular recognition and/or adhesion. The gametes of opposite mating types pair and five to ten minutes later a tubelike protoplasmic extension is produced by one or both gametes and serves to hold the fusing gametes together (Fig. 7-10B–D). The gametes swim freely for a period of time by the activity of the flagella of only one gamete (Fig. 7-10E). During swimming, the gamete apices are in contact (Fig. 7-11A), and fusion proceeds, resulting in a single-celled zygote (Fig. 7-11B, C). After cytoplasmic fusion, chloroplast and nuclear fusion follow (Fig. 7-11D), and the zygote wall is formed, with the resulting $2n$ cell becoming a resting stage.

When transferred to fresh mineral medium in the light (after maturing in the dark), the zygotes germinate within 24 to 48 hours. The heavy wall expands, and the diploid zygote nucleus undergoes meiosis followed by mitosis. In species with two mating types, half the progeny are one mating type and half the other (Fig. 7-9L–O). Because the zygote is the only diploid stage, the results of any recombination and segregation are immediately apparent in the haploid offspring.

In *Volvox*, only a few large cells divide vegetatively (Fig. 7-12A). When cell division is completed, the cells of the new colony are oriented with the apical, or flagellar, end pointing toward the inside of the new colony. Thus, the developing colony must literally turn inside out so the flagella are on the outside of the colony (**inversion**; Fig. 7-12B–G). As this occurs, the flagella are elongated and begin to expand in size (Fig. 7-12G, H). There are several ways by which inversion proceeds; essentially it is a folding back on itself of a hollow ball. The new colony remains in the parental colony (Fig. 7-12A, H) until breakdown of the parental colony.

Sexual reproduction in *Volvox* is oogamous, with a large, nonmotile egg fertilized by a small, elongate, biflagellated sperm. Studies using controlled cultures and culture conditions show that some species produce inducer substances that trigger gamete production and/or differentiation. The inducer can cause certain vegetative cells to act as female gametes (eggs), or it can trigger cell division in certain cells so that a new colony containing some eggs is formed. In addition, the production of male gametes by a number of divisions producing 64 or 128 sperms in a unit (sperm packet) can result (Fig. 7-13A). The sperm packets swim as a unit, and upon contact with an egg-containing colony, they dissolve a hole (Fig. 7-13A–C). The packet dissociates, and individual, biflagellated sperms burrow through to eggs (Fig. 7-13D–F), ultimately effecting fertilization. The diploid zygote develops a heavy

Figure 7-6 Flagellar arrangement and structure in Chlorophyceae of motile cells, electron microscope. **A, B,** symmetrical cells with cross-shaped flagellar roots. **A,** longitudinal section at apex showing flagella (*f*), flagellar insertion, basal body (*bb*), *Tetraspora*, ×1700; **B,** tangential section just below flagellar insertion near basal bodies, showing flagellar roots (*fr*), *Fritschiella*, ×22,500. **C, D,** asymmetrical cells sectioned at flagellar insertion, showing single broad band (*b*), *Klebsormidium*. **C,** ×34,500; **D,** ×28,000. Note also mitochondrion (*m*). **E,** shadow-cast preparation of flagellar scales (*sc*), ×11,000. (**A,** courtesy of J. Pickett-Heaps and the Linnean Society of London, from Duckett, J. G., and Racey, P. A., eds., *The Biology of the Male Gamete*, 1975, with permission of the Academic Press; **B,** courtesy of M. Melkonian and *Protoplasma*; **C, D,** courtesy of J. D. Pickett-Heaps, H. Marchant, K. Jacobs, and *Cytobios*; **E,** courtesy of Ø. Moestrup and the Linnean Society of London, from Ø. Moestrup, *Biological Journal of the Linnean Society*, 6:111–25, 1974, with permission of Academic Press.)

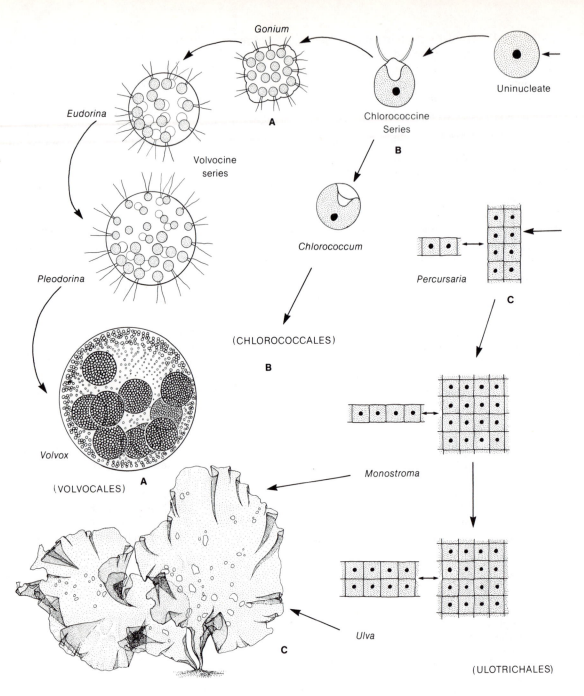

Gonium

Uninucleate

Chlorococcine Series

A

Eudorina

Volvocine series

B

Chlorococcum

Percursaria

C

Pleodorina

(CHLOROCOCCALES)

B

Volvox

Monostroma

(VOLVOCALES) **A**

Ulva

C

(ULOTRICHALES)

Figure 7-7 Groups in Chlorophyceae based on developmental series resulting from planes of nuclear and cell divisions (order names in *parentheses*). **A,** motile, or volvocine, series, flagella not shown for *Volvox* (Volvocales). **B–G,** nonmotile series culminating in different plant types. **B,** chlorococcine series (Chlorococcales, Tetrasporales); **C, D,** trichocine series with symmetrical zoospores (Ulotrichales); **E,** trichocine series with asymmetrical zoospores, possibly leading to land plants; **F, G,** trichocine series side-

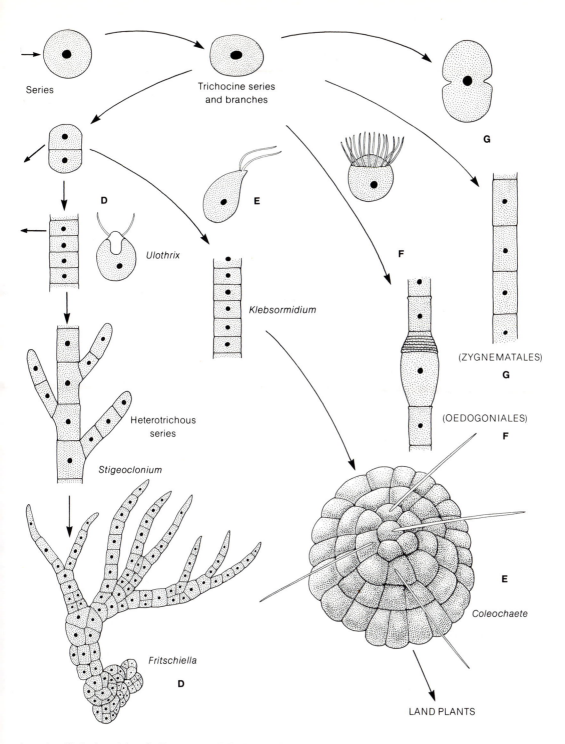

Series

Trichocine series
and branches

G

D

E

Ulothrix

Klebsormidium

F

(ZYGNEMATALES)

G

(OEDOGONIALES)

F

Heterotrichous
series

Stigeoclonium

Fritschiella

D

E

Coleochaete

LAND PLANTS

branches (**F**, Oedogoniales; **G**, Zygnematales). Surface and transverse section views are
shown for parenchymatous series **C.**

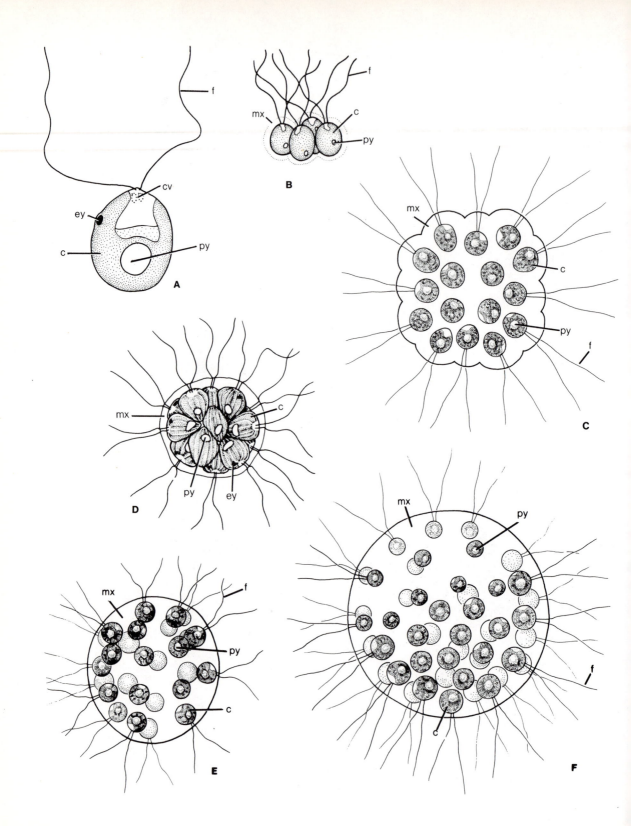

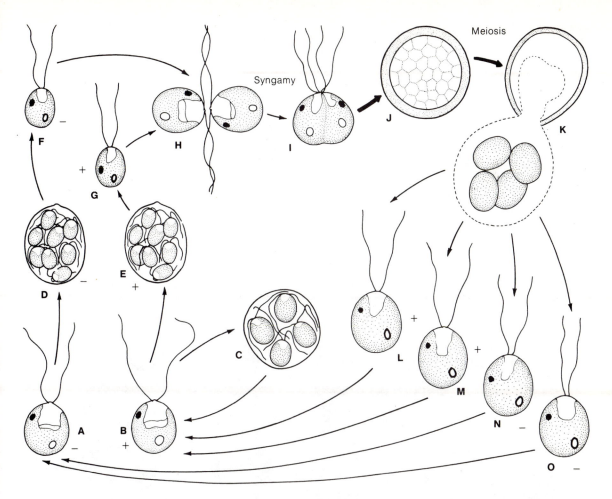

Figure 7-9 Predominantly haploid life history, *Chlamydomonas.* **A, B,** vegetative condition. **C,** asexual (or vegetative) reproduction. **D, E,** gamete production (by mitosis). **F, G,** gametes. **H,** fusion of isogametes (syngamy). **I,** motile zygote (or planozygote); **J,** resting zygote with heavy wall (zyospore). **K,** germinating zygote (after meiosis). **L–O,** motile meiospores (+ and − indicate mating types; *thin lines* indicate haploid phase, *thick lines* indicate diploid phase).

Figure 7-8 Morphological diversity in volvocine series (Volvocales). **A,** unicellular, *Chlamydomonas,* showing eyespot (*ey*), contractile vacuoles (*cv*), pyrenoid (*py*) in chloroplast (*c*), and flagella (*f*), ×2800. **B–F,** coenobial; **B, C,** *Gonium* in side view (**B**) and surface view (**C**), showing colony matrix (*mx*), ×800; **D,** *Pandorina,* ×800; **E,** *Eudorina,* ×400; **F,** *Pleodorina,* ×450.

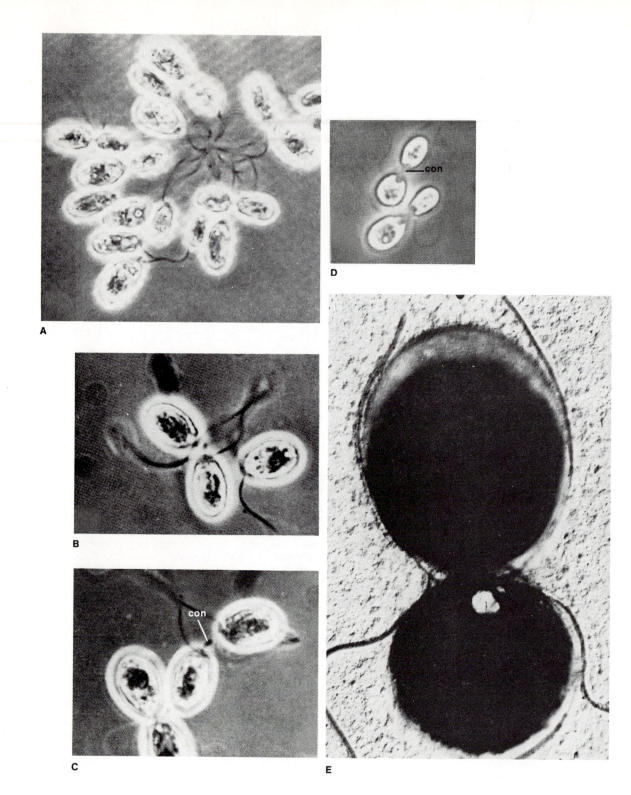

wall (Fig. 7-14A) and remains in the parental colony. In the laboratory, zygotes can be induced to germinate approximately three weeks after formation by transferring them to fresh growth medium. Germination is similar to that described for *Chlamydomonas* (p. 263); however, generally only one meiotic product is viable. The meiospore is a biflagellated, spherical, pale cell (Fig. 7-14B) that swims for a short period before dividing to produce a small colony of 8–128 cells (Fig. 7-14C). The new, haploid colony contains primarily vegetative cells and a few reproductive cells, which will produce new colonies.

Chlorococcine Group The morphology of the nonmotile, nonfilamentous algae of the chlorococcine group is more variable than in the volvocine series. Diversity ranges from simple unicellular to more complex, multicellular plants; the cells and plants are of varied shapes, sizes, and forms, but never filamentous or parenchymatous as in the trichocine group. When present, motile cells are a reproductive stage, and many are similar to the chlamydomonad form.

Chlorococcales is the main order illustrating this unicellular to multicellular line (Fig. 7-7B). Basic chlorococcalean forms are unicellular *Chlorococcum* and *Chlorella* (Fig. 7-15A–C). *Chlorococcum* produces motile reproductive cells, whereas *Chlorella* produces only nonmotile spores, or aplanospores. Some colonial genera have a definite number of cells in a colony. In some of these, such as the coenobial *Scenedesmus*, *Pediastrum*, and *Hydrodictyon* (Fig. 7-15D–F), the zoospores or aplanospores are retained within parental cells until a new colony having the parental form is produced.

Some Chlorococcales have been effectively exploited for laboratory studies of photosynthesis and other biochemical research. These include *Chlorella* and *Scenedesmus* (Fig. 7-15C, D), which were used in early investigations on the splitting of water and fixation of carbon in photosynthesis. Because of their small size and simple nutrient requirements (only minerals, no vitamins or growth factors), these algae are easy to grow and maintain in the laboratory with minimal space and equipment.

Chlorococcales are probably an evolutionary sideline in green algae, much like Volvocales. Their great diversity and wide distribution possibly indicate that it is a heterogeneous order.

Trichocine Group Morphological variation in the large, diverse trichocine group is almost limitless. The trichocine series is composed primarily (although not exclusively) of multicellular forms that are filamentous or parenchymatous (Fig. 7-7C–E). Motile cells are present only as reproductive cells and can be symmetrical or asymmetrical (Fig. 7-6A, C). However, not all representatives produce motile cells. The main order is Ulotrichales (in the broadest sense, this includes the Ulvales and Chaetophorales of many authors). These algae regularly have vegetative divisions in the plant in a regular manner, and they also produce reproductive cells. At one time, members were placed in a single line, but this is now subdivided on the basis of cytological features (primarily ultrastructural).

In addition to the ulotrichalean branches (Fig. 7-7C–E), there are two possible side branches (Fig. 7-7F, G) containing the specialized orders Oedogoniales and Zygnematales. Some authors would not consider these to be in the trichocine series; however, they do have morphological features similar to ulotrichalean algae.

Ulotrichales. If a nonmotile, *Chlorococcum*like cell divides vegetatively (not to produce spores), and the cell wall is involved, its derivatives dividing in the same plane will produce a simple **uniseriate** (*uni* = one; *seri* = series, row) filament of cells. Essentially, this is a simple type of division in a linear fashion occurring anywhere in the filament, as in *Ulothrix* or *Klebsormidium*

Figure 7-10 Sexual reproduction in *Chlamydomonas (C. moewusii)*. **A,** clumping of gametes by flagellar tips. **B–D,** mating pairs held by cytoplasmic connection (*con*). **A, D,** ×1200; **B, C,** ×1550. **E,** platinum-carbon shadowed electron microscope preparation showing free-swimming pair; gamete at top is passive partner, ×7500. (Courtesy of R. M. Brown, Jr. and *Journal of Phycology*.)

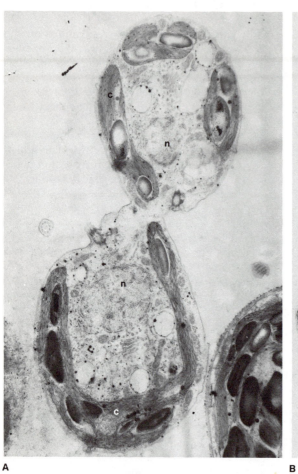

A

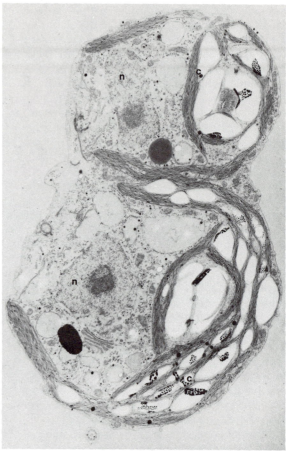

B

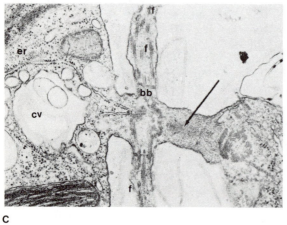

C

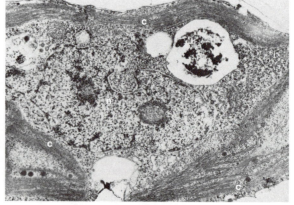

D

(Fig. 7-16A, B). Repeated cell divisions in a second plane will produce a multiseriate plant (Fig. 7-7C). If cells in a primary, unbranched, uniseriate axis divide once in a second plane, a **biseriate** (*bi* = two) plant comparable to marine *Percursaria* results (Fig. 7-16C). And if, in addition, division occurs repeatedly in the second plane, then a single-layered plant typical of *Monostroma* is derived (Fig. 7-16D–F). It is also possible for early cell divisions to result in **foliose** (*foli* = leaf) plants in the form of a monostromatic tube, as in *Enteromorpha*, or a distromatic blade, in which the two cell layers are usually in close contact, as in *Ulva* (Fig. 7-16H). Both the foliose types can be considered parenchymatous, as cell divisions occur somewhat randomly in the growing plant.

It is possible to achieve other types of plants by supplementing cell divisions of the uniseriate filament with occasional cell divisions in a second plane (Fig. 7-7D, E). Thus a branch initial is established. A uniseriate, branched filament, such as the freshwater *Stigeoclonium* (Figs. 7-17A and 7-18A), can achieve growth in two directions: upright and horizontal (Fig. 7-7D). This growth habit with erect and prostrate parts is known as **heterotrichy** (*hetero* = different; *trich* = thread, filament) (Figs. 7-17B and 7-18A, B), and is considered a necessary precursor for the development in land plants of any erect assimilative system and prostrate attaching system. In some species of *Stigeoclonium*, both prostrate and erect systems are almost equally well represented

under natural conditions. In *Fritschiella* (Figs. 7-17B and 7-18B), both systems develop. The prostrate portion futher differentiates into multicellular, colorless **rhizoids** (*rhiz* = root; *oid* = like) which anchor the plant in mud and parenchymatous, photosynthetic portions on the surface. In other genera, one or the other of these systems has been reduced or even eliminated. The prostrate system is completely suppressed in *Draparnaldia* (Fig. 7-17C), whereas in *Coleochaete* (Figs. 7-17D and 7-18C), the prostrate system is usually better-developed, and the upright system is reduced, sometimes to only a few colorless hairs.

Representatives of the basic group described here are interesting, in that one series, *Ulothrix* to *Ulva*, occurs primarily in the marine environment (Fig. 7-7C), and another, *Ulothrix* to *Fritschiella*, occurs primarily in the freshwater habitat (Fig. 7-7D). Some species of *Ulothrix* grow in fresh water and others in seawater, indicating that it is either a primitive algal form or has secondarily returned to the marine habitat. A third series (Fig. 7-7E), *Klebsormidium* to *Coleochaete*, which also occurs in fresh water, is probably more important with regard to the origin of land plants but is still incompletely known. The significance of this last line is considered in the section on the relationships of Chlorophyta (see p. 306). However, at this point the important feature is the occurrence of cell divisions in three planes with growth in two directions. These are basic for true parenchymatous tissue, which makes possible the development of complex tissues in higher plants. This is not proof that any developmental series or line led directly to higher plants; rather, it shows that the genetic potential for the development of higher plants is present in ulotrichalean algae. They contain the potential for the basic form and tissue of land plants.

Of the many representatives, relatively few are known in any detail. *Ulothrix* (Fig. 7-16A) is unbranched except for the basal holdfast cell, with each cell of the filament capable of producing isogametes or zoospores. The zoospores are biflagellated or quadriflagellated, and gametes are biflagellated. Freshwater species of *Ulothrix* form a resting zygote following syngamy, as is typical of freshwater algae (e.g., *Chlamydomonas*, Fig. 7-9). In contrast, marine species, which are

Figure 7-11 Stages in syngamy in *Chlamydomonas*. **A, B,** *C. reinhardtii*, ✕ 15,200. **A,** early stage showing apical contact and nuclei (*n*). **B,** later stage showing two nuclei (*n*) and chloroplasts (*c*). **C, D,** *C. moewusii*, showing bridge formation (*arrow*). Note basal body (*bb*), contractile vacuole (*cv*), endoplasmic reticulum (*er*) and flagella (*f*) ✕23,250. **D,** chloroplast and nuclei undergoing fusion, ✕ 14,700. (**A, B,** courtesy of E. I. Friedman, A. L. Colwin, L. H. Colwin, and *Journal of Cell Science;* **C,** courtesy of R. E. Triemer and *Protoplasma;* **D,** courtesy of R. M. Brown, Jr. and *Journal of Phycology.*)

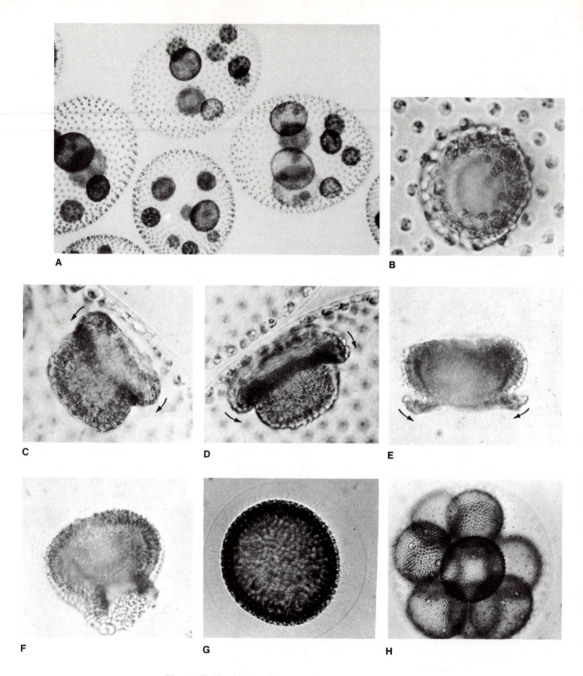

Figure 7-12 *Volvox.* **A,** vegetative cells, showing new colonies (coenobial), ×80. **B–H,** new colony development showing a type of inversion; arrows indicate movement of cells during inversion. **B,** end view of young colony to show opening through which colony will invert; **C,** beginning of inversion, with new colony starting to fold back; **D–F,** further stages in inversion; **G,** inversion complete, with flagella extended; **H,** several young colonies within parental colony (**B–D,** ×500; **E–G,** ×560; **H,** ×110). (**A,** courtesy of R. Starr; **B–H** courtesy of W. H. Darden and the *Journal of Protozoology.*)

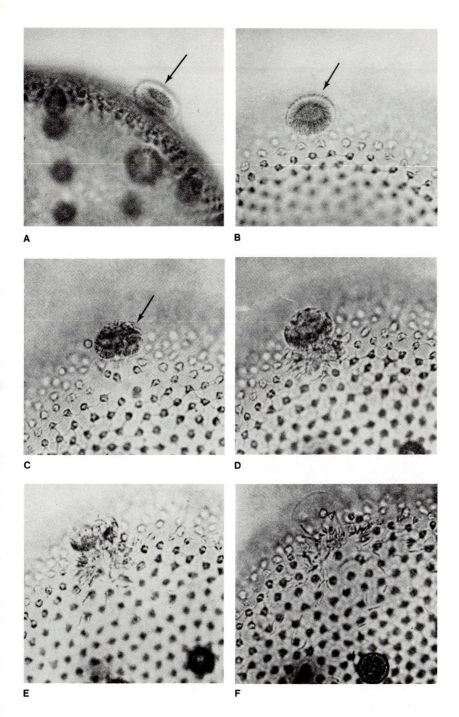

Figure 7-13 Sexual reproduction in *Volvox,* ×225. **A–C,** sperm packet (*arrow*) attachment to female colony. **D, E,** breakdown of sperm packet and penetration of individual sperm. **F,** sperm migration through colony, ×230. (Courtesy of M. McCracken and *Archiv für Protistenkunde.*)

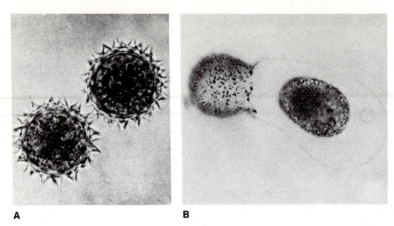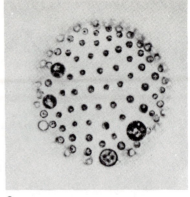

A **B** **C**

Figure 7-14 Sexual reproduction in *Volvox*. **A,** zygotes (*V. rousseletii*), ×310. **B, C,** zygote germination (*V. gigas*). **B,** meiospore (*left*) in transparent vesicle (flagella not shown), ×570; **C,** first colony formed from meiospore; large cells will be reproductive and form new colonies, ×220. (**A,** courtesy of M. McCracken and *Archiv für Protistenkunde;* **B, C,** courtesy of W. J. Vande Burg and *Archiv für Protistenkunde*.)

usually larger, have two generations (isomorphic or heteromorphic), and the zygote develops immediately into the diploid phase. Similar in appearance to *Ulothrix*, (except lacking a holdfast cell), is *Klebsormidium* (previously known as *Hormidium* and *Chlorhormidium;* Fig. 7-16B). It is known only from the freshwater habitat and grows readily on moist soil or in cool, fast-moving water. Both algae have a single ring-shaped, peripheral chloroplast per cell with at least one pyrenoid. *Ulothrix* readily produces symmetrical motile cells with apical flagella and cruciately arranged flagellar roots (Fig. 7-6B). *Klebsormidium* does not produce motile cells readily, and when it does, they are asymmetrical. Their flagella are subapically attached and have a single, broad band root (Fig. 7-6C, D). Ultrastructural differences between the two genera involve the manner of cytokinesis and wall deposition (cell wall formation).

The freshwater *Fritschiella* (Figs. 7-17B and 7-18B), which grows naturally on mud banks beside streams or ponds, will produce an extensive basal portion, complete with colorless rhizoids and parenchymatous patches, when grown on a solid substrate with a thin, liquid layer on top (as

in the laboratory on mineral agar). *Stigeoclonium* (Figs. 7-17A and 7-18A) has a similar plasticity in form, depending upon available water. Growth in liquid culture enhances the upright system at the expense of the basal system. *Fritschiella* and *Stigeoclonium*, like *Ulothrix*, are isogamous, and in *Fritschiella* there are reports on an alternation of isomorphic phases.

Coleochaete (Figs. 7-17D and 7-18C), which is postulated as a possible "link" to land plants, grows prostrate on the substrate as a flat disc. Basal filaments are usually closely appressed, giving the appearance of parenchymatous development. In vegetative cell division, there are persistent, longitudinal microtubules following the mitotic spindle (Fig. 7-5C), similar to those in land plants. However, the new wall forms first at the outside of the cell. *Coleochaete* produces biflagellated, asymmetrical sperms with subapical flagellar insertion, scaly flagella, and a single, broad, striated flagellar root. The egg is retained on the vegetative plant, and syngamy occurs *in situ*. The zygote then becomes a resting stage that develops within haploid vegetative tissue produced by the surrounding cells.

Ulva grows in the marine habitat (Fig.

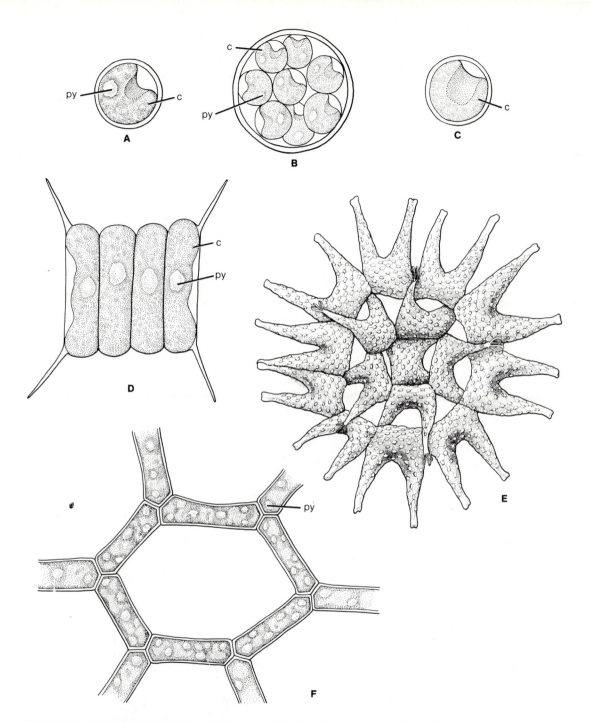

Figure 7-15 Morphological variation in chlorococcine series (nonfilamentous, nonmotile; Chlorococcales). **A, B,** *Chlorococcum,* ×1000. **A,** vegetative cell; **B,** zoospore production (flagella not shown): *c,* chloroplast; *py,* pyrenoid. **C,** *Chlorella,* ×1000. **D–F,** coenobial forms; **D,** *Scenedesmus,* ×1000; **E,** *Pediastrum,* ×400. **F,** *Hydrodictyon,* ×200.

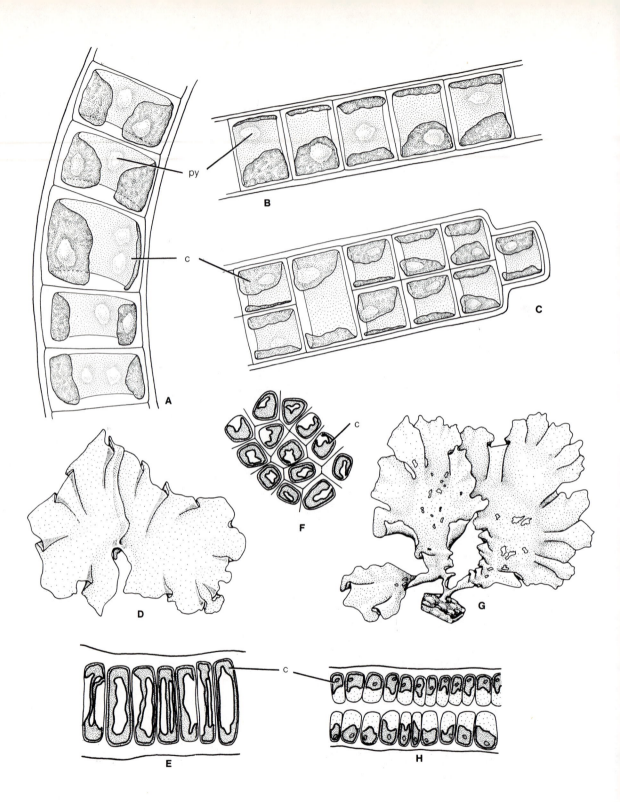

7-18D) and produces a juvenile vegetative plant that starts as a uniseriate filament but quickly becomes biseriate. Divisions occur in both planes, eventually resulting in a flat, foliose plant (Fig. 7-16G). The two layers that result are usually in close contact (Fig. 7-16H). Basal cells are rhizoidal and aggregate to form a compact holdfast. *Ulva* has an alternation of isomorphic phases, with no apparent morphological difference between haploid gametophyte (male and female) plants and diploid sporophytic plants. In some species, the mature male plant appears more orange or yellow due to the abundance of carotenoid pigments in the smaller male gametes. Except for basal cells, which usually remain sterile, all cells are potentially capable of producing gametes, starting at the plant margin.

Ulva undergoes sexual reproduction readily, with uninucleate cells of haploid gametophytes (Fig. 7-19A, B) dividing mitotically to produce 8, 16, or 32 terminally biflagellated gametes that are released through a pore or rupture in the wall (Fig. 7-19C–F). Syngamy occurs in water after a period of swarming. As in *Chlamydomonas* (Fig. 7-10A), gametes of opposite mating types clump before pairing (Fig. 7-20A). Cytoplasmic contact near the anterior end is established within ten seconds after pairing, and fusing cells swim away from light. Fusion is completed in three minutes (Fig. 7-20B), and quadriflagellated zygotes (Fig. 7-20C) start to settle after approximately five minutes (Fig. 7-19G–I). The zygote attaches to the substrate by the anterior end, and the flagella (which lie along the cell) are ab-

sorbed in 15 minutes; nuclear fusion is complete 30 minutes from the start of mating.

The zygote, which secretes a cellulose wall material before germinating, produces a multicellular filament (Fig. 7-19J, K) by repeated mitotic divisions. Subsequently, the typical, flat, foliose plant is formed (Fig. 7-19L). This plant is morphologically identical to the gametophyte, but cytologically it is different, as the cells are diploid. At maturity, all cells of this diploid plant (except those near the base) are potentially capable of producing meiospores, although usually only the marginal cells undergo meiosis. Meiosis is followed by several mitotic divisions to produce 4, 8, or 16 terminally quadriflagellated meiospores (Fig. 7-19M–O). Spores are released in the same way as gametes, and after a period of swimming, they settle down, losing their flagella and secreting a cellulose wall (Fig. 7-19P, Q). Cell divisions form a filament (Fig. 7-19R–U) and then the typical foliose plant, which is the new gametophytic phase (Fig. 7-19A, B). Both gametes and zoospores are similar to those of *Ulothrix, Stigeoclonium,* and *Fritschiella*.

Other Related Chlorophyceae The trichocine line can be considered to have two side branches, or offshoots (Fig. 7-7F, G), each comprising a single order: Oedogoniales and Zygnematales. These orders are restricted to fresh water and are specialized morphologically and reproductively. Both are ubiquitous and have the predominantly haploid life history characteristic of freshwater algae. The order Oedogoniales contains only three filamentous genera, whereas in Zygnematales both unicellular and multicellular genera are represented. Some workers consider both orders distinct from trichocine algae, and in some instances they recognize these orders as separate classes, subclasses, or even divisions.

Oedogoniales. The order Oedogoniales is set apart on the basis of its cell division and reproductive cell morphology. The vegetative cells contain a single, peripheral, netlike chloroplast (Fig. 7-21A). The motile reproductive cells characteristically have a ring, or crown, of flagella at the anterior end of the cell (Fig. 7-21B); zoospores have more flagella (approximately 128) than male gametes (about 30). Cell division is

Figure 7-16 Morphological variation in filamentous and parenchymatous trichocine series (Ulotrichales). **A,** *Ulothrix,* showing unbranched, uniseriate filament; note single chloroplast (*c*) with many pyrenoids (*py*), ×400. **B,** *Klebsormidium,* morphologically similar to *Ulothrix,* but with only one pyrenoid per chloroplast, ×500. **C,** *Percursaria,* showing biseriate form, ×400. **D–F** *Monostroma.* **D,** habit, ×0.5; **E,** transverse section showing single layer of cells, ×530; **F,** surface view, ×750. **G, H,** *Ulva.* **G,** habit, ×0.5; **H,** transverse section showing two layers of cells, ×500; surface view of *Ulva* similar to that of *Monostroma* (**F**).

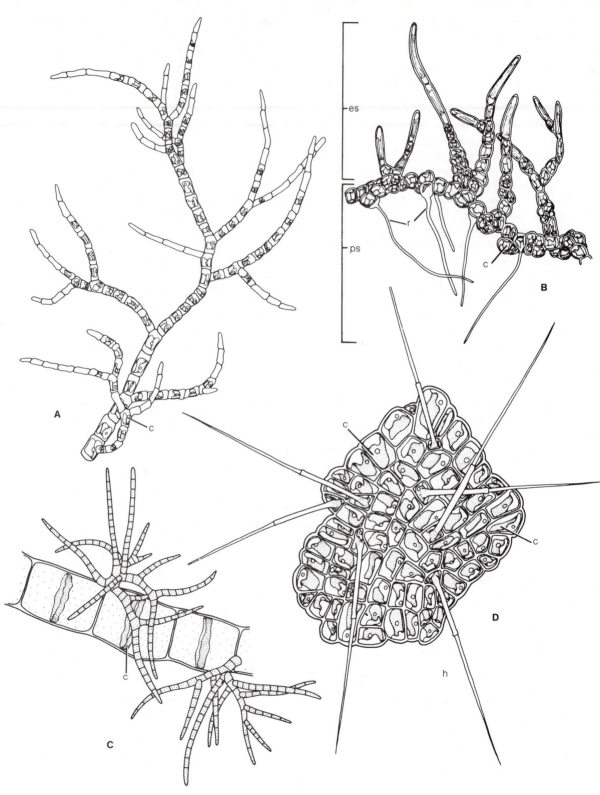

unique for this order, in that there are cellulose rings produced at the anterior end of some vegetative cells (Fig. 7-21A). These rings are often very obvious and help in the recognition of genera, especially the ubiquitous *Oedogonium* (Fig. 7-21A).

All three genera of Oedogoniales are oogamous, with both gametes produced in specialized gametangia that are easily recognized (Fig. 7-21C, D). In syngamy, the sperm is attracted to the egg, which is retained in the female gametangium on the haploid filament (Fig. 7-22A, B). The fusion of gametes occurs in less than a minute (Fig. 7-22C–N), and the zygote develops into a resting stage with a heavy, often ornamented, cellulose wall. In germination of the zygote, the diploid nucleus undergoes meiosis, producing one or two motile, haploid meiospores with the typical ring of flagella. Upon settling, the meiospore forms a filamentous, haploid plant.

Zygnematales. This order includes filamentous forms, as *Mougeotia, Zygnema,* and *Spirogyra* (Figs. 7-23A–C and 7-24) and unicellular forms, as the desmids *Staurastrum, Micrasterias, Cosmarium,* and *Closterium* (Fig. 7-23D–G). One of the most distinctive features of Zygnematales is the complete lack of flagellated cells at any time. Sexual reproduction results from conjugation of somewhat amoeboid gametes, one or both of which move (Fig. 7-25). Vegetative reproduction is by fragmentation in multicellular forms and cell division in unicellular forms. In desmids (most of which are unicellular or may be considered so), the vegetative cell is composed of two halves that are mirror images if split longitudinally or horizontally (Fig. 7-26A). Cell division results in each half-cell regenerating a new half-cell and producing two new, identical cells (Fig. 7-26H). This

process begins by nuclear division in the area between the two half-cells (Fig. 7-26A), followed by a gradual formation of new half-cells to form two new identical cells (Fig. 7-26B–H).

Class Bryopsidophyceae

The recognition that multinucleate Chlorophyta should be considered as a separate class results from cytological, biochemical, and life-history differences that set them apart. Essentially, this class encompasses the siphonous type of development of Chlorophyta. The Bryopsidophyceae is poorly represented in fresh water; most occur in warmer marine waters, and many are macroscopic.

Cell Structure

Cells of the Bryopsidophyceae are multinucleate at maturity, and many genera form transverse walls only when producing reproductive structures. Thus, each protoplasmic unit contains many nuclei, chloroplasts, and other organelles. A striking feature of siphonous forms is the large central vacuole occupying a large part of the cell (Fig. 7-27A, B), with the cytoplasm only a thin layer appressed to the cell wall. In some cells there are two zones in the cytoplasm, with numerous chloroplasts in the inner zone adjacent to the vacuole (Fig. 7-27C).

Generally, the small, disclike chloroplasts are similar to those of Chlorophyceae, except pyrenoids can be absent, and thylakoids are present in larger bands, similar to grana in land plants. Some Bryopsidophyceae have, in addition to chloroplasts, a second, colorless plastid containing starch but no thylakoids. In these algae, starch is also present in the cytoplasm. The pigments of Bryopsidophyceae are the same as in Chlorophyceae, but some contain additional carotenoids not reported for other algae. In addition to mitochondria, golgi, and endoplasmic reticulum, other cytoplasmic inclusions, such as protein bodies, are present.

Figure 7-17 Morphological variation in heterotrichous, trichocine series (Ulotrichales): *c*, chloroplast. **A,** *Stigeoclonium*, showing branched growth, ×150. **B,** *Fritschiella*, showing heterotrichy with prostrate (*ps*) and erect (*es*) systems and rhizoids (*r*), ×375. **C,** *Draparnaldia*, showing main axis, erect form, and smaller branches; note chloroplast (*c*). **D,** *Coleochaete*, prostrate species with erect hairs (*h*), ×435.

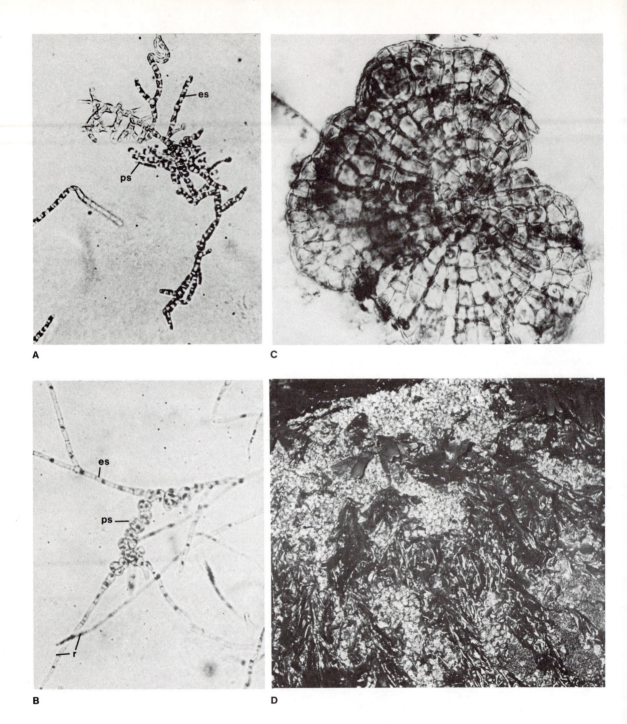

Figure 7-18 Heterotrichous habit. **A,** *Stigeoclonium,* showing basal prostrate system (*ps*) and start of erect filaments (*es*), ×150. **B,** *Fritschiella,* young plant showing rhizoids (*r*) and start of prostrate and erect system, ×150. **C,** *Coleochaete,* prostrate plant, top view. ×1000. **D,** *Ulva,* habitat on intertidal rocks. ×0.5. (**D,** courtesy of M. Higham.)

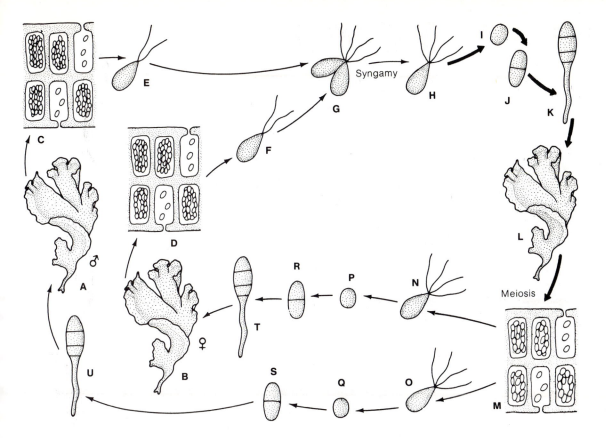

Figure 7-19 Life history of *Ulva* with alternation of isomorphic haploid (*thin lines*) and diploid phases (*thick lines*). **A, B,** mature gametophytes (pale margin indicates region of gamete discharge). **C, D,** gamete production by mitosis. **E, F,** gametes (biflagellated). **G,** fusion of isogametes (syngamy). **H,** planozygote. **I,** zygote. **J, K,** filamentous juvenile sporophyte. **L,** mature sporophyte (pale margin indicates region of spore discharge). **M,** zoospore production (by meiosis). **N, O,** motile meiospores (quadriflagellate). **P–U,** filamentous juvenile gametophytes.

Cell walls are generally thick and contain other polysaccharides than cellulose; the most common are mannan, composed of mannose, and xylan, composed of xylose. Interestingly, xylan maintains its crystalline nature only when it is wet, whereas cellulose and mannan are equally crystalline whether dry or wet. The chemical nature of the cell wall can vary during the life history of a given species. For example, in *Bryopsis* and *Derbesia* (Fig. 7-30A–D), the cell walls of the diploid plant contain mannan, but the haploid plant contains cellulose or xylan. Walls have alternating layers of parallel-oriented microfibrils, which spiral around the length of the plant and form helices. This probably imparts strength to the wall, which is perhaps necessary for a large, multinucleate plant. Many tropical species have calcium carbonate in the wall and are important in coral reef formation.

Motile reproductive cells are uninucleate, generally resembling the symmetrical reproductive cells of the Chlorophyceae. Contractile vacuoles (a feature of many freshwater motile cells) are usually not present, and the occurrence of eyespots is variable. Motile cells can have two equal, anteriorly attached, whiplash flagella simi-

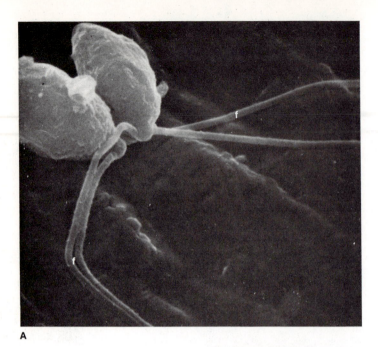

A

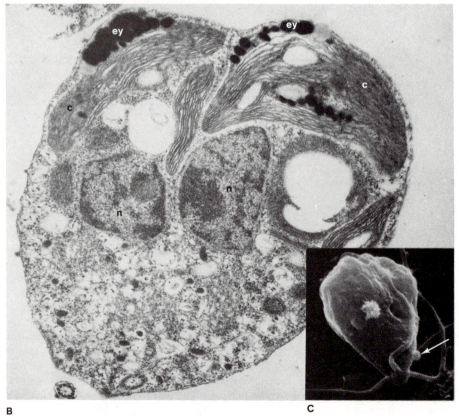

B

C

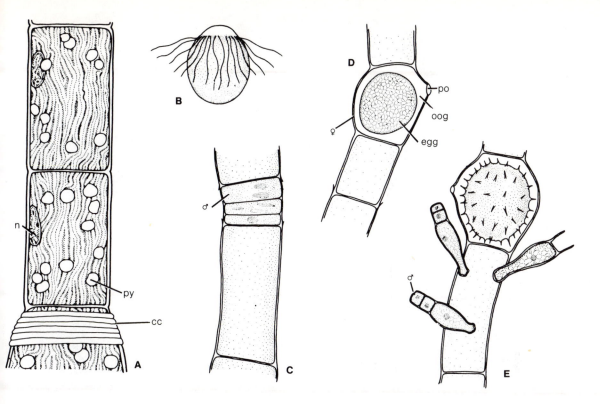

Figure 7-21 Morphology of *Oedogonium* (Oedogoniales). **A,** part of vegetative filament, showing netlike chloroplast with pyrenoids (*py*) and cell wall rings comprising cellulose cap (*cc*): *n*, nucleus ×260. **B,** multiflagellate sperm, with anterior crown of flagella, ×500. **C,** filament with narrow, immature, male gamete-producing cells (♂), ×700. **D,** filament with mature female (♀) gametangium (*oog*) containing single egg (*egg*) and subapical fertilization pore (*po*), ×700. **E,** filament with spiny resting zygote, showing three epiphytic, dwarf male plants; two plants immature, each with two male gametangia (*left*); third plant (*right*) mature with discharged male gametangia, ×825.

Figure 7-20 Sexual reproduction in *Ulva*. **A, C,** scanning electron microscope micrograph. **A,** mating gametes, fixed prior to cytoplasmic contact while held together by flagella (*f*), ×10,000. **B,** young zygote (transmission electron micrograph) five minutes after start of syngamy: *n*, two nuclei; *c*, chloroplasts; *ey*, eyespots ×24,000. **C,** young zygote showing four flagella and extruding vacuole (*arrow*), ×8000. (Courtesy of T. Braten and *Journal of Cell Science*.)

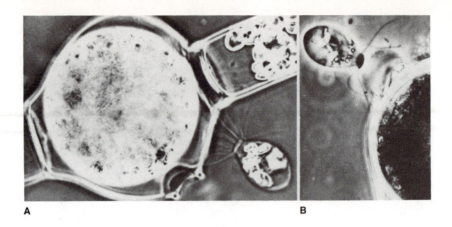

A   B

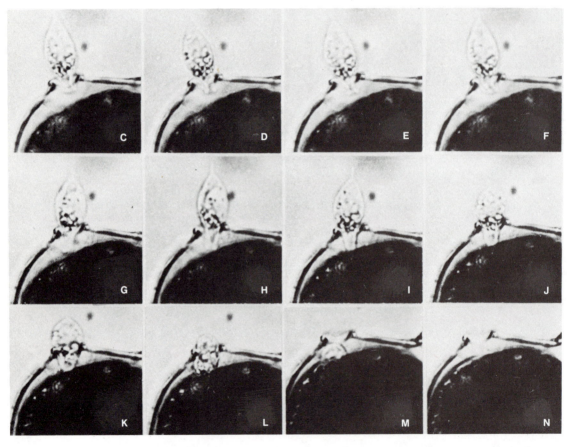

Figure 7-22 Sexual reproduction in *Oedogonium* (Oedogoniales). **A, B,** sperm cell attached near opening of female gametangium. **B,** pore substance holding sperm, ×850. **C–N,** sequence of syngamy with cine film (16 frames/sec, ×960); time from **C–N:** 53.5 seconds. Note change in sperm shape as syngamy proceeds (especially compare, **B, C, I**). (Courtesy of L. R. Hoffman and *Journal of Phycology*.)

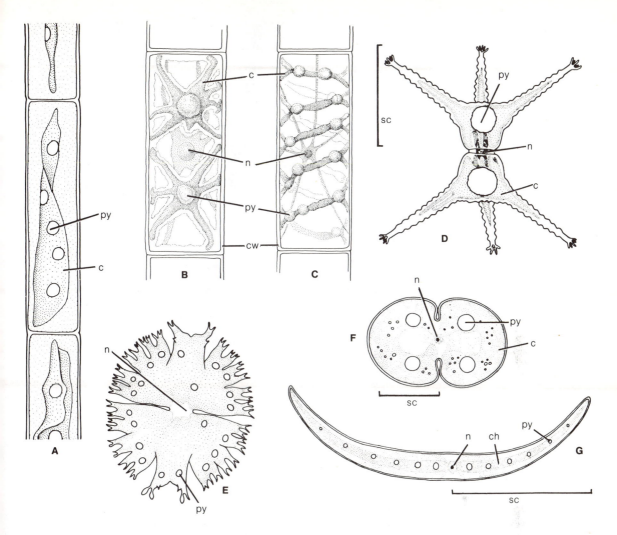

Figure 7-23 Morphological diversity in Zygnematales. **A–C,** filamentous forms showing cell wall (*cw*), nucleus (*n*), and chloroplast (*c*) with pyrenoid (*py*). **A,** *Mougeotia,* ×510; **B,** *Zygnema;* **C,** *Spirogyra,* ×700. **D–G,** unicellular desmids, showing the half-cells (*sc*), nucleus (*n*), chloroplast (*c*), and pyrenoid (*py*). **D,** *Staurastrum;* **E,** *Micrasterias,* ×270; **F,** *Cosmarium,* ×220; **G,** *Closterium,* ×525.

lar to *Chlamydomonas* (Fig. 7-8A) or a ring of anterior, whiplash flagella, as in Oedogoniales (Fig. 7-21B). The arrangement of flagellar roots is a cross, as in many Chlorophyceae (Fig. 7-6B).

Nuclear division and division of other organelles occurs regularly. Some genera (in Cladophorales and Siphonocladales) produce transverse walls that create multinucleate protoplasmic units. In other genera (in Codiales and Dasycladales), transverse wall material is formed, but these walls are not complete, and cytoplasmic continuity is maintained.

Reproduction

Life histories of many Bryopsidophyceae are predominantly diploid, gametes being the only haploid cells (see Fig. 2-8B). Other members have

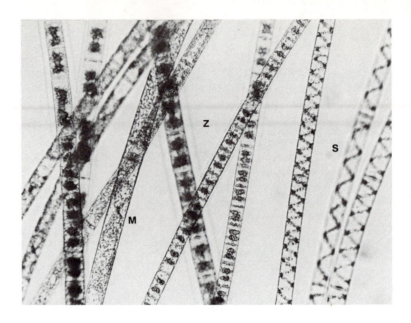

Figure 7-24 Vegetative cells of filamentous Zygnematales, common in shallow bodies of water from spring through autumn. Note characteristic chloroplast shapes: spiral (*Spirogyra, S*), band (*Mougeotia, M*), and star (*Zygnema, Z*), ×200.

two different morphological phases (Fig. 2-9A). The predominantly haploid life history (see Fig. 2-8A) common in Chlorophyceae is rare in this class. Asexual reproduction is usually by fragmentation of vegetative parts. Asexual spores can be produced but are irregular; their production is possibly restricted to conditions of environmental stress.

Gametes of many Bryopsidophyceae are anisogamous, with small, almost colorless, biflagellated male gametes. In contrast, the female gamete contains a great deal of cytoplasm and is often motile for only a short time. After syngamy, the zygote produces a new plant without any resting period.

Morphological Variation and Classification

Bryopsidophyceae includes at least four orders separated on the basis of the location and the time in development in the life history when transverse walls (or *septa*) are formed (Table 7-3). Whether the morphological variation present can be derived from a single progenitor is uncertain. A possibile progenitor is an alga similar to a living chlorophycean genus in the chlorococcine group (Fig. 7-7B), since many of these unicellular forms become multinucleate prior to spore formation. If the alga remains multinucleate but does not produce spores, then a subsequent in-

Figure 7-25 Sexual reproduction (conjugation) in Zygnematales (*Spirogyra*), light microscope. **A, B,** phase-contrast microscope. **A,** contraction of male gametes (lower filament) and early contraction of female gametes (*arrows*) in upper filament; **B,** migrating gametes. **C,** section, similar to **B; D,** section showing final stages in migration of male gamete (*arrow*). Note also young zygotes, ×400. (Courtesy of L. C. Fowke and *Journal of Phycology*.)

A

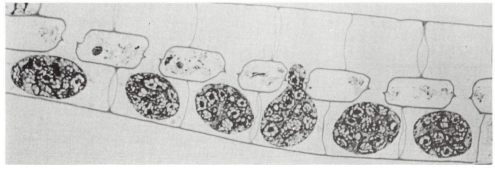

B

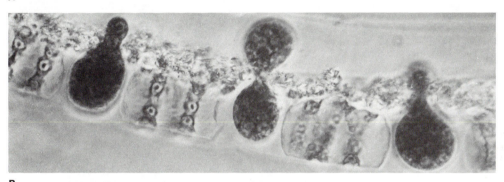

C

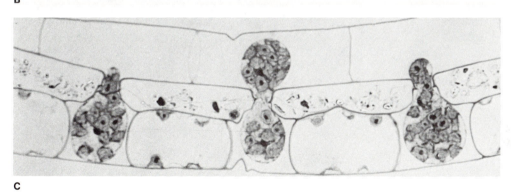

D

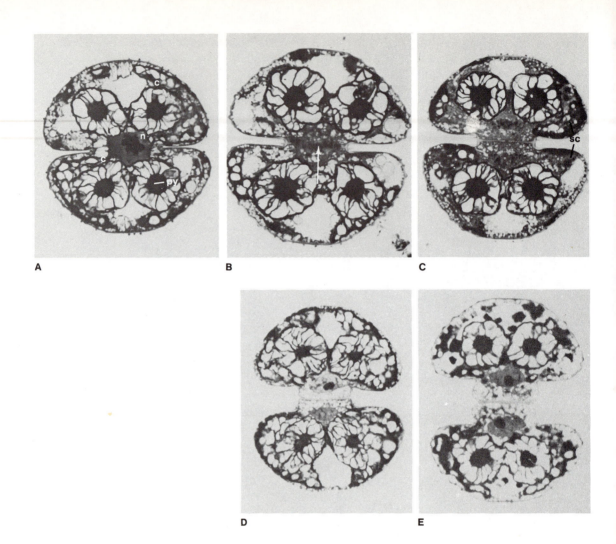

A   B   C

D   E

crease in size could produce the bryopsido-phycean form. For example, in the freshwater chlorococcalean *Hydrodictyon* (Fig. 7-15F), nuclear division occurs while the cells are increasing in size, so that each cell of the large net is multi-nucleate for a period of time before motile cells are formed.

The Cladophorales shows a series in which there is regular (and complete) transverse wall formation of multinucleate units (Fig. 7-28A). The simplest form is the unbranched filament, as in *Urospora* (Fig. 7-29A, B). As in the fil-amentous, uninucleate trichocine series (Chlo-rophyceae), occasional divisions produce branch initials and result in a branched plant

with multinucleate cells, such as *Cladophora* (Fig. 7-29C).

Development similar to that in Clado-phorales occurs in Siphonocladales (Fig. 7-28B); however, transverse wall formation is irregular, and the plant—for example, *Valonia* (Fig. 7-29D, E)—is never filamentous. The alga usually con-sists of many large, multinucleate units (up to 4 cm) often surrounding a large, central vacuole.

Complete transverse walls are not always produced; thus, another developmental series is possible, as shown by Codiales (Fig. 7-28C). In these, further elongation produces a branched, multinucleate tube, or **coenocyte** (*coeno* = com-mon; *cyte* = cell) (Fig. 7-30A, D). There can be an

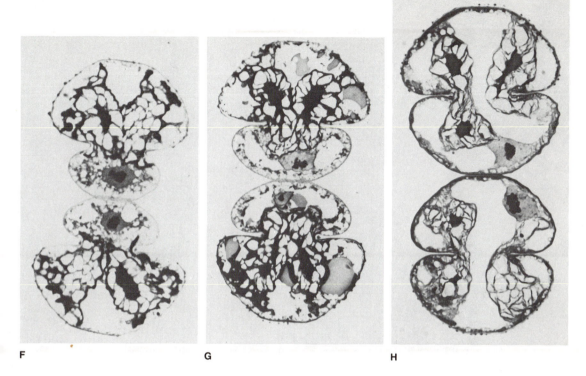

F **G** **H**

Figure 7-26 Cell division in desmids (Zygnematales, *Cosmarium*). **A,** before division; note central nucleus (*n*), two chloroplasts (*c*) with pyrenoids (*py*), ×1100. **B,** division with metaphase chromosomes at center (*arrow*), ×1300. **C,** after wall formation (*sc*) between new half-cells (*sc*). **D–G,** expansion of new half-cell after transverse wall formation. **H,** wall deposition with symmetry restored, ×1100. (Courtesy of J. Pickett-Heaps and *Journal of Phycology*.)

aggregation of dense, intertwined coenocytes, as in *Codium* or *Halimeda* (Fig. 7-30E–G), the appearance of which superficially resembles parenchymatous tissue.

Finally, in this siphonous series, the Dasycladales have extensive plant development while uninucleate (Fig. 7-28D) and become multinucleate only late in development. The multinucleate stage occurs after extensive growth and differentiation.

As can be seen, the Bryopsidophyceae cannot be considered a linear series. In fact, they are more like the trichocine series of Chlorophyceae (Fig. 7-7C–G), composed of several branches with

a possible series of parallel lines, each consisting of one or more orders.

Cladophorales and Siphonocladales The algae in the orders Cladophorales and Siphonocladales contain many multinucleate cells. Cladophorales are present in both freshwater and marine habitats; some freshwater species are nuisance algae in lakes or rivers. In contrast, Siphonocladales are restricted to tropical and subtropical marine waters. The similarites between the two orders are indicated in Table 7-3. Their life histories are different, but those in Cladophorales are so vari-

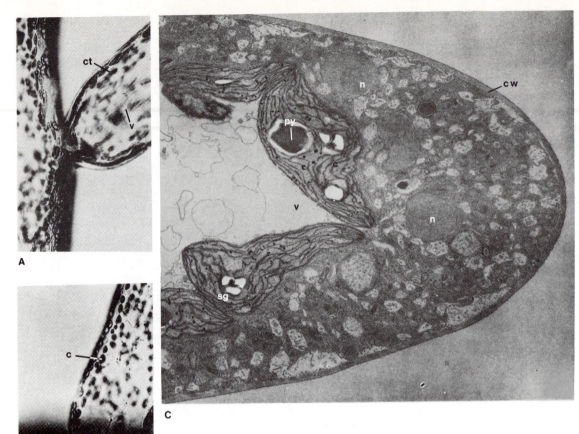

Figure 7-27 Cell structure in Bryopsidophyceae (*Bryopsis*). **A, B,** light microscope view, showing central vacuole (*v*), peripheral cytoplasm (*ct*) and chloroplasts (*c*) at point of lateral branch attachment, ×300. **C,** electron microscope view of growing tip, showing cell wall (*cw*), several nuclei (*n*), central vacuole, chloroplasts with pyrenoids (*py*) and starch (*sg*); note that chloroplasts are in a zone adjacent to vacuole, ×4000. (**A, B,** courtesy of F. A. Burr, J. A. West, and *Journal of Ultrastructure Research,* from F. A. Burr , J. A. West, "Protein bodies in *Bryopsis hypnoides:* their relationship to wound-healing and branch septum development," *Journal of Ultrastructure Research* 35:476–98, 1971; **C,** courtesy of F. A. Burr, J. A. West, and *Phycologia.*)

able that the Siphonocladales could conceivably have developed from a closely related form, if not from a simple cladophoralean alga.

Variability in life history is used frequently to divide Cladophorales into a number of orders. Some species (especially marine *Cladophora* species; Fig. 7-29C) have a life history identical to that of *Ulva* (Fig. 7-19), in which there is an alternation of isomorphic phases. In others, there are two phases of different appearance; one generation is large and predominant, and the other is small and almost microscopic. In most freshwater *Cladophora* species, sexual reproduction is lacking, but biflagellated and quadriflagellated zoospores are readily produced.

Codiales Algae in the order Codiales are char-

Table 7-3 Orders of Bryopsidophyceae, Including Selected Genera

Orders and Genera	Cell Walls		Pigments*	Starch Plastids	Dominant Life History†
	Transverse	Constituent			
Cladophorales *Aegarophila, Cladophora, Rhizoclonium,* *Spongomorpha, Urospora*	regular	cellulose	regular	absent	haploid/diploid, (haploid)
Siphonocladales *Dictyosphaeria, Siphonocladus, Valonia*	irregular	cellulose	regular	absent	diploid
Codiales *Bryopsis, Caulerpa, Codium, Derbesia (Halicystis),* *Halimeda, Udotea*	rare	xylan, mannan, (CaCO₃)	extra carotenoids	present	diploid, haploid/diploid
Dasycladales *Acetabularia, Acicularia, Batophora, Dasycladus,* *Neomeris*	rare	mannan, cellulose, (CaCO₃)	regular	absent	diploid

*Regular pigments refer to those characteristic for the division.
†See footnote † in Table 7-1.

acterized by specific differences in cell wall chemistry and pigmentation (Table 7-3). When present, cellulose is a minor constituent in the cell wall, with mannan and xylan the predominant components. In addition, the chloroplasts contain two xanthophylls specific to the order. Codiales are almost exclusively restricted to ocean shores; in some areas they are the dominant vegetation. As a result of the morphological and reproductive variability of the representatives, the order is not considered by all phycologists to be a natural grouping. Morphological similarites to land plants are not present, although some have the same life history, in which there are two different plants—one diploid and the other haploid (see Figs. 2-9A and 7-31).

An example is *Derbesia*, in which the sporophytic (diploid) phase is a coenocytic, multinucleate, irregularly branched plant (Figs. 7-30A and 7-31I, J). Depending on the species, the gametophyte (haploid) phase is either a globose, coenocytic plant (e.g., *D. tenuissima*, Figs. 7-30B and 7-31A, B), or a regularly branched, coenocytic plant (e.g., *D. neglecta*). In both species, gametes are biflagellated (Fig. 7-33C, D), whereas the motile meiospores (zoospores) have a crown of flagella of equal length (Fig. 7-31L, M), similar to those in Oedogoniales.

Codium is a genus in which gametes are the only haploid cells (Fig. 7-32). Thus the plant is diploid. There are definite male and female gametes (Fig. 7-32E, F) released into the water. After syngamy (Fig. 7-32G), the zygote remains motile (Fig. 7-32H) for several minutes before rounding up and becoming attached (Fig. 7-32I). There is no resting period before germination occurs (Fig. 7-32J). For a period of time (depending upon environmental conditions), the germling can be an unorganized mass of coenocytic threads before forming the massive, diploid plant (Figs. 7-30E and 7-32A).

Sexual reproduction is unknown for some *Codium* populations. The effectiveness of asexual reproduction is shown by the massive increase in the area occupied by one species (*C. fragile*) on the northeastern coast of the United States. The plants produce large, motile cells resembling female gametes, but no male gametes or indications of syngamy have been seen. In winter, transverse walls form on the terminal branches. These tips fall off and act as vegetative units that grow into new plants.

Dasycladales Finally, in the order Dasycladales, cells are uninucleate until maturity. The

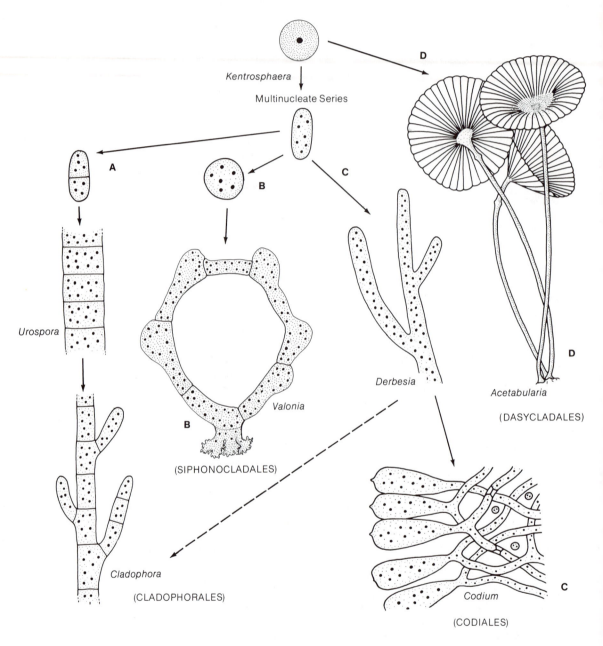

Figure 7-28 Groups of Bryopsidophyceae forming developmental series based on cell number (orders names in *parentheses*). **A, B,** multinucleate, multicellular, with transverse walls (**A,** Cladophorales; **B,** Siphonocladales). **C,** multinucleate (primary one cell), lacking transverse walls (Codiales.) **D,** multinucleate only at maturity, uninucleate during development (Dasycladales).

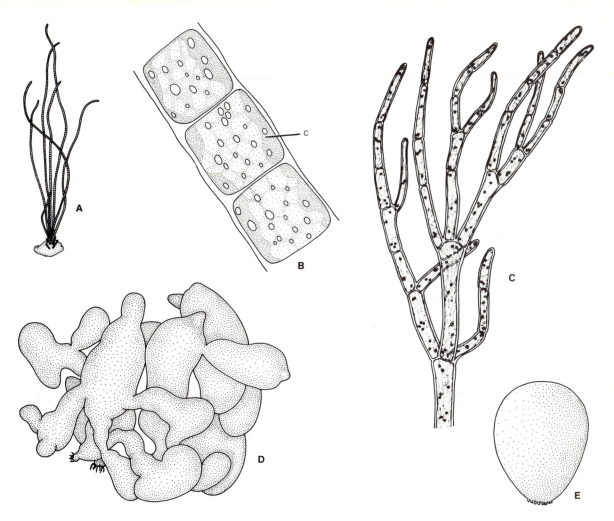

Figure 7-29 Morphological variation of multinucleate forms with transverse walls (Cladophorales, Siphonocladales). **A–C,** Cladophorales. **A, B,** *Urospora,* showing unbranched, uniseriate form. **A,** habit, ×9; **B,** vegetative cells *c*, chloroplast, ×450; **C,** *Cladophora,* showing branched, uniseriate form, ×75. **D, E,** Siphonocladales; habit of *Valonia* spp. (cell wall detail not shown). **D,** ×1; **E,** ×0.5.

diploid nucleus is in the basal, rhizoidal region of the elongate cylindrical plant and remains there throughout growth and differentiation of the plant and production of the apical cap (as in *Acetabularia,* Fig. 7-30H). If the plant is damaged during development, including cap production, it heals and can produce a new cap. *Acetabularia* has been used extensively for studies of nuclear-cytoplasmic interactions on development. By replacing the cap of one species with that of another, it has been shown that the nucleus controls the type of cap differentiated. Upon formation of the cap, the primary nucleus produces derivatives, probably as a result of meiosis, that migrate to the cap. Haploid anisogametes, produced within individual units in the cap, are released, and syngamy occurs in the water. The diploid zygote grows into the new *Acetabularia* plant without further nuclear divisions until after cap production.

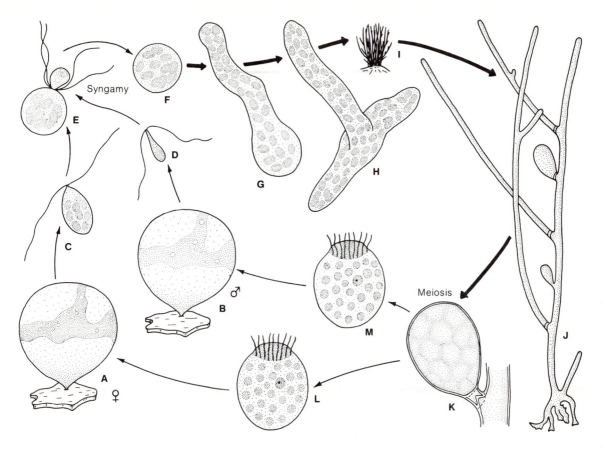

Figure 7-31 Life history of *Derbesia* with alternation of heteromorphic phases (*thin lines* indicate haploid phase, *thick lines* indicate diploid phase). **A, B,** globose, coenocytic stage. **A,** mature female gametophyte; **B,** mature male gametophyte, **C, D,** biflagellate gametes (produced by mitosis). **E,** fusion of anisogametes (syngamy). **F,** zygote. **G, H,** coenocytic juvenile sporophyte. **I,** habit of mature sporophyte. **J,** mature, coenocytic sporophyte with two meiosporangia. **K,** mature meiosporangium. **L, M,** multiflagellated zoospores (produced by meiosis).

Figure 7-30 Morphological variation of siphonous, multinucleate forms, lacking transverse walls (Codiales, Dasycladales). **A–G,** Codiales. **A, B,** *Derbesia:* **A,** coenocytic stage, ×110; **B,** globose stage, showing differentiated gametangial region (*g*), ×4.5. **C, D,** *Bryopsis:* **C,** habit, ×1.5; **D,** coenocytic branch, ×110. **E, F,** *Codium:* **E,** habit. ×1; **F,** coenocytic branch tip (*arrow*) with gametangia (*g*), ×1000. **G,** habit of *Halimeda* (cellular detail not shown), ×0.5. **H,** Dasycladales, *Acetabularia,* showing apical cap, ×1.5.

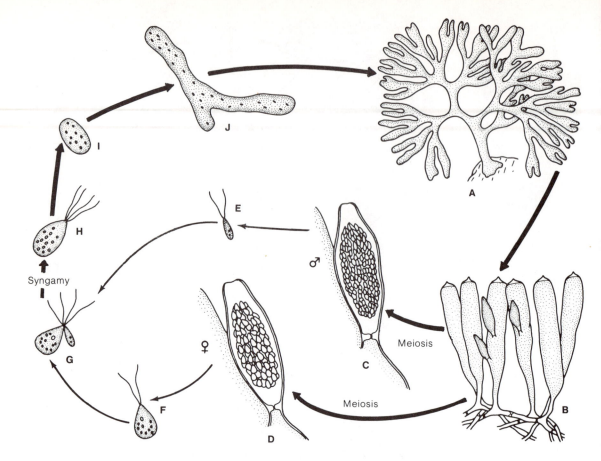

Figure 7-32 Predominantly diploid life history of *Codium*. **A,** mature diploid plant. **B,** coenocytic tips with gametangia. **C,** male gametangium after meiosis. **D,** female gametangium after meiosis. **E,** biflagellated male gamete. **F,** biflagellated female gamete. **G,** fusion of anisogametes (syngamy). **H,** planozygote. **I,** zygote. **J,** coenocytic juvenile stage. (*Thin lines* indicate haploid phase, *thick lines* indicate diploid phase).

Class Charophyceae

The class Charophyceae, as delimited in this text, is a small, distinctive group occurring submerged and attached to the bottom, primarily in fresh and brackish waters. Although there are only six living genera with approximately 250 species, many genera and species are known only as fossils. Most members of Charophyceae have heavily calcified (calcium carbonate) walls—thus the common name, *stoneworts*.

The members of this group are so distinctive morphologically (Figs. 7-33A and 7-34B) that some botanists place them in a separate division. They have been considered taxonomically remote from other groups of green algae, but their biochemical, cytological, and some morphological characteristics indicate that they may have a common evolutionary origin with some Chlorophyceae. Because some features of reproduction and life history are more like those of bryophytes (see Chapter 12), it is interesting to note that some botanists have excluded charophytes from algae and considered them closer to bryophytes.

A

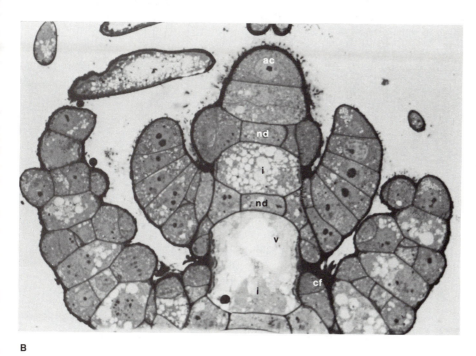

B

Figure 7-33 Charophyceae, *Chara*. **A,** habit view, showing characteristic whorled branches, ×5. **B,** cell and plant structure; longitudinal section of apical region, showing apical cell (*ac*), internode cells (*i*) with large central vacuole (*v*), and node cells (*nd*); enclosing internodal filaments (*cf*) at right (chloroplasts not obvious), ×200. (**B,** courtesy of J. Pickett-Heaps.)

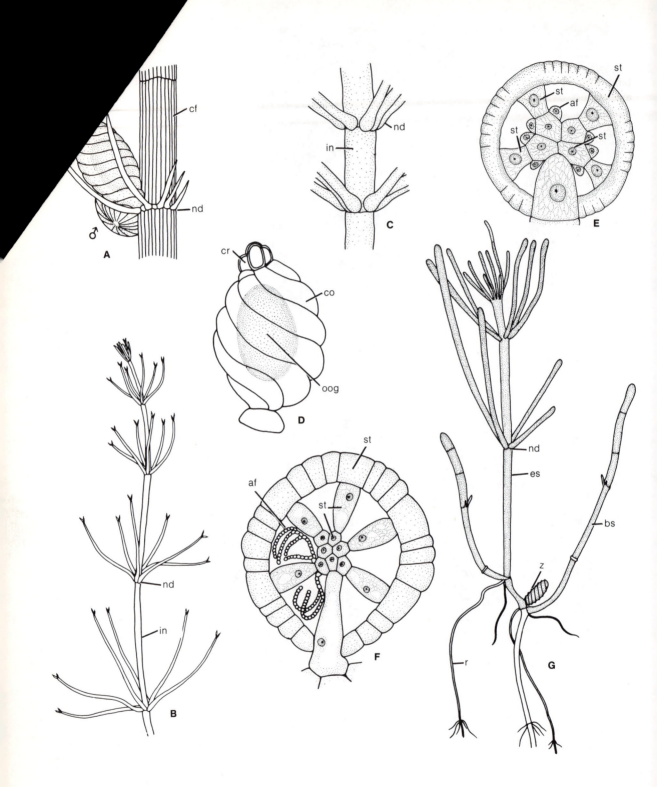

Structure and Reproduction

Cells of Charophyceae are similar to those of land plants; they have a large, central vacuole and numerous, disc-shaped chloroplasts in the peripheral cytoplasm (Fig. 7-33B). The cell organelles are like those of Chlorophyceae; starch is present in the chloroplast, although pyrenoids are absent. Cytoplasmic streaming is especially evident in large cells, and chloroplasts remain stationary.

Charophyceae are distinguished from most Chlorophyceae by producing plants differentiated into short **node** (*nod* = knot, swelling) and long **internode** (*inter* = between) regions (Fig. 7-34B). Longitudinal growth is primarily apical and involves a single, large, apical cell for each branch (Fig. 7-33B). From the nodes, whorls of branches of limited growth arise. In *Chara*, the long internodal region is covered by enclosing filaments of cells, which arise at the nodes and extend up and down over the internodal cells (Fig. 7-34A). In *Nitella* (Fig. 7-34B, C), covering filaments are absent, and the long internodal cells, which are often several centimeters long, are obvious. The cells of the nodal region are uninucleate, but the large internodal cells become multinucleate. At the base of the plant there is a colorless, branched rhizoidal system.

Vegetative reproduction is uncommon but occurs by fragmentation of specialized groups of cells. Zoospores are not produced. Sexual reproduction is the prime method of reproduction, with definite male and female gametes produced. The zygote is the only diploid stage. The female gametangium is borne at a node (Fig. 7-34A) and contains a single egg (Fig. 7-34D). Surrounding the nonmotile egg is a sterile, vegetative sheath, which originates beneath the gametangium and grows up spirally to form a protective layer of cells. At maturity, this sheath covers the tip of the gametangium with one or two rows of crown cells (Fig. 7-34D). A sperm must penetrate this crown to contact the egg. The female structure, termed **nucule** (*nucl* = small nut), appears to be multicellular but is unicellular, with the sheath cells produced secondarily, (as happens in the trichocine alga *Coleochaete*; see p. 274). The immature female gametangium is white, but after syngamy it becomes black.

The male reproductive structure, or **globule** (*glob* = ball), is the most complex one in all the algae. It also occurs at a node (Fig. 7-34A) and is generally a spherical, orange structure attached by a short, stalklike cell. Sterile cells surround the fertile, sperm-producing cells (Fig. 7-34E, F). The formation of sterile and fertile cells occurs simultaneously so that the male reproductive unit is considered multicellular and is interpreted as a complex branch. Within the globule are cells in a linear series, termed *fertile filaments* (Fig. 7-34F). Each cell in the filament produces a single, coiled, asymmetric, biflagellated sperm. The coiled flagella extend backward from the point of attachment, which is some distance from the anterior tip of the sperm (Fig. 7-35A). The sperm body and flagella are both covered with scales, and microtubular flagellar roots form a flat band. The sperm is comparable to that of some higher plants and to motile cells of *Klebsormidium* and *Coleochaete* (Fig. 7-6C, D).

The female gametangium containing the nonmotile egg is retained on the plant after fertilization, and the zygote becomes a resting stage within a thick, protective covering of vegetative cells. The zygote is the only diploid phase in the life history (see Fig. 2-8A), and at germination the nucleus divides by meiosis, producing four haploid nuclei. Three of these nuclei degenerate, and the functional nucleus divides to produce a filamentous stage (Fig. 7-34G), which is similar to the **protonema** of mosses (see Chapter 12). Sub-

Figure 7-34 Morphology and reproduction of the Charophyceae (Charales). **A,** *Chara,* showing enclosing filaments (*cf*) and reproductive structures (gametangia) at nodes (*nd*), ×30. **B, C,** *Nitella.* **B,** habit showing characteristic nodes and internodes (i), ×0.5; **C,** axis, ×30. **D–G,** *Chara.* **D,** female structure with twisted cover cells (*co*) and one row of crown cells (*cr*), protecting single gamete (*oog*), ×750; **E, F,** developmental stages of male reproductive structure, showing sterile cells (*st*) and reproductive filaments (*af*), ×750; **G,** juvenile, filamentous plant from germinating zygote (*z*), showing basal system (*bs*), rhizoids (*r*), and erect systems (*es*), ×225. (**E–G,** after Smith, with permission of McGraw-Hill Book Co.)

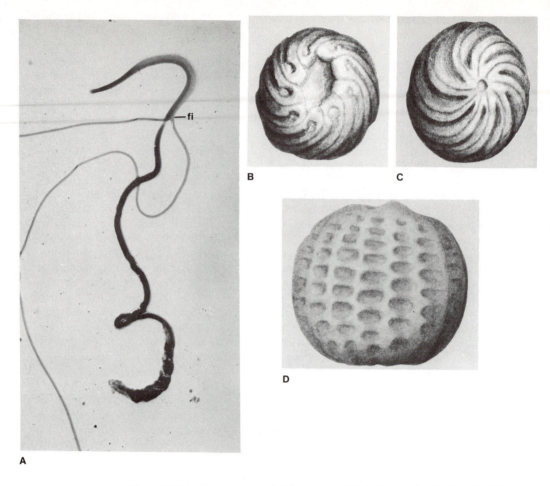

Figure 7-35 Charophyceae. **A,** *Chara* sperm (dried), showing flagellar insertion (*fi*) away from apex, ×3000. **B–D,** fossil female reproductive structures (gyrogonites), ×50. (**A,** courtesy of Ø. Moestrup and *Planta Berlin;* **B–D,** courtesy of R. E. Peck and *Micropaleontology*.)

sequently this develops into the typical plant with apical growth (Fig. 7-35A).

Morphological Variation and Diversity

There is little diversity within this class compared to other Chlorophyta. All living genera of Charophyceae are placed in a single order, Charales, but extinct representatives are placed in four orders, including Charales. Living genera are separated on the basis of gametangial features and gametophyte morphology, including the presence of enclosing filaments at the internodes. Extinct genera are distinguished mostly on features of the female reproductive structure that have survived as calcified fossils (known as **gyrogonites:** *gyro* = round, turning; *gon* = seed; Fig. 7-35B–D).

In 1975, the class Charophyceae was expanded to include those green algae with per-

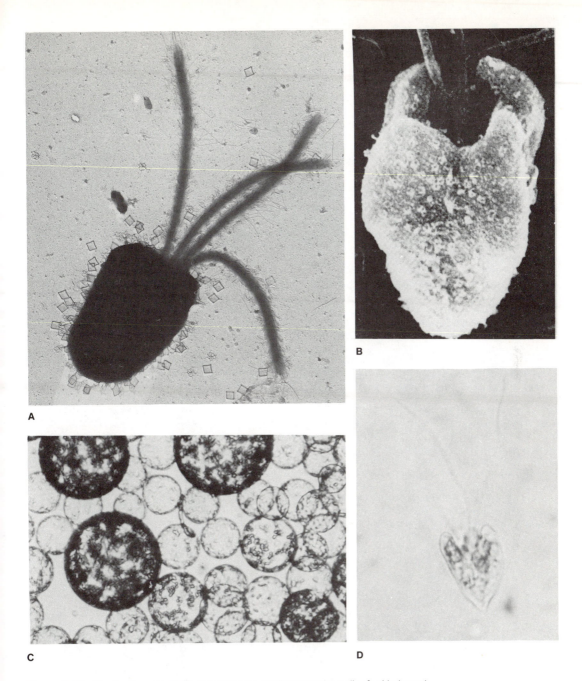

A

B

C

D

Figure 7-36 Prasinophyceae. **A, B,** *Pyramimonas,* motile vegetative cells. **A,** dried specimen, showing body and flagellar scales, ×8000; **B,** scanning electron microscope preparation, showing apical depression, ×6000. **C, D,** *Halosphaera,* nonmotile genus. **C,** nonmotile state, ×200; **D,** reproductive cell, showing apical depression, ×1400. (**A,** courtesy of D. Hibberd; **B,** courtesy of R. E. Norris and *Archiv für Protistenkunde;* **C, D,** courtesy of M. Parke, from "Three species of *Halosphaera," Journal of the Marine Biological Association, U.K.* 45:537–57, 1965, with permission of Cambridge University Press.)

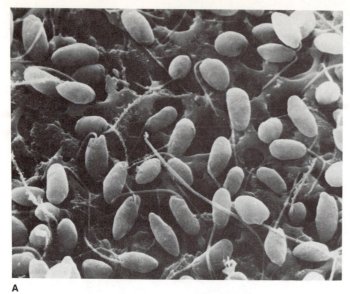

A

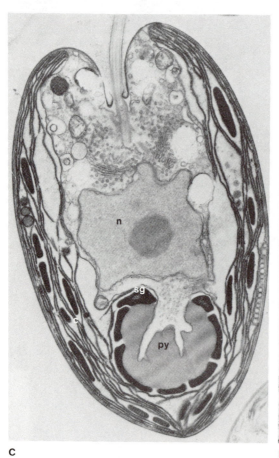

C

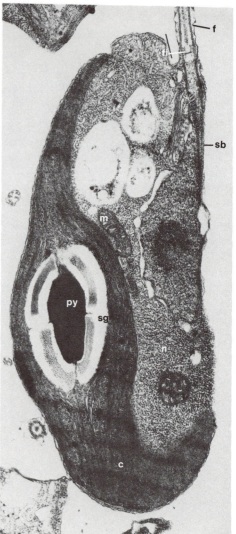

B

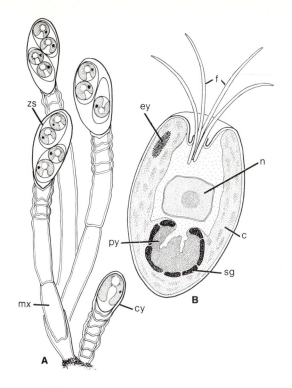

Figure 7-38   Nonmotile Prasinophyceae, *Prasinocladus*. **A,** vegetative stage, showing tubular stalks (*mx*) and containing cysts (*cy*) or zoospores (*zs*), ×40. **B,** motile stage, ×400; note chloroplast (*c*) with pyrenoid (*py*) and starch (*sg*), eyespot (*ey*), nucleus (*n*), and flagella (*f*).

Figure 7-37   Cell structure of Prasinophyceae, electron microscope. **A, B,** *Pedinomonas*. **A,** scanning electron micrographs of cells, showing long flagellum and flagellar insertion (*arrow*), ×2000; **B,** longitudinal section of vegetative cell, showing chloroplast (*c*) with pyrenoid (*py*) and starch (*sg*). Note nucleus (*n*), flagellar insertion (*fi*), attachment of flagellum (*f*) to striated band (*sb*), and mitochondrion (*m*), ×2125. **C,** *Platymonas*, longitudinal section of vegetative cell with off-center pyrenoid; note nucleus and apical depression, ×9000. (**A, B,** courtesy of J. D. Pickett-Heaps and *Cytobios;* **C,** courtesy of I. Manton and M. Parke, from "Observations on the fine structure of two species of *Platymonas* with special reference to flagellar scales and mode of origin of the theca," *Journal of the Marine Biological Association, U.K.* 45:743–54, 1965, with permission of Cambridge University Press.)

sistent spindle and microtubules during cyto-kinesis, and asymmetric motile cells in which the flagellar basal bodies are associated with a single, broad band (Fig. 7-7E). *Klebsormidium* (Fig. 7-16B) and *Coleochaete* (Figs. 7-17D and 7-18C) are included, as well as the Zygnematales (which lack any motile cells). Since then, with the completion of further cytological and physiological studies, a few other genera have been placed in the class. Until further research elucidates which genera should also be included, this text considers the Charophyceae as a restricted, well-defined morphological group.

Class Prasinophyceae

The class Prasinophyceae presently consists of those uninucleate, unicellular, or multicellular green algae with flagellated cells different from the "typical" Chlorophyceae (see p. 258; compare Fig. 7-6A, B with Fig. 7-37B, C). The class includes forms with scaly flagella arising from an apical depression or groove (Figs. 7-36A, B and 7-37A). Motile cells have one, two, or more equal or unequal flagella (Figs. 7-36A, B, D and 7-37A). Although most genera are flagellated in the vegetative phase, there are nonmotile, nonfilamentous representatives (Figs. 7-36C and 7-38). Most genera are less than 30 μm, with more representatives known from marine waters. As most are small, and can be easily overlooked, it is possible that prasinophytes are present but not well-documented in freshwater habitats. As constituted by some phycologists, the Prasinophyceae are a heterogenous class; however, here they are considered in a restricted sense.

In addition to similar morphological features of the motile cell, some prasinophytes are similar to certain Chlorophyceae and Charophyceae in cell division. In these, newly formed nuclei are separated some distance from each other, and the mitotic spindle fibers can remain throughout mitosis. It is thought that some prasinophytes are similar to progenitors of various groups of Chlorophyta and may be in an evolutionary line to land plants (see p. 306 and Fig. 7-39B).

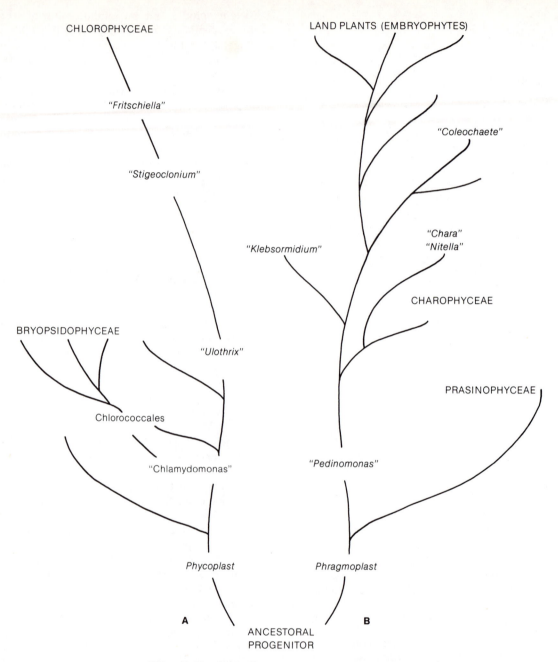

Figure 7-39 Possible derivation of Chlorophyta and land plants from flagellated, unicellular progenitors. **A,** primarily Chlorophyceae, starting with symmetrical *Chlamydomonas*like motile cell; possibly Bryopsidophyceae is an offshoot from chlorococcalean forms. **B,** selected Chlorophyceae starting from asymmetrical, *Pedinomonas*like, motile cell and leading toward land plants (embryophytes), with a sideline to Prasinophyceae and Charophyceae; generic names in quotation marks indicate genera similar to present-day forms. (The Chlorophyceae line might be further subdivided, as noted on page 306, into two lines, the Chlorophyceae and Ulvaphyceae).

Ecology of Chlorophyta

Few other groups of algae have such a wide range of distribution and variety of habitat as Chlorophyta. They can occur wherever some moisture and light are available. Green algae are common, floating free or swimming in **plankton** (*plank* = wandering) of lakes, ponds, rivers, and streams. To a lesser extent, green algae occur in oceans and can be abundant in tidepools. There are many larger, attached forms, especially in the marine habitat, where Chlorophyta have a broad, vertical distribution through the intertidal zone into deep water. In cold waters, they are more conspicuous in shallow regions and in the intertidal zone, but in warm waters they can extend to depths exceeding 100 meters.

Most genera (even orders) are restricted to either the marine or freshwater habitat. However, as noted previously, some genera (e.g., *Ulothrix*, Fig. 7-16A; *Chlamydomonas*, Fig. 7-8A; *Cladophora*, Fig. 7-29C) are represented in both. Rarely is the same species an inhabitant of both; however, there are some green algae that can live in estuarine waters, where salinity fluctuates widely.

Although most green algae are aquatic, some are terrestrial and occur almost everywhere on damp surfaces, such as tree trunks, moist walls, and leaf surfaces. Some green algae occur in soil or in lichens as the algal component or phycobiont (see Chapter 6). The commonest cause of *red snow* in areas of permanent or semipermanent snow and ice is the presence of *Chlamydomonas* species and related genera containing excess carotenoid pigments in the cells.

Some green algae occur epiphytically on larger algae or aquatic seed plants. Others, especially filamentous Zygnematales (*Zygnema, Mougeotia, Spirogyra*, Fig. 7-24), occur in large, free-floating or submerged mats in freshwater ponds, ditches, and slow streams. Some Chlorophyta grow inside the shells of molluscs or other invertebrates, or within tissues of higher plants. A few Chlorophyta lack chloroplasts and occur as saprobes. One colorless genus (*Prototheca* in Chlorococcales) is the cause of a skin disease (*protothecosis*) in humans.

Almost any container of water that receives some light will become colonized by green algae.

The water in fish tanks can have any one of a number of Chlorococcales as **phytoplankton** (*phyt* = plant); whereas on the aquarium sides, the filamentous *Oedogonium* (Fig. 7-21A) often occurs. Some Chlorophyta are present in large numbers in domestic water supplies, giving a disagreeable flavor to the water or clogging filters. Some occur in great masses, as does *Cladophora* (Fig. 7-29C), which grows in some nutrient-rich waters. One example is *Cladophora glomerata* in the Laurentian Great Lakes between Canada and the United States (Lakes Ontario, Erie, Michigan). Evidently growth is favored by high light intensity, high water turbulence, and water containing large amounts of calcium and carbonate, as well as nitrogen, phosphorus, silicon, and sometimes thiamin. The algal masses interfere with boating, swimming, and fishing, and can be responsible for fish kills by depleting the oxygen level. The odor and appearance of this alga when washed ashore and rotten make it even more unpopular.

Relationships of Chlorophyta

The division Chlorophyta contains representatives that are more like higher plants than those of any other algal division. The fossil history goes back to the Precambrian, with the oldest supposed green algae being microfossils (1.3 billion years old) in sediments in eastern California. The interpretation of these microfossils is subject to question. It has been shown in some instances that their internal contents may not be remnants of eukaryotic organelles, but are degraded prokaryotic cells in which cell wall and sheath are intact, but protoplasm is a globular mass. More recent data on the ultrastructural features and cell size of similar microfossils support their recognition as eukaryotes, possibly green algae.

In general, the fossil record for most Chlorophyta is poor, as they are morphologically simple and do not contain material that withstands decomposition. The best-known fossil Chlorophyta are representatives of Bryopsidophyceae (Dasycladales, Codiales) and Charophyceae. These fossils are calcified remnants resulting from calcium carbonate deposition in the cell walls. There are indications of some filamentous forms in Cambrian and younger rocks as well as geochemical evidence of biological compounds attributable to green algae. At present, the main speculation concerning the phylogeny of Chlorophyta is based primarily on living organisms.

It is thought that possibly the four classes of Chlorophyta had a common origin but that at least three separate lines developed very early. Because of physiological and biochemical similarities, Chlorophyta are accepted as obvious progenitors of land plants. Exactly how this has occurred is unknown and will probably remain so. However, it is possible to construct a series based on present-day algae. This is a **phenetic** (*pheno* = appear) series; it rarely considers ancestral forms or actual **phylogeny** (*phyl* = tribe; *gen* = origin) because of the poor fossil record. Thus, the starting point of the "series" is a unicellular, flagellated alga, and it continues through a filamentous, heterotrichous form, ultimately producing a "prototype" of a land plant adapted to the harsher terrestrial environment.

Such a series includes not only biochemical features but also morphological and cytological features and the genetic potential implicit in sexual reproduction. The obvious biochemical features are pigmentation (chlorophyll *a, b,* carotenoids), cell walls (cellulose as a substance that maintains its structure when out of water), food manufacture (typical carbon fixation into simple carbohydrates), and storage products (starch). The morphological features include specialized tissues for anchoring, assimilation, storage, and reproduction, as well as cytological features. These cytological features can be divided into (1) those of vegetative cells, such as type of cell division (origin and orientation of microtubules in cell wall formation; see Fig. 7-5A, C), direction of cytokinesis (from center out, or toward center; Fig. 7-4), and presence of cytoplasmic connections or **plasmodesmata** (*plasmo* = mold, form; *desma* = bond) between cells; and (2) those of motile cells, such as asymmetry of cell, presence of a single, broad flagellar root, presence of flagellar scales, and lack of a firm cell wall (Fig. 7-6). The importance of sexual reproduction involves not only the type of gametes but also an understanding of the alternation of phases (genera-

tions) which necessitates multicellular haploid and diploid stages.

The occurrence of more than one, somewhat parallel series in Chlorophyceae may explain the diversity of forms in green algae and in land plants. One series (Fig. 7-39A), with a *Chlamydomonas*like form as the flagellated unicell (Fig. 7-8A), could lead through a simple filament, like *Ulothrix* (Fig. 7-16A), to something similar to the heterotrichous genera *Stigeoclonium* and *Fritschiella* (Figs. 7-17A, B and 7-18A, B). This series did not evolve to land plants, although certain forms (e.g., *Fritschiella*) are somewhat adapted to fluctuating moisture conditions. A second series (Fig. 7-39B) culminating in land plants may have a *Pedinomonas*like, flagellated cell (Fig. 7-37A, B). The simple, filamentous form would be *Klebsormidium*like (Fig. 7-16B). Intermediate stages could be similar to *Coleochaete* and *Chara* (Figs. 7-19D, 7-18C, and 7-33A), although neither should be considered in a "direct line" to land plants. This second series is very similar to the expanded concept of the Charophyceae proposed in 1975. Current research shows there may be even a third series involving species presently classified in the ulotrichalean genera *Ulothrix, Monostroma,* and *Ulva* (Fig. 7-16A, D, G). In these species, microtubules at cytokinesis are lacking, and motile cells have a cruciate root system and body scales. Thus, some features of each of the two lines shown in Figure 7-39 occur in these species. In addition, some members of the Bryopsidophyceae have been shown to belong to this third series, which has been named the Ulvaphyceae (not shown in Fig. 7-39). As more taxa are examined thoroughly (primarily with regard to ultrastructure), the validity of the phenetic schemes may be clarified. There will also probably be taxonomic changes at both specific and generic levels for some of the taxa involved.

In establishing such phenetic series, it is important to remember that present-day green algae are not as morphologically complex as representatives of some other algal divisions (Phaeophyta, Rhodophyta; Chapters 10, 11). However, possibly the most highly evolved green algae are actually land plants, such as mosses or ferns! The series in Figure 7-39 is based primarily on cytological features, with some consideration of the habitat. This is especially true for *Klebsor-*

midium, which often grows in moist, temporary habitats. *Coleochaete* occurs submerged and usually on other plants. Some species have an open type of branching, but all lack the well-defined, upright system that occurs in *Stigeoclonium* or *Fritschiella*. Thus, further studies are needed of other heterotrichous and parenchymatous green algae to fill the blank space on the land plant phenetic series. It is apparent that this series could lead to a *Chara*like organism as an offshoot terminating in rather complex forms. However, these advanced branches are actually dead ends in an evolutionary series to land plants. The similarities of the Charophyceae to higher plants and their long fossil history indicate early divergence (probably 500 million years ago) in the early Cambrian period.

Termination of such a line as charophytes may be self-imposed, due to the complexities produced. This, then, can be natural selection in one form. Similar dead ends occur in various groups of Chlorophyceae (e.g., Volvocales) and Bryopsidophyceae.

References

For other publications concerned with the Chlorophyta, see References at the end of Chapter 2.

Bold, H. C., and Wynne, M. J. 1978. *Introduction to the algae*. Prentice-Hall, Englewood Cliffs, N.J.

Bourrelly, P. 1966. *Les algues d'eau douce*, vol. 1, *Les algues vertes*. N. Boubée & Cie., Paris.

Cox, E. R., ed. 1980. *Phytoflagellates*. Elsevier/North-Holland: New York. (Unicellular chlorophytes, pp. 5–39; prasinophytes, pp. 85–145; colonial chlorophytes, pp. 147–163).

Dodge, J. D. 1979. *The phytoflagellates: fine structure and phylogeny*. In Levandowsky, M., and Hutner, S. H., eds., *Biochemistry and physiology of protozoa*, vol. 1, pp. 7–57. 2nd ed. Academic Press, New York.

Fritsch, F. E. 1935. *The structure and reproduction of the algae*, vol. 1. Cambridge Univ. Press, Cambridge.

Moestrup, Ø. 1978. On the phylogenetic validity of the flagellar apparatus in green algae and other chlorophyll A and B containing plants. *BioSystems* 10:117–44.

Pickett-Heaps, J. D. 1979. Electron microscopy and the phylogeny of green algae and land plants. *American Zoologist*, 19:545–54.

————. 1975. *Green algae: structure, reproduction and evolution of selected genera*. Sinauer Associates, Sunderland, Mass.

Round, F. E. 1971. The taxonomy of the Chlorophyta II. *Brit. Phycol. J.* 6:235–64.

Stewart, K. D., and Mattox, K. R. 1975. Comparative cytology, evolution, and classification of the green algae, with some consideration of the origin of other organisms with chlorophyll A and B. *Bot. Rev.* 41:104–35.

————. 1978. Structural evolution in the flagellated cells of green algae and land plants. *Biosystems* 10:142–52.

Wood, R. D., and Imahori, K. 1965. *Revision of the Characeae*. 2 vols. J. Cramer, Weinheim.

As noted in Chapter 2 (Table 2-2), there are several algal divisions. The largest forms in the aquatic environment belong to Chlorophyta, Phaeophyta, and Rhodophyta. Representatives of other divisions are generally less prominent and consist of many small, often motile forms. These are divided into two groups for ease of discussion: (1) the yellow-green to yellow-brown forms classified as Chrysophyta (Table 8-1); and (2) the primarily motile divisions—often thought of as "photosynthetic animals," since most are flagellated. These two groups are treated in separate chapters.

8

CHRYSOPHYTA (CHRYSOPHYTES)

Introduction

The yellow-green to yellow-brown algae considered in this chapter make up the algal classes Xanthophyceae, Raphidophyceae, Eustigmatophyceae, Chrysophyceae, Bacillariophyceae, and Prymnesiophyceae. Although they are treated as one division (Table 2-3), other authors recognize at least two divisions. The classes have several important features in common:

1. Presence of accessory chlorophyll and carotenoid pigments.

2. Chloroplast composed of many units of three associated but unfused thylakoids.

3. Storage product not starchlike but laminaranlike, termed **chrysolaminaran,** as well as fats and oils.

4. Storage product stored outside the chloroplast.

5. At least one tinsel flagellum (except in Prymnesiophyceae).

6. Cell covering, when present, outside the plasma membrane, sometimes not a wall and not cellulose.

To facilitate consideration of these classes, only three are discussed in detail: Xanthophyceae, Chrysophyceae, and Bacillariophyceae (Table 8-1). These contain algae that are most likely to be encountered in aquatic or damp environments. From an evolutionary viewpoint, many algae in these classes are known only as extant forms, so their interrelationships are based

Table 8-1 Characteristics of Main Classes of Chrysophyta

Characteristics	Class		
	Xanthophyceae	Chrysophyceae	Bacillariophyceae
Vegetative Stage			
Form (range)	nonmotile (amoeboid, flagellate)	flagellate (amoeboid, nonmotile)	nonmotile "box-shaped"
Cell covering	cellulose	scales—silica wall—cellulose	wall—organic + silicon
Chloroplast			
number	many	one, two	one, two, many
color	yellow-green	yellow-brown (golden)	yellow-brown
Motile Stage			
Flagellation			
vegetative cell	rare	primarily	none
reproductive cell	yes	yes	male gametes
number	2; unequal	1, 2; unequal	1 (2, unequal)
type	longer—tinsel	longer—tinsel	one tinsel
Eyespot	in chloroplast	in chloroplast	none
Pigmentation			
Chlorophyll *c*	many (?)	all	all
Fucoxanthin	absent	present	present

only on living features (phenetic relationships). With regard to the evolution of land plants, these classes are definitely a side branch. Whether this side branch may have a relationship to other algae or even to animals is discussed later.

Class Xanthophyceae (xanthophytes, or yellow-green algae)

There are approximately 100 genera with some 550 species referred to as Xanthophyceae (*xantho* = yellow). They occur predominantly in freshwater habitats and soil, but some are marine. Most species are nonmotile, small, and easily overlooked. Many xanthophytes are ephemeral and delicate—so delicate that they die when collected in soft (soda) glass containers.

Cell Structure and Cell Division

The xanthophyte cell is superficially similar to other eukaryotic algal cells, with the usual organelles present. One to several disc-shaped, smooth-edged, green or yellow-green chloroplasts occur peripherally in the cell (Fig. 8-1A, B). The main green pigment is chlorophyll *a*, with chlorophyll *c* present in some species. Carotenoid pigments mask the chlorophylls to some extent and include at least three xanthophylls, also characteristic of some related classes. The chloroplasts are composed of bands of three unfused thylakoids (Fig. 8-2A, B). When pyrenoids are present, they are always traversed by thylakoids. Pyrenoids can be embedded in the chloroplast or protrude from it, and they have a different appearance from the pyrenoids in Chlorophyceae. Pyrenoids are always associated with

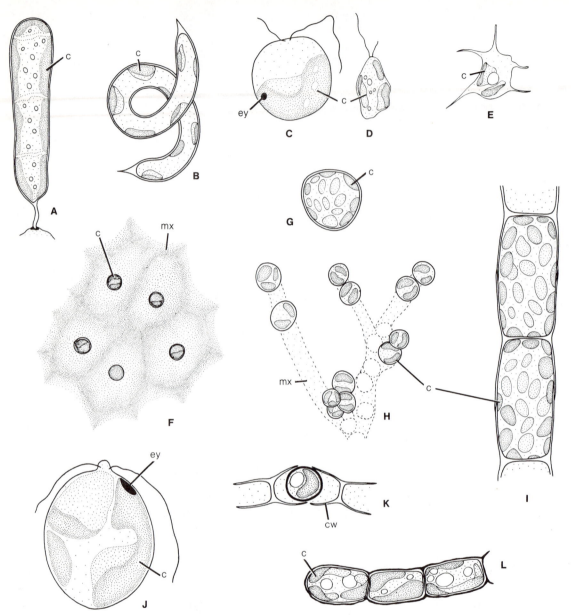

Figure 8-1 Cell structure, morphological diversity, and reproduction of Xanthophyceae. **A, B,** coccoid forms (*Ophiocytium* spp.), showing peripheral, disc-shaped chloroplasts (*c*); **A,** ×235; **B,** ×500. **C, D,** motile, flagellated form. **C,** *Xanthophycomonas*, with eyespot (*ey*), ×2150; **D,** *Heterochloris,* ×560. **E,** amoeboid forms (*Rhizochloris*), ×830. **F,** palmelloid form (*Chlorosaccus*), portion of colony, ×650. **G, H,** coccoid forms. **G,** *Pleurochloris,* ×940; **H,** *Mischococcus,* within matrix (*mx*), ×855. **I,** filamentous form (*Tribonema*); vegetative filament, showing disc-shaped chloroplasts and overlapping cell walls (*arrow*), ×915. **J–L,** reproduction. **J,** heterokont zoospore (*Tribonema*), ×8000; **K,** internal cyst (*Tribonema*) in cell wall (*cw*), ×400; **L,** akinete formation (*Bumilleria*), ×450. (**D, E, K, L,** after Pascher, in Rabenhorst, with permission of Akademische Verlagsgesellschaft, Geest & Portig K.-G., Leipzig.)

photosynthate storage, which always occurs outside the chloroplast.

Some storage material is a laminaranlike compound similar to, if not identical with, that in Chrysophyceae (**chrysolaminaran,** see p. 317). Studies of some Xanthophyceae indicate that other polysaccharides are present (although the exact identity is still unknown) as well as fatty acids and sterols. Storage materials can accumulate adjacent to the pyrenoid, when present. Since starch is lacking in Xanthophyceae, the chloroplast is smooth and almost shiny in contrast to the rough-edged and granular chloroplasts of Chlorophyceae.

The main cell wall materials are cellulose, pectic substances, and sometimes silicon. In some xanthophytes, the cell wall is composed of two overlapping equal or unequal halves (Figs. 8-1I and 8-2A). A central vacuole is present in some cells. Generally, xanthophytes are uninucleate when young, although at maturity they are often multinucleate. Two of the most commonly occurring genera (*Botrydium* and *Vaucheria*) produce multinucleate, coenocytic thalli. The endoplasmic reticulum is continuous from nuclear envelope to chloroplast (Fig. 8-2B), a feature also occurring in Chrysophyceae (see p. 319) and Phaeophyta (Chapter 10).

As there are few filamentous genera and no parenchymatous representatives, vegetative cell division, as described for Chlorophyceae, is not of particular significance in Xanthophyceae. In most xanthophytes, nuclear division precedes spore formation. Prior to the time of cell division, centrioles are often present.

Movement

Few xanthophytes are motile in the vegetative stage, and those that are can move by either flagella, amoeboid movement, or a combination of both. Thus, motile cells are generally either zoospores or gametes and have two unequal, anteriorly inserted flagella (Figs. 8-1C, D, J and 8-3A). The shorter flagellum, which can be trailing, is of the whiplash type; the longer one is an-

teriorly directed and contains a double row of tubular tinsels (Fig. 8-3A). This longer flagellum is considered responsible for forward propulsion. In some instances, the uniflagellated condition occurs, and the shorter whiplash flagellum is absent. Some uniflagellated forms have been referred to Eustigmatophyceae (see p. 335).

In the coenocytic genus *Vaucheria*, an unusually large, multinucleate, and multiflagellated zoospore is produced. The flagella occur in pairs, with one member of each pair slightly longer than the other; both are whiplash. It is thought that this zoospore is a compound one, composed of many flagellar units (Fig. 8-4C, E). Sperms of *Vaucheria* have typical xanthophyte flagellation, except the longer flagellum is whiplash (Figs. 8-3C and 8-4I).

Reproduction

The chief method of reproduction is by spore production. Zoospores and aplanospores are produced singly in the cell, or the protoplast can divide to produce several spores. Zoospores (Fig. 8-1J) are generally bilaterally symmetrical, with two dorsiventrally arranged chloroplasts. Near the base of the shorter flagellum is a swelling associated with the eyespot (Fig. 8-3B). In the *Vaucheria* zoospore (Fig. 8-4D), each flagellar pair is associated with a nucleus and several chloroplasts.

In multicellular species, fragmentation also serves as a means of vegetative reproduction. Some genera produce internal, or **endogenous** (*endo* = within) cysts (*cyst* = bag) (Fig. 8-1K), similar to those of Chrysophyceae (see p. 317). In some genera, the entire cell (including cell wall) is enclosed in a heavy wall to form an **akinete** (Fig. 8-1L).

Sexual reproduction is not widespread in Xanthophyceae, although it is common in *Vaucheria* (see p. 315). Isogamy, anisogamy, and oogamy are all represented in the class. In the freshwater forms studied, meiosis apparently occurs on germination of the zygote. Thus, the haploid phase predominates, and the life history is haploid (see Fig. 2-8A).

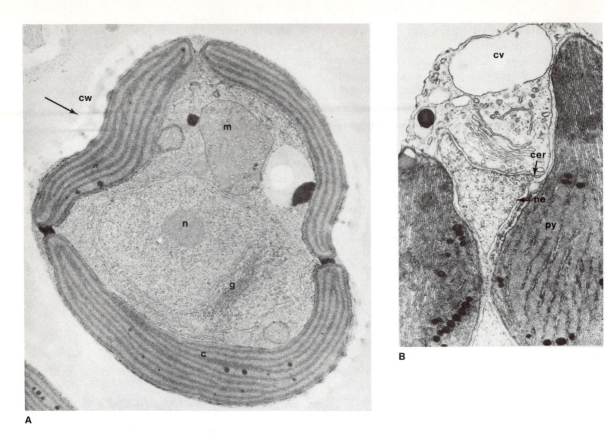

Figure 8-2 Cell structure of Xanthophyceae, electron microscope. **A,** cell (*Tribonema*), showing nucleus (*n*), chloroplast (*c*), golgi (*g*), mitochondrion (*m*), and overlapping cell wall (*cw*) at *arrow,* ×10,000. **B,** section of anterior of zoospore (*Bumilleria*), showing relationship between nucleus and chloroplasts; nuclear envelope (*ne*) continuous with endoplasmic reticulum of chloroplast (*cer*); note pyrenoid (*py*) and contractile vacuole (*cv*), ×40,000. (**A,** courtesy of T. Bisalputra; **B,** courtesy of A. Massalski and *British Phycological Journal*.)

Figure 8-3 Motile cells of Xanthophyceae, electron microscope. **A, B,** zoospores (*Bumilleria*). **A,** shadow-cast to show flagella; forward projecting one with tinsels, ×4000; **B,** longitudinal section showing parietal chloroplasts (*c*), with eyespot (*ey*), nucleus (*n*), flagellar insertion (f_1, f_2), and flagellar swelling (*fs*), ×10,000. **C, D,** sperm of *Vaucheria*. **C,** entire cell, showing lateral insertion of flagella; forward flagellum with tinsels (*arrow*), ×6000. **D,** longitudinal section showing nucleus, mitochondrion (*m*), and lateral insertion of flagella (*f*), ×19,500. (**A, B,** courtesy of A. Massalski and *British Phycological Journal;* **C, D,** courtesy of Ø. Moestrup, from "On the fine structure of the spermatozoids of *Vaucheria sescuplicaria* and the later stages in spermatogenesis," *Journal of the Marine Biological Association, U.K.* 50:513–23, 1970, with permission of Cambridge University Press.)

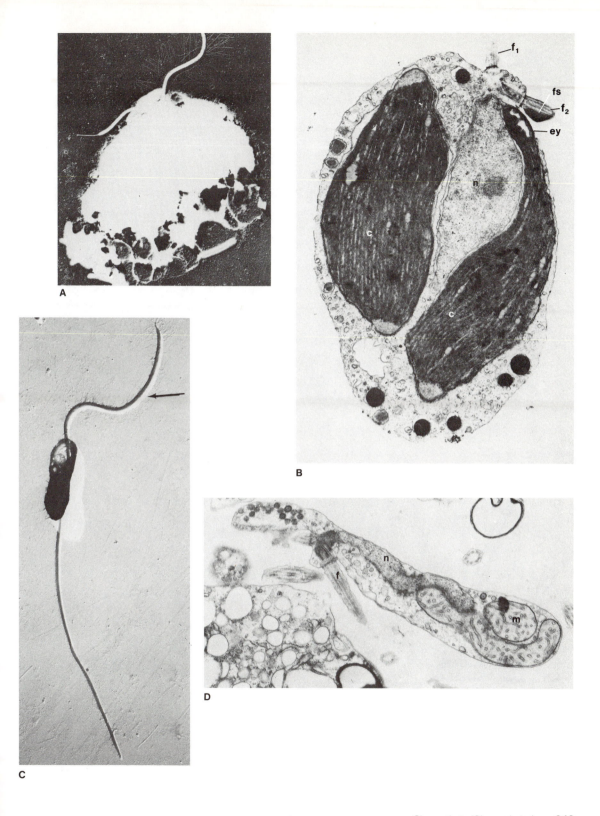

A

B

C

D

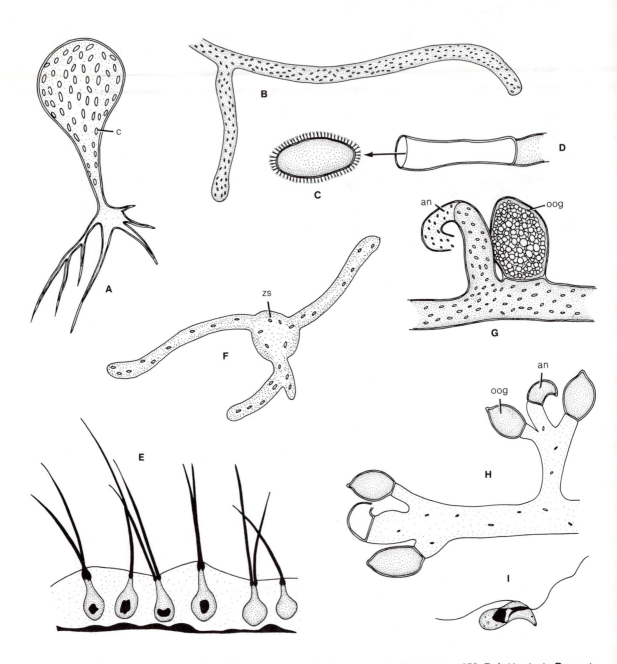

Figure 8-4   Coenocytic Xanthophyceae. **A,** *Botrydium,* ×250. **B–I,** *Vaucheria.* **B,** vegetative coenocyte, ×60. **C–E,** zoospore (**C,** after release from zoosporangium, ×200; **D,** empty sporangium, ×200; **E,** detailed structure of edge of zoospore, showing flagellar apparatus, ×1240). **F,** young germling from zoospore (*zs*), ×400. **G, H,** gametangial arrangement (*an,* antheridium; *oog,* oogonium), ×140. **I,** sperm, ×1250. (**E, I,** courtesy of W. J. Koch and *Journal of the Elisha Mitchell Scientific Society.*)

Morphological Variation and Classification

The same range of morphological variation in Chlorophyceae is present in Xanthophyceae. However, most xanthophytes are nonmotile, unicellular, and small (less than 50–75 μm). Some are of common occurrence, such as *Ophiocytium* (Fig. 8-1A, B). However, most are easily overlooked because of their size and because they are often **epiphytes** and occur in specialized habitats, such as alpine pools, ditches, or sphagnum bogs. Motile, biflagellated xanthophytes include *Xanthophycomonas* and *Heterochloris* (Fig. 8-1C, D). There are a few unicellular amoeboid forms, such as *Rhizochloris* (Fig. 8-1E), as well as nonmotile colonies, such as *Chlorosaccus* (Fig. 8-1F). Typical coccoid forms include the unicellular *Pleurochloris* (Fig. 8-1G) and the multicellular *Mischococcus* (Fig. 8-1H). Unicellular genera are separated by cell shape, number of chloroplasts, and production of aplanospores or zoospores (similar to the Chlorococcales of the Chlorophyceae; see Chapter 7). The most ubiquitous and commonly encountered Xanthophyceae are species belonging to three genera: the filamentous *Tribonema* (Fig. 8-1I) and the coenocytic *Botrydium* and *Vaucheria* (Fig. 8-4A, B). These three genera occur floating freely or attached in water, or on moist soil. Thus there is diversity in the class but few species of any one form. This text treats in detail only two of the commonly encountered morphological types: *Tribonema* and *Vaucheria*.

Tribonema The unbranched, filamentous *Tribonema* is generally present in cool waters (around 12 °C or less), usually in the spring, as light green or almost colorless floating masses. The cells are often thick in the middle and slightly constricted at the transverse walls. This is the result of **H**- or **I**-shaped transverse walls that overlap at the center of the cell (Figs. 8-1I and 8-2A). This transverse wall is formed from the center outward, and gradual separation of the parental half-cells follows. Sometimes yellow- to brown-colored materials settle irregularly on older **H**-shaped pieces.

One or two zoospores (Fig. 8-1J) are formed in ordinary vegetative cells. They are **pyriform** to ovoid, with typical xanthophyte flagellation.

Numerous chloroplasts occur, with an eyespot present in one of the anterior ones. Zoospores show some plasticity when swimming; ultimately, they settle to form an elongated holdfast cell.

Sexual reproduction is not commonly observed but is isogamous, with one gamete settling prior to syngamy. The zygote is a resting stage, and it is assumed that meiosis occurs during germination.

Vaucheria For many years it was uncertain whether the common, often mud-inhabiting alga *Vaucheria* was a member of siphonous greens or xanthophytes. Anomalies regarding pigmentation and flagellation fostered the confusion. It has now been shown conclusively that *Vaucheria* is indeed a member of the Xanthophyceae. Chlorophyll *b* is not present, and small amounts of chlorophyll *c* are. In addition, another chlorophyll (chlorophyll *e*) has been reported present in certain reproductive stages of some species. Besides chlorophylls, there are carotenoids characteristic of other xanthophytes. As noted previously, flagellation of the zoospore differs from that of the sperm (see p. 317); however, the presence of at least one tinsel flagellum on the sperm supports the xanthophycean relationships (Fig. 8-3C and 8-4I).

Vaucheria is a coenocyte (Fig. 8-4B) that produces transverse walls only when injured or when forming reproductive cells. The zoospore is produced at the end of a somewhat swollen tube and ultimately becomes separated from the rest of the plant. The zoosporangium produces a single, large zoospore with many flagellar units, each associated with a nucleus. This multiflagellated zoospore is released through an opening at the end of the zoosporangium (Fig. 8-4C, D) and swims sluggishly for a few hours. It is over 100 μm long, and the surface is covered with many pairs of unequal whiplash flagella (Fig. 8-4E), with more at the anterior end. Each flagellar pair and associated nucleus has associated chloroplasts. When the zoospore settles, it grows into a new coenocyte by retracting the flagella and forming a cell wall. The coenocytic tubes produced are narrower than the zoospores, so it is easy to recognize the origin of the new plant (Fig. 8-4F).

In sexual reproduction, gametangia can be

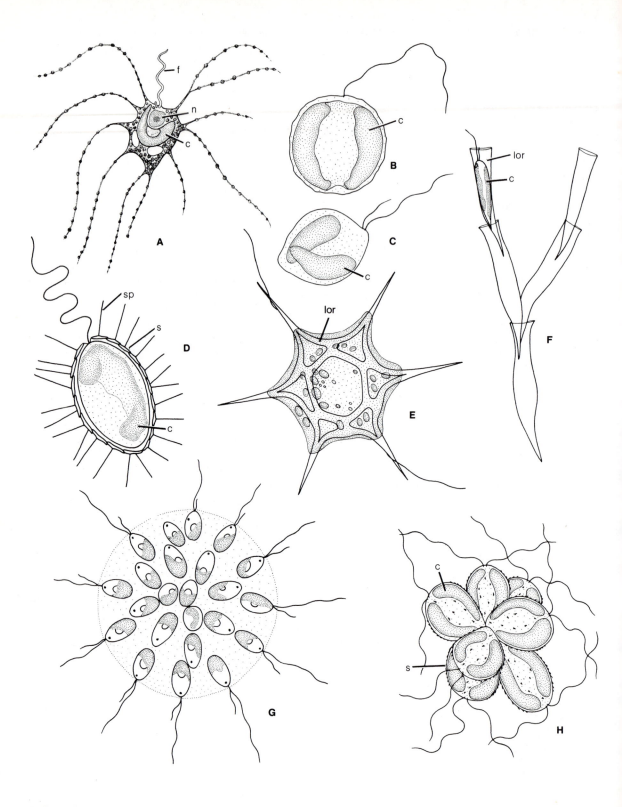

adjacent to one another and are easily recognized (Fig. 8-4G, H). Material grown on a solid mineral medium (agar) or just barely immersed in nutrient solution will produce gametangia in seven to ten days when placed in light (constant or diurnal) at room temperature. A single large, nonmotile, green egg is retained within the female gametangium. A number of chloroplasts and storage-product granules seem to accumulate prior to syngamy. Many small, elongate, colorless male gametes are produced in the male gametangium and are released from an apical pore. They are approximately 5 μm in size and lack chloroplasts (Figs. 8-3C, D and 8-4I). Each sperm contains only a single elongate nucleus and three or four elongate mitochondria (Fig. 8-3D). The flagella are laterally inserted; the forward, short flagellum has tinsels, and the longer flagellum is smooth (Fig. 8-3C). Interestingly, the flagellation of the *Vaucheria* sperm is very much like that of the sperm of the brown alga *Fucus,* which differs from that of other phaeophytes (see Chapter 10).

The sperm fertilizes the egg *in situ,* and a heavy-walled, resting zygote results. Zygote germination occurs under favorable conditions of light, temperature, and moisture. Meiosis is believed to occur upon germination, with only one meiotic nucleus surviving.

Many details of the life history of *Vaucheria* are lacking. Whether meiosis does occur and what is the fate of the products are only two questions still to be answered. Additional ones are: Assuming the egg is uninucleate, how does it become so? Does self-fertilization occur? How is the sperm attracted to the egg?

Figure 8-5 Cell structure and morphological diversity of flagellated Chrysophyceae. **A,** amoeboid form (*Chrysamoeba*), with flagellum, chloroplast (*c*), and nucleus (*n*), ×500. **B–E,** unicellular, motile forms, showing chloroplasts (*c*), scales (*s*), and spines (*sp*). **B,** *Chromulina,* ×2700; **C,** *Ochromonas,* ×3200; **D,** *Mallomonas,* optical section to show cell contents and side view of scales and spines; **E,** *Distephanus,* a silicoflagellate with siliceous endoskeleton (*arrow*), ×1200. **F–H,** multicellular, motile forms. **F,** *Dinobryon,* with lorica (*lor*) around individual cells, ×1000; **G,** *Uroglena,* ×600; **H,** *Synura,* with scales, ×2700. (**A,** after D. J. Hibberd; **E,** after R. E. Norris.)

Ecology

Xanthophyceae are primarily freshwater organisms, often occurring in small ponds or pools. Because of their delicate nature and small size, their ecology is not well known. A common inhabitant of swimming pools is the unicellular *Pleurochloris* (Fig. 8-1G), which is known as the *mustard alga.* The common *Tribonema* occurs in cool waters in the spring, and *Vaucheria* grows in muddy areas, often on the edge of drying ponds or puddles. Some species of *Vaucheria* are extensive colonizers of brackish water marshes and embayments. Most xanthophytes occur in acid ponds, sphagnum bogs, or alpine and subalpine pools. Because they are small and superficially resemble Chlorophyceae, these algae are easily overlooked.

Class Chrysophyceae (Golden Algae)

Over 100 genera with approximately 500 species are referred to Chrysophyceae (*chryso* = golden), which are primarily motile forms. Most representatives occur in fresh water, but a number of species appear to be significant in the marine **plankton** (*plank* = wandering) and **nannoplankton** (*nan* = dwarf).

Cell Structure and Cell Division

The typical golden algal cell contains one or two smooth-edged, golden to yellow-brown chloroplasts (Figs. 8-5B, C and 8-6A). The color results from the predominance of the specific xanthophyll **fucoxanthin,** which masks chlorophyll *a.* Additional pigments include chlorophyll *c* and other carotenoids. The photosynthate is chrysolaminaran, which differs from the laminaran occurring in Phaeophyta (Chapter 10): it lacks the sugar alcohol **mannitol.** Chrysolaminaran often accumulates as a large, refractive, cytoplasmic granule at the posterior of the cell (Fig. 8-6A) or adjacent to an excentric pyrenoid. In some motile forms, an eyespot is present near the anterior

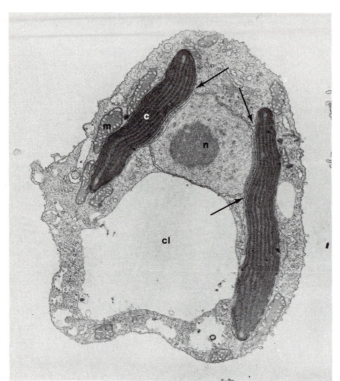

A

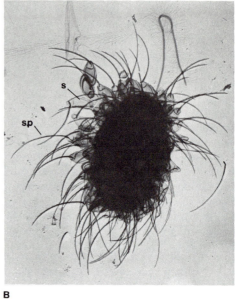

B

C

part of the chloroplast. Contractile vacuoles occur in some species. The nuclear membrane and endoplasmic reticulum of the chloroplast are associated (Fig. 8-6A) as in Xanthophyceae (p. 311; Fig. 8-2B).

Most species are unicellular or colonial flagellates (Fig. 8-5), which generally lack a definitive cell wall. Instead there are scales (Figs. 8-5D and 8-6B, C) or a loose type of covering known as a **lorica** (Figs. 8-5F and 8-7A). The lorica (*lorica* = armor) is essentially a shell to which the cell may or may not be attached by cytoplasmic processes. One or more openings are present, depending upon the shape of the opening and whether there are protruding cytoplasmic processes. The lorica is microfibrillar or scalar in nature, sometimes cellulosic, although minerals (e.g., iron, magnesium, and carbonate) can be deposited. Lorica formation appears to result from longitudinal microfibrils being laid down in a circular fashion, possibly as the swimming cell rotates. When scales are present, they are inorganic, usually composed of silica. They are produced in golgi vesicles adjacent to the cell exterior. In silicoflagellates (Fig. 8-5E), there is an extensive siliceous endoskeleton. Nonmotile forms have a cellulosic wall with pectic substances. In a few species, this wall is composed of two overlapping halves (as in Xanthophyceae; see p. 311).

In the large number of golden algae that are motile, cell division usually occurs while the cells are moving. Prior to nuclear division in motile forms, flagellar basal bodies replicate, and the chloroplast is constricted into two lobes. Early in division, the golgi apparatus replicates, and new flagella are synthesized. Some chloroplast endoplasmic reticulum, which surrounds the nucleus, becomes incorporated into the nuclear envelope.

Movement

In motile cells, flagella are anteriorly inserted, sometimes in a small depression, or pit. In the biflagellated condition, the longer flagellum is tinsel and the other whiplash. If the two flagella are markedly unequal in length, the whiplash one is reduced, and reduction may be so complete that only the tinsel flagellum protrudes. In some biflagellated species, the second flagellum is very delicate and is shed easily (e.g., *Mallomonas*, Fig. 8-6B). Many loricate forms are motile although only the flagellum projects. Other loricate species are stationary even when functional flagella are present (Fig. 8-7A).

The longer, tinsel flagellum has been shown in some species to have a ribbon-shaped root originating from the basal body. This continues to the nucleus, where the root lies in the area between the golgi apparatus and nuclear envelope. During mitosis this root, which has been duplicated, forms poles for the spindle microtubules. A similar structure is rarely present in Chlorophyta and is generally unknown in other algal classes. At the base of the shorter flagellum is a swelling, generally adjacent to the chloroplast. In uniflagellated forms, this swelling and a second basal body are present, indicating reduction of the whiplash flagellum.

Reproduction

As noted previously, reproduction in flagellated forms can occur while they are still motile. In many loricate forms, following cell division, one of the new cells remains in the parental lorica, and the other swims out and forms a new lorica. In nonmotile forms, the zoospores produced resemble either biflagellated or uniflagellated

Figure 8-6 Cell structure of Chrysophyceae, electron microscope. **A,** longitudinal section (*Ochromonas*) showing two lobes of chloroplast (*c*), nucleus (*n*), nuclear-chloroplast endoplasmic reticulum (*arrows*), mitochondrion (*m*), and chrysolaminaran vacuole (*cl*), ×9000. **B, C,** dried cells (*Mallomonas*), showing scales (*s*) with spines (*sp*). **B,** whole cell with obvious tinsel flagellum (*arrow*), smooth flagellum not present, ×2500; **C,** group of scales, ×7000. (**A,** courtesy of T. Slankis and S. Gibbs and *Journal of Phycology;* **B, C,** courtesy of H. Belcher and *Nova Hedwigia.*)

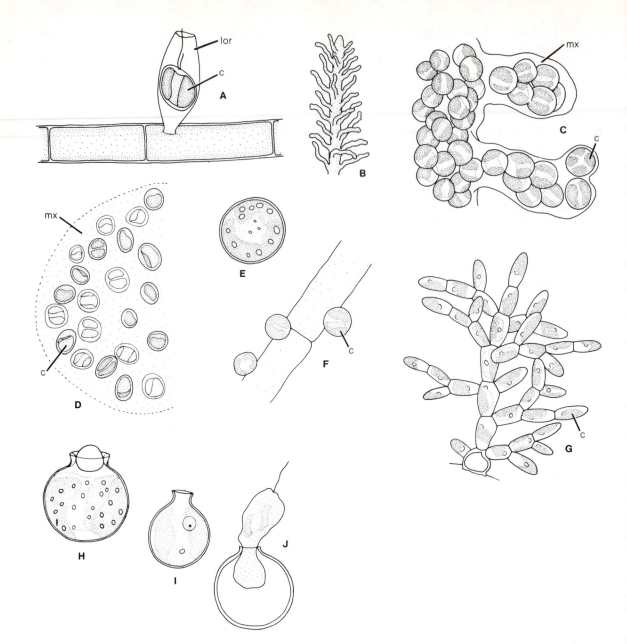

Figure 8-7 Morphological diversity and reproduction of Chrysophyceae. **A–G,** nonmotile forms. **A,** attached, loricate (*lor*) form, *Epipyxis* (on algal filament), with chloroplast (*c*), and two flagella for creating water currents, × 1000. **B–E,** palmelloid forms. **B, C,** *Hydrurus,* showing habit (**B,** × 50) and detail of colony of matrix (*mx*); **C,** × 1000. **D, E,** *Chrysocapsa* with portion of colony in matrix (**D,** × 500) and a single cell (**E,** × 1000). **F,** coccoid form (*Epichrysis*), epiphytic on algal filament, × 930. **G,** filamentous form (*Phaeothamnion*), × 5000. **H–J,** statospores (**H,** with plug, × 3060; **I,** without plug, × 1000); **J,** statospore germination with single zoospore being released, × 1000. (**J,** after Smith, with permission of McGraw-Hill Book Co.)

vegetative cells (Fig. 8-5B, C). Other nonmotile genera form aplanospores.

In many Chrysophyceae, a resting, internal cyst, the **statospore** (*stat* = standing), is produced. A single statospore is formed endogenously in a cell, often excluding parts of the cytoplasm. The cyst wall becomes heavily silicified throughout, except for a small opening containing a plug, evidently composed of silicon and pectic substances, that closes the statospore (Fig. 8-7H, I). Under favorable conditions, this resting stage germinates, generally liberating one or two motile cells, or zoospores, which escape through the pore left after the plug dissolves (Fig. 8-7J). Some statospores are the result of syngamy.

Sexual reproduction is uncommon and usually isogamous, with flagellated gametes. In loricate forms, gametes move to the edge of the lorica but do not become free. Thus the zygote, which is a resting cell, remains attached to the two loricae or caught in a matrix.

Morphological Variation and Classification

Chrysophyceae are almost as diverse in form as Chlorophyceae (see Chapter 7) and Xanthophyceae (p. 315), although most genera and species of Chrysophyceae are motile in the vegetative state. This includes unicellular amoeboid forms, such as *Chrysamoeba* (Fig. 8-5A); unicellular flagellates, such as *Ochromonas*, *Mallomonas*, and *Chromulina* (Fig. 8-5B–D); and multicellular flagellates, as *Dinobryon*, *Synura*, and *Uroglena* (Fig. 8-5F–H). Amoeboid forms, which can have a very short (2 µm) flagellum in the amoeboid phase, can have a flagellated stage with a long tinsel flagellum and a short flagellar stub. Other amoeboid forms lack flagella, and many are within a lorica. Also included in the motile, unicellular Chrysophyceae are silicoflagellates, which have a siliceous endoskeleton (Fig. 8-5E). These algae are known mainly from their siliceous remains. Loricate forms can be multicellular, such as *Dinobryon* (Figs. 8-5F and 8-13A), which also occurs as a single cell, or *Epipyxis* (Fig. 8-7A), which is usually epiphytic on other algae. Nonmotile golden algae include colonies, such as *Hydrurus* and *Chrysocapsa* (Fig. 8-7B–E); coccoid types, such as *Epichrysis* (Fig. 8-7F); and

filamentous forms, such as *Phaeothamnion* (Fig. 8-7G). These nonmotile forms are not commonly collected and comprise less than half the genera in the class. When motile reproductive cells are produced, they resemble either *Chromulina* (uniflagellated) or *Ochromonas* (biflagellated) cells.

Ecology

Freshwater flagellated species of Chrysophyceae are common in the plankton of lakes, particularly during colder seasons of the year, when these algae are abundant. Sometimes the cells are in a mucilaginous envelope on the upper surface of the water. Zoospores are released into the matrix and can then swim to another place on the surface film. Many nonmotile, freshwater species occur commonly in cold (8–10 °C), fast-running streams, springs, or spring-pools, and shallow ponds. Acid habitats, such as those present in sphagnum bogs, support a large variety of smaller, delicate forms. These can be motile, freefloating, epiphytic, or endophytic. Some golden algae occur in tidepools and salt marshes, often in large enough numbers to color the water a golden-brown. Motile Chrysophyceae in the nannoplankton of colder waters, in both the Northern and Southern Hemispheres, are considered important primary food producers.

A few Chrysophyceae are **phagotrophic** (*phag* = eat), often ingesting bacteria or bluegreen algae. This occurs in species containing chloroplasts as well as in those without. Ingestion occurs by cytoplasmic engulfment of anything lying on the cell surface. Not surprisingly, algal cells "prefer" to ingest bacteria rather than carbon particles when given the choice! Thus, some golden algae are opportunistic with regard to their energy source.

Class Bacillariophyceae (Diatoms)

Bacillariophyceae (*bacill* = little stick) comprise nonflagellated, vegetative cells. They are conspicuous in both marine and freshwater habitats and well-represented in the fossil record. Outstanding features are the boxlike form of the cell and the presence of silicon in the cell wall. Approximately 200 genera with over 5600 living species belong to this class. Including fossil forms, there are probably over 10,000 species described.

Cell Structure and Cell Division

The cells, or **frustules** (*frust* = little piece), are complex. They consist of two overlapping halves (Fig. 8-8), each termed a **theca** (*thec* = case, box). Because of these two parts, the common name for the class is *diatom*, meaning "cut in two." As much as 95% of the cell wall is silica, which is deposited in an organic framework composed of sugars, lipids, and amino acids. The siliceous shell is formed within the cell in a membrane-bound vesicle. The pattern of silica deposition is, in general, species specific, although sometimes affected by environmental factors. The outer part of the cell wall (comparable to the top of a box) is the **epitheca** (*epi* = above), and the inner part of the cell (or bottom of the box) is the **hypotheca** (*hypo* = below), as shown in Figure 8-8. Each theca consists of a flattened or convex valve with a connecting band attached along its edge. When the top or bottom surface of the cell is viewed, it is the **valve view** (Figs. 8-8 and 8-9A, B). The overlapping walls of the cell form a girdle region, and when the cell is observed from the side, it is the **girdle view** (Figs. 8-8 and 8-9B). The girdle area is often wide because of the presence of intercalary bands (Fig. 8-9A, B).

The pattern of silica deposition on frustules forms characteristic markings. Electron microscope studies show that these markings are more intricate than can be resolved with the light microscope. Thus it has been necessary to redefine and standardize the terms used to describe them. For example, silica depositions that form openings are known as **areolae** (*areol* = little space) (Figs. 8-9C, D and 8-10A). They contain a thin, siliceous layer with slits or holes (Fig. 8-9D). Perforations in the thin, siliceous layer are **puncta** (*punct* = sting, prick) and may be of unknown structure (Fig. 8-9E). Elongate thickenings can be solid or form chambers that are open at both ends. Current usage terms the solid thickenings

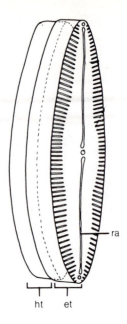

Figure 8-8   Cell structure of Bacillariophyceae (*Pinnularia*), showing girdle (*left*) and valve (*right*) and to illustrate hypotheca (*ht*), epitheca (*et*), raphe (*ra*), and linear markings, ×500.

Figure 8-9   Frustule structure of Bacillariophyceae. **A, B,** pennate diatoms, scanning electron microscope. **A,** theca (*Tabellaria*), showing valve (*lower*) and intercalary band (*upper*): note rows of punctate striae on valve, ×3300. **B,** external features of two cells (*Didymosphaenia*), showing valve (*lower*) and girdle views, ×650. **C,** light microscope of centric diatom, *Coscinodiscus*, in valve view, showing areolae, ×1200. **D, E,** transmission electron microscope. **D,** areolae with perforate siliceous plate with puncta (*Triceratium*), ×5000. **E,** carbon replica of frustule (*Cymbella*), showing raphe (*ra*) from both outside (*straight*) and inside (*curved*) of the valve, ×3600. (**A,** courtesy of J. D. Koppen and *Journal of Phycology;* **B,** courtesy of P. A. Dawson and the *British Phycological Journal;* **C,** courtesy of D. Walker; **D,** courtesy of R. Ross and the *British Phycological Journal;* **E,** courtesy of R. W. Drum, from "Electron microscope observations of diatoms," *Oesterrich Botanische Zeitschrift* 116:321–30, 1969, with permission of Springer-Verlag.)

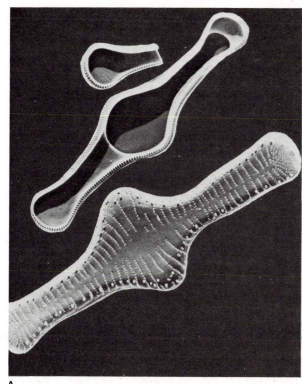

A

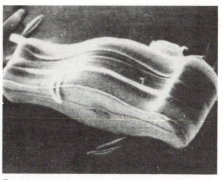

B

C

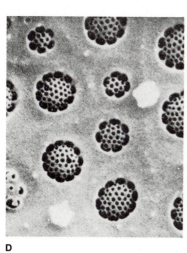

D

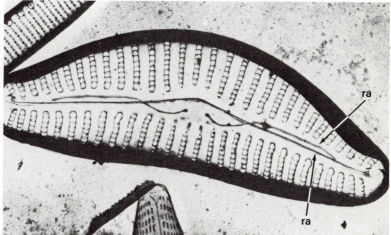

E

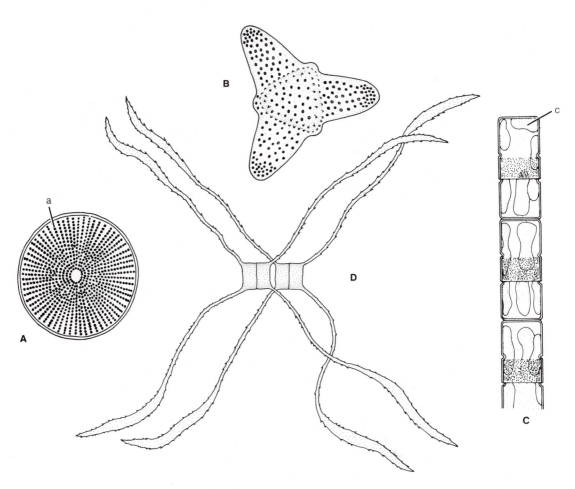

Figure 8-10 Morphological diversity and wall markings of Bacillariophyceae. **A–D,** centric diatoms, **A,** *Coscinodiscus,* valve view, showing rows of areolae (*a*), ×125; **B,** *Triceratium,* valve view, ×545; **C,** *Melosira* cells, girdle view, showing chloroplasts (*c*), ×640; **D,** *Chaetoceros* cells, girdle view, with long spines, ×500. **E–H,** pennate diatoms. **E,** *Asterionella* colony, girdle view, ×1000; **F,** *Fragilaria* colony, girdle view, ×750; **G,** *Gomphonema,* valve view (note transverse asymmetry), ×1250; **H,** *Cymbella,* valve view (note longitudinal asymmetry), ×545.

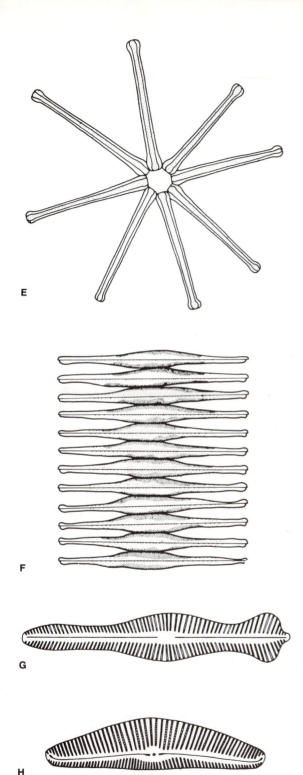

E

F

G

H

costae (*cost* = rib) and the chamber an **alveolus** (*alveol* = cavity). Areolae, puncta, and alveoli, when arranged linearly, form **striae** (*stria* = furrow) (Fig. 8-9A). Sometimes it is difficult to resolve the individual puncta or alveoli of a stria using the light microscope. In addition to silica deposition patterns, projections, spines, or granules of various types can be present (Fig. 8-10D).

Cell shape and valve markings are radial (including triangular), rectangular, or asymmetrical. The radial form delimits the **centric** diatoms (Figs. 8-9C and 8-10A–D), whereas rectangular and asymmetric forms delimit the **pennate** diatoms (Figs. 8-8, 8-9A, B, E, and 8-10E–H). Several orders are included in both centric and pennate groups.

Many pennate diatoms have an additional valve marking, termed the **raphe** (*raph* = seam, suture) (Figs. 8-8 and 8-9E), which is essential to the motility of the diatom. This is an unsilicified groove or channel that can be straight, undulate, or S-shaped. The raphe extends through the frustule and can be straight or curved, so that the outer and inner faces of the valve differ in appearance (Fig. 8-9E). Part of the raphe or the raphe channel can contain cytoplasm, which is enclosed by the plasma membrane. Sometimes the raphe also lies in a raised canal.

The uninucleate diatom cell has one, two, or many yellow- or golden-brown peripheral chloroplasts, which are disc-shaped or ribbonlike. The centric diatoms generally have many chloroplasts, whereas pennate forms have only one or two large ones. The ultrastructure of the chloroplast (Fig. 8-11) is similar to that of other Chrysophyta as well as that of Pyrrhophyta (Chapter 9) and Phaeophyta (Chapter 10). The chlorophylls (a, c_1, c_2) are masked by carotenoids, especially fucoxanthin and other xanthophylls that also occur in Chrysophyceae. Fucoxanthin is the most abundant carotenoid pigment present and is responsible for the characteristic color.

The large, well-organized nucleus is located either centrally or peripherally in the cell. In many pennate diatoms, the nucleus is often located on a cytoplasmic bridge across the center of the large central vacuole (Fig. 8-11A). Food reserves are stored as fats, oils, and the carbohydrate chrysolaminaran; fat droplets occur in chloroplasts or vacuoles (Fig. 8-11A). Pyrenoids are present in the chloroplasts of some diatoms.

The photosynthate can be adjacent to the pyrenoids but is always external to the chloroplast envelope, as in other Chrysophyta.

Cell division in diatoms involves formation of two new valves, somewhat analogous to the formation of new half-cells in the desmids of green algae (Chapter 7). Nuclear and cytoplasmic division precede frustule formation. There is a concomitant increase in the cytoplasmic vesicles originating from golgi, which evidently contribute to the vesicle in which the new valve is formed. The new valves, each of which is a hypotheca, are completed before the girdle or intercalary bands are formed (Fig. 8-11B). Silicon is an essential element for diatom wall construction.

The valves of the parental cell always become the epithecae of the new cells, putting a limit on the size of the diatom. Thus, there is a gradual diminution in cell size of the population. Certain diatom populations do not show this diminution in size, since weakly silicified cells expand and regain their original size and shape after cell division. In laboratory cultures, cell size can often be maintained by nutritional and cultural changes. Many natural diatom populations gradually diminish in overall cell size, and in these a large cell termed an **auxospore** (*aux* = grow, enlarge) is produced (see Reproduction, this page).

Movement

Only those pennate diatoms with a raphe (Figs. 8-8 and 8-9E) move by a smooth or somewhat jerky gliding motion. Rates of movement have been calculated at 0.2 to 25 μm/second. The path taken seems to depend on the shape of the raphe, with the rate of movement possibly light-dependent. The mechanism is believed to involve cytoplasmic streaming within the raphe system. One hypothesis is that motile cells apparently possess a fibrillar system under the raphe and produce a mucous material that is secreted through the raphe and adheres to the substrate. Another hypothesis postulates capillary forces and pressure gradients that play a part in cell locomotion. Other hypotheses propose a series of dynamic oscillations of protoplasmic protein fibrils.

In centric diatoms, motile uniflagellated or biflagellated swarmers are produced. In most instances, these function as male gametes. Swarmers have at least one tinsel flagellum, which can be anterior or posterior in position.

Reproduction

The main method of reproduction in diatoms is asexual, or vegetative, by cell division. Their rapid rate of division under optimum conditions can result in dense blooms, which give the water a distinct brown color over an extensive area. Silicon is an absolute requirement not only for cell wall formation but also for cell division, as mitosis is inhibited by its absence. Germanium (as germanium dioxide), which belongs to the same series of elements as silicon, is an inhibitor of silicate utilization.

Sexual reproduction varies throughout the class and can be isogamous, anisogamous, or oogamous. In contrast to those of most algae, the vegetative cell is diploid (see Fig. 2-8B for diploid life history). Meiosis occurs during gamete formation. The resulting zygote, often different from vegetative cells in appearance, enlarges immediately into an auxospore, or it can undergo a resting period as a round cell.

Details of sexual reproduction have been studied in a few diatoms. In pennate forms, gametes are usually isogamous and lack flagella. Parental cells pair along the girdle, and meiosis results in one or two viable gametes. In contrast, centric diatoms have oogamy; the tiny, almost colorless, flagellated male gametes have at least one tinsel flagellum. In some species, the central two flagellar microfibrils are lacking, but this does not seem to impede motility. Sperms are released from the girdle region, and penetration of the female gamete, which is retained within the frustule, is often by way of the girdle.

Gamete nuclei either fuse immediately or later. When the auxospore is formed it undergoes two nuclear divisions, with one nucleus aborting during late telophase. The auxospore differs from the vegetative cell (Fig. 8-12A, B), and when cell division occurs, the first cells are "atypical," since the "typical" mold for the new valves is lacking.

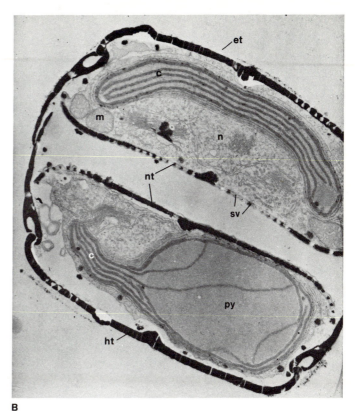

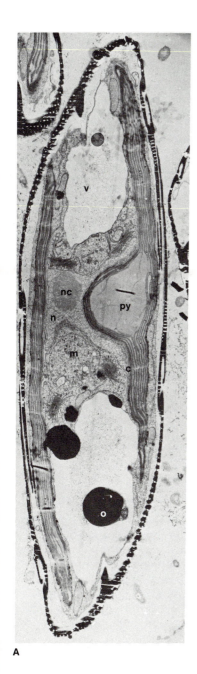

Figure 8-11 Cell structure and division of Bacillario-phyceae (*Amphipleura*), transmission electron microscope. **A,** longitudinal section, slightly oblique to valve, showing chloroplasts (*c*) with pyrenoids (*py*), central cytoplasmic bridge with nucleus (*n*) and nucleolus (*nc*), mitochondria (*m*), vacuole (*v*), and oil bodies (*o*), ×6750. **B,** transverse section of recently divided cell, showing chloroplast with pyrenoid, nucleus, mitochondria, silicon deposition vesicles (*sv*), new hypotheca (*nt*), and parental thecae (hypotheca, *ht;* epitheca, *et*), which become epithecae of new cells (note girdle bands for new thecae not formed), ×5500. (Courtesy of E. F. Stoermer and *American Journal of Botany*.)

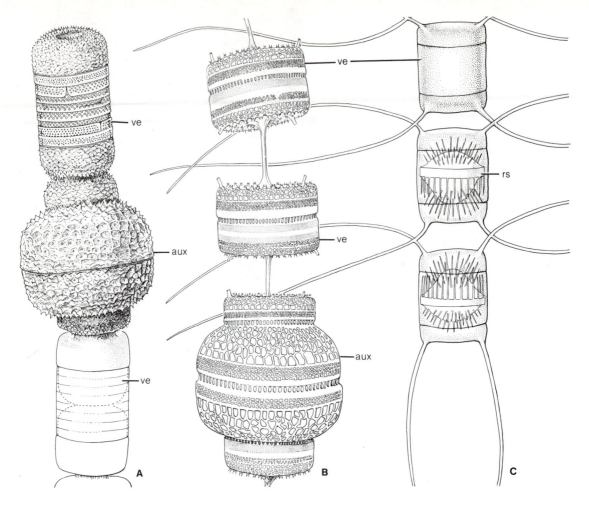

Figure 8-12 Reproductive cells and morphological forms of Bacillariophyceae. **A, B,** auxospores (*aux*) in filamentous colonies with vegetative cells (*ve*). **A,** *Melosira;* **B,** *Thalassiosira;* ×500. **C,** resting spores (*rs*), orcysts, in filamentous colonies (*Chaetoceros*) with vegetative cells, ×500. (After Hendey.)

This probably indicates a certain amount of genetic plasticity in thecal structure.

Auxospore formation is reported in natural populations, and in some genera it occurs only every two to five years, following a prolonged period of vegetative reproduction. Some marine diatoms that grow in water of low salinity form auxospores upon dilution of the medium. In others, sexual reproduction and subsequent auxospore formation can be controlled nutritionally

(e.g., with manganese) or influenced by light regime, a temperature change, or additional carbon dioxide. Most diatoms studied are self-fertilizing. Therefore, sexual reproduction can actually involve little if any genetic segregation and recombination, since fusing protoplasts are produced by the same parental cell. Even though meiosis occurs, segregation and recombination do not, and there is no new genotype. Thus, auxospore formation can be considered a type of reproduc-

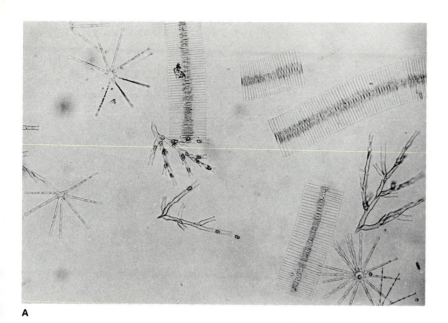

A

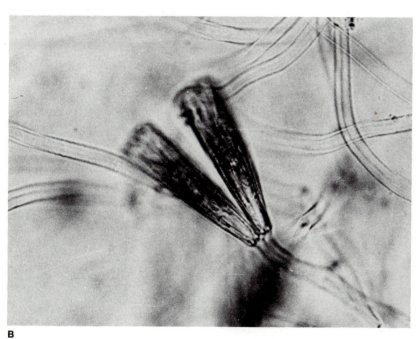

B

Figure 8-13 Light microscope preparations of Bacillariophyceae. **A,** phytoplankton from freshwater lake, showing predominance of diatoms, including *Fragilaria* (chains of elongate cells) and *Asterionella* (star-shaped colony); also present is the motile *Dinobryon* (Chrysophyceae—dendroid colonies), ×150. **B,** *Gomphonema,* showing mucilaginous stalks and vegetative cells in girdle view, ×1000.

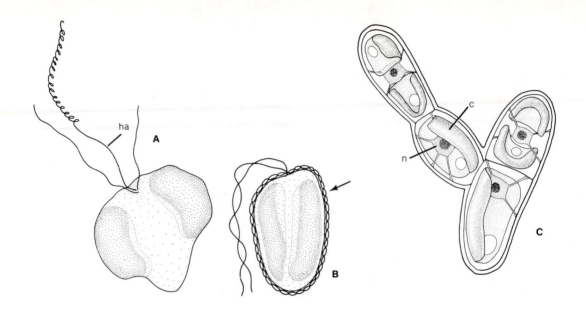

Figure 8-14 Cell structure and morphological diversity of Prymnesiophyceae. **A, B,** motile forms. **A,** *Chrysochromulina* with coiled haptonema (*ha*), ×700; **B,** *Hymenomonas* with surface coccoliths (*arrow*), ×1000. **C,** nonmotile form, *Pleurochrysis* (*Apistonema*), ×1200: *c*, chloroplast; *n*, nucleus.

tion that maintains a high degree of species uniformity and allows for reestablishment of maximum vegetative cell size.

Cysts, or resting cells, are produced by diatoms. These are smaller than the cell producing them and have a thickened wall. The shape is unlike that of the parental cell (Fig. 8-12C). Cysts are generally produced asexually.

Morphological Variation and Classification

Several orders of diatoms are recognized, but all are equivalent to the nonmotile, nonfilamentous order of other algal classes (e.g., Chlorococcales of Chlorophyceae). The main morphological separation into centric and pennate diatoms is adequate to show the diversity within the Bacillariophyceae, although, as noted previously, there are several orders of each type. In addition to wall markings and symmetry, centric and pennate diatoms are separated on the basis of shape,

chloroplast number, and motility. Centric diatoms are radially symmetrical, usually with circular valves, as in *Coscinodiscus* (Fig. 8-9C); sometimes, however, they are triangular, as in *Triceratium* (Fig. 8-9B), or elongate, as in *Melosira* (Fig. 8-10C). Valve markings radiate from a central point of the valve, and spines and external appendages are present (Fig. 8-10C), as well as intercalary bands in the girdle region. Pennate diatoms have bilateral symmetry with elongate valves, as in *Tabellaria* (Fig. 8-9A). Some genera are asymmetrical in girdle view, as in *Gomphonema* and *Didymosphaenia* (Figs. 8-9B and 8-13B). Valve markings occur in two longitudinal series on either side of a median line (Fig. 8-8), and the valve and/or girdle is longitudinally or transversely symmetrical or asymmetrical (Fig. 8-10G, H). Most pennate diatoms have a raphe in the valve (Fig. 8-9E).

Most diatoms occur as solitary cells. Others are loosely aggregated in irregular filamentous chains, with adjacent cells interconnected by mucilaginous pads that join the cells into a zigzag colony. Species of *Gomphonema* form stalked, treelike colonies, often with large amounts of stalk material (Fig. 8-13B), as does *Cymbella*. Some diatoms form macroscopic, branched strands resembling filamentous brown algae (Chapter 10). The stalk and mucilaginous tubes are formed of carbohydrate, or chitin similar to that in fungal cell walls, or sulfated polysaccharides, as in brown and red algae (Chapters 10, 11). Very little is known about the material or its production. Other colonial species are formed by adhesion of valves, as in the starlike *Asterionella* (Figs. 8-10E and 8-13A) and the long, chainlike *Melosira* and *Fragilaria* (Figs. 8-10C, F and 8-13A). Long, spinelike valve processes assist in forming other colonies, as in *Chaetoceros* (Fig. 8-10D).

Ecology

Since the vegetative cells of planktonic diatoms are not flagellated, they tend to sink from the illuminated surface waters. However, the presence of lipids and elaborate cell extensions helps keep them afloat. The presence of spinelike extensions, as in *Chaetoceros* (Fig. 8-10D), increases the surface area of the cell immensely and tends to trap small bubbles of oxygen liberated during photosynthesis. Similarly, filamentous, chainlike colonies may serve the same function.

Bacillariophyceae are widely distributed throughout freshwater and marine environments; centric species are primarily marine, and pennate diatoms occur more often in fresh water (Fig. 8-13A). Diatoms occur free-floating in plankton, as well as on the surfaces of solid substrates, on mud-flat surfaces, in salt marshes, and attached as epiphytes to other algae and aquatic plants. Diatoms also occur in soil, on old brickwork or rock walls, and in almost any moist place, including human livers and kidneys! Diatoms occur in vast numbers in phytoplankton, often forming extensive blooms. In arctic waters, diatoms often produce extensive masses on the lower ice surface.

Some diatoms can tolerate only a narrow range of environmental conditions and are used as indicators of the physical-chemical characteristics of some waters. The composition of diatom communities is also used as an indicator of varied ecological conditions. For example, rivers free from pollution have many different diatom species, each present in relatively small numbers. In contrast, polluted waters have few diatom species, often in very large numbers. Lake sediment cores contain diatom frustules that indicate the history of the lake with regard to the nutrients in its waters.

Other Classes of Chrysophyta

Representatives of the remaining three classes are often overlooked because of their small size and paucity of numbers. Of these, Prymnesiophyceae are the most important in the food chain of some marine waters, as they are the prime food source for the animals (e.g., zooplankton, fish).

Class Prymnesiophyceae (Prymnesiophytes)

Until the early 1960s, the Prymnesiophyceae (also known as Haptophyceae), comprising approximately 60 genera with nearly 200 species, were classified as an order in Chrysophyceae. Until then, little was known about most members, except that they were small, biflagellated, and golden-brown. With increasing interest in the productivity of the seas and the refinement of techniques for culturing and the study of ultrastructure, a better understanding has become possible. Their separation from Chrysophyceae is made primarily on the basis of cellular differences. In addition, there are life-history differences; many prymnesiophytes show an alternation of nonmotile stages with motile ones. Thus, what was once considered as a group of primarily motile forms may well consist also of a large number of forms with alternating motile and nonmotile phases.

Cell Structure, Cell Division, and Movement
Motile cells lack cell walls, but do have scales of organic materials and/or calcium carbonate (Figs. 8-14B and 8-15A–C). The calcium carbonate scales are large enough to be seen with the light micro-

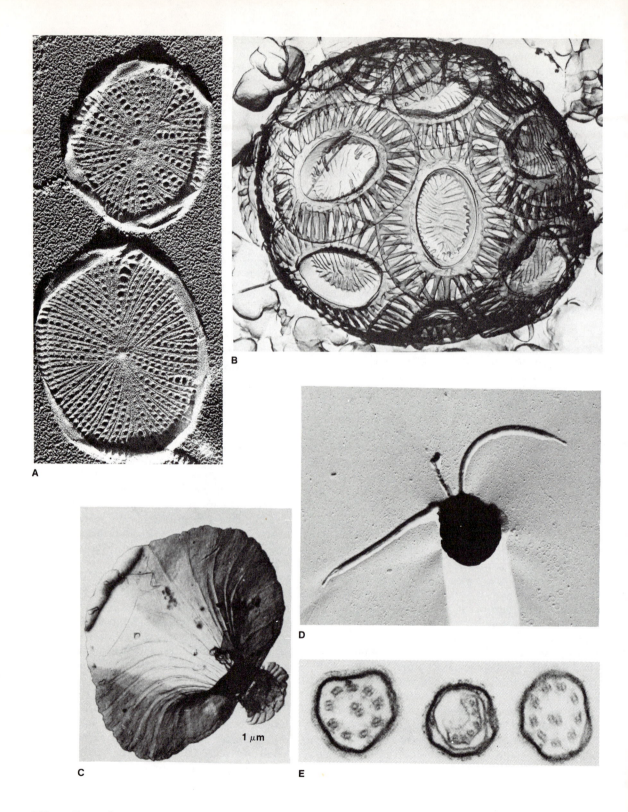

A

B

C

1 μm

D

E

scope and, in fact, were first known from microscopic examination of chalk deposits. These scales are termed **coccoliths** (Fig. 8-15B, C); those organisms bearing them are the *coccolithophorids*. Scales are formed within the golgi apparatus by the deposition of calcium carbonate on an unmineralized base, which then becomes coated with an organic membrane. In addition to coccoliths, there are organic scales present on coccolithophorids as well as other Prymnesiophyceae (Fig. 8-15A). Organic scales also are formed within the golgi apparatus and are primarily polysaccharide, some being cellulosic. Neither scale type is present in all stages of the life history. An example is *Pleurochrysis*, which produces two types of swarmers—one bearing coccoliths and the other bearing organic scales—from a nonmotile, coccolith-free, colonial thallus (Fig. 8-14C).

The internal structure of the cell is similar to that described for Chrysophyceae (Fig. 8-6A; p. 317). However, one outstanding difference is the presence on many forms of a unique, apical organelle: the **haptonema** (Figs. 8-14A, 8-15D, and 8-16C). The name means "fastening thread," which was considered originally to be its function, although at present the exact function is unknown. The haptonema, which often looks like a third flagellum, is located between the two flagella, sometimes lying opposite the direction of the flagella. As with a flagellum, the cell plasma membrane is continuous around the haptonema. However, in contrast to the flagellum, with its typical 9 + 2 arrangement of micro-

tubules (see Fig. 2-2B), the haptonema contains extensions of endoplasmic reticulum surrounding the six or seven longitudinally oriented microtubules (Fig. 8-15E). In some instances, the haptonema is long and coiled (Fig. 8-14A), whereas in others it is short (Figs. 8-15D and 8-16C). The surface can be covered with fine, organic scales. In some short, uncoiled haptonema, the tip is swollen, and more endoplasmic reticulum is associated with it. In the coiled haptonema, a refringent inclusion moves along the length during coiling and uncoiling. Coiling itself seems to be automatic and responds in less than $1/50$ of a second to minor shock (e.g., chemicals). The haptonema is also not present in all life history phases of an organism.

Cell division occurs in either motile or nonmotile phases. Generally, duplication of organelles occurs prior to the separation of the cells. In attached or sessile forms, new cells remain adjacent, increasing the size of the filamentous or colonial plant.

A second important difference between Chrysophyceae and Prymnesiophyceae is in the flagella. Prymnesiophytes have two smooth, whiplash flagella which are equal in length (Fig. 8-15D). Their insertion is similar to that in Chrysophyceae and is either apical or subapical. Often the flagella trail behind or seem to be "pushing" the cell.

Reproduction Reproduction in Prymnesiophyceae is varied, with vegetative reproduction being the main form. For unicellular forms, asex-

---

Figure 8-15 Cell structure of Prymnesiophyceae, electron microscope. **A–C,** scales. **A,** organic scales (*Chrysochromulina*), ×29,000. **B, C,** coccoliths made of calcium carbonate. **B,** whole cell (*Coccolithus*), ×15,000; **C,** single scale, ×13,000. **D, E,** flagella and haptonema. **D,** zoospore showing central haptonema with bulbous tip and two flagella, ×6000; **E,** transverse section of two flagella and central haptonema (Phaeocystis), ×60,000. (**A, D,** courtesy of I. Manton, from I. Manton and M. Parke, "Preliminary observations on scales and their mode of origin in *Chrysochromulina polylepis* sp. nov.," *Journal of the Marine Biological Association, U.K.* 42:565–78, 1962, with permission of Cambridge University Press; **B, C,** courtesy of A. McIntyre, from A. McIntyre and A. W. H. Bé "Modern Coccolithophoridae of the Atlantic Ocean. I. Placoliths and cryoliths," *Deep-Sea Res.* 14:561–97, 1967, with permission of Pergamon Press Ltd.; **E,** courtesy of J. C. Green and I. Manton, from M. Parke, J. C. Green, and I. Manton, "Observations on the fine structure of zoids of the genus *Phaeocystis* (Haptophyceae)," *Journal of the Marine Biological Association, U.K.* 51:927–41, 1971, with permission of Cambridge University Press.)

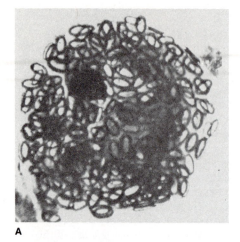

A

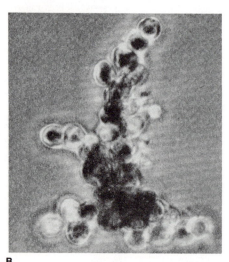

B

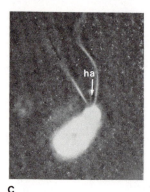

C

ual reproduction usually occurs by longitudinal splitting of the cells; this produces two new cells, although sometimes four cells are formed. Fragmentation of multicellular forms also occurs. Resting cells have been reported, but little is known concerning their formation.

Sexual reproduction as such has been rarely observed, and the tiny gametes are isogamous and almost colorless. It is possible that other motile stages function as gametes, but this has not been confirmed. Meiosis has not been confirmed for any species. Chromosome counts of one nonmotile, coccolith-forming plant indicate that it is diploid and alternates with a nonmotile, attached form that is haploid. Meiosis in this organism may occur in swarmer production. The fact that there are several forms in the life history indicates the possibility of sexual reproduction (Fig. 8-16).

Morphological Variation and Classification
Diversity in form is similar to that in Chrysophyceae (see p. 321). However, in Prymnesiophyceae, many diverse forms apparently are stages in the life history of a species. For example, in one species the motile, coccolith-bearing stage (*Hymenomonas*-stage, Fig. 8-16A) is diploid and gives rise to motile and nonmotile, coccolith-bearing spores. In addition, motile, coccolith-bearing cells produce biflagellated cells lacking coccoliths but having a haptonema. Zoospores (Fig. 8-16C) may be the result of meiosis. They can grow into a uniseriate, filamentous thallus (*Apistonema*-stage, Fig. 8-16B) lacking coccoliths. The filamentous stage, which is also produced by motile, coccolith-bearing cells, produces biflagellated, haptonema-bearing but coccolith-free cells. These zoospores are capable of forming either the *Apistonema*- or *Hymenomonas*-

Figure 8-16 Life history of Prymnesiophyceae. **A,** coccolith-bearing stage (*Hymenomonas*-stage), ×3000. **B,** nonmotile thallus (*Apistonema*-stage), ×750. **C,** coccolith-free swarmer from nonmotile thallus, with short haptonema (*ha*), ×2000. (Courtesy of B. S. C. Leadbeater and the *British Phycological Journal*.)

stage. It is thought that in some species, these bi-flagellated cells from the *Apistonema*-stage are gametes, and the coccolith-bearing, motile stage is a zygote. It is also possible for motile, coc-colith-bearing cells to form nonmotile, coccoid cells lacking coccoliths. This is the *Ocrosphaera*-stage, which produces biflagellated, coccolith-and haptonema-free zoospores that settle, pro-ducing nonmotile, attached forms that ultimately bear coccoliths and then become flagellated. Thus, many different morphological stages are part of one life history:

1. Coccolith-bearing, motile cells (Fig. 8-16A).
2. Coccolith-free, motile cells with haptonema (Fig. 8-16C).
3. Coccolith-free, motile cells without hap-tonema.
4. Coccolith-bearing, nonmotile cells.
5. Coccolith-free, nonmotile cells.
6. Filamentous plants (Fig. 8-16B).

Ecology   Most prymnesiophytes are apparently marine, and some are important in marine nan-noplankton. Coccolithophorids are often pre-dominant in warmer waters of the tropics and subtropics. Certain living species serve as good indicators of water temperature. Fossils with coc-coliths identical to those of living ones serve as indicators of prehistoric climatic conditions.

Some nonmotile phases become free-floating and occur in large masses that adversely affect herring migration. Other prymnesiophytes (some noncoccolithophorids) are responsible for marine fish kills and mass mortalities of inver-tebrates in brackish waters. The effect on the ani-mals evidently is primarily on the permeability of biological membranes.

Class Eustigmatophyceae (Eustigmatophytes)

The class Eustigmatophyceae was established in 1970 to include some forms previously classifed as Xanthophyceae. These forms, when examined with the electron microscope, were seen to be definitely different. Because of their small size—usually less than 50 μm—it was necessary to em-ploy the electron microscope to establish the dif-ferences. Presently, the number of Eustig-matophyceae is small, because only a few of the possible forms have been studied in detail.

The main difference is in the zoospore struc-ture, although there are also differences in vege-tative cells. The class name is derived from the conspicuous eyespot in the zoospore (Fig. 8-17A), which is outside the single, elongate chloroplast (Fig. 8-17C). In addition, the posteriorly flattened zoospore generally has only a single, emergent, tinsel flagellum (Fig. 8-17B), although there are indications (basal body presence) of a second, rudimentary flagellum. In contrast to Xantho-phyceae and Chrysophyceae (and Phaeophyta, Chapter 10), there is no connection of the nucle-ar envelope with the endoplasmic reticulum of the chloroplast (see pp. 311, 319; Figs. 8-2B and 8-6A). Additional differences in pigments—pri-marily xanthophylls and possibly chlorophyll *c*—also separate the classes.

Class Raphidophyceae (Raphidophytes)

The group of unicellular flagellates known as Raphidophyceae contains just over a dozen gen-era placed in a single order, of which half are colorless. Our knowledge of raphidophytes and their affinities is poor. In the past they have been allied with Cryptophyceae (Chapter 9) and Xanthophyceae or separated into their own group. Few species have been completely stud-ied; however, present data indicate possible af-finities to Xanthophyceae. They are also known as the chloromonads, as they contain many dis-coid, green chloroplasts.

The cell is relatively large (50–100 μm) for a flagellated organism. It lacks a cell wall but has a delicate covering and is plastic in shape. In some forms, flagellated cells quickly become non-motile by ejecting mucilaginous threads that form a matrix. The flagellated cell is somewhat circular in side view but is usually compressed, frequently with a longitudinal groove from which flagella emerge (Fig. 8-18B). In some forms, small organelles occur in the outer cyto-plasm, either throughout the cell (Fig. 8-18B) or in localized areas (Fig. 8-18A). These are released

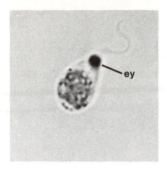

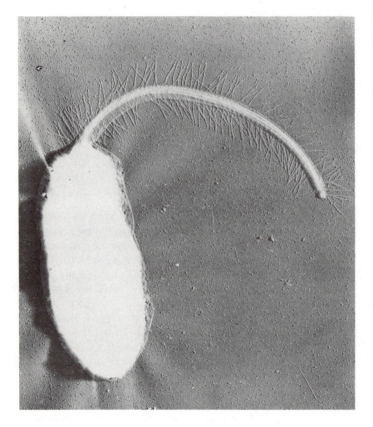

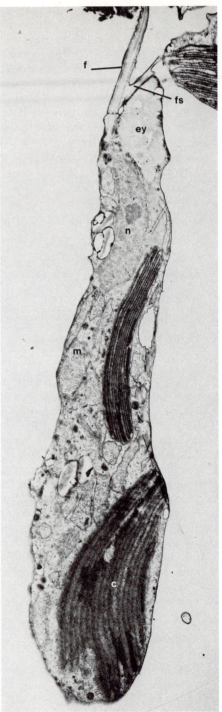

Figure 8-17 Zoospores of Eustigmatophyceae. **A,** light microscope view showing single flagellum and prominent anterior eyespot (*ey*), ×2000. **B,** shadow-cast cell preparation showing bilateral tinsels on single flagellum, ×6750. **C,** longitudinal section of cell showing flagellum (*f*), flagellar swelling (*fs*), large eyespot, nucleus (*n*), elongate chloroplast (*c*), and mitochondrion (*m*), ×9000. (Courtesy of D. J. Hibberd and *Annals of Botany* Company.)

C

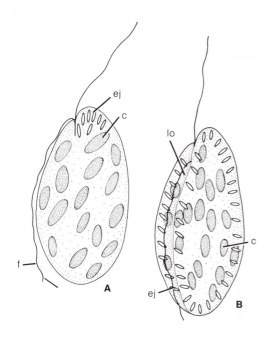

Figure 8-18 Morphological diversity of Raphidophyceae. *Gonyostomum* with ejectosomes (*ej*), disc-shaped chloroplasts (*c*), trailing flagellum (*f*), and longitudinal groove (*lo*). **A,** ×460; **B,** ×500.

as fine threads when the organism is disturbed and may be the source of the mucilaginous matrix that forms macroscopic masses. These rodlike organelles **(ejectosomes)** are the source of the class name Raphidophyceae (*raph* = needle). Usually, the anteriorly biflagellated cell has the swimming flagellum bearing tinsels directed anteriorly. The posteriorly directed second flagellum is smooth, lies close to the cell, and is often difficult to discern. Many contractile vacuoles are present, but no eyespot has been noted. Photosynthetic forms have many disc-shaped chloroplasts.

Relationships of Chrysophyta

Interrelationships of the classes constituting Chrysophyta, as presented in this text, are complex. Pigmentation, cell covering, and storage

products serve to unify the classes. Ultrastructural differences of Prymnesiophyceae (flagellation, haptonema; pp. 331–335) and Eustigmatophyceae (eyespot, endoplasmic reticulum; p. 335) are features that cannot be overlooked. The basic question concerning the importance of such features needs to be considered when establishing evolutionary relationships. It is probable that each should be placed in a division separate from, but closely related to other Chrysophyta.

The fossil record for Chrysophyta (in the broadest sense) includes primarily coccolithophorids and diatoms. A few cysts allied to Chrysophyceae and Xanthophyceae have been reported. Some diatoms occur in Mesozoic strata about 150 million years old. However, they are present in much larger numbers since the beginning of the Cenozoic era (65 million years ago). Traces of Chrysophyceae, primarily as cysts and also as scales, are in the Cenozoic. It is becoming evident that these microfossils will be useful in determining past distributions and environmental conditions.

It is likely that all classes discussed in this chapter could have originated from a phaeophycean ancestor (Chapter 10). As the characteristic wall of the diatoms sets them apart, it is conceivable that early in the evolution of a prototype diatom, the "box construction" was successful in establishing diatoms in the aquatic environment—probably as plankton in the upper surface waters. The incorporation of silicon into the cell wall may have originated in a time when silicon was more abundant in the water than carbon, although at present it is difficult to imagine such a situation. The diversity of prymnesiophytes may also represent a line that diverged early but occupied a different part of the aquatic environment, namely that of the intertidal and subtidal habitat.

The lack of masking carotenoids, such as fucoxanthin, in xanthophytes (as well as eustigmatophytes and raphidophytes) seems to set them apart. At present, there is not enough information available to indicate the relationship of fucoxanthin to other xanthophyll pigments in these groups. Possibly the fact that most representatives of these groups are freshwater forms may have some bearing on their evolutionary relationships and the development of pigments. Whether these freshwater forms could have evolved from predominantly marine brown algae is speculative. Yet, in contrast, Chrysophyceae have many phaeophycean (as well as xanthophycean) features but occur abundantly in the freshwater environment.

From this brief discussion (and speculation), it is clear that many features and factors must be considered in proposing evolutionary patterns. Sometimes it is not possible to draw really meaningful conclusions; however, that there is diversity of organisms in similar environments as well as similarity of organisms in diverse environments is of significance as well as interest.

References

Aaronson, S. 1980. Descriptive biochemistry and physiology of the Chrysophyceae (with some comparisons to Prymnesiophyceae). In Levandowsky, M., and Hutner, S. H., eds., *Biochemistry and physiology of protozoa*. Vol. 3. 2nd ed. Academic Press, New York and London.

Boney, A. D. 1975. *Phytoplankton*. Edward Arnold, London.

Bourrelly, P. 1968. *Les algues d'eau douce*. Vol. 2. *Les algues jaunes et brunes, Chrysophycées, Phaeophycées, Xanthophycées et Diatomées*. N. Boubée & Cie., Paris.

Cox, E. R., ed. 1980. *Phytoflagellates*. Elsevier/North Holland, New York. (Chapters on chrysophytes, xanthophytes, prymnesiophytes, eustigmatophytes, silicoflagellates, and chloromonads.)

Dodge, J. D. 1979. The phytoflagellates: fine structure and phylogeny. In Levandowsky, M., and Hutner, S. H., eds., *Biochemistry and physiology of protozoa*. Vol. 1. 2nd ed. Academic Press, New York and London.

Hibberd, D. J. 1976. The ultrastructure and taxonomy of the Chrysophyceae and Prymnesiophyceae (Haptophyceae): a survey with some new observations on the ultrastructure of Chrysophyceae. *Bot. J. Linn. Soc.* 72:55–80.

———. 1981. Notes on the taxonomy and nomenclature of the algal classes, Eustigmatophyceae and Tribophyceae (synonym Xanthophyceae). *Bot. J. Linn. Soc.* 82:93–119.

Klaveness, D., and Paasche, E. 1979. Physiology of coccolithophorids. In Levandowsky, M., and Hutner, S. H., eds., *Biochemistry and physiology of protozoa*. Vol. 1. 2nd ed. Academic Press, New York and London.

U.S. Department of Interior. 1966. *A guide to the common diatoms at water pollution surveillance system stations*. Federal Water Pollution Control Administration, Cincinnati.

Werner, D., ed. 1976. *The biology of diatoms*. Univ. of California Press, Berkeley and Los Angeles.

The three divisions considered in this chapter have very few similarities that hold them together, except that most representatives in each are flagellated. However, this similarity does not include the insertion and appearance of the flagella. Pigmentation, storage products, cell covering, and ultrastructure are varied, although there are similarities among these divisions (see Table 9-1). Thus, the discussion treats each division separately.

Division Pyrrhophyta (Class Dinophyceae: Pyrrhophytes or Dinoflagellates)

The Pyrrhophyta (*pyrrh* = red, burning) include over 1000 species in about 125 genera. Most are unicellular flagellates, but there are nonmotile forms that produce flagellated cells typical of the division (Fig. 9-1A). The Pyrrhophyta are here considered a single class, the Dinophyceae (*dino* = whirling), known as dinoflagellates. The mode of nutrition is varied, with colorless, phagotrophic forms as well as autotrophs occurring.

Cell Structure, Cell Division, and Movement

Nonmotile forms (filamentous or nonfilamentous) have a cellulose wall. However, most dinoflagellates are biflagellated cells (Fig. 9-1A, B) lacking a cellulose wall but usually containing cellulose plates of varied thickness within the plasma membrane (Fig. 9-2B, C). Forms in which plates are obvious with the light microscope (Fig. 9-3C) are known as armored, or thecate (*thec* = box or case), dinoflagellates. Many supposedly unarmored (or naked) species have been shown to have very delicate plates. By using the electron microscope, it can be seen that the cell covering involves a system of *vesicles* (*vesicul* = little bladder) that can contain cellulose plates adjacent to the outer cell membrane (Fig. 9-2B, C). The number of plates in a species is generally fixed, with a correlation of plate number to thick-

9

ALGAL POTPOURRI (PYRRHOPHYTES, EUGLENOPHYTES, AND CRYPTOPHYTES)

Table 9-1   Characteristics of Pyrrhophyta, Euglenophyta, and Cryptophyta

Characteristics	Division (Class)		
	Pyrrhophyta (Dinophyceae)	Euglenophyta (Euglenophyceae)	Cryptophyta (Cryptophyceae)
Flagellation			
number	2	2 (1, 3)	2
form	ribbon, smooth	tinsel, unilateral	tinsel, uni- or bilateral
insertion	lateral grooves	apical groove	subapical groove
Chlorophyll a +	chlorophyll c_2	chlorophyll b	chlorophyll c_2
dominant pigments	chlorophylls + carotenoids	chlorophylls	carotenoids + phycobiliproteins
Percent colorless (approx.)	20	66	10
Storage product	starch	laminaranlike (paramylon)	starch
Cell covering			
form	plates	strips (spiral),	strips (straight)
chemical nature	cellulose	proteinaceous	proteinaceous

ness—the thicker the plates, the fewer their number. The arrangement of plates is used as a generic and specific characteristic.

Photosynthetic dinoflagellates have a pigment complex and chloroplast structure similar to that of most Chrysophyta (Chapter 8). In addition to chlorophyll a and c_2, there are dominant xanthophylls, some specific to the dinoflagellates, which usually give them a characteristic yellow to brown color. Thylakoids are in threes; pyrenoids, when present, are varied in position. The storage reserve is starch, but unlike the starch in Chlorophyta, it is stored outside the chloroplast. Sometimes fats and oils accumulate in the cell in large quantities and are red or orange. Eyespots are varied and may or may not be associated with chloroplasts.

The cells are generally uninucleate, with the chromosomes condensed and visible throughout cell division and interphase (Figs. 9-2A and 9-3A). During mitosis, the nuclear membrane and nucleolus persist. Cytoplasmic microtubules separate new nuclei in division but are not involved in the actual separation of chromosomes, which appear attached to the nuclear envelope. The dinoflagellate nucleus is unlike that of other eukaryotic cells, including euglenophytes (see p.

345) and cryptophytes (see p. 350), which also have condensed chromosomes during interphase. It is different in its appearance and in having only trace amounts of specific nuclear proteins (*histones*) characteristic of eukaryotic nuclei. A saclike structure is present in many motile forms that consists of a system of tubes or vesicles connected to the outside. This structure may serve in excretion, in osmoregulation, or for buoyancy. Some dinoflagellates also have organelles that are ejected (ejectosomes or trichocysts, Fig. 9-2A), possibly as a defense mechanism.

The typical dinoflagellate cell has two flagella laterally inserted (Fig. 9-1A, B). One flagellum encircles a transverse groove, the **girdle**, whereas the other trails posteriorly in a second groove, the **sulcus** (*sulc* = groove), as shown in Figures 9-1A–D and 9-2D. The flagella can emerge from a common pore or two separate ones, and there can be a series of plates in the grooves. The girdle (transverse) flagellum is ribbonlike owing to the presence of a striated band of material within the flagellar membrane. This band is shorter than the flagellar microtubules, causing a "buckling" of the flagellum similar to that of a hoop within a skirt hem (Fig. 9-2E). Fine hairs can be present on this flagellum, but they are unlike the tubular tinsels of chrysophytes (Chapter 8) and brown algae (Chapter 10). The trailing flagellum is whiplash and is responsible for most forward and rotating movement of the cell. This motion explains the derivation of the name *dinoflagellates,* or "whirling" flagellates. The location of the girdle with regard to the overall cell appearance is used as a taxonomic criterion. In one order, the flagella are apically inserted, often with one encircling the other.

Reproduction

Vegetative reproduction in Pyrrhophyta is varied. Following nuclear and organelle division in motile forms, separation of new cells can include the parental cell plates or vesicles and associated membranes. The general situation seems to involve an invagination, or pinching in half, of the parental outer layers. In some forms with thick plates, there is a diagonal splitting of the parental plates, with each new cell receiving some parental plates. A third type of cell division occurs completely within the parental cell, with new cells being released as two or more spores and forming new plates, vesicles, and external membranes.

Nonmotile species with a cellulosic wall produce motile cells, which may or may not have plates. These swarmers often resemble described motile species of dinoflagellates and have the lateral flagellation typical of dinoflagellates.

Some dinoflagellates produce cysts within the cell; they may resemble the parental cell or differ radically (Fig. 9-3D, E). Only since the 1960s has it been possible to show that some elaborate microfossils are dinoflagellate cysts.

Sexual reproduction is not well known in Pyrrhophyta. In some genera (*Peridinium, Gymnodinium, Woloszynskia;* Figs. 9-1A, 9-2D, and 9-3A), isogametes resembling the parental cell are produced by mitosis. These gametes fuse to form a motile zygote having two trailing flagella and sometimes a single girdle flagellum. After a swimming period of up to several days, the planozygote (*plano* = wandering) forms a heavy-walled cell (or cyst) and undergoes a resting period. In germination, the zygote forms one, two, or four swarmers by a prolonged meiosis, lasting several days. In the armored, photosynthetic *Ceratium* (Figs. 9-1C and 9-3B), anisogametes are produced by haploid cells. The larger (female) gamete evidently absorbs the smaller gamete, which is considered male. In dinoflagellates, the zygote is the only diploid state. However, in the colorless *Noctiluca,* dinoflagellatelike isogametes produced by meiosis fuse to form the zygote, which is the diploid, typical *Noctiluca* cell.

Morphological Variation and Classification

The most commonly occurring type of dinoflagellate is a photosynthetic, motile unicell with lateral flagellar insertion (Figs. 9-1A–C, 9-2D, and 9-3A–C). There are also genera with apical flagella, in which one flagellum (the ribbonlike one) encircles the other (*Prorocentrum,* Fig. 9-1E). Other genera, such as *Polykrikos* (Fig. 9-1D), are colonial. Not all motile forms are flagellated. *Noctiluca,* for example, has a motile "tentacle," although dinoflagellatelike swarmers are produced (see preceding). There are also unicellular, amoeboid forms, such as *Dinamoebidium* (Fig. 9-1F, G), which apparently produce unarmored, dinoflagellate zoospores. Nonmotile forms include nonfilamentous genera, such as *Rufusiella, Cystodinium, Pyrocystis,* and *Stylodinium* (Figs. 9-1H–J and 9-3F), and filamentous *Dinothrix* (Fig. 9-1K). In these nonmotile genera, motile cells are "typical"—that is, dinoflagellatelike (compare Figs. 9-1I, J and 9-3G with Figs. 9-1A, B, 9-2D, and 9-3A). Note that, as in many other algal classes (Chlorophyceae, Xanthophyceae, Chrysophyceae), the range of morphological variation includes motile and nonmotile as well as filamentous and nonfilamentous forms.

Classification is based primarily on motility. Species with apical flagellation are in an order separate from the laterally flagellated species. Amoeboid species are in still another order. Nonmotile forms are classified in separate orders, based on their morphology (filamentous, nonfilamentous, etc.).

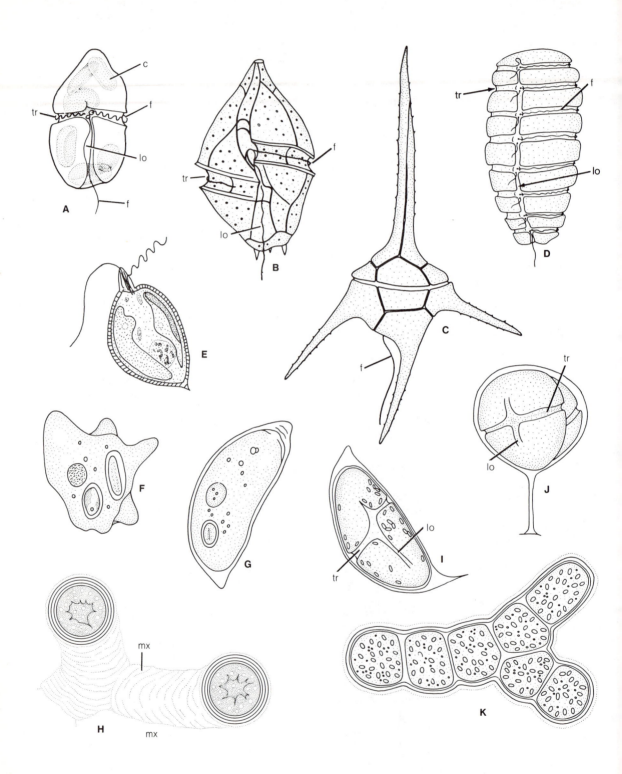

Ecology

Dinoflagellates occur in both freshwater and marine habitats. In abundance, they are second only to diatoms (Chapter 8) in marine phytoplankton, and in tropical waters they can be the dominant phytoplankters. They can occur in numbers sufficient to color the water a reddish-brown, often termed a *red tide*. This red color explains the division name Pyrrhophyta. Under optimum growth conditions, rapid asexual reproduction occurs, producing a bloom similar to that occurring in Cyanophyceae (Chapter 3). Generation time is as short as one or two days, producing 20,000 to 50,000 cells per ml. Some dinoflagellates (e.g., *Gonyaulax*) are bioluminescent, giving a glow (or *phosphorescence*) to sea water when it is disturbed at night. In general, armored forms are more common in coastal waters, whereas unarmored forms occur chiefly in open oceanic waters.

Some Pyrrhophyta (pigmented or not) are symbiotic in tissues of marine invertebrates. In some sea anemones, various tentacle colors result from the presence of pigmented dinoflagellates. Other dinoflagellates (photosynthetic or not) ingest particulate food. A few dinoflagellates contain a chrysophycean symbiont with a typical eukaryotic nucleus.

The great variety in ornamentation of the dinoflagellate cell, particularly the development from the plates of horns, spines, and saillike or winglike structures (Fig. 9-3B) may increase its ability to stay in the upper water layers. For photosynthetic forms, it is vital that the cell remain in the illuminated zone. However, non-photosynthetic forms can have elaborations and flotation mechanisms in order to assist in feeding. Thus, unpigmented dinoflagellates remain in an area with smaller plankton; their elaborations enhance the flow of water for a specialized type of feeding. Some colorless, unicellular dinoflagellates contain symbiotic bluegreen or green algae, which may provide food or, in the bluegreen algae, actually fix nitrogen for the host (Chapter 3).

Some dinoflagellates (*Gonyaulax*, *Gymnodinium*; Figs. 9-1A, B and 9-3A) produce substances extremely poisonous to man and many animals. Poisoning occurs indirectly when people eat shellfish that have fed on dinoflagellate cells. These toxic blooms are most frequent in waters north or south of 30°. The toxin does not usually affect shellfish (although some will not feed during massive blooms); however, in some vertebrates the toxin acts to inhibit the diaphragm and causes respiratory failure. Dinoflagellate blooms often color the water and are the reason for the red tides that cause concern in coastal waters, especially of North American coasts (e.g., New England, Florida, Southern California, Washington, British Columbia). Not all

Figure 9-1 Cell structure and morphological diversity of Dinophyceae. **A–D,** motile forms viewed from ventral side (**A, B, D**) to show orientation of flagella (*f*) and grooves (transverse, *tr;* longitudinal, *lo*). **A,** *Gymnodinium*, unarmored form, showing chloroplasts (*c*), ×250; **B,** *Gonyaulax*, armored form, ×750; **C,** *Ceratium*, armored form with horns (dorsal view), ×550; **D,** *Polykrikos*, a colonial form, ×75. **E,** apical flagellation (*Prorocentrum*), ×400. **F–K,** nonmotile forms. **F, G,** *Dinamoebidium*, an amoeboid type in vegetative stage (**F,** ×500) and cyst stage (**G,** ×625); **H,** *Rufusiella*, nonfilamentous type with extensive mucilaginous matrix (*mx*), ×245. **I, J,** coccoid forms, each containing characteristic dinoflagellate swarmers. **I,** *Cystodinium*, ×650; **J,** *Stylodinium*, ×900. **K,** *Dinothrix*, filamentous type, ×720. (**B, D, K,** after Schiller, in Rabenhorst, with permission of Akademische Verlagsgesellschaft, Geest & Portig K.-G., Leipzig; **F, G,** after Fott's *Algenkunde*, with permission of VEB G. Fischer Verlag, Jena; **H,** from R. H. Thompson, "Algae," in W. T. Edmonson, ed., *Freshwater Biology*, 2d ed., copyright © 1959, by John Wiley & Sons. Reprinted by permission of the author and John Wiley & Sons, Inc.)

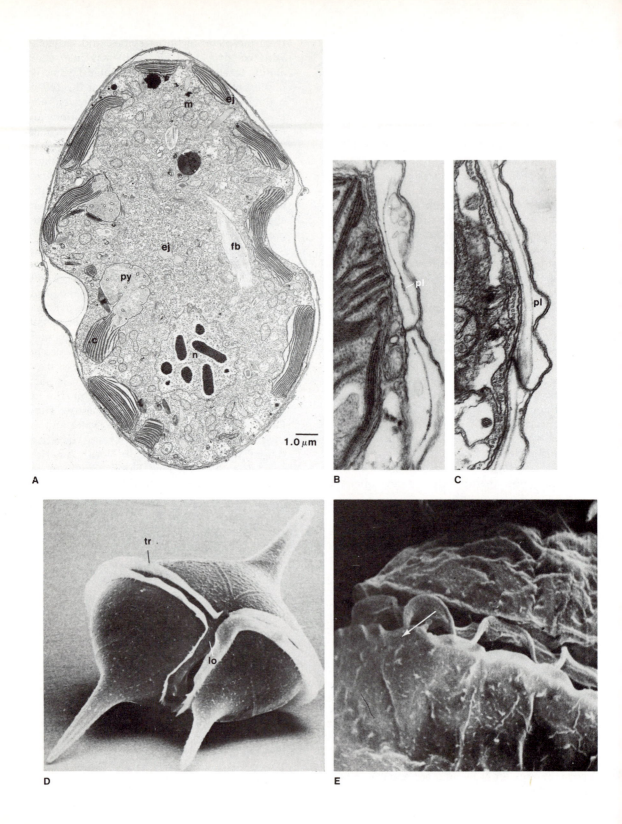

A

B

C

D

E

toxic, but enough are that they are important to a large number of people who eat various shellfish and other seafoods. Also, not all red tides are caused by dinoflagellates.

Division Euglenophyta (Class Euglenophyceae: Euglenophytes or Euglenoids)

There are approximately 40 genera and 450 species of algae or algalike organisms in the Euglenophyta. All euglenoid genera are unicellular and motile, with only one sessile genus (Fig. 9-4F). This group has both plantlike and animallike features, and representatives can be considered equally well as algae or protozoa.

Cell Structure, Cell Division, and Movement

The cell lacks a firm wall; however, the outer part of the protoplast consists of a **pellicle** (*pell* = skin) covered by the plasma membrane. The pellicle consists of articulating, proteinaceous, ridged strips that spiral from the posterior to the apical end of the cell and are sometimes conspicuous (Fig. 9-5A). Associated with these strips are

Figure 9-2 Cell structure of Dinophyceae, electron microscope. **A–C,** transmission EM. **A,** entire cell (*Cachonia*) with transverse flagellum groove in middle, showing many chloroplasts (*c*) with pyrenoid (*py*), mitochondria (*m*), ejecting organelles ejectosomes (*ej*), fibrillar band (*fb*), golgi (*arrow heads*), and distinctive interphase dinoflagellate nucleus (*n*) with interphase chromosomes, ×6190. **B, C,** section through periphery of cell, showing plates (*pl*); **B,** thin plates (*Katodinium*); **C,** thicker plates and junction (*Glenodinium*), ×4900. **D, E,** scanning EM (*Peridinium*). **D,** entire cell, ventral view, showing plates, pores, and furrows (both transverse, *tr*, and longitudinal, *lo*), ×14,000; **E,** portion of girdle flagellum, showing straight, striated strand (*arrow*) within flagellar membrane, ×13,000. (**A,** courtesy of E. M. Herman and *Journal of Phycology*; **B, C,** courtesy of J. D. Dodge, from J. D. Dodge and R. M. Crawford, "A survey of thecal fine structure in the Dinophyceae," *Botanical Journal of the Linnaean Society* 63:53–67, 1970, with permission of Academic Press; **D, E,** courtesy of F. J. R. Taylor.)

mucilage-producing bodies and microtubules. These are involved in the characteristic plasticity of some euglenoid cells, which involves sliding of the pellicular strips. In addition to the pellicle, some forms have a lorica (*lorica* = armor) (Figs. 9-4C and 9-6).

At the anterior end of the cell is an invagination, or *groove*, from which flagella emerge (Fig. 9-4A). The invagination consists of a narrow *canal* that opens into a larger *reservoir*; pellicular strips occur only in the canal. Contractile vacuoles empty into the reservoir, and flagella are inserted on the reservoir surface (Fig. 9-4A). Adjacent to the reservoir is a large eyespot (Figs. 9-4A and 9-5B) that is independent of chloroplasts (contrary to many algae). The eyespot probably acts as a light screen to a photoreceptive flagellar swelling at the base of one flagellum (Fig. 9-4A).

The basic flagellar arrangement is two flagella, with unilateral tinsels present on at least one. In most euglenoids, the second flagellum is very short and may not extend beyond the canal (Fig. 9-4A). The emergent flagellum has tinsels and is often easily seen with the light microscope, as it can have additional thickening material within the membrane. In photosynthetic forms, the flagellar swelling is near the base of the emergent flagellum.

Approximately one-third of the euglenophytes are photosynthetic. The bright-green chloroplasts contain both chlorophyll *a* and *b* and additional carotenoids. One, two, or many disc-shaped, elongate, platelike or ribbon-shaped chloroplasts are present. Thylakoids are arranged generally in groups of 3 to 12. Pyrenoids are present in some euglenoids, their occurrence and position being regular features of certain chloroplast types. The photosynthate is like laminaran in brown algae and is known as **paramylon** (*para* = near; *amyl* = starch). The photosynthate is always outside the chloroplast (Fig. 9-6) and sometimes associated with the pyrenoid. With the light microscope, paramylon appears as refractive granules of various sizes and shapes (Fig. 9-5A) and is deposited in a helical fashion.

Most euglenoids are colorless and thus obligate heterotrophs (Fig. 9-4D, E). Many secure food saprobically, as do slime molds and fungi (Chapters 4–6). However, a few euglenoids are phagotrophic and ingest particulate material,

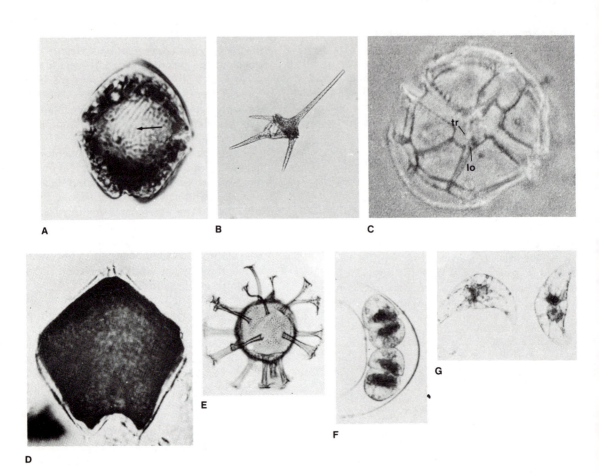

Figure 9-3 Cell structure, morphological diversity, and reproduction of Dinophyceae, light microscope. **A–C,** motile unicellular forms. **A,** *Gymnodinium,* showing characteristic dino-flagellate nucleus (*arrow*), ×1000; **B,** *Ceratium,* with plates and horns, ×500; **C,** *Peridinium* theca, showing plate arrangement and transverse (*tr*) and posterior longitudinal (*l*) flagellar grooves, ×800. **D, E,** resting cysts. **D,** living cyst resembling parent cell, ×450; **E,** fossil cyst differing from parent cell, ×360. **F, G,** nonmotile species (*Pyrocystis*) in vegetative stage (**F,** ×400) and producing motile cells (**G,** ×500). (**A,** courtesy of H. A. von Stosch and *British Phycological Journal;* **D,** courtesy of W. R. Evitt and *Stanford University Publications, Geological Sciences;* **E,** courtesy of D. Wall and *Journal of Phycology;* **F, G,** courtesy E. Swift, E. G. Durbin, and *Journal of Phycology.*)

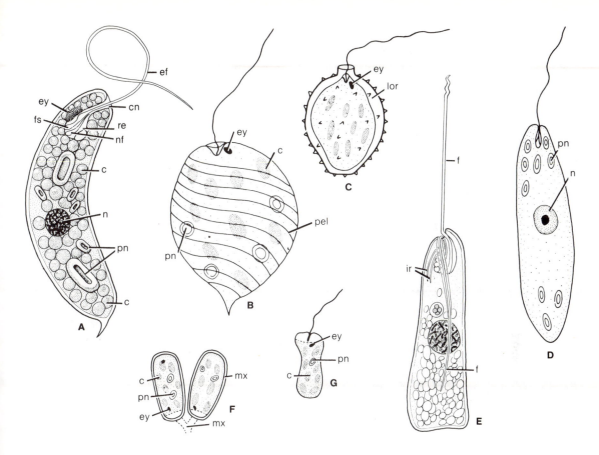

Figure 9-4 Cell structure and morphological diversity of Euglenophyceae. **A,** *Euglena*, showing disc-shaped chloroplasts (*c*), paramylon (*pn*), nucleus (*n*), reservoir (*re*), eyespot (*ey*), canal (*cn*), flagellar swelling (*fs*), emergent flagellum (*ef*), and nonemergent flagellum (*nf*), ×1400. **B,** *Phacus*, with pellicle striations (*pel*), ×1000. **C,** *Trachelomonas*, with lorica (*lor*), ×155. **D,** *Astasia*, a colorless form, ×1915. **E,** *Peranema*, with two flagella (*f*), showing ingestion rods (*ir*), ×1000. **F, G,** *Colacium*, showing attached stage with matrix (*mx*) (**F,** ×800) and motile stage (**G,** ×530). (**A, E,** after Leedale.)

sometimes with a special organelle that captures and helps ingest the food (Fig. 9-4E). Evidently the reservoir is not involved in ingestion.

The euglenophyte cell contains a single, large, centrally-located nucleus. Chromosomes remain condensed throughout interphase (Fig. 9-6), and a nucleolus and nuclear envelope are present during division. Centromeres and spindle fibers are lacking, but microtubules along the division axis within the nucleus have been demonstrated. Nuclear structure is only superficially similar to that of the Dinophyceae (see p. 340; Figs. 9-2A and 9-3A).

In cell division, the nucleus migrates to the anterior end of the cell. At this time, eyespot, chloroplast, and other cytoplasmic organelles, including reservoir and canal, can divide. The flagella may be formed from recently replicated basal bodies, or the parental apparatus may replicate itself. In the latter instance, one new cell retains the parental apparatus. Upon completion of organelle replication, cell cleavage occurs by in-

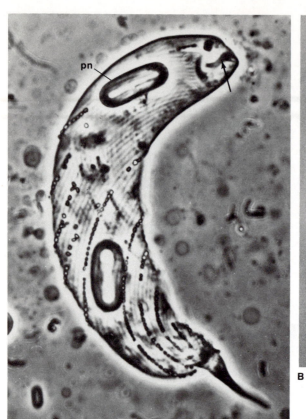

A

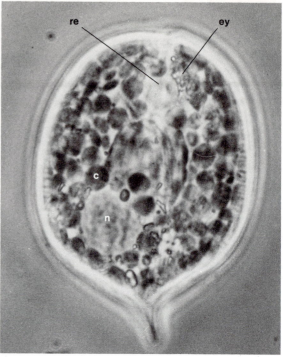

B

Figure 9-5 A, B, cell structure of Euglenophyceae, light microscope. **A,** external view of *Euglena,* showing pellicle striations with warts, canal (*arrow*), and two paramylon (*pn*) granules (anoptral phase-contrast), ×1500; **B,** cell contents of *Phacus,* showing chloroplasts (*c*), nucleus (*n*), and eyespot (*ey*) adjacent to reservoir (*re*) (phase-contrast), ×2000. **C,** Habitat of nonmotile Euglenophyceae, *Colacium* (*arrows*), sessile on the zooplankter *Daphnia,* ×400. (**A, B,** courtesy of G. F. Leedale and Prentice-Hall, Inc.)

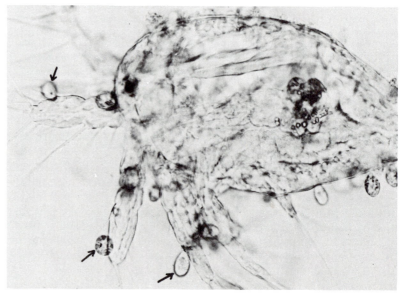

C

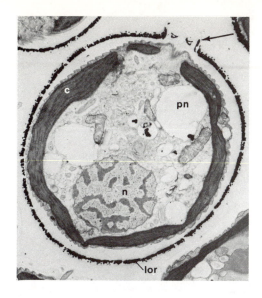

Figure 9-6 Cell structure of Euglenophyceae, electron microscope; oblique section of *Trachelomonas*, showing chloroplast (*c*), paramylon (*pn*), nucleus (*n*), and lorica (*lor*) with pore and collar (*arrow*), ×4000. (Courtesy of G. F. Leedale and *British Phycological Journal*.)

Reproduction

Only asexual reproduction is known in Euglenophyceae, and it generally occurs in the motile cells. Under some conditions, the cell becomes nonmotile and undergoes repeated divisions. Under unfavorable conditions, a cell can encyst, becoming resistant to extremes in the environment. The wall of the cyst is sometimes stratified and often ornamented. Some encysted forms are red because they accumulate large amounts of carotenoids, often referred to as **hematochrome** (*hema* = blood; *chrome* = color),

although it is the xanthophyll *astaxanthin*. These cysts usually float on the water surface. Germination of thick-walled cysts results in only one motile cell; in thin-walled cysts, the contents often divide into a number of cells.

Sexual reproduction evidently is absent, although there are a few records reporting union of gametes. These reports are unsubstantiated, and some are apparently due to the presence of other organisms or to misinterpretation. There are reports of the occurrence of meiosis in some euglenoids, but this seems to be rare and follows a nonsexual process of nuclear fusion.

Morphological Diversity, Classification, and Ecology

Classification of Euglenophyceae is currently based on flagellar organization, although morphological form (motile or nonmotile) or mode of nutrition (photosynthetic, saprobic, or phagotrophic) have been used. The use of flagellar organization as the prime feature is acceptable to those who consider euglenoids as being plantlike or animallike.

All genera but one are motile in the vegetative stage. This exception is the photosynthetic *Colacium*, which retains two flagella within the reservoir and attaches at its anterior end (Figs. 9-4F and 9-5C). It can undergo cell division in this attached, nonmotile, or the motile state (Fig. 9-4F, G). As noted previously, only a third of the Euglenophyceae contain chlorophyll. This includes such common genera as *Euglena*, *Phacus*, and the loricate *Trachelomonas* (Figs. 9-4A–C and 9-5A, B). Some more common, colorless, saprobic forms are *Astasia* and *Peranema* (Fig. 9-4D, E). *Peranema* also contains ingestion rods and thus also has a phagotrophic mode of nutrition.

As two-thirds of the genera are colorless, it is interesting to note that even photosynthetic forms require at least one vitamin, generally cyanocobalamin (B_{12}). Euglenoids grow in waters in which some decaying organic matter is present. This includes barnyard and pasture ponds and ditches. Most are freshwater organisms, although a few occur in brackish or marine habitats—especially on mudflats and in areas where organic material collects. A few euglenoids occur

vagination of the pellicle at the anterior end of the cell between the openings of the two new canals. This cleavage line progresses posteriorly following the helix of the pellicle, until the new cells are connected by a narrow, posterior, protoplasmic bridge. New cells can remain attached for varied periods of time.

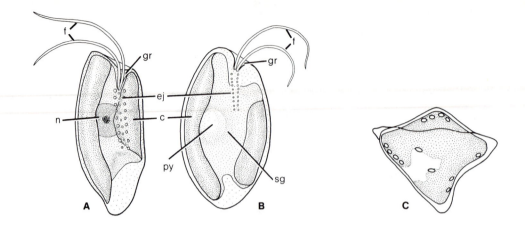

Figure 9-7 Cell structure and morphological diversity of Cryptophyceae. **A, B,** motile forms, ×2850. **A,** *Cryptomonas*, showing chloroplast (*c*), groove (*gr*), flagella (*f*), and groove ejectosomes (*ej*); ×2850; **B,** *Chroomonas*, also showing starch (*sg*) surrounding pyrenoid (*py*); compare ejectosome distribution with Figure 9-7A. **C,** nonmotile form, *Tetragonidium*, ×2000. (**A, B,** after Butcher; **C,** from R. H. Thompson, "Algae," in W. T. Edmonson, ed., *Freshwater Biology*, 2d ed., copyright © 1959 by John Wiley & Sons. Reprinted by permission of John Wiley & Sons, Inc.)

endozoically in invertebrates such as rotifers, nematodes, flatworms, oligochaetes, and copepods. One lightly pigmented genus occurs only in the lower intestinal tract of tadpoles.

Division Cryptophyta (Class Cryptophyceae: Cryptophytes or Cryptomonads)

The division Cryptophyta contains approximately 100 species in less than 25 genera, which occur in both fresh and marine waters. Most cryptophytes are flagellated, and a single class, Cryptophyceae, is recognized (Fig. 9-7).

Cell Structure, Cell Division, Reproduction, and Movement

The typical cryptophyte is a dorsiventrally flattened, oblong or ovoid, motile unicell. It is biflagellated, generally with an anterior groove from which the flagella arise (Figs. 9-7A, B and 9-8). As in Dinophyceae and Euglenophyceae (see pp. 339, 345), a true cell wall is lacking. The outer cell layer, the **periplast** (*peri* = around; *plast* = form), differs from the pellicle of Euglenophyceae and the plates of Dinophyceae. The cryptophyte periplast is composed of a series of regularly aligned, longitudinal, protein plates delineated by a number of shallow ridges extending the length of the cell. In addition, lateral furrows are present, resulting in an alternating pattern (Fig. 9-8A). The longitudinal plates do not extend into the flagellar grooves.

At the junction of ridges and lateral furrows are located small ejecting organelles, which are more numerous in the cell posterior (Fig. 9-8A, B). The ejecting organelles are analogous to the trichocysts of dinoflagellates (p. 340). In the flagellar groove region are located larger, somewhat more complex ejecting organelles approximately twice the size of the peripheral ones. These larger ones are visible with the light microscope (Fig. 9-7A, B) and are used as a taxonomic characteristic. Both types of ejecting organelles are probably produced by the golgi apparatus. Their function is uncertain but may involve protection of the cell against other organisms.

The cell contains one or two elongate chloroplasts (Figs. 9-7A, B and 9-8B) of varied color—green, bluegreen, golden, brown, or red—depending in part on environmental conditions. The main pigments include chlorophyll a, c_2, and specific carotenoids unique to the division. However, also present are red and blue masking pigments: the phycobiliproteins. These phycobiliproteins do not differ significantly from those of Cyanophyta (Chapter 3) and Rhodophyta (Chapter 11). The chloroplast is composed of pairs of unfused thylakoids (Fig. 9-8B) with phycobiliproteins in the space within a thylakoid, not in the form of discrete units as in Cyanophyta and Rhodophyta. Pyrenoids can be present, with the starch storage product adjacent to but outside the chloroplast (Fig. 9-8B), as in most algal divisions (except the Chlorophyta). The starch can be enclosed within the nuclear-chloroplast endoplasmic reticulum (Fig. 9-8B). Eyespots are uncommon; when present, they are in the chloroplast, usually near the nucleus and adjacent to a flagellar base.

The two flagella are unequal in length, and both have tinsels along their length (Fig. 9-8A). The shorter flagellum generally has tinsels arranged unilaterally; the longer flagellum has bilateral tinsels. The flagella are attached near the base of the groove and emerge subapically or almost laterally. A flagellar swelling is often present on the longer flagellum near the point where it emerges from the groove.

The nucleus is generally posteriorly located and irregular in outline, often with cytoplasmic invaginations. As with Pyrrhophyta and Euglenophyta, the interphase nucleus has large **chromatin** (*chrom* = color) areas and a prominent nucleolus. However, the nuclear envelope, unlike that of Pyrrhophyta and Euglenophyta, breaks down during division, and microtubules extend into the chromosome mass as in some golden algae (see Chapter 8).

Reproduction is only asexual, with no confirmed reports of sexuality. The cell can divide while motile, or in a nonmotile condition when it comes to rest, and becomes embedded in a mucilaginous matrix. In some genera, thick-walled, internally-formed cysts are reported. In nonmotile genera, cell division results in packet-formation, or "typical," motile, cryptophyte cells.

Morphological Diversity, Classification, and Ecology

Cryptophyceae are classified on the basis of vegetative morphology—motile versus nonmotile. Only two genera are nonmotile, and these are rarely collected. Motile genera are separated on the basis of cell shape, appearance and location of the flagellar groove, number and location of large ejecting organelles, and chloroplast number. Previous systems incorporating color are unsatisfactory, since color appears to vary depending on environment.

Motile genera include *Cryptomonas* and *Chroomonas* (Fig. 9-7A, B), which are pigmented, whereas *Tetragonidium* (Fig. 9-7C) is one of the two pigmented, nonmotile genera. The nonmotile phase for some species has previously been treated as a separate genus, but motility is evidently also a response to environmental conditions.

Cryptophyta occur more commonly in fresh water than in marine waters. Some species of *Cryptomonas* and *Chroomonas* occur regularly in saltmarsh pools with salinities varying from 5 to 30 parts per thousand, and at all times during the year. In fresh water, cryptophytes are often common when some organic material is present. They are sometimes the dominant phytoflagellates in early spring blooms of temperate lakes. The few species studied in culture are obligate photoautotrophs and require the vitamin cyanocobalamin (B_{12}). Some marine cryptophytes are endosymbionts in radiolarians, sponges, and corals, and are referred to as **zooxanthellae** (or, more properly, as *zoocryptophytes*; *zoo* = animal; *xanth* = yellow). In addition, cryptophytelike chloroplasts (as seen with the electron microscope) occur in a few marine ciliates. These ciliates can be present in such large numbers as to create red tides similar to those caused by Dinophyceae (see p. 343). How the ciliates acquire these is unknown, but the chloroplasts are photosynthetically active.

A

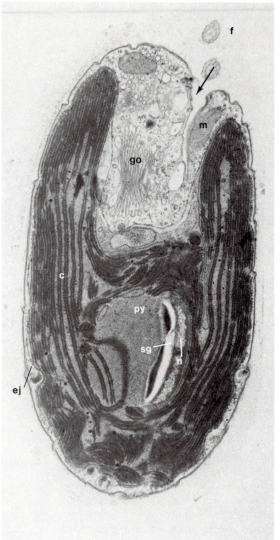

B

Relationships

As mentioned at the beginning of this chapter, there are few similarities among the three divisions discussed. Only the Pyrrhophyta have any fossil record. This record, which goes back possibly to late Precambrian time (650 million years ago), consists primarily of cysts and the cells that produced them. Neither the Euglenophyta nor Cryptophyta are known much before 200,000,000 years ago (late Mesozoic).

The nuclear morphology of the three divisions seems superficially similar—all show persistence of condensed chromosomes in interphase and persistence of some nuclear envelope and nucleolus during division. However, the paucity of certain key proteins and the lack of intranuclear microtubules during division separate pyrrhophytes from euglenophytes. Interestingly, nuclear division in cryptophytes is similar to that in Chrysophyta (Chapter 8) and not to that in euglenophytes, dinoflagellates, or the phycobiliprotein-containing rhodophytes (Chapter 11). It has been proposed that all three classes represent endosymbiotic relationships of a photosynthetic cell within a colorless, flagellated cell. This hypothesis is based on ultrastructural features, especially of the chloroplast. Thus, a *Euglena* would contain a chlorophycean symbiont, and some dinoflagellates would contain a prymnesiophyte (see Chapters 7 and 8 for details of chloroplast structure).

The different life histories in the Pyrrhophyta—colorless *Noctiluca* with its diploid vegetative phase and some photosynthetic forms (*Peridinium, Gymnodinium, Woloszynskia*) with haploid vegetative phases—are intriguing. Possibly the Pyrrhophyta, as presently circumscribed, is composed of diverse forms that appear similar; as more organisms are studied, it may be shown that the dinoflagellates are of diverse origins.

All three divisions resemble chrysophytes (Chrysophyta, Chapter 8) and brown algae (Phaeophyta, Chapter 10) more closely than green algae (Chlorophyta, Chapter 7). However, specific differences for each division set them apart. In Pyrrhophyta, these are the nuclear condition, variation in xanthophylls, and storage of starch. In Euglenophyta, the presence of chlorophyll *b* and a laminaranlike storage product are outstanding features. In the Cryptophyta, the presence of phycobiliproteins and starch are noteworthy. The possibility that cryptophytes are closely related to either Cyanophyta (Chapter 3) or Rhodophyta (Chapter 11) based on pigmentation is untenable because of more basic differences, such as the lack of flagellated cells in either bluegreen or red algae.

At this time, it is not (and probably never will be) possible to indicate positive relationships among these three divisions of flagellates. Some have suggested that the organisms considered here, plus most of those in the preceding chapter (excluding Prymnesiophyceae, and perhaps Eustigmatophyceae) belong in a loosely-organized grouping along with Oomycota and Hypochytridiomycota of the fungi (Chapter 5). Certainly, information derived from future investigations of any of these organisms should be exciting and provide information on possible relationships.

Figure 9-8 Cell structure of Cryptophyceae, electron microscope (*Chroomonas*). **A,** platinum-carbon replica of cell, showing unequal tinsel flagella emerging from groove region (*top*), and surface pattern formed by plate areas with small bulges of the ejecting organelles (*arrows*), ×18,400. **B,** longitudinal section showing chloroplast (*c*) with pyrenoid (*py*) and thylakoids, starch (*sg*), nucleus (*n*), Golgi (*go*), mitochondrion (*m*), small peripheral ejecting organelles (ejectosomes, *ej*), and groove (*arrow*); note transverse section of the two flagella (*f*) above groove (*top*), ×17,850. (Courtesy of E. Gantt and *Journal of Phycology*.)

References

Bourrelly, P. 1979. *Les algues d'eau douce*. Vol. 3. *Les algues bleues et rouges, Eugléniens, Peridiniens et Cryptomonadines*. N. Boubée & Cie., Paris.

Buetow, D. E., ed. 1968. *The biology of* Euglena. Vols. 1 and 2. Academic Press, New York.

Cox, E.R., ed. 1980. *Phytoflagellates*. Elsevier/North-Holland, New York. (Chapters on euglenoid flagellates; photosynthetic cryptomonads; free-living dinoflagellates.)

Dodge, J. D. 1979. The phytoflagellates: fine structure and phylogeny. In Levandowsky, M., and Hutner, S. H., eds. *Biochemistry and physiology of protozoa*. Vol. 1. 2nd ed. Academic Press, New York.

Leedale, G. F. 1967. *Euglenoid flagellates*. Prentice-Hall, Englewood Cliffs, New Jersey.

Sargeant, W. A. S. 1974. *Fossil and living dinoflagellates*. Academic Press, New York.

Wolken, J. J. 1967. Euglena, *an experimental organism for biochemical and biophysical studies*. 2nd ed. Appleton-Century-Crofts, New York.

The division Phaeophyta (brown algae) is a large, conspicuous group of algae containing about 260 genera and about 2000 species. They are restricted to the sea, with the exception of four freshwater genera that are rare and inconspicuous. Phaeophyta are generally classified (Table 10-1) in 11 to 14 orders, which can be grouped in one class or in three separate classes. Their separation into classes based on life history has been traditional. However, many brown algal genera are not fully known, especially in their life-history details, and recent studies suggest that a separation into classes on this basis is highly artificial. Their uniformity with respect to many biochemical characteristics supports the concept of a single class, Phaeophyceae, which is the system adopted in this text. At the same time, increasing knowledge concerning the ultrastructural features of flagella clearly indicates that there are two main evolutionary series within the brown algae, and these are placed in two distinct subclasses, Phaeophycidae and Cyclosporidae. Hence, in this text, 12 orders of brown algae are recognized and placed in one class, Phaeophyceae, with two subclasses: Phaeophycidae (with 11 orders), comprising the ectocarpalean line of evolution, and Cyclosporidae (with 1 order), comprising the fucalean line of evolution.

10

PHAEOPHYTA (BROWN ALGAE)

Cell Structure

Cell Wall

Except for reproductive cells (gametes and spores), which have naked protoplasts, the cells of brown algae are bounded by a well-developed cell wall. In general, the cell wall is composed of an inner, firm, **fibrillar,** cellulose layer and an outer, amorphous, mucilaginous, **pectic** or pectinlike layer. The cellulose resembles that occurring in the Ulvales (Chlorophyceae) in containing glucose and xylose units. Various carboxylated **(alginic acid)** and sulphated **(fucoidan)** polysaccharides, termed **phycocolloids,** also occur in the matrix of the cell walls or in the intercellular spaces of brown algae. These colloidal substances help prevent the plant from becom-

Table 10-1 Orders of Division Phaeophyta and Representative Genera

Class	Subclass	Order	Representative Genera
Phaeophyceae	Phaeophycidae (Ectocarpalean line)	Ectocarpales	*Ectocarpus, Pilayella* (= *Pilayella*), *Ralfsia*[1], *Streblonema*
		Chordariales	*Analipus* (= *Heterochordaria*)[1], *Haplogloia, Leathesia, Myrionema*
		Sporochnales	*Sporochnus*
		Desmarestiales	*Desmarestia*
		Sphacelariales	*Sphacelaria*
		Cutleriales	*Cutleria* (= *Aglaozonia*), *Zanardinia*
		Tilopteridales	*Tilopteris*
		Dictyotales	*Dictyota, Padina, Syringoderma*[2], *Zonaria*
		Dictyosiphonales	*Phaeostrophion, Punctaria, Soranthera*
		Scytosiphonales	*Hydroclathrus, Petalonia, Scytosiphon*
		Laminariales	*Agarum, Alaria, Cymathere, Egregia, Laminaria, Macrocystis, Nereocystis, Pelagophycus, Pterygophora, Thalassiophyllum*
	Cyclosporidae (Fucalean line)	Fucales	*Ascophyllum, Ascoseira*[3], *Bifurcariopsis, Cystophora, Cystoseira, Durvillea*[4], *Fucus, Halidrys, Himanthalia, Hormosira, Pelvetiopsis, Sargassum*

[1] The crustose *Ralfsia*, parenchymatous *Analipus*, and certain other genera are sometimes placed in a separate order, Ralfsiales.
[2] The genus *Syringoderma* should probably be placed elsewhere, possibly in a separate order.
[3] The genus *Ascoseira* is sometimes placed in a separate order, Ascoseirales.
[4] The genus *Durvillaea* is sometimes placed in a separate order, Durvillaeales.

ing dehydrated, a characteristic that is especially significant for algae living in the intertidal area, where they are subjected to varied degrees of exposure and desiccation. Alginic acid and fucoidan occur in some larger genera of kelps (Laminariales) and rockweeds (Fucales) in sufficient quantity to be commercially important. Alginic acid can represent 10–40% of the dry weight of the plant. Both alginic acid and fucoidan have been found in all species of brown algae so far examined. Alginic acid is present in the form of an **alginate** (a salt of alginic acid), which is insoluble in water and commonly known as **algin.** Algin is chiefly bound tightly to the cell wall, whereas fucoidan occurs primarily in the intercellular matrix. However, both algin and fucoidan can also occur to some extent in the outer, mucilaginous cell wall matrix.

Alginic acid is a polyuronide, a linear polymer of β-1, 4-linked D-mannuronic and L-guluronic acids, although the proportion of the two types of residues present varies. In structure it is somewhat similar to cellulose and pectic acid, but the fine details of the structure of the macromolecule have not yet been fully worked out. Fucoidan is typified by the presence mainly of L-fucose (a methyl pentose sugar) residues esterified by sulphuric acid. Lesser amounts of D-xylose, galactose, and D-glucuronic acid are also present. The structure of fucoidan is also still uncertain.

Calcification ($CaCO_3$) of the cell wall occurs in one genus, *Padina* (Fig. 10-1A, B), and small amounts of silica occur in some brown algae.

Chloroplast

The pigments are present in one or more usually peripheral chloroplasts, which vary to some extent in size and shape (Fig. 10-2). Where they are numerous in the cell, chloroplasts are peripheral and usually small and discoid, but they can be elongate, ribbon-shaped, netlike, **laminate,** or irregular. In a few instances they are **stellate** and **axile.** Electron micrographs show that the chloroplasts are bound by two membranes and contain photosynthetic lamellae surrounded by a granular matrix, the **stroma**. These photosynthetic lamellae are formed by associated but unfused thylakoids in stacks of three (Fig. 10-3), which

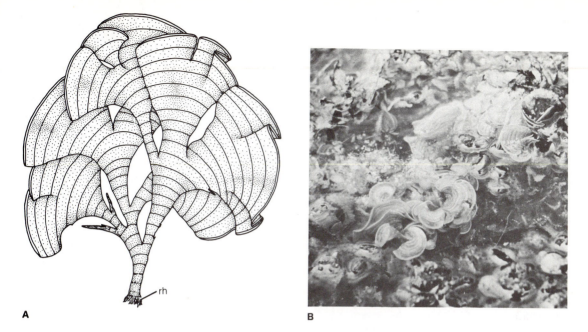

A

B

Figure 10-1 A, B, *Padina,* a tropical member of the Dictyotales with calcified cell walls. **A,** the plant is attached to the substratum by tufts of rhizoids (*rh*); concentric zones on the fanlike branches are due to rows of microscopic filamentous "hairs," ×0.5. **B,** habit view of group of plants, ×0.25.

form parallel bands running the full length of the plastid. The bands are separated by a uniform space, but there can be some exchange of thylakoids between them. It appears that a peripheral, or girdle, band encircles the other bands, following the contour of the chloroplasts. Small, membrane-free DNA regions occur between the peripheral and parallel bands, forming a ring at the ends of the chloroplast. Longitudinal division of the chloroplast is preceded by division and separation of these DNA regions. The chloroplast lies within a double-membrane sac, the chloroplast endoplasmic reticulum, which is continuous with the outer membrane of the nuclear envelope.

The characteristic brownish color of these algae results from the masking of chlorophyll pigments by an excess of accessory carotenoid pigments, especially the xanthophyll pigment **fucoxanthin** and β-carotene. The chlorophylls present are chlorophyll *a* and *c*. In addition to

the dominant fucoxanthin and β-carotene, additional xanthophyll pigments, such as **violaxanthin,** are usually present in smaller amounts. Pigments occur in the photosynthetic lamellae (thylakoids). Fucoxanthin is actively involved in the absorption of light energy used in photosynthesis. It absorbs light at wavelengths in which chlorophyll is not particularly effective and then transfers the energy to the chlorophyll *a*.

Although there are epiphytic and endophytic species of brown algae, some of which have had parasitic features attributed to them, all have chloroplasts and are apparently partially, if not wholly, autotrophic. However, knowledge of the nutritional requirements of brown algae is still meager. According to the conditions in which they occur, brown algae range in color from light yellow-brown or olive-brown through a rich, golden brown to practically black. This variation is a reflection of the relative proportions of chlorophylls to other pigments present.

Other Cell Components and Storage Products

A central vacuole can be present in the cell, but usually there are numerous, small vacuoles scattered throughout the cytoplasm (Fig. 10-2B). Typical dictyosomes, ribosomes, endoplasmic reticulum, and mitochondria are present in the cytoplasm. The principal product of photosynthesis that is stored as food reserves is the polysaccharide **laminaran,** which is a polymer of D-glucose, characterized by β-1,3 and 1,6 linkages. The sugar alcohol, D-mannitol, is a terminal residue of the polymer. **Mannitol,** which is a hexahydric alcohol, can also be found free in extracts of these algae. Small amounts of other carbohydrates have been reported. Fat globules and oil droplets can also be stored. In electron micrographs, laminaran appears as opaque granules scattered throughout the cytoplasm, external to the chloroplast and frequently adjacent to the pyrenoid, when present.

Pyrenoids are apparently not universally present in the brown algae. They are common in Ectocarpales, Scytosiphonales, and Dictyosiphonales, but they are vestigial or absent in Desmarestiales, Laminariales, Dictyotales, and Fucales. However, although apparently absent from the vegetative cells of members of the Fucales, a vestigial pyrenoid has been reported in the egg cell of several genera in this order. It has been suggested that the occurrence of pyrenoids in the brown algae may be influenced by environmental or developmental conditions. When present, the pyrenoid is excentric, projecting from either an end or a side of the chloroplast (Fig. 10-3), apparently outside the chloroplast, capped by and separated from the cytoplasm by a dilated sac. The pyrenoid is not penetrated by the thylakoids.

Numerous small, highly refractive vesicles, commonly referred to as **fucosan vesicles,** or **physodes,** originate within the chloroplast. They are extruded from the chloroplast into the cytoplasm, and are usually aggregated about the nucleus; they contain a tanninlike substance. They are sometimes referred to as **phaeophyte tannin.** It has been suggested that fucosan vesicles represent a waste product of metabolism; however, the substance is not a carbohydrate,

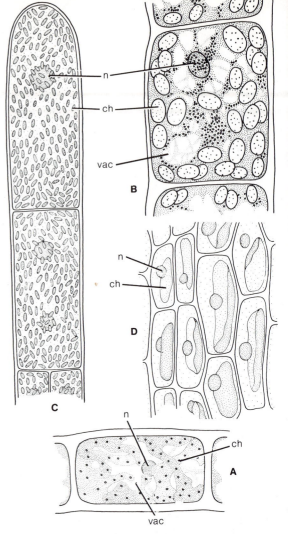

Figure 10-2 Chloroplast (*ch*) types in Phaeophyta. **A,** reticulate (*Ectocarpus*), ×1500. **B, C,** discoid. **B,** *Pilayella*, ×690; **C,** *Sphacelaria*, ×320. **D,** laminate (*Scytosiphon*), ×300: *n*, nucleus; *vac*, vacuole.

and it appears to be most abundant in regions of high metabolic activity, such as dividing cells. Sterols, such as **fucosterol** and **sargasterol,** are also reported in brown algae. Saturated and unsaturated hydrocarbons have been found in a number of algae, with the saturated *n*-pentadecane predominating in brown algae. Several of the unsaturated hydrocarbons have already been shown to function as **gamones,** including *ectocarpen* (in *Ectocarpus* and *Cutleria*), *multifiden*

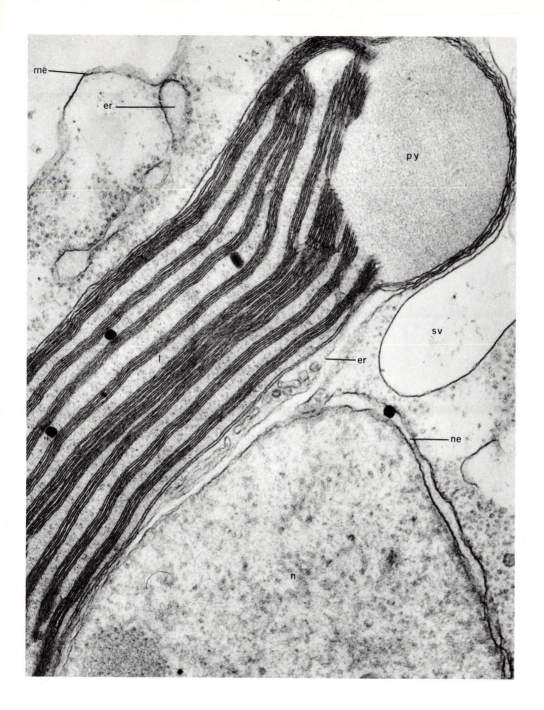

Figure 10-3 Electron micrograph showing part of chloroplast of *Pilayella* near pyrenoid (*py*), with characteristic compound lamellations, each composed of fused thylakoids in stacks of three (*l*), and dense lipid droplets lying between them: *n*, nucleus; *ne*, nuclear envelope; *er*, layer of endoplasmic reticulum surrounding chloroplast; *me*, membrane; *sv*, part of storage vesicle, ×54,600. (Photograph courtesy of L. V. Evans.)

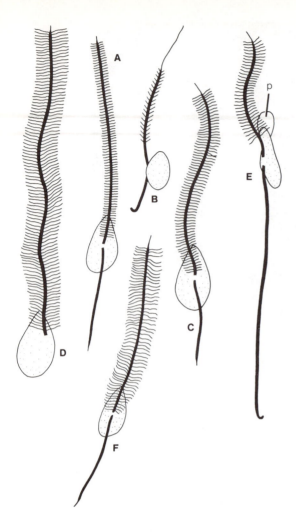

Figure 10-4 Arrangement of flagella of various Phaeo-phyta (anterior end toward top of each diagram). **A–C,** zoospores. **A,** *Pilayella;* **B,** *Chordaria;* **C,** *Laminaria.* **D–F,** sperms. **D,** *Dictyota;* **E,** *Fucus;* **F,** *Alaria: p,* proboscis.

certain holdfast cells of some kelps) they can be multinucleate. The nucleus, with a well-developed nuclear envelope, contains one or more nucleoli. Interphase chromatin is granular, and chromosomes contain DNA and histones similar to that in higher plants. Mitosis and meiosis are similar to that in higher plants, although there is some doubt about persistence of the nuclear envelope during division. It has been reported that the nuclear envelope persists during division, except at the poles, where spindle microtubules originate. Divisions complete with centrosomes, centromeres, and spindles have been reported. The spindle is reported to be **intranuclear,** whereas the centrosomal apparatus is **extranuclear.** In cell division, a centrally located membrane apparently extends **centrifugally** and becomes thickened with a fibrillar layer to form the cell plate. It is thought that dictyosomes give rise to vesicles from which the wall components alginic acid and fucoidan are secreted. However, the details of wall formation are still not well known in brown algae. Although chromosomes are usually small and morphologically similar, in some Laminariales an XY chromosomal sex determination has been reported; the larger X chromosome appears in the female gametophyte, and the smaller Y chromosome in the male gametophyte. Chromosome numbers reported in brown algae range from $n = 8$ to $n = 32$. Polyploidy has been suggested in a few instances, but the genetics of brown algae on the whole is a neglected field of study.

Flagellated Cells

No brown alga has a motile vegetative state, although motile cells (gametes and zoospores) are produced (Figs. 10-4 and 10-5). Motile cells often have a reddish eyespot, are almost always biflagellate and somewhat beanshaped or pyriform, and have flagella of unequal length. In the ectocarpalean line, the longer flagellum is of the tinsel type, with two rows of mastigonemes, and is directed anteriorly; the shorter flagellum is of the whiplash type and is directed posteriorly. Usually, mastigonemes extend almost the full length of the flagellum. However, in some instances (*Chordaria,* Fig. 10-4B), the anterior

and *aucanten* (in *Cutleria*), and *fucoserraten* (in *Fucus*).

Nucleus and Other Organelles

The cells are generally uninucleate (Fig. 10-2); however, in certain regions of the thallus (as in

flagellum does not have mastigonemes along the full length of the flagellum; the distal end terminates in a fairly extensive smooth point. In Dictyotales, the posterior flagellum is absent (Figs. 10-4D and 10-5A).

In the fucalean line, the longer flagellum is usually directed anteriorly, and the shorter one trails posteriorly (Fig. 10-4E). In addition, the motile cell (sperm) is described as being somewhat bell-shaped and usually has a striking, transparent, striated beaklike **proboscis** (Figs. 10-4E, 10-5C, and 10-6). This structure is intimately associated with the anterior flagellum and a strand passing from the basal body of the anterior flagellum. It has been suggested that the proboscis may serve to detect an exudate from the female gamete. The proboscis is well developed in some fucalean genera, such as *Fucus* (Fig. 10-5C) and *Pelvetiopsis* (Fig. 10-6). But it is less well developed in others, such as the fucalean genus *Himanthalia*, where the fibrillar root of the flagellum bears a superficial resemblance to a reduced proboscis. Other modifications occur in *Himanthalia*, in which the relative length of the flagella differs from that in other fucoids. Also, the mastigonemes on the anterior flagellum do not cover the full length of the flagellum; they stop about one-quarter of the way from the distal end, at which point a single, conspicuous spine occurs, attached to a single fibril of the peripheral series.

Basal bodies of the flagella are joined laterally, and both are in close contact with the nucleus. The anterior flagellum is attached to the nucleus by a fibrous connection **(rhizoplast)**. The posterior flagellum is attached to the cell surface immediately above the eyespot, and the flagellar membrane is dilated at this point. In Dictyotales, despite the absence of a posterior flagellum, there are two basal bodies present in the sperm. The posterior part of the one without a flagellum projects as a fibrillar root to the surface of the nucleus. It would appear that the sperm is a reduced type of motile cell derived from a biflagellate prototype through loss of the emergent portion of a posterior whiplash flagellum.

In many brown algae, the nucleus occupies only a small part of the motile cell; in others, as in *Fucus* sperm, most of the cell is occupied by the nucleus.

Classification and Morphological Diversity

The 12 orders of Phaeophyta in the class Phaeophyceae are placed in two evolutionary lines, based on their biochemical and flagellar characteristics. One of these, the fucalean line (subclass Cyclosporidae), comprising the single, oogamous order Fucales, also has a distinctive type of life history (gametic) common to all its genera[1]. The other 11 orders, placed in the ectocarpalean line (subclass Phaeophycidae), can be arranged along several lines of development, in which the main features emphasized relate to life history and alternation of generations, vegetative construction (filamentous or parenchymatous), and reproductive characteristics (isogamy, anisogamy, and oogamy) (see Fig. 13-14).

The Ectocarpalean Line

The orders in the ectocarpalean line have an alternation of isomorphic or heteromorphic generations, sexual reproduction ranging from isogamy through anisogamy to oogamy, and morphological diversity ranging from simple, filamentous types to complex, parenchymatous forms. In this line, there are some orders that are still very poorly known. Only representative genera of the better-known orders are dealt with in this text.

Since there is no motile or nonmotile unicellular or colonial brown alga, the simplest organization is an unbranched or sparingly branched, uniseriate type. In more specialized members, well-developed tissues are present, and there can be considerable differentiation of cells in the thallus and of the method of growth from one phase to another in the life history. No other group of algae has attained the diversity of form, complexity of vegetative construction, or size of Phaeophyta. They range from microscopic, filamentous forms (less than 1 mm) to massive plants 50 meters or more in length. Their diversity and size are achieved largely as a result of the great variety of types of growth occurring in brown algae.

---

[1]see footnote on p. 383

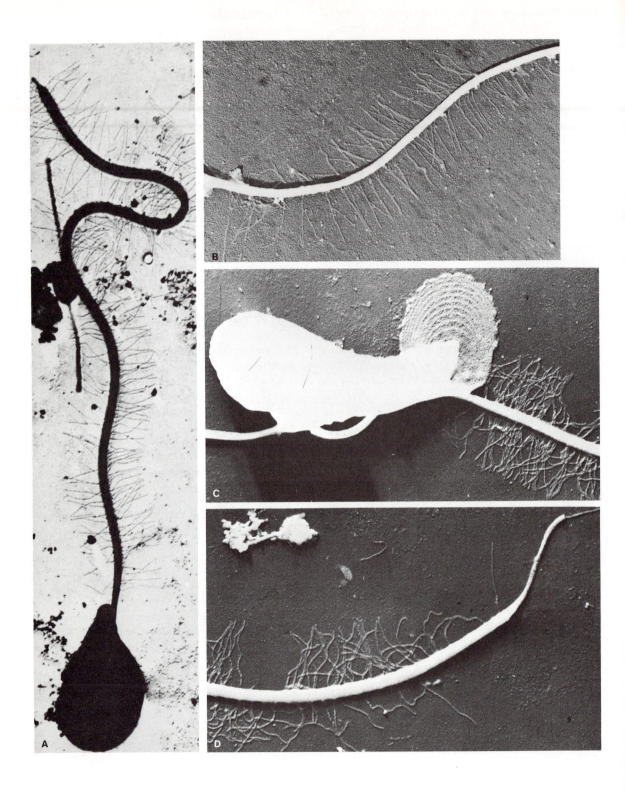

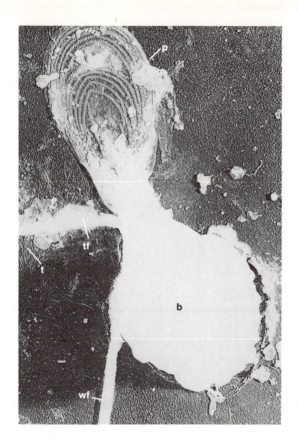

Figure 10-6 Electron micrograph of sperm of *Pelvetiopsis*, showing body (*b*) of sperm with anterior proboscis (*p*) and proximal regions and insertion of flagella, ×46,200: anterior flagellum with tinsel (*t*); *wf*, posterior whiplash flagellum. (Photograph courtesy of E. Henry.)

Filamentous and Pseudofilamentous Oganization In the ectocarpalean line, the simplest type of growth, which results from **intercalary** cell divisions occurring at any point in the thallus in one plane, produces an unbranched, uniseriate filament (Fig. 10-7A, B). An occasional division in a second plane, on the same basic theme, results in a sparingly branched, **uniseriate** filament, as in *Pilayella* (Figs. 10-7C and 10-8G). With more frequent, but still intermittent division in the second plane, a profusely branched, loosely filamentous plant is derived, as in *Ectocarpus* (Figs. 10-7B and 10-8C). Erect, unbranched filaments arising from a prostrate basal system of branched filaments are characteristic of some genera, such as *Myrionema* (Fig. 10-8K). This type of growth habit, referred to as **heterotrichy,** is common in brown algae and also occurs in some red and green algae. In some brown algae, there can be an aggregation of only slightly branched, erect filaments, which adhere laterally to form a compact, encrusting layer, as in *Ralfsia* (Fig. 10-8I, J). In other instances, a **pseudoparenchymatous** thallus can be formed (Fig. 10-9) by an aggregation or intertwining of branched filaments that are loosely organized, as in *Leathesia* (Fig. 10-9C, D), or densely compacted, as in *Analipus* (Fig. 10-9A, B). In the mature thallus of *Leathesia*, the central portion becomes hollow.

Figure 10-5 Electron micrographs of flagellar structure of Phaeophyta. **A,** *Dictyota* sperm, showing tinsel flagellum, ×7500. **B,** anterior portion of tinsel flagellum of *Chordaria* zoospore, ×20,000. **C, D,** *Fucus* sperm, showing (**C**) body of sperm, proboscis, and insertion of posterior whiplash and anterior tinsel flagella, and (**D**) distal end of tinsel flagellum; **C,** ×26,000; **D,** ×28,000. (**A,** courtesy of I. Manton, from I. Manton, B. Clarke, and A. D. Greenwood, "Further observations with the electron microscope on spermatozoids in the brown algae," *Journal of Experimental Botany* 4:319–29, 1953, with permission of Oxford University Press; **B,** photograph courtesy of J. B. Hansen, J. B. Petersen, B. Caram, and *Botanisk Tidsskrift;* **C, D,** photographs courtesy of I. Manton, from Manton and Clarke, "An electron microscope study of the spermatozoid of *Fucus serratus*," *Annals of Botany*, N.S, 15:461–71, 1951, with permission of Oxford University Press.)

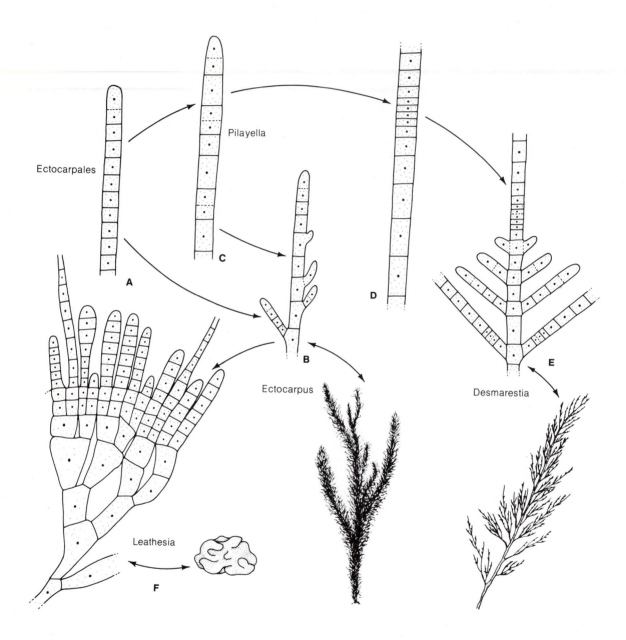

Figure 10-7 Growth types of filamentous and pseudoparenchymatous genera of Phaeophyta with possible evolutionary lines. **A–C,** unbranched and branched filamentous forms. **A, B,** with apical growth; **C,** with intercalary cell division in addition to apical growth. **D, E,** trichothallic growth. **E, F,** pseudoparenchymatous forms. **E,** trichothallic growth; **F,** apical growth.

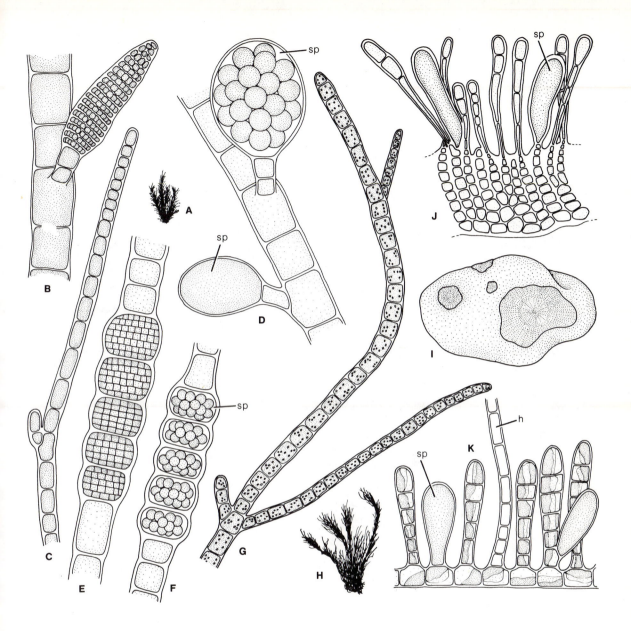

Figure 10-8 Filamentous forms of Phaeophyta with reproductive structures. **A–D,** *Ecto-carpus.* **A,** habit, ×0.5; **B,** portion of filament bearing plurilocular structure, ×250; **C,** vegetative portion of thallus, showing initiation of a branch, ×235; **D,** portion of filament, showing immature (*below*) and mature (*above*) unilocular meiosporangia (*sp*), ×250. **E–H,** *Pilayella.* **E,** fertile portion of a filament, showing intercalary plurilocular structure, ×110; **F,** fertile portion of filament, showing intercalary unilocular meiosporangia, ×110; **G,** vegeta-tive portion of thallus, ×110; **H,** habit, ×0.5. **I, J,** *Ralfsia.* **I,** habit, showing several encrusting thalli growing on a rock, ×0.5; **J,** transverse section through portion of encrust-ing thallus, showing erect compacted filaments and two unilocular meiosporangia (*sp*), ×225. **K,** *Myrionema,* showing prostrate basal system, hair (*h*), erect vegetative filaments of cells, and two unilocular meiosporangia (*sp*), ×800.

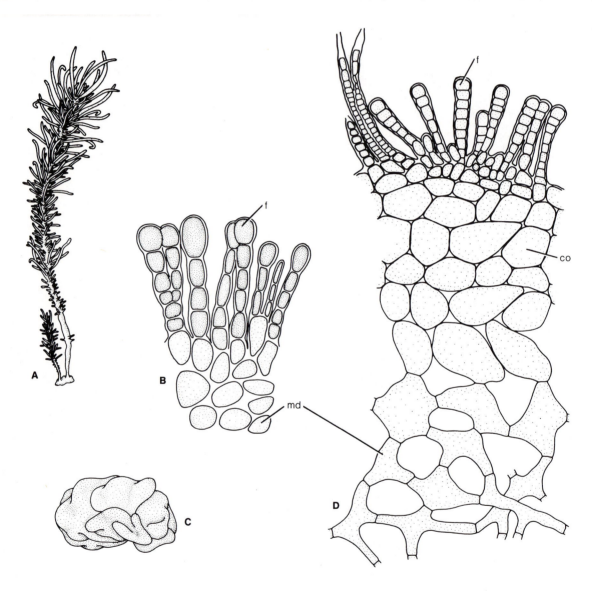

Figure 10-9 Pseudoparenchymatous forms. **A, B,** *Analipus*. **A,** habit, ×0.5; **B,** transverse section through outer portion of thallus, showing pseudoparenchymatous medulla (*md*) and free surface filaments (*f*), ×300. **C, D,** *Leathesia*. **C,** habit, ×0.5; **D,** transverse section through outer portion of thallus, showing loose filamentous medulla, pseudoparenchymatous cortical region, and free surface filaments (*f*), ×155.

Parenchymatous Growth In some brown algae, the cells of a primary, uniseriate axis can divide in more than one plane (Fig. 10-10) to produce a **corticated** thallus, as in *Sphacelaria* (Figs. 10-10 C–E and 10-11), suggesting an incipient parenchymatous development. In other instances, diffuse parenchymatous growth occurs as a result of divisions only in surface cells. **Periclinal** divisions in these surface cells increase the thallus thickness, and occasional **anticlinal** divisions permit surface growth to keep pace with the increase in girth (Fig. 10-12D, F). The resulting thallus can be a solid, foliose one, as in *Phaeostrophion* (Fig. 10-12E, F) and *Petalonia*, or it can become hollow, as in *Scytosiphon* (Fig. 10-12C, D). In some instances, one phase in the life history can be an erect parenchymatous thallus and can alternate with another free-living phase, which may be a prostrate pseudoparenchymatous disc or a branched filament.

Apical Growth Growth is usually intercalary, although in Sphacelariales (*Sphacelaria*, Figs. 10-10C–E and 10-11C), Dictyotales (*Dictyota*, Figs. 10-10F–J and 10-13B; *Syringoderma*, Figs. 10-10K, L and 10-12A, B), and Fucales there is a marked apical development of the thallus. There can be a single apical cell, as in *Sphacelaria* (Figs. 10-10C and 10-11C). Most genera of Fucales, such as *Fucus*, also have a single apical cell, but in some there can be a group of several apical cells at a branch apex. Some of the Dictyotales also have a single apical cell, as in *Dictyota* (Fig. 10-13B, C). In still others, there is a margin of apical cells, as in *Padina* (Fig. 10-1), *Zonaria*, and *Syringoderma*[2] (Figs. 10-10K, L and 10-12A, B). The apical cells found in brown algae are simpler than those found in most higher plants in that they usually contribute cell derivatives from a single (posterior) cutting face.

Intercalary Meristems The ultimate in size and complexity in brown algae is achieved through **meristematic** activity in specific, lo-

calized regions. Some meristems have potentialities somewhat similar to those in vascular plants. One type of meristem, unique to brown algae, results in **trichothallic growth** (Figs. 10-7D, E and 10-14C, D), which is the result of a special type of cell division in a filament. This region can be recognized as a subapical band of narrow actively dividing cells in the filament. In this subapical position, the meristem is somewhat protected from injury. In its simplest form, this type of localized, meristematic activity occurs in some of the loosely branched, uniseriate, filamentous forms in Ectocarpales. However, it reaches an advanced stage of development in some pseudoparenchymatous forms and is especially characteristic of Desmarestiales (Fig. 10-14).

In *Desmarestia*, cell divisions are confined to these localized, subapical portions of the thallus apex at the base of hairlike, uniseriate filaments and result in trichothallic growth. Cells are added both **distally** and **proximally** to these apical filaments (Fig. 10-14C, D). Those added distally increase the length of the terminal filament to only a limited extent, as older cells are gradually eroded away. However, those added proximally increase the length of the thallus, and from this axial row of cells arise lateral orders of branching (Fig. 10-14G). These lateral branches also establish similar subapical meristems, thus increasing the girth of the thallus by the same type of development. Despite the relative simplicity of this type of growth, the combined system of branched, aggregated, and compacted filaments leads to the development of a mature, pseudoparenchymatous organization (Fig. 10-14F–H) that permits a variety of morphological types and the attainment of great size. Some straplike species of *Desmarestia* (Fig. 10-14A) are up to a meter in width and four or more meters in length.

The most highly developed type of meristematic activity in brown algae occurs in the parenchymatous kelps, or Laminariales, such as *Laminaria*, *Nereocystis*, *Macrocystis*, and *Pterygophora*. (Figs. 10-10M–O, 10-15, 10-17, 10-19, 10-23, 10-25, 10-29, and 10-32). A primary meristem called the **transition zone** (Fig. 10-10N–O) is an intercalary meristem located between the bladelike distal portion and the narrow, stemlike proximal region of the plant. This meristem contributes cells distally, increasing the length and

---

[2]On the basis of recent studies, reproduction and life history details suggest that *Syringoderma* should be removed from the Dictyotales and possibly referred to the Sphacelariales or to a new order.

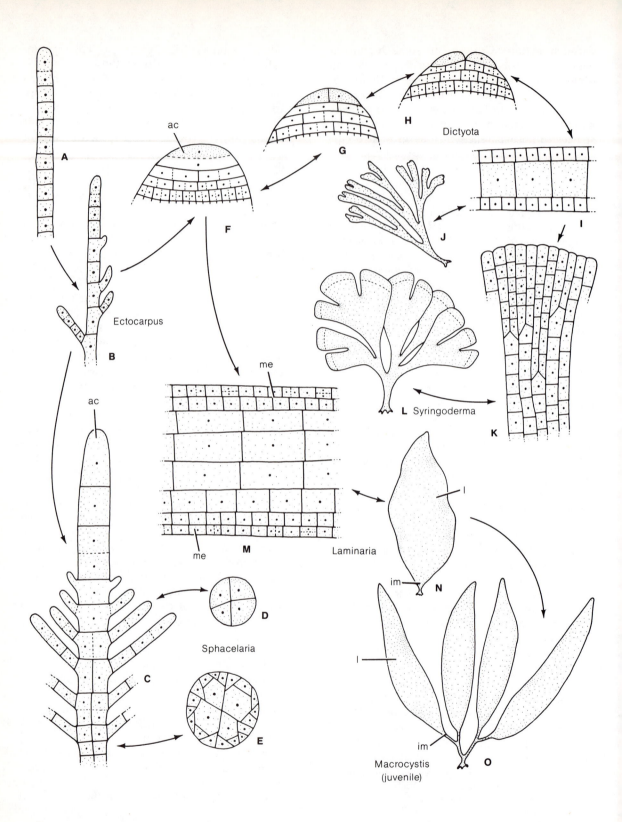

ac

A

B

Ectocarpus

F

ac

G

Dictyota

H

J

I

ac

me

me

M

Laminaria

K

L Syringoderma

C

Sphacelaria

D

E

l

im

N

l

im

Macrocystis
(juvenile)

O

development of the blade; proximally, increasing the length and development of the basal portion of the plant; and in the second and third planes, increasing the girth of the plant. In addition, surface meristematic activity (Figs. 10-10M and 10-15B, D) occurs in the epidermal region, which is referred to as the **meristoderm.** It has also been suggested that there is a deeper, internal, meristematic region in the stemlike portion of kelps that is somewhat similar (geometrically) to the cambium of vascular plants.

Most large, conspicuous brown algae have these well-developed intercalary meristems, which account for their great size. However, in two additional ways brown algae can attain macroscopic dimensions: (1) forms with **diffuse** intercalary meristematic regions form an extensive multilayered, or parenchymatous, thallus by undergoing cell divisions in three planes, as in *Phaeostrophion* (Fig. 10-12E, F) and *Scytosiphon* (Fig. 10-12C, D); (2) simple intercalary cell divisions, trichothallic growth, strict apical growth, or a combination of these can result in a basically filamentous thallus. However, an aggregation, lateral fusion, or intertwining of these branched filaments results in a large pseudoparenchymatous thallus, as in *Leathesia* (Figs. 10-7F and 10-9C, D) and *Desmarestia* (Figs. 10-7E and 10-14). This second type is fundamentally simpler than the first, but neither generally results in forms as large as those with intercalary meristems, as in Laminariales.

In *Fucus*, as well as other Fucales, there is also a highly developed type of meristematic activity, and a variety of different types of growth occur during **ontogeny** of the conspicuous diploid phase. The embryo develops a definite polarity at an early stage of development (Fig. 10-16B–F). After the developing embryo elongates to produce a short filament, filamentous rhizoids develop at one end, and a parenchymatous apex is soon established at the other (Fig. 10-16G). Although surface cells continue to divide, following the development of a prominent apical depression (Fig. 10-16K), trichothallic growth becomes established (Fig. 10-16I). However, this method of growth soon ceases. The establishment of the first apical cell (Fig. 10-16K) in the apical depression initiates apical growth, which continues throughout the life of the mature plant.

Cellular and Structural Differentiation Differentiation in more complex brown algae, such as Laminariales (Figs. 10-17, 10-18D, 10-19, 10-21, 10-22, 10-29, and 10-32) and Fucales (Figs. 10-18A–C, 10-20, 10-22, 10-30, and 10-31), is such that structures somewhat comparable to higher plant organs can be distinguished. Superficially, these parts can resemble leaves, stems, and roots, but they do not function in exactly the same way as these organs do in higher plants, because vascular tissues are lacking. The leaflike part of a brown alga is referred to as a **lamina,** or **blade;** the stemlike part as a **stipe;** and the attachment part as a **holdfast,** often with rootlike structures, the **haptera.** Despite the absence of vascular tissues in these structures, they are well adapted for some of the functions of leaves, stems, and roots: blades have expanded surfaces that increase their photosynthetic efficiency, stipes are frequently stiffened or almost woody and provide excellent support, and holdfasts are effective anchors.

Although these structures are not as complex as vascular plant organs, they are not necessarily composed of simple tissues. Definite anatomical regions can be differentiated (Figs. 10-21A–E and 10-22) as meristoderm (epidermis), **cortex,** and innermost **medulla.** In the cortex, pitted elements (Figs. 10-21A and 10-22) and **mucilage canals** with adjacent **secretory cells** (Figs. 10-15C and 10-21C) can occur, whereas phloemlike elements (Figs. 10-15E and 10-21D, E) can occur in the medulla. Pitted walls with **plasmodesmata**

Figure 10-10 Growth types of parenchymatous genera of Phaeophyta, with possible evolutionary lines from filamentous prototypes. **A, B,** filamentous forms. **A,** unbranched filament; **B,** branched filament. **C–J,** apical growth followed by intercalary and parenchymatous cell divisions. **C–E,** with a single apical cell (*ac*) at the tip of axis; **D, E,** transverse views of axis at points indicated in **C** by *arrows.* **F–J,** also with a single apical cell, but with dichotomous branching as shown in **G, H,** and **J; I,** transverse section of thallus, showing three cell layers. **K, L,** marginal row of apical cells; **K,** detail of portion of **L. M–O,** growth by meristoderm (*me*) and intercalary meristem (*im*); **M,** detail of transverse section of lamina (*l*) of **N** or **O,** showing intercalary meristem (transition zone) (*im*) and meristoderm (*me*).

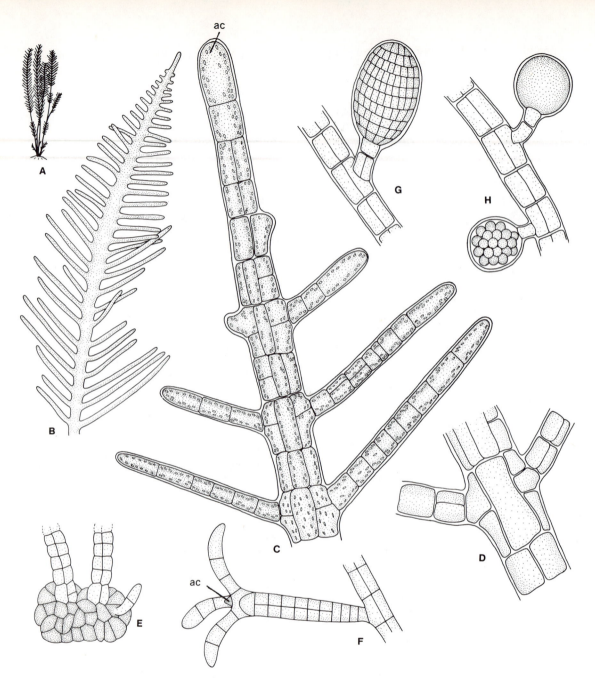

Figure 10-11 Morphology and reproduction of parenchymatous form with apical growth (*Sphacelaria,* Sphacelariales). **A,** habit, ×1. **B,** apical portion of thallus, showing pinnate branching, ×30. **C,** apical portion of thallus, showing apical cell (*ac*), branches, and parenchymatous development of main axis, ×150. **D,** more mature portion of axis, showing cells in surface view, ×160. **E,** basal attachment region of a young thallus, showing prostrate basal system and erect axes, ×120. **F,** portion of axis, showing asexual reproductive unit, the propagule, ×120. **G, H,** branches with reproductive structures. **G,** plurilocular structure, ×180; **H,** immature (*above*) and mature (*below*) unilocular meiosporangia, ×180.

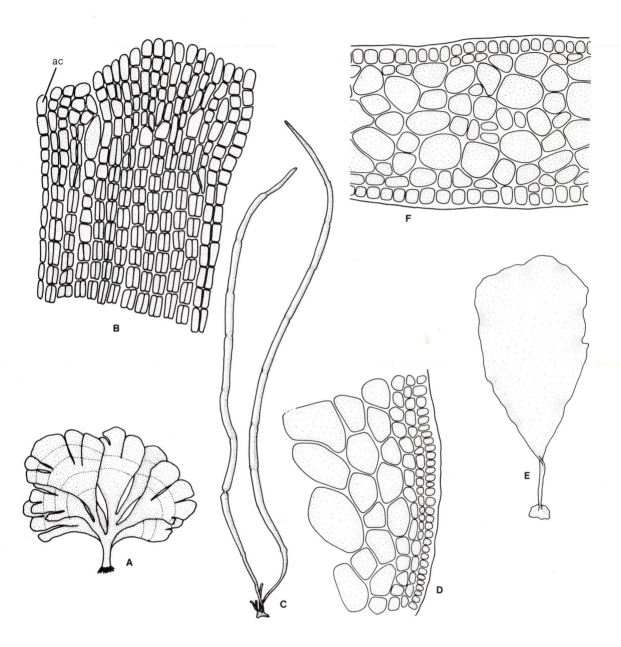

Figure 10-12 Parenchymatous forms of Phaeophyta. **A, B,** *Syringoderma*. **A,** habit of fan-shaped thallus, ×2.5; **B,** outer region of monostromatic thallus, showing marginal row of apical cells (*ac*), ×125. **C, D,** *Scytosiphon*. **C,** habit of hollow thallus, ×0.5; **D,** transverse section of outer portion of thallus, showing how anticlinal and periclinal divisions lead to parenchymatous organization of cells, ×500. **E, F,** *Phaeostrophion*. **E,** habit, ×1; **F,** transverse section through portion of thallus, showing parenchymatous organization, ×300.

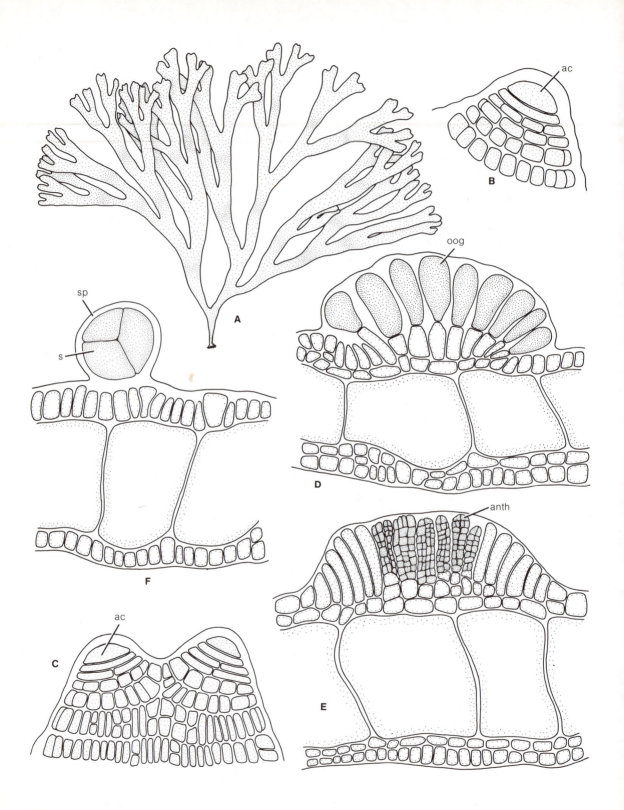

(Fig. 10-22) appear to be widespread in brown algae. In Laminariales plasmodesmata are common between adjacent cells in the outer cortex, between adjacent secretory cells, and between secretory cells and adjacent cortical cells. Morphologically and chemically, the phloemlike elements are very similar to the sieve tube elements of vascular plants. Kelp **callose** (Fig. 10-21D) similar to that present in the sieve tube elements of higher plants has been demonstrated. Evidence using radioactive tracers shows that these sievelike elements function in the conduction of photosynthate.

Some regions of the stipe can become hollow and develop into floats, or **pneumatocysts** (Figs. 10-15A, 10-17, and 10-23P), which hold the plant more or less erect in the water. These pneumatocysts contain a mixture of gases, including nitrogen, oxygen, and carbon dioxide (e.g., *Macrocystis*) or carbon monoxide (e.g., *Nereocystis*).

Despite the localized nature of meristematic activity in the normal vegetative condition of the thallus, some complex brown algae show remarkable regenerative powers. It has been demonstrated in a species of *Laminaria* that vegetative proliferation can occur from old haptera of plants that are completely devoid of their old blades, transition zones, and stipes. These regenerated blades soon establish a normal transition zone and develop to maturity in typical fashion. Also, in *Fucus* (Fig. 10-19D), complete new plants develop from fragments of almost any part of the plant and grow to sexual maturity.

Reproduction and Life Histories

Although many brown algae are complex vegetatively, they all have relatively simple reproductive structures. Sexual reproduction varies greatly, especially in the ectocarpalean line, where isogamy, anisogamy, and oogamy are represented. In this group of orders, there is typically an alternation of two multicellular phases or generations. The fucalean line is entirely oogamous, with a gametic type of life history[3] in which the diploid plant is the conspicuous and only multicellular phase in the life history.

Vegetative or Asexual Reproduction

Vegetative reproduction can occur by simple thallus fragmentation in simpler filamentous types or in somewhat specialized multicellular portions such as the **propagules** of *Sphacelaria* (Fig. 10-11F). It has been shown that these propagules are produced when plants are subjected to a specific temperature and light regime at which reproductive structures are apparently unable to develop. Thallus fragmentation is the only method of reproduction known in some free-floating species of *Sargassum*. Species of this same genus found attached (Fig. 10-18A–C) to the ocean bottom in coastal areas reproduce sexually; but unattached, free-floating species have apparently lost this ability and are perpetuated vegetatively (as in the Sargasso Sea). **Mitospores** (also termed **neutral spores**) from many-chambered **(plurilocular)** mitosporangia can also occur as an accessory method of reproduction. Meiosis occurs in an unpartitioned **(unilocular)** meiosporangium (Figs. 10-8D, F, 10-11H, and 10-13F). Meiospores (zoospores or aplanospores) are produced by **free-nuclear division.**

Sexual Reproduction

In sexual reproduction, plants with isogamy, anisogamy, and oogamy occur, and apparently these various types have evolved repeatedly along diverging evolutionary lines. Motile gametes (Fig. 10-4F) are morphologically similar to the zoospores (Fig. 10-4A–C) of brown algae,

---

Figure 10-13 Morphology and reproduction of parenchymatous form with apical growth (*Dictyota*, Dictyotales). **A,** habit, ×0.5. **B, C,** apical region of branch with apical cells (*ac*), showing dichotomous division in **C; B,** ×300; **C,** ×215. **D–F,** transverse sections through portions of different thalli, showing reproductive structures at surface. **D,** mature oogonia (*oog*), ×140; **E,** mature antheridia (*anth*), ×145; **F,** mature meiosporangium (*sp*) with three of four meiospores (*s*) visible.

[3]see footnote on p. 383

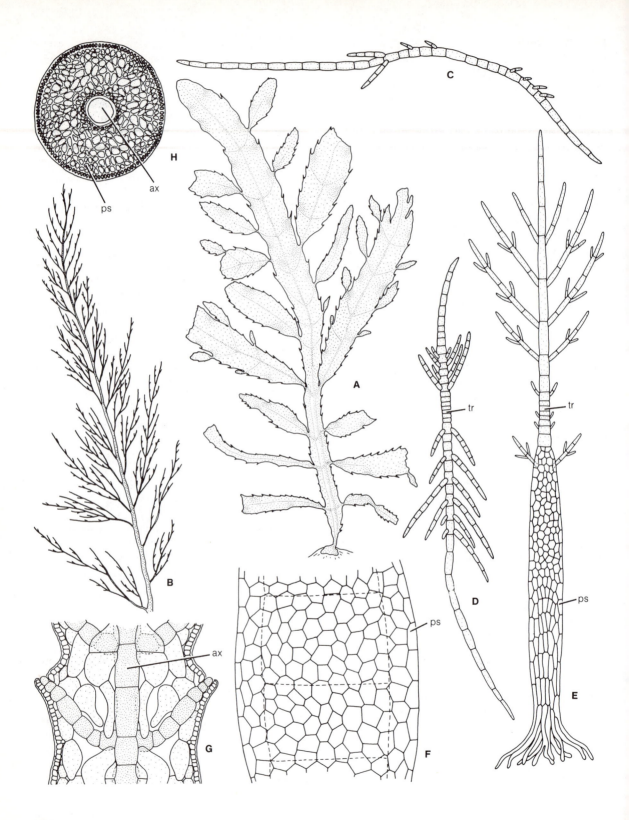

although often smaller. Isogamous forms produce gametes morphologically alike and of the same size. Anisogamous gametes are also morphologically alike, but one gamete (female) is consistently larger than the other (male). In oogamous forms, a nonmotile egg is fertilized by a small, motile sperm. Generally, syngamy occurs in the water, although an egg can remain fastened on the haploid female plant. In oogamous species, such as those of Dictyotales, Laminariales, and Fucales, one to several eggs or sperms form in a gametangium. However, in some isogamous and anisogamous species (e.g., Ectocarpales and Sphacelariales), gametangia are subdivided into many compartments, each of which produces a single, motile gamete. These gametangia are referred to as being *plurilocular* (Figs. 10-8B, E and 10-11G).

Life Histories

Sexual plants of brown algae generally produce either male or female gametangia. In some instances, male and female gametophytes are morphologically distinguishable, as in *Nereocystis* (Fig. 10-23). In other brown algae, gametophytes are indistinguishable, as in some species of *Ectocarpus*. Less commonly, other species of brown algae produce both kinds of gametes on the same plant.

Alternation of Isomorphic Generations
Brown algae, except in Fucales, usually have an alternation of free-living, *multicellular*, gametophytic and sporophytic generations. In simpler filamentous brown algae, such as Ectocarpales, there is usually an alternation of **isomorphic** generations (Fig. 10-24), in which haploid and diploid phases are morphologically identical and physiologically independent in the vegetative condition. However, even in the same genus, as in *Ectocarpus* and *Sphacelaria*, there are some species that have an alternation of isomorphic generations and others in which there are slightly **heteromorphic** generations.

Ectocarpus. In *Ectocarpus* (Ectocarpales), motile isogametes (Fig. 10-24E, F) are produced in many-chambered (plurilocular) gametangia (Fig. 10-24C, D). It has been shown in one species that a highly volatile sexual attractant, or gamone, called *ectocarpen* (S-1-but-1'-enyl-2,5-cycloheptadiene) is produced by female gametes and is responsible for the clumping of male gametes. The female gamete settles to the substratum shortly after release from the gametangium, withdraws its flagella, and at this point apparently releases the highly volatile gamone, which attracts male gametes to it. The male gamete has a longer period of motility (about 24 hours) and, when attracted to the female gamete, attaches to it by the smooth tip of the tinsel flagellum. The flagella are then retracted, and the male gamete fuses with the female gamete to form the diploid zygote (Fig. 10-24H), which immediately divides to produce the diploid sporophytic generation.

The sporophytic generation produces many-chambered (plurilocular) mitosporangia (Fig. 10-24T), which in turn produce diploid mitospores capable of maintaining the diploid generation by asexual means. Meiosis occurs in one-chambered (unilocular) meiosporangia (Fig. 10-24K–M). The resulting meiospores grow into two types of haploid gametophytes (Fig. 10-24A, B), each producing one type of gamete. In recent studies it has been shown that haploid gametophytes can also produce many-chambered mitosporangia, which in turn produce mitospores capable of maintaining the haploid generation by asexual means.

Dictyota. The life history of *Dictyota* (Dictyotales) is very similar to that of *Ectocarpus*. However, in this oogamous genus, only four

Figure 10-14 Morphology and anatomy of pseudoparenchymatous form (*Desmarestia*, Desmarestiales). **A, B,** species diversity; except for symmetry of thallus, anatomical detail is similar in all species. **A,** flattened, strap-shaped species; **B,** terete species (distal portion only shown), ×0.3. **C–E,** juvenile sporophyte plants. **C, D,** early filamentous phases, ×150; **D,** showing initiation of trichothallic growth (*tr*), ×150; **E,** more mature phase, showing trichothallic growth and pseudoparenchymatous habit (*ps*) of cortical filaments, ×110. **F,** enlarged surface view of basal portion of **E,** showing pseudoparenchymatous cells surrounding central axis (*broken lines*), ×200. **G,** enlarged longitudinal section of more mature portion of the thallus, showing axial row of cells (*ax*), ×250. **H,** transverse section of mature portion of plant, ×150.

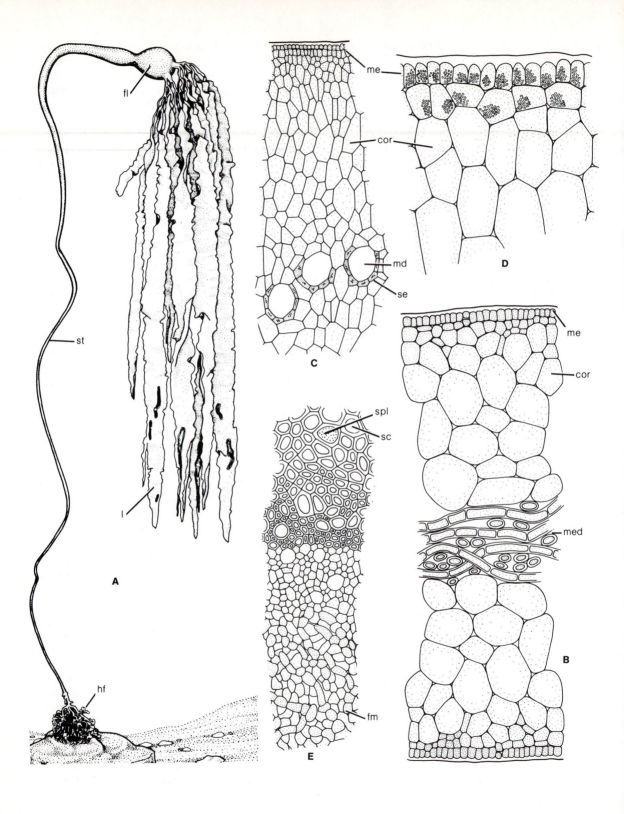

nonmotile meiospores (Fig. 10-13F) are produced in the meiosporangium. It has been demonstrated that **genotypic sex determination** takes place in the meiosporangium, and that half (two) of the spores develop into male and the other half (two) into female gametophytic plants.

Alternation of Heteromorphic Generations
In more specialized orders, such as Laminariales and Desmarestiales, there is typically an alternation of heteromorphic generations (Fig. 10-25), in which the haploid phase is a free-living, physiologically independent, diminutive (although multicellular) generation and the diploid phase is the conspicuous, relatively large generation. Some of these large, more complex, sporophytic generations are annual, as in *Nereocystis*, but they still reach a length of 35 meters or more in less than four months. Others are perennial, such as *Pterygophora* (Fig. 10-19), and can reach a length of about four meters or more in 10 to 15 years. Some evidence indicates that free-living, gametophytic generations (Fig. 10-23F, G) of Laminariales can also be perennial; however, their fate has not been followed in the sea. Gametophytes of some Laminariales have been grown continuously in the laboratory for up to six years.

◄──────

Figure 10-15 Morphology and anatomy of parenchymatous form (*Nereocystis*, Laminariales). **A,** habit view showing branched holdfast (*hf*) of rootlike structures, erect stalklike stipe (*s*), enlarged spherical float (*fl*), and dense cluster of terminal, bladelike laminae (*l*), ×0.1. **B–E,** anatomical details of portions of **A. B,** transverse section through sterile portion of lamina, showing meristoderm (*me*), cortex (*cor*), and medulla (*med*), ×350. **C–E,** transverse sections through portions of stipe. **C, D,** outer portion of stipe, showing meristoderm (*me*), compact arrangement of cells in outer cortex (*cor*), mucilage ducts (*md*), and secretory cells (*se*); **C,** ×150; **D,** ×600. **E,** inner portion of stipe, showing compact cells in transition region between inner cortex region and outer medulla region with sieve cells (*sc*), sieve plates (*spl*), and central filamentous medulla (*fm*), ×150.

Nereocystis. In the typical kelp (Laminariales), such as *Nereocystis,* the haploid gametophytes are microscopic. The filamentous female gametophyte is made up of somewhat larger cells than the male, and the filaments are slightly less branched (Figs. 10-23F, G and 10-25A, B). The female gametophyte produces one egg (per oogonium) which is extruded from but held at the open end of the oogonium (Figs. 10-23F and 10-25D). The male gametophyte is made up of smaller cells, and the filaments are profusely branched (Figs. 10-23G and 10-25A). It produces many motile sperms (Fig. 10-23H) (one per antheridium usually), which swim to the egg (Fig. 10-25E).

On fertilization, the resulting zygote (Figs. 10-23I and 10-25F), as in Ectocarpales, undergoes mitosis to form the embryonic sporophyte (Figs. 10-23L–M and 10-25G–I). The sporophyte continues growing into the typical kelp plant (Figs. 10-23O, P and 10-25J). In specialized areas on the sporophyte blade, meiosis occurs in one-chambered (unilocular) meiosporangia (Figs. 10-22F and 10-25K), each of which produces 64 meiospores (Fig. 10-16F). In other Laminariales, the number produced is 16; in still others it is 32, or 128. Motile meiospores (Figs. 10-23A and 10-25M, N) swim for a brief period, then settle to the substratum, retract their flagella, and secrete a cell wall. The mucilaginous cell wall cements the settled zoospores to the substratum, and they develop into the two types of haploid gametophytes. Genotypic sex determination has been demonstrated during meiosis, with half the zoospores developing into male and the other half into female gametophytes.

Fucus. In *Fucus* (Fucales), the diploid generation is the only multicellular, conspicuous, and free-living phase (Figs. 10-20D and 10-26G). The diploid phase is **perennial,** lasting in some species up to four years. In other genera in Fucales, plants are reported to reach an age of at least 19 years. The haploid generation has been reduced essentially to single cells that become the gametes themselves. In other words, meiosis occurs at gamete formation in the diploid plant. Thus, gametangia are comparable to the unicellular one-chambered (unilocular) sporangia of other brown algae. By comparison with other brown algae, spores function essentially as

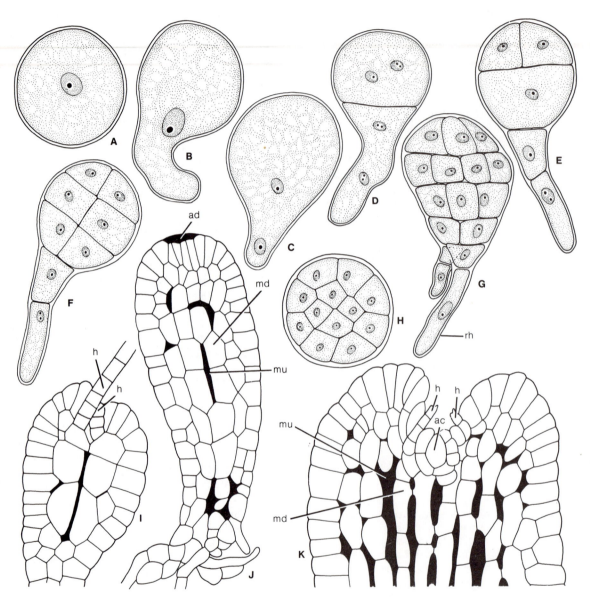

Figure 10-16 Early developmental stages of *Fucus* (Fucales.) **A,** zygote, ×600. **B–G,** stages in development of juvenile thallus, showing nuclei, ×600. **H,** transverse section view of terminal portion of **G,** ×600. **I,** upper portion of more mature juvenile, showing two terminal hairs (*h*), ×325. **J,** juvenile, showing early development of apical depression (*ad*), filamentous medulla (*fm*), and mucilage (*mu*), ×250. **K,** upper portion of more mature juvenile thallus, showing apical depression, remnants of terminal hairs (*h*), and initiation of apical cell (*ac*), ×325. (**I,** after Nienburg; **J, K,** after Oltmanns.)

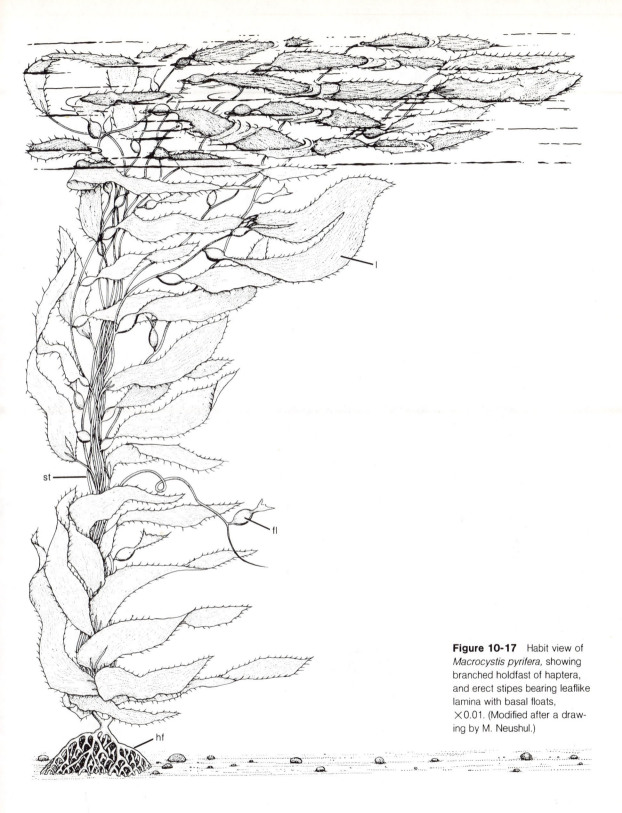

Figure 10-17 Habit view of *Macrocystis pyrifera*, showing branched holdfast of haptera, and erect stipes bearing leaflike lamina with basal floats, ×0.01. (Modified after a drawing by M. Neushul.)

Figure 10-18 **A–C,** *Sargassum muticum,* a species accidentally introduced from Japan to the Pacific Coast of North America with oyster spat. **A,** attached plant, ×0.1; **B,** basal portion of plant, showing straplike ''leaves'' and discoid holdfast, ×1; **C,** distal portion of plant, showing leaflike branches and spherical pneumatocysts (floats), ×1. **D,** plant of *Macrocystis integrifolia,* attached by holdfast, at low tide in northern British Columbia, ×0.1.

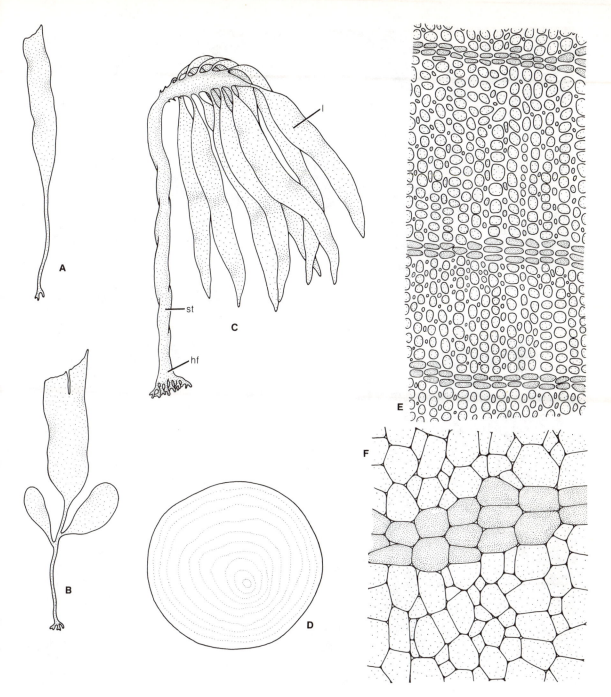

Figure 10-19 Morphology and anatomy of parenchymatous form (*Pterygophora,* Laminariales). **A, B,** young sporophyte plants, ×0.2. **C,** mature sporophyte, showing stipe (*st*), holdfast (*hf*), and laminae (*l*), ×0.07. **D,** transverse section of stipe, showing "growth" rings formed by concentric zones of differentiated cells, ×0.5. **E,** enlarged view of a portion of stipe, showing distinct zones of differentiated cells, ×50. **F,** the same as **D** and **E,** but at higher magnification, ×150.

Figure 10-20 **A, B,** *Durvillaea* in South Australia. **A,** a single large plant, ×0.1; **B,** appearance of a bed at low tide. **C,** *Hormosira* in South Australia; close-up showing conceptacles on swollen, beadlike receptacles, ×0.5. **D,** *Fucus* growing in the intertidal zone in British Columbia, with wartlike appearance of conceptacles on swollen receptacles, ×0.3. (Photographs **A–C,** courtesy of H. B. S. Womersley.)

gametes, fusing immediately instead of developing into free-living gametophytes.[4]

The zygote is spherical when first formed, but soon develops a definite **polarity** (Fig. 10-18E). The factors influencing polarity are very complex, involving a number of environmental factors. The direction of the light source seems to be especially important. Rhizoids (Fig. 10-16G) always develop at the end of the zygote directed away from the light, and the apex develops toward the light. However, other factors including temperature gradient, pH gradient, presence of other (unfertilized) eggs in close proximity, electric current, growth substances, and point of entry of the sperm all may contribute to the establishment of this polarity. The embryo undergoes a complex development (Fig. 10-16A–J) before it becomes the typical, dichotomously branched, diploid thallus (Fig. 10-26G).

At maturity, swollen tips (receptacles, Fig. 10-26H) on the thallus produce numerous small cavities, called **conceptacles** (Figs. 10-26I and 10-27A), in which gametangia are formed (Figs. 10-26I and 10-27A). Meiosis occurs (Fig. 10-27B–E) in the one-chambered female gametangium (oogonium), resulting in eight functional eggs (Figs. 10-26K and 10-27H–K), and in the one-chambered male gametangium (antheridium), resulting in 64 sperms (Figs. 10-26L, M and 10-27M–O). Gametes are usually released from the diploid plant following a period of water loss during a low tide period. Shrinkage of tissues surrounding the conceptacle results in detachment of the oogonia and antheridia. Swelling of

the mucilaginous material within the conceptacle as water is taken in during a high tide period results in the release and extrusion of eggs (Figs. 10-26B and 10-27I–K) and sperms (Figs. 10-26A and 10-27P) through the **ostiole** of the conceptable into the water, where fertilization occurs (Fig. 10-26C, D).

Motile sperms are attracted (Fig. 10-27L) to the egg by the gamone, fucoserraten (a conjugated hydrocarbon, 1,3,5-octatriene, C_8H_{12}). The proboscis on the *Fucus* sperm is thought to be involved in the response, although the tip of the longer flagellum apparently is the first point of physical contact in the process. This type of life history is somewhat comparable to that occurring in animals, with the organism diploid and gametes the only haploid cells.[5] The zygote sinks to the bottom, develops a firm, mucilaginous wall, which adheres to the substratum, and germinates soon after fertilization. There are no known examples of the diploid generation in the brown algae being reduced to only a single cell, the zygote, as in many green algae.

Other Life History Complexities In some brown algae, isogametes or anisogametes develop directly into haploid plants, thus behaving like mitospores. This suggests that vegetative reproduction is more primitive, that fusion of cells has occurred secondarily, and that sexuality is perhaps relative. For example, in some isogamous species of algae, both gametes are produced by the same plant; in other isogamous forms, gametes fuse only with those from different plants. The next step in the development of sexual differentiation was probably anisogamy. The most advanced type of sexual reproduction is oogamy. One of the advantages of oogamous sexual reproduction is that the large, nonmotile female gamete, or egg, can survive longer before being fertilized without exhausting its stored food reserves. Also, because of the egg's relatively large contribution to it, the zygote too can survive longer before developing without having to photosynthesize.

Detailed examples used in the foregoing discussion, *Ectocarpus*, *Dictyota*, *Nereocystis*, and

---

[4]An early interpretation by Strasburger (1906) of the life history of *Fucus* as being sporic, which has been adopted by other authors (Smith 1938), has been reinforced by a recent study (Jensen 1974) of *Bifurcariopsis*, another member of the Fucales. However, until the order has been more thoroughly investigated, we prefer to follow the interpretation of Kylin (1933), Fritsch (1945), and Papenfuss (1951) and consider the typical life history as being gametic. In the Strasburger hypothesis regarding *Fucus*, now elaborated by Jensen in *Bifurcariopsis*, the stage immediately following meiotic division of the diploid nucleus in the "sporangium" (= gametangium of the interpretation followed in this text) is regarded as a much-reduced, endosporic "gametophyte" retained within the "unilocular sporangium."

---

[5]*See* footnote on p. 383 for alternate interpretation.

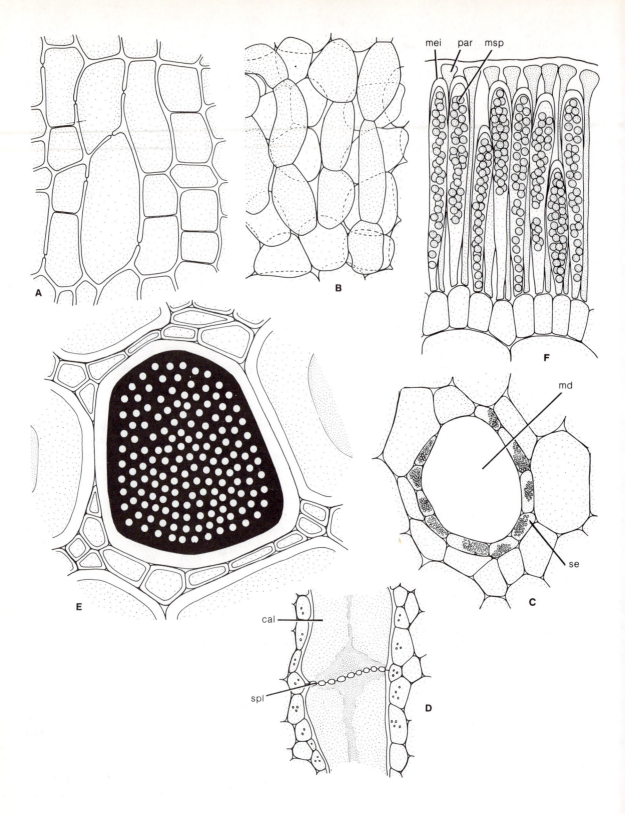

Figure 10-21 Anatomy and reproduction of parenchymatous form (*Nereocystis*, Laminariales). **A, B,** longitudinal sections of portions of stipe, showing (**A**) arrangement of cortical cells with "pits" between some of the adjacent cells and (**B**) filamentous organization of medulla region; **A,** ×300; **B,** ×450. **C,** transverse section through mucilage duct, showing secretory cells (*se*) surrounding duct (*md*), ×400. **D,** longitudinal section through region of sieve plate (*spl*), showing adjacent cells (note heavy deposit of callose, *cal*, in sieve cells), ×650. **E,** enlarged transverse section of sieve plate, ×1200. **F,** transverse section through outer portion of fertile region of lamina, showing almost mature unilocular meiosporangia (*mei*) with meiospores (*msp*) and paraphyses (*par*), ×720.

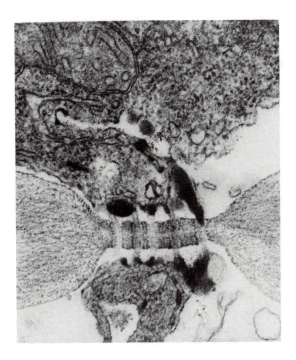

Figure 10-22 Electron micrograph of portion of wall between adjacent outer cortical cells of *Nereocystis* stipe. Transverse section, showing thinner region of wall in the pit-field area, and plasmodesmata, ×37,600. (Photograph courtesy of D. C. Walker.)

Fucus, are well known with respect to their life histories and reproduction. However, the number of brown algae that are equally well known represents only a small fraction of this large and heterogeneous group of algae. It is almost 100 years since Reinke and Falkenberg connected two morphologically different genera (*Aglaozonia* and *Cutleria*, order Cutleriales) of brown algae as phases in the life history of the same plant with an alternation of heteromorphic generations. Still, recent culture studies continue to yield similar surprises among genera whose life histories are completely unknown or incompletely known. In one instance, a *Ralfsia*like stage has been reported in cultures initiated with *Scytosiphon;* in another, a *Scytosiphon*like stage has been reported in cultures initiated with *Streblonema.* Some species of *Ectocarpus* and *Sphacelaria* have been shown to have an alternation of heteromorphic generations, although they are in orders traditionally considered as comprising forms having only an alternation of isomorphic generations. These studies, and the absence of information on all stages in the life history of many of the larger brown algae, lead one to suspect that many genera whose life histories are still incompletely known will be found to be linked to other entities presently considered as separate genera. Among Ectocarpales, genera such as *Myrionema, Ralfsia,* and other simple filamentous and encrusting types are especially open to suspicion in this respect.

Lines of Evolution

Repeatedly, in the highly evolved groups of algae, we find a trend from isogamy through anisogamy to oogamy. In the brown algae, evidence indicates that this trend has occurred along several distinct lines of evolution, especially within orders having an alternation of heteromorphic generations. Three of these evolutionary lines are particularly well distinguished.

The order Ectocarpales contains the simplest forms of brown algae. A number of genera in other orders, especially in Dictyosiphonales, Scytosiphonales, and Chordariales, merge with this basic group, both in respect to their vegetative structure and life histories. This is the chief

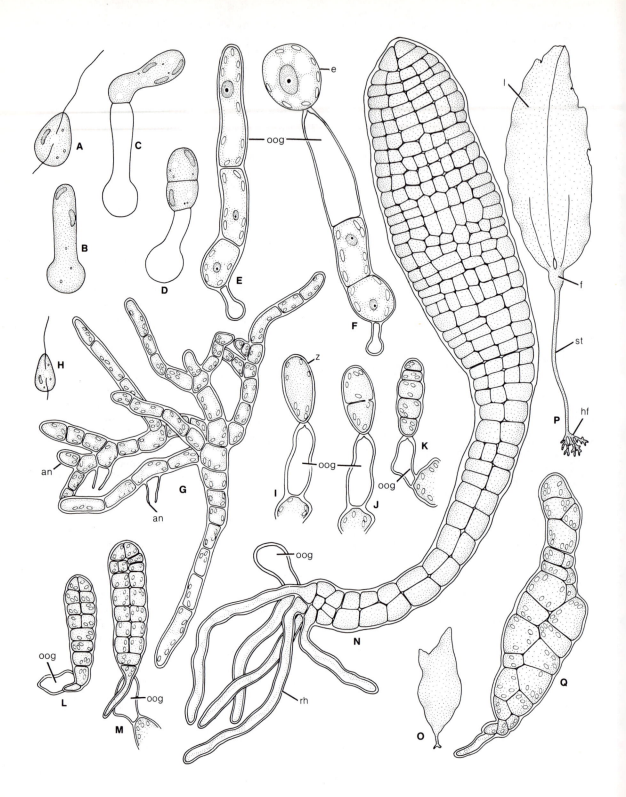

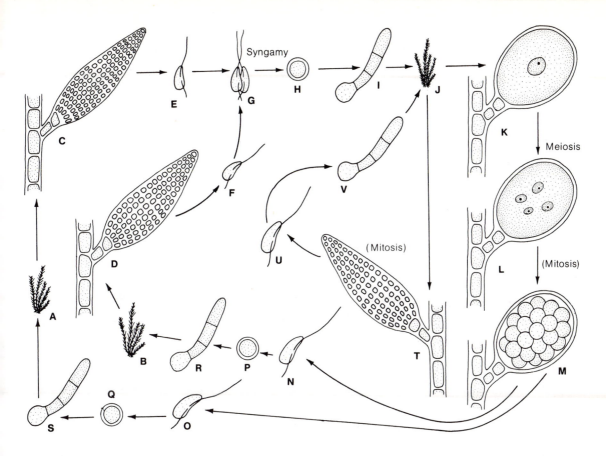

Figure 10-24 Life history of *Ectocarpus* (Ectocarpales). **A, B,** habit of gametophyte plants. **C, D,** plurilocular gametangia. **E, F,** gametes. **G,** syngamy. **H,** zygote. **I,** developing sporophyte. **J,** habit of mature sporophyte. **K–M,** maturation of meiosporangium. **N, O,** meiospores. **P–S,** developing gametophytes. **T,** plurilocular sporangium on sporophyte. **U,** mitospore. **V,** developing sporophyte. (For structural details and relative sizes of structures refer to Fig. 10-8B–D.)

Figure 10-23 Stages in life history of *Nereocystis* (Laminariales). **A,** zoospore (meiospore), ×1000. **B,** germinating zoospore, ×1000. **C–E,** stages in development of female gametophyte, ×1000. **F,** mature female gametophyte with empty oogonium (*oog*) and discharged, unfertilized egg (*e*), ×1000. **G,** mature male gametophyte with immature and mature (*empty*) antheridia (*an*), ×1000. **H,** sperm, ×1000. **I–K,** zygote (*z*) and early divisions in young sporophyte attached to empty oogonium, ×1000. **L–N,** stages in development of young sporophyte (**L, M,** ×800; **N,** ×400). **O, P,** later stages (macroscopic) of young sporophyte (**O,** ×0.5; **P,** ×0.3): *f,* float; *hf,* holdfast; *l,* lamina; *rh,* rhizoid; *st,* stipe.

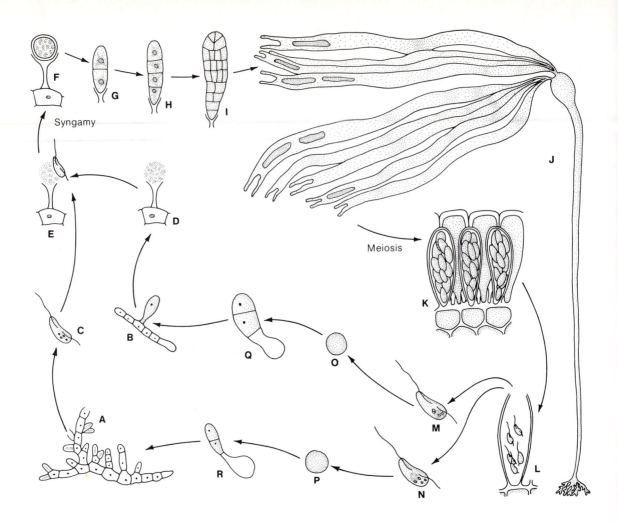

Figure 10-25 Life history of *Nereocystis* (Laminariales). **A,** male gametophyte with antheridia. **B,** female gametophyte with oogonium. **C,** sperm. **D,** unfertilized egg. **E,** syngamy. **F,** zygote. **G–I,** development of young sporophyte on female plant. **J,** mature sporophyte with dark, fertile sori on distal portion of lamina. **K,** meiosporangia. **L,** release of meiospores. **M, N,** meiospores. **O–R,** development of gametophytes. (For details and relative sizes of structures refer to Figs. 10-15A, 10-23A–P, and 10-30.)

reason why an arbitrary main division into several classes based primarily on the type of life history does not result in a tenable phylogenetic (phenetic) arrangement. *Ectocarpus* illustrates well the generalized features of the brown algae and, at the same time, serves to illustrate a possible prototype from which all other groups of brown algae (with the possible exception of Fucales) may be derived (see also Fig. 13-14).

From the *Ectocarpus* type of life history, forms with anisogamy and an alternation of isomorphic generations can readily be derived, as in

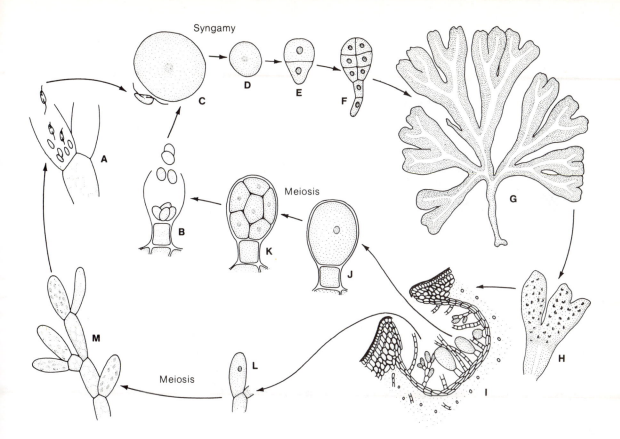

Figure 10-26 Life history of *Fucus* (Fucales). **A,** mature antheridium liberating sperm. **B,** mature oogonium liberating eggs. **C,** mature sperm and egg prior to fusion. **D,** zygote. **E, F,** early development of thallus. **G,** mature thallus. **H,** fertile receptacle with numerous conceptacles. **I,** conceptacle with male and female structures. **J, K,** development within conceptacle of oogonium and eggs. **L, M,** development within the conceptacle of branch with several antheridia. (For structural details and relative sizes of structures refer to Figs. 10-16, 10-20, and 10-27.)

Sphacelaria. Similarly, a more advanced type of sexual reproduction, oogamy, still with an alternation of isomorphic generations, as in *Dictyota,* is in a direct line of evolution from this basic group. Although vegetatively *Nereocystis* has a reduced, *Ectocarpus*like, filamentous gametophytic stage, the conspicuous generation has become a massive, complex, and highly developed sporophytic stage. *Fucus,* on the other hand, has the ultimate in reduction in the haploid (*n*) phase, namely the gametes, but also has a complex, highly developed diploid (2*n*) phase.

Relationships of Phaeophyta

There are no fossils that can be referred to brown algae with any degree of certainty from periods earlier than the Triassic, although some plants resembling *Fucus* apparently occurred in the early Paleozoic Period. Some **phylogeneticists** have suggested a possible Precambrian origin for brown algae in the Proterozoic or late Archeozoic periods. However, except for the **extant**

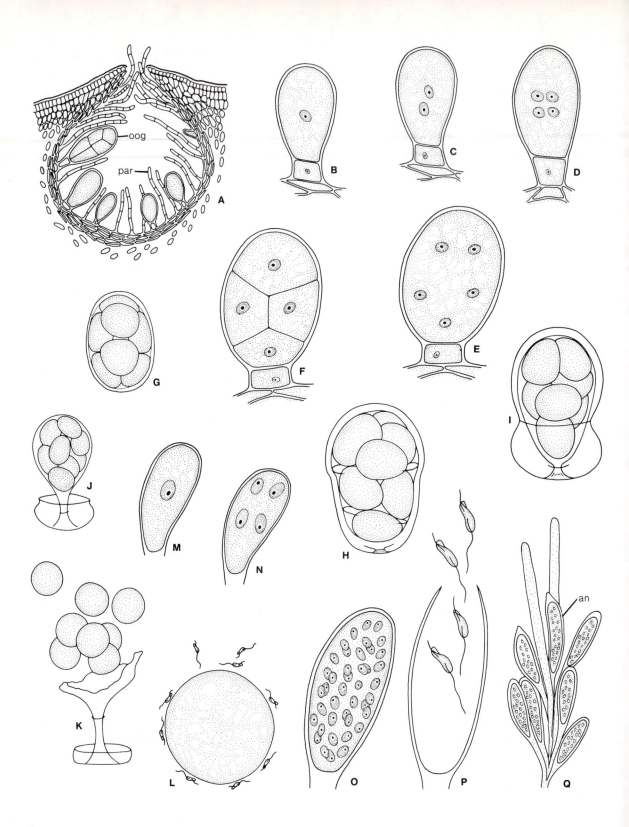

genus *Padina* (Fig. 10-1), which is only lightly calcified, all extant brown algae are noncalcareous. Hence, the fossil record for the Phaeophyta, in contrast to that for the Chlorophyta and Rhodophyta, in which certain groups are heavily calcified, is relatively poor. There is thus little paleobotanical evidence of a geological history to support phylogenetic considerations in the Phaeophyta.

From Paleozoic rocks in the Silurian and Devonian, certain fossil plants have been reported that bear a similarity to modern members of Fucales, Dictyotales, Chordariales, and Sphacelariales. These observations suggest, at least, that Phaeophyta had evolved before late Devonian time.

However, a number of less dubious Cenozoic records of several genera and species from Miocene **diatomite** deposits in southern California provide some of the best fossil evidence known. Certain of these, such as *Julescraneia* (Fig. 10-28C), have been assigned to Laminariales. Most have been referred to Fucales, and because of their similarity in appearance to the extant genera *Halidrys*, *Cystoseira*, and *Cystophora*, they bear the names *Paleohalidrys* (Fig. 10-28A), *Cystoseirites*, and *Paleocystophora* (Fig. 10-28B). *Julescraneia*, the largest brown algal fossil known, bears some remarkable similarities to the extant kelps *Pelagophycus* and *Nereocystis* (Fig. 10-15A). It must be emphasized that although these noncalcareous forms have been perfectly preserved under anaerobic conditions in deep marine deposits, in most instances it is impossible to determine anything of the nature of their cell contents or their reproductive characteristics.

Brown algae are probably not closely related to any other group of living algae, although they share some characteristcs with Chrysophyta. One can assume that in evolutionary sequence, nuclear condition, flagellation, and biochemical characteristics (pigments and nature of food reserves) followed one another. Most likely, brown algae are derived from some laterally biflagellated ancestral stock comparable to the characteristic motile cells of brown algae. However, the fact that there are no unicellular flagellates referred to Phaeophyta does not mean that they do not exist in the sea; a prototype may still be found among living flagellates when the marine nannoplankton are better known. Because of the similarity in flagellation to some Chrysophyta, a common ancestry has been suggested for the brown algae and Chrysophyta. This suggestion is further supported by the presence of chlorophyll *c* and fucoxanthin in Chrysophyta and the storage of food reserves in Chrysophyta as a laminaranlike polysaccharide. A less likely relationship to Pyrrhophyta has been suggested because of the presence of chlorophyll *c* in that group. If these three divisions had a common origin, it is likely that the brown algae had some prechrysophyte ancestor.

The most generalized order of brown algae is thought to be Ectocarpales. From this assemblage, or an *Ectocarpus*like ancestry, all other orders of brown algae, except for Fucales, may have evolved (see also Fig. 13-14). Biochemical characteristics, including the pigment complex, support a common origin for the Fucales within the same division (Phaeophyta) and class (Phaeophyceae), possibly at some early stage of evolution. However, in view of the emphasis placed on flagellation in algal classification in general, it would appear that Fucales, because of the strikingly different form and flagellation of some of their sperms and their unique type of life history, comprise an evolutionary series (the fucalean line) distinct from the ectocarpalean evolutionary line.

Within the ectocarpalean line there have evolved two distinct evolutionary series, based on vegetative construction, from the simple uniseriate prototype. One of these, comprising filamentous forms with various degrees of aggregation of filaments, culminates in complex pseudoparenchymatous types. The other, comprising parenchymatous forms with highly organized meristems and differentiated tissues,

Figure 10-27 Reproduction in *Fucus* (Fucales). **A,** transverse section of fertile portion of thallus, showing conceptacle containing immature and mature oogonia (*oog*), and paraphyses (*par*), ✕ 125. **B–K,** stages in development and release of eggs in oogonium; **B–F,** ✕500; **G,** ✕400; **H, I, K,** ✕500; **J,** ✕350. **L,** sperms swimming about unfertilized egg, ✕1500. **M–P,** stages in development of antheridia and sperm release, ✕1800. **Q,** branch bearing several antheridia (*an*), ✕800. (**H–L,** after Thuret.)

A

B

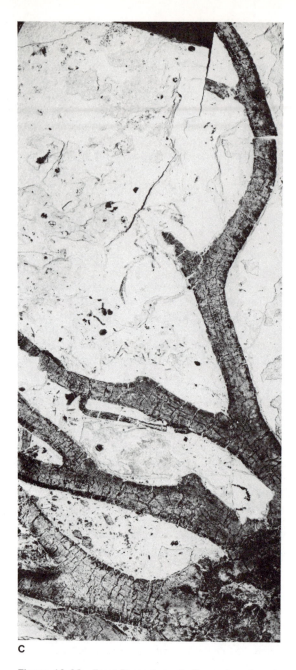

C

Figure 10-28 Fossil Phaeophyta. **A, B,** species referred to the family Cystoseiraceae (Fucales). **A,** *Paleohalidrys lompocensis,* ×1.0; **B,** *Paleocystophora delicatula,* ×1.3. **C,** *Julescraneia grandicornis,* a species referred to the family Lessoniaceae (Laminariales) and believed to be closely related to *Pelagophycus,* ×0.38. (Photographs courtesy of B. C. Parker, with permission of *Nova Hedwigia.*)

Figure 10-29 Dense bed of *Nereocystis* at low tide in Queen Charlotte Islands, British Columbia.

Figure 10-30 Intertidal view of plants of *Ascophyllum* at low tide in Iceland.

culminates in the largest and most complex aquatic plants (Laminariales). The fucalean line is entirely oogamous, but within the ectocarpalean line, there is an advance from the primitive, isogamous condition through anisogamy to oogamy in both filamentous and parenchymatous series.

Distribution and Ecology

Brown algae are almost entirely marine in distribution, including brackish water and salt marshes. Some species of *Sargassum* can occur free-floating far from shore in the North Atlantic (Sargasso Sea). In the Sea of Japan, other species of the same genus can occur either attached or free-floating in salt marshes and lagoons or embedded in sand. Floating populations of *Fucus* and *Macrocystis* have also been reported. In the unattached condition, these algae generally reproduce only by vegetative propagation. A number of large genera, such as *Nereocystis* and *Macrocystis*, which grow attached in deeper water (to 35 meters or more), form dense subtidal forests. At times the distal branches of these large kelps float in dense mats at the sea surface (Figs. 10-29 and 10-32).

The most conspicuous brown algae belong to Laminariales and Fucales, which have representatives in both the Northern and Southern Hemispheres. However, Laminariales are richest in genera and species in the North Pacific. In contrast, Fucales are most abundant in genera and species in the Southern Hemisphere, appearing to have had a center of distribution in Southern Australia and New Zealand. Certain genera in each of these groups occur in only one or the other region. In Fucales, for example, the large *Durvillaea* (Fig. 10-20A, B) does not occur in the Northern Hemisphere; it is found only in colder waters of Australia, New Zealand, Chile, and other subantarctic regions. *Fucus* (Fig. 10-20D), on the other hand, occurs only in the Northern Hemisphere, although it is present in both the

North Atlantic and North Pacific. However, *Ascophyllum* (Fig. 10-30) is restricted in the Northern Hemisphere to the North Atlantic. The ecological niche occupied by *Fucus* in the Northern Hemisphere is occupied in Australia and New Zealand by another member of Fucales, *Hormosira* (Figs. 10-20C and 10-31). In Laminariales, *Macrocystis* (Fig. 10-17) is found in colder water in both the Northern and Southern Hemispheres; it does not occur in the western Pacific or in the North Atlantic Ocean.

Apart from *Sargassum* (Fig. 10-18A–C) and a few other members of Fucales, which are especially widespread in tropical oceans, brown algae tend to be small and inconspicuous in the tropics. However, there are some forms, such as *Padina* (Fig. 10-1A, B) and *Hydroclathrus*, that are often fairly conspicuous and restricted to the tropics and subtropics.

Importance and Uses

The direct importance to humans of the brown algae in coastal regions is related to their potential as **primary producers.** The plankton algae—such as diatoms, dinoflagellates, and others—are probably the most important primary producers in the sea as a whole; a much greater area of the globe and volume of water is available for the support of these free-floating or free-swimming microscopic plants than for the **benthic** algae. The benthic, or attached, marine algae (the seaweeds) cover a small area—a relatively limited area of **continental shelf** or water shallow enough to allow the penetration of sufficient light for plant growth.

Less than a tenth of the oceans, including adjacent seas, have a depth shallower than about 225 meters, and algae seldom grow to this depth. In coastal areas, however, seaweeds such as kelps are important contributors to the economy of the sea and can be more significant than the phytoplankton in areas where dense growth of attached algae occurs. Their large plant bodies provide a substrate for small, sessile animals; among their rootlike branches, the holdfasts (Fig. 10-18D) can house many diverse small animals. Dense subtidal forests can provide protection and **spawning** areas for larger, free-swimming

Figure 10-31 Dense bed of *Hormosira* at low tide in South Australia. (Photograph courtesy of H. B. S. Womersley.)

animals, including fish. Some marine animals, including some fish, browse directly on plant tissues of larger seaweeds, and others indirectly make use of these large algae by feeding upon detritus originating from decaying plant tissues. Specimens of *Nereocystis* (Fig. 10-15A) individually reach a length of over 35 meters and can weigh over 125 kilograms (fresh weight). Such plants provide an abundant source of detritus in coastal regions for filter-feeding animals. The larger algae also contribute significantly to the phytoplankton at certain stages in their life history through their motile stages (zoospores and gametes). In dense beds of *Nereocystis* (Fig. 10-29), through a water column having an average depth of about 20 meters, an estimated 3 million zoospore cells (about 5 μm in diameter) may be produced per liter per day from June to September. The total primary production in such an area also includes the reproductive cells liberated by other, smaller kelps and other algae growing around the holdfasts of the *Nereocystis* plants.

The larger brown algae are also directly important to humans. Species of Laminariales (especially *Laminaria*) have been used as food

Figure 10-32 Harvesting barge at work in dense bed of *Macrocystis pyrifera* in California. (Photograph courtesy of the Kelco Company.)

(called *kombu*) in the Orient since at least 1730. Most kombu is produced in China and Japan; although much of the domestic production in Japan is consumed by the Japanese, a large amount is exported to China and elsewhere.

Algin, or alginic acid, is produced exclusively from brown seaweeds. Valuable algin industries have developed in Britain and California, and to a limited extent in Japan, Australia, Canada, and elsewhere. The greatest production of algin is from *Macrocystis* in California. Dense stands are harvested by mowers on floating barges (Fig. 10-32). These mechanical harvesters can collect several hundred tons of seaweed a day by cutting through beds at a depth of about one meter.

Although alginic acid as such is limited in use, its salts—the alginates—in both soluble and insoluble form, have a great variety of uses. Alginic acid can absorb 10 to 20 times its own weight of water, but when dried it becomes so hard that it can be turned on a lathe. It produces viscous, gummy sols with water. Being a weak acid, it forms a wide variety of alginates having interesting properties, as well as complex compounds of the plastic type. For example, ammonium alginate is used in fireproofing fabrics; calcium alginate in plastics and as a laundry starch substitute; and sodium alginate as a stabilizer in ice cream and other dairy products.

There are many other uses of alginates and their derivatives in industrial processes. They are used as a binder in printer's ink; in soaps and shampoos; in molding material for artificial limbs; and in the manufacture of buttons. They are used in photography for film coatings; in paints and varnishes; in dental impression materials; in leather finishes; and in insecticides. Pharmaceutical preparations such as toothpaste, shaving cream, lipsticks, medicines, and tablets are manufactured from alginates. As mentioned previously, they are widely used in the food industry as stabilizers; they are also used as a clarifier in the manufacture of beer.

Brown algae have been used at times for fertilizer. Much of the value of seaweed as fertilizer lies in its provision of trace nutrients and organic growth substances. Marine algae accumulate certain substances, such as potassium and iodine salts, in significantly higher concentrations than those present in sea water. Because the great number of elements present in sea water are found also in algae, deficiency diseases in land plants can often be remedied with seaweed fertilizer. The total amount of fertilizing constituents in fresh seaweed is said to be comparable to the amount present in barnyard manure. However, seaweed contains more potassium, less phosphorous, and about the same proportion of nitrogen as barnyard manure. Iodine was reclaimed from seaweeds on a large scale in the last century until cheaper sources of supply were discovered. This industry has largely disappeared, although it still exists to a small extent in Japan and northern Europe.

References

Abbott, I. A., and Dawson, E. Y. 1978. *How to know the seaweeds.* Wm. C. Brown Co. Pubs., Dubuque, Iowa.

Abbott, I. A., and Hollenberg, G. J. 1976. *Marine algae of California.* Stanford Univ. Press, Stanford, Calif.

Bold, H. C., and Wynne, M. J. 1978. *Introduction to the algae.* Prentice-Hall, Englewood Cliffs, N.J.

Boney, A. D. 1965. Aspects of the biology of the seaweeds of economic importance. *Advances Mar. Biol.* 3:105–253.

Bouck, G. H. 1969. Fine structure and organelle associations in brown algae. *J. Cell. Biol.* 26:523–37.

Chapman, V. J. 1970. *Seaweeds and their uses.* 2nd ed. Methuen & Co., London.

Chapman, V. J., and Chapman, D. J. 1973. *The algae.* 2nd ed. Macmillan & Co., London.

Druehl, L. D. 1968. Taxonomy and distribution of northeast Pacific species of *Laminaria. Can. J. Bot.* 46:539–47.

Druehl, L. D., and Wynne, M. J. 1971. Bibliography on Phaeophyta. In Rosowski, J. R., and Parker, B. C., eds. *Selected papers in phycology.* Dept. of Botany, Univ. of Nebraska, Lincoln, Neb.

Edelstein, T.; Chen, L. C-M.; and McLachlan, J. 1970. The life cycle of *Ralfsia clavata* and *R. borneti. Can. J. Bot.* 48:527–31.

Evans, L. V. 1963. A large chromosome in the laminarian nucleus. *Nature* 198:215.

———. 1965. Cytological studies in the Laminariales. *Ann. Bot.* 29:541–62.

———. 1966. Distribution of pyrenoids among some brown algae. *J. Cell. Sci.* 1:449–54.

Fritsch, F. E. 1945. *The structure and reproduction of the algae.* Vol. 2. Cambridge Univ. Press, Cambridge.

Hoek, C. van den, and Flinterman, A. 1968. The life-history of *Sphacelaria furcigera* Kütz. (Phaeophyceae). *Blumea* 16:193–242.

Hoek, C. van den, and Jahns, H. M. 1978. *Algen.* Georg Thieme Verlag, Stuttgart.

Hollenberg, G. J. 1969. An account of the Ralfsiaceae (Phaeophyta) of California. *J. Phycol.* 5:290–301.

Jensen, J. A. 1974. Morphological studies in Cystoseiraceae and Sargassaceae (Phaeophyceae) with special reference to apical organization. *Univ. Calif. Publ. Bot.* 68:1–63.

Kylin, H. 1933. Über die Entwicklungsgeschichte der Phaeophyceen. *Lund Univ. Årsskr.,* N. F., Avd. 2, 29(7):1–102.

Lee, R. E. 1980. *Phycology.* Cambridge Univ. Press, Cambridge.

Levring, T.; Hoppe, H. A.; and Schmid, O. J. 1969. *Marine algae: a survey of research and utilization.* Cram, DeGruyter & Co., Hamburg.

Manton, I., and Clarke, B. 1951a. Electron microscope observations on the zoospores of *Pylaiella* and *Laminaria. J. Exper. Bot.* 2:242–46.

———. 1951b. An electron microscope study of the spermatozoid of *Fucus serratus. Ann. Bot.* 15:461–71.

Manton, I.; Clarke, B.; and Greenwood, A. A. 1953. Further observations with the electron microscope on spermatozoids in the brown algae. *J. Exper. Bot.* 4:319–29.

Markham, J. W. 1973. Observations on the ecology of *Laminaria sinclairii* on three northern Oregon beaches. *J. Phycol.* 9:336–41.

Mathieson, A. C. 1967. Morphology and life history of *Phaeostrophion irregulare* S. et G. *Nova Hedwigia* 13:293–318.

Müller, D. G. 1967. Generationswechsel, Kernphasen-wechsel und Sexualität der Braunalge *Ectocarpus siliculosus* im Kulturversuch. *Planta* 75:39–54.

———. 1979. Genetic affinity of *Ectocarpus siliculosus* (Dillw.) Lyngb. from the Mediterranean, North Atlantic, and Australia. *Phycologia* 18:312–18.

Müller, D. G.; Jaenicke, L.; Donike, M.; and Akintobi, T. 1971. Sex attractant in a brown alga: chemical structure. *Science* 171:815–17.

Nakamura, Y. 1965. Development of zoospores in *Ralfsia*-like thallus, with special reference to the life cycle of the Scytosiphonales. *Bot. Mag.* 78:109–10.

Nizamuddin, M. 1970. Phytogeography of the Fucales and their seasonal growth. *Bot. Mar.* 13:131–39.

Papenfuss, G. F. 1951. *Phaeophyta.* In Smith, G. M., ed. *Manual of phycology.* Chronica Botanica, Waltham, Mass.

———. 1955. Classification of the algae. In *A century of progress in the natural sciences, 1853–1953.* California Academy of Sciences, San Francisco.

Parker, B. C. 1965. Translocation in the giant kelp *Macrocystis.* I. Rates, direction, and quality of C^{14}-labeled organic products of fluorescein. *J. Phycol.* 1:41–46.

Parker, B. C., and Dawson, E. Y. 1965. Non-calcareous marine algae from California Miocene deposits. *Nova Hedwigia,* 10:273–95.

Parker, B. C., and Huber, J. 1965. Translocation in *Macrocystis.* II. Fine structure of the sieve tubes. *J. Phycol.* 1:172–79.

Russell, G. 1973. The Phaeophyta: a synopsis of some recent developments. *Oceanogr. Mar. Biol. Ann. Rev.* 11:45–88.

———. 1978. Environment and form in the discrimination of taxa in brown algae. In Irvine, D. E. G., and Prince, J. H., eds. *Modern approaches to the taxonomy of red and brown algae.* Academic Press, New York.

Scagel, R. F. 1966. The Phaeophyceae in perspective. *Oceanogr. Mar. Biol. Ann. Rev.* 4:123–94.

Setchell, W. A., and Gardner, N. L. 1925. The marine algae of the Pacific Coast of North America. III. Melanophyceae. *Univ. Calif. Publ. Bot.* 8:383–898.

Smith, G. M. 1938. *Cryptogamic botany.* Vol. 1. *Algae and Fungi.* McGraw-Hill Book Co., New York.

Strasburger, E. 1897. Kernteilung und Befruchtung bie *Fucus. Jahrb. Wiss. Bot.* 30:351–74.

———. 1906. Zur Frage eines Generationswechsel bei Phaeophyceen. *Bot. Zeit.* 64(2):1–7.

Tatewaki, M. 1966. Formation of a crustaceous sporophyte with unilocular sporangia in *Scytosiphon lomentaria. Phycologia* 6:62–66.

Taylor, W. R. 1957. *Marine algae of the northeastern coast of North America.* 2nd ed. Univ. of Michigan Press, Ann Arbor.

———. 1960. *Marine algae of the eastern tropical and subtropical coasts of the Americas.* Univ. of Michigan Press, Ann Arbor.

Tokida, J., and Hirose, H. 1975. *Advance of phycology in Japan.* Junk, The Hague.

Wynne, M. J. 1969. Life history and systematic studies of some Pacific North American Phaeophyceae (brown algae). *Univ. Calif. Publ. Bot.* 50:1–88.

Wynne, M. J., and Loiseaux, S. 1976. Recent advances in life history studies of the Phaeophyta. *Phycologia* 15:435–52.

11

RHODOPHYTA (Red Algae)

The division Rhodophyta (red algae) is an extremely large and diverse group of algae with over 800 genera and over 5200 species. The group is predominantly marine, with less than four percent of the species (in about 35 genera) occurring in fresh water. With few exceptions, the genera are either strictly marine or exclusively fresh water in occurrence. In this text, the division is considered in two classes, Bangiophyceae and Florideophyceae (Table 11-1). Each class has several orders, some of which are still poorly known, and will undoubtedly be further subdivided when the constituent genera have been more fully studied.

Flagellated cells are absent in red algae. Although the vegetative plant is relatively simple, most of these algae have a complex reproductive system. This is especially true of the female reproductive system, postfertilization processes, and **embryogeny.** Characteristics associated with the female reproductive system provide the primary basis for distinguishing orders of Florideophyceae.

Red algae as a whole are distinguished by having oogamous sexual reproduction, usually with a regular alternation of syngamy and meiosis. A nonmotile female gamete, retained on the haploid gametophyte, is fertilized *in situ* by a nonmotile male gamete carried by water currents. Various types of spores are also produced— some as a result of mitosis and others as a result of meiosis. Male and female reproductive structures can occur on the same thallus; they can also be on separate thalli, with male and female plants vegetatively indistinguishable. In some instances, there are **unisexual** and **bisexual** species in the same genus. Most genera have an alternation (or sequence) of free-living generations; in addition, there is generally a morphologically and cytologically distinct diploid phase or generation, which may be essentially parasitic, remaining attached to the haploid female gametophyte.

In red algae, the **antheridium** (or male gametangium) is referred to as a **spermatangium**; and the nonmotile male gametes are called **spermatia**. Each spermatangium usually produces a single spermatium. The **oogonium**, or female gametangium, called a **carpogonium**, usually has an elongate, emergent, threadlike, receptive portion called the **trichogyne**. The uni-

Table 11-1 Orders of Division Rhodophyta and Representative Genera

Class	Order	Representative Genera
Bangiophyceae	Porphyridiales	*Asterocystis, Goniotrichum, Porphyridium*
	Bangiales	*Bangia, Porphyra (= Conchocelis)*
	Rhodochaetales	*Rhodochaete*
	Erythropeltidales	*Compsopogon, Erythrocladia, Erythrotrichia, Smithora*
Florideophyceae	Acrochaetiales	*Audouinella*
	Nemaliales	*Batrachospermum, Bonnemaisonia (= Trailliella), Cumagloia, Gelidium, Pseudogloiophloea (Acrochaetium), Nemalion (Acrochaetium)*
	Cryptonemiales	*Bossiella, Callophyllis, Choreocolax, Cryptosiphonia, Dudresnaya, Gloiopeltis, Gloiosiphonia (Cruoriopsis), Harveyella, Lithothamnium, Melobesia, Mesophyllum, Thuretellopsis (= Erythrodermis), Weeksia*
	Gigartinales	*Neoagardhiella, Ahnfeltia (Porphyrodiscus), Chondrus, Eucheuma, Gigartina (Petrocelis), Gracilaria, Gymnogongrus, Iridaea, Halarachnion (Cruoria), Turnerella (Cruoria)*
	Rhodymeniales	*Rhodymenia*
	Palmariales	*Palmaria, Halosaccion*
	Ceramiales	*Antithamnion, Ceramium, Nitophyllum, Odonthalia, Polyneura, Polysiphonia, Rhodomela, Spermothamnion*

nucleate protoplast of the carpogonium functions as the female gamete. The carpogonium can be borne at the end of a special filament, the carpogonial branch.

The spermatium fuses with the trichogyne, and the male nucleus enters it and passes down into the enlarged base of the carpogonium, where the male nucleus fuses with the female nucleus, thus completing the process of fertilization. The cytoplasm in the carpogonium does not become a discrete female gamete, nor does the cytoplasm of the fertilized carpogonium form a discrete zygote. Following fertilization, the zygote nucleus undergoes further divisions *in situ*, or it (or a derivative of it) is transferred to another cell where further nuclear divisions occur.

Division in the zygote nucleus is mitotic and leads directly or indirectly to the production of **carposporangia**, which produce *2n* **carpospores**. Carposporangia can occur on filaments called **gonimoblast filaments**. Gonimoblast filaments, including the carposporangia, make up the **carposporophyte**, which is also referred to as a **gonimoblast**.[1]

In simpler members of Florideophyceae, gonimoblast filaments arise directly from the fertilized carpogonium, or from a cell very closely associated with it, into which a diploid nucleus has been transferred. In more specialized members, a diploid nucleus is generally transferred by a **connecting filament** to one or more specialized cells, called **auxiliary cells**, which are often remote from the carpogonium. Gonimoblast filaments then emanate from the auxiliary cell. Where a derivative of the primary diploid nucleus is transferred directly or indirectly to more than one auxiliary cell, several carposporophytes (one from each "diploidized" auxiliary cell) can arise as a result of a single sexual fusion.

Carpospores generally give rise directly to a free-living, sporophyte generation—the **tetrasporophyte**. The tetrasporophyte is generally morphologically similar, but can be distinctly dissimilar, to the free-living gametophytes in the life history, and it is in the free-living, diploid phase that meiosis occurs. The meiosporangium of Florideophyceae is referred to as a **tetrasporangium**, and each produces four **tetraspores**. Tetraspores then germinate into free-living gametophytes.

---

[1]See footnote on p. 415.

Figure 11-1 *Lithothamnium* (Florideophyceae), an encrusting coralline alga, is heavily calcified and grows firmly "cemented" to rocky surfaces, ✕ 1.

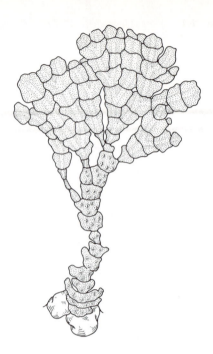

Figure 11-2 *Bossiella* (Florideophyceae), an articulated coralline alga, ✕ 1.

Cell Structure

Cell Wall Composition

The red algal cell wall is typically differentiated into an inner, cellulosic layer and an outer, pectic layer. Cellulosic fibrils are generally randomly scattered throughout a granular matrix. A xylan (containing both β-1,3 and 1,4 linkages) forms a significant portion of the matrix. A number of complex, colloidal, mucilaginous substances (**galactans**), such as **agar, carrageenan, xylan, mannan, porphyran, odonthalan, aeodan, funoran, ahnfeltan,** and **furcellaran**, can also occur in the cell wall and intercellular spaces of many red algae. These complex compounds are mixtures of sulfated polysaccharides (mainly galactan sulfate,

containing 1,3-linked galactose together with certain proportions of 3,6-anhydrogalactose residues and sulfate). Other mucilages or gelans, such as dulsin (a xylan), can also occur. In certain genera, such as coralline algae (Corallinaceae, Cryptonemiales), there is a heavy calcification of the outer part of the cell wall, which is composed chiefly of calcium carbonate but can also contain some magnesium carbonate. In some coralline genera, the whole thallus is encrusted with lime (**encrusting** forms, such as *Lithothamnium*, Fig. 11-1). In other genera, such as *Bossiella* (Fig. 11-2), certain regions of the plant remain uncalcified (**articulated** forms). In many calcified Nemaliales, the deposition is light, and the plants retain a good deal of flexibility in contrast to the rigid condition of the encrusting coralline algae and most of the plant body in the articulated forms.

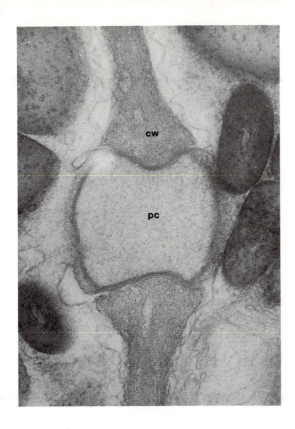

Figure 11-3 Section through plugged pit connection (*pc*) in cell wall (*cw*) between cells of conchocelis stage of *Porphyra* (Bangiophyceae), ×65,000. (Photograph courtesy of K. Cole, with permission of *Phycologia*.)

Nucleus, Chromosomes, and Cell Division

Cells in simpler forms are usually uninucleate. In more advanced forms, they are frequently multinucleate, except for very young or reproductive cells. In some instances, large cells of red algae can contain hundreds, even thousands, of nuclei. The nucleus in red algae has a well-defined nuclear envelope and a nucleolus. Mitosis is essentially similar to that in higher plants, with spindle fibers, distinct chromosomes, and **kinetochores**. However, the nuclear envelope can still be intact at metaphase, except for fenestrations (gaps) at the poles in some. In some in-

stances, a structure referred to as a **polar ring** occurs during division at the poles. However, no true centrioles have been confirmed. Not only the nuclei, but the chromosomes also are very small in red algae. Chromosome numbers from $n = 2$ to $n = 90$ have been reported. As chromosomes separate, the nucleus elongates and constricts in the middle, resulting in two daughter nuclei. Division of the nucleus is usually followed by the annular ingrowth of a **septum** from the lateral walls, resulting at first in a rimmed aperture through which the cytoplasm is continuous from cell to cell. This is followed by the development of a highly structured proteinaceous plug (Fig. 11-3), which usually completely blocks the aperture and apparently eliminates all vestiges of cytoplasmic continuity. This unique red algal structure, which is characteristic of all Florideophyceae and is present in certain stages of some Bangiophyceae, is referred to as a **primary pit connection**.

Chloroplasts and Other Cell Organelles

Chloroplasts can be central (axile) or peripheral (parietal) in position. In some simple forms, there can be only one chloroplast per cell. Generally, more complex forms have numerous, small, discoid chloroplasts in each cell, although irregular or bandlike ones also occur. Pyrenoids frequently occur within the chloroplast in simpler forms (Bangiales, Nemaliales) but are absent in more advanced forms (Ceramiales). The pyrenoid has a dense matrix and is generally intimately associated with penetrating thylakoids of the chloroplast (Fig. 11-4). The chloroplast is limited by a double membrane and has a homogeneous stroma traversed by a number of widely spaced thylakoids lying separately (never stacked) (Fig. 11-5). DNA has been demonstrated in the chloroplast matrix; small, dense granules and larger, lipidlike granules have been reported. Thylakoids usually occur as single, unassociated, unstacked, photosynthetic lamellae, roughly parallel to each other (Fig. 11-5). The unstacked thylakoid feature may be a primitive characteristic linking Cyanophyta and Rhodophyta. Just within the bounding membrane, a

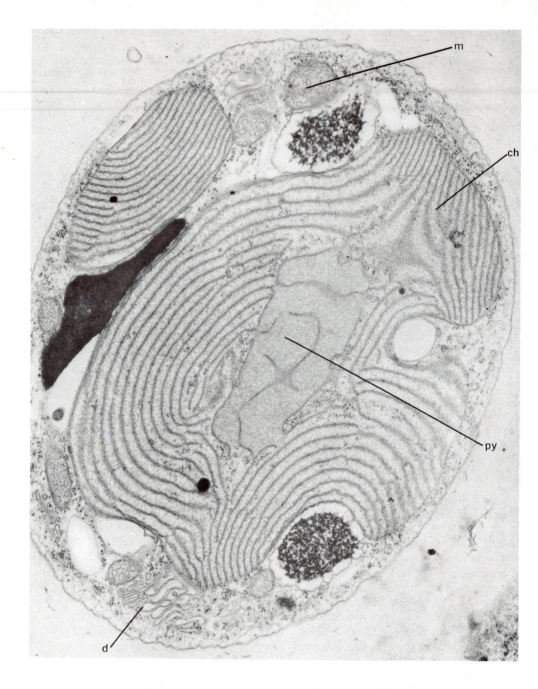

Figure 11-4 Ultrastructure of whole cell of *Porphyridium* (Bangiophyceae), ×25,000. Note close association of pyrenoid and penetrating thylakoids: *ch,* chloroplast; *d,* dictyosome; *m,* mitochondrion; *py,* pyrenoid. (Photograph courtesy of A. D. Greenwood.)

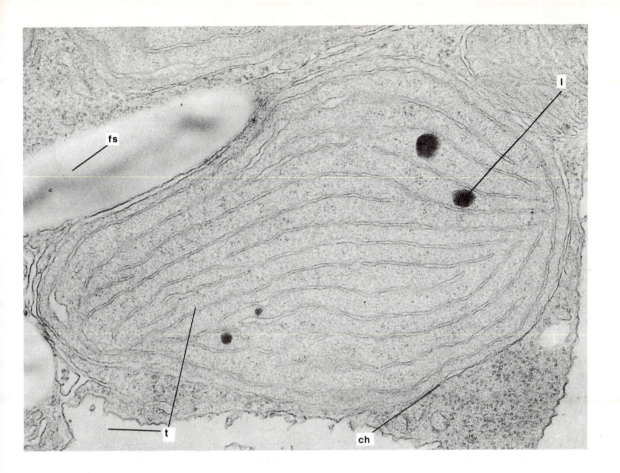

Figure 11-5 Ultrastructure of chloroplast of cortical cell in *Ceramium* (Florideophyceae), ×53,000. Note that thylakoids (*t*), consisting of single membrane pair, are not elaborately layered as is typical of other algal chloroplasts (see Figs. 7-2, 8-6A, 9-2A); also compare this chloroplast ultrastructure with photosynthetic lamellae in Cyanophyta (see Fig. 3-46B). Large translucent areas are floridean starch granules (*fs*) outside the chloroplast (*ch*): *l*, lipoid globule. (Photograph courtesy of A. D. Greenwood.)

single thylakoid sometimes passes all the way around the chloroplast (Fig. 11-6A), and others seem to arise from it by invagination. In other instances, this outer, encircling thylakoid is absent (Fig. 11-6B). Dictyosomes, endoplasmic reticulum, mitochondria, ribosomes, and starch granules occur in the cytoplasm, and one or more vacuoles can be present.

Pigments

Red algae owe their characteristic color to the presence of accessory, water-soluble, proteinaceous **phycobilin** pigments in the chloroplasts. These bilin pigments (phycoerythrobilin and phycocyanobilin) are present as prosthetic

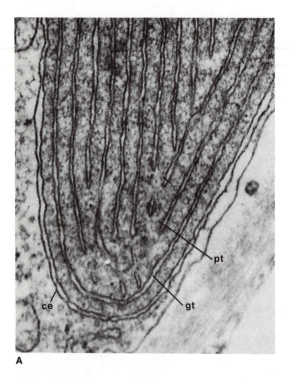

A

groups of proteins (biliproteins), to which they are firmly bound. The **phycoerythrins** that occur in red algae include *R*-phycoerythrin and *B*-phycoerythrin; the **phycocyanins** include *C*-phycocyanin, *R*-phycocyanin, and allophycocyanin. A nonproteinaceous red pigment, **floridorubin**, of unknown structure and function also occurs in the Rhodomelaceae (Florideophyceae, Ceramiales). Accessory carotenoid pigments are also present. These include two carotenes, α-carotene and β-carotene, and the xanthophylls lutein and zeaxanthin. β-carotene is the principal carotene, although in some instances α-carotene predominates. Lutein and zeaxanthin seem to be almost universal in red algae, with lutein usually more abundant. Green pigments include universal chlorophyll *a* and, in some instances, chlorophyll *d*. The green pigments are generally masked by accessory pigments. However, depending on environmental conditions, red algae exhibit a variety of shades from green, through red-brown, bright red, blue, bluegreen, purple-red, and even black. These shades depend on the relative amounts of pigments present. Plants living in subdued light or shaded locations tend to have a higher proportion of phycoerythrin.

Biliproteins are aggregated in small, regularly arranged granules (**phycobilisomes**) attached to the outer surface of the thylakoid membrane. Visible light energy is absorbed by both phycoerythrin and phycocyanin and transferred to chlorophyll *a* in the photosynthetic lamellae. Carotenes also occur in the chloroplast, where light energy absorbed by β-carotene is transferred to chlorophyll *a*. Xanthophylls also occur in the chloroplast, but they are apparently

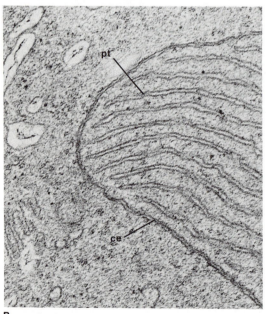

B

Figure 11-6 Thylakoid arrangement in *Porphyra* (Bangiophyceae). **A,** portion of chloroplast of vegetative cell of conchocelis stage, with characteristic girdle thylakoid (*gt*) inside chloroplast envelope (*ce*) and encircling inner parallel thylakoids (*pt*), $\times$80,000. **B,** portion of chloroplast of vegetative cell of foliose *Porphyra* thallus with parallel thylakoids (*pt*) terminating at chloroplast envelope (*ce*) (note absence of encircling thylakoid), $\times$80,000. (Photographs courtesy of K. Cole, with permission of *Phycologia*.)

not involved in energy transfer. Isolated phycobilisomes have been shown to contain 80% phycoerythrin, 10% phycocyanin, and 10% allophycocyanin, with the latter forming a core surrounded by a shell of phycoerythrin.

Polysaccharide and Other Storage Products

Food reserves are stored chiefly as a polysaccharide, generally referred to as **floridean starch** (α- 1,4- and 1,6-linked D-glucose). The floridean starch occurs as granules outside the chloroplast (Fig. 11-5), often associated with the outer surface of the nucleus or, when present, the pyrenoid. This starch is identical to the branched, or amylopectin, fraction of starches that occur in green algae and other plants; however, a slightly different staining reaction with iodine-potassium-iodide solution is obtained when testing for the presence of floridean starch (a red-violet color is obtained). Other polysaccharide reserves may be present, including reserve substances of low molecular weight such as the disaccharides trehalose and sucrose and glycosides of glycerol, such as floridoside, isofloridoside, and mannoglyceric acid. Certain sugar alcohols, such as mannitol, are abundant in some forms. Phytosterols stored include cholesterol, β-sitosterol, and fucosterol.

Among the saturated hydrocarbons that occur in red algae, n-heptadecane predominates. Of the polyunsaturated hydrocarbons found, red algae contain only C_{19} polyolefins. Observations suggest a possible connection between reproductive chemistry and the olefin content of benthic algae.

Mode of Nutrition

Most red algae are autotrophic. A number occur as epiphytes or endophytes; some are very pale, or completely white, and it is presumed that they are in part, if not wholly, dependent on the host for their nutrition. In the small, colorless, red algal parasite *Choreocolax*, double-membrane limited chloroplasts are present, but they lack any internal thylakoid development. In its outer cortical cells, floridean starch granules are absent, but they are abundant in the inner medullary cells of the parasite. On the other hand, adjacent cells of the autotrophic host (*Polysiphonia*) have little stored floridean starch. In *Harveyella* (Fig. 11-7), a diminutive, almost colorless, red algal parasite that occurs on *Odonthalia* and *Rhodomela*, a physiological dependence of the parasite on host tissues has been demonstrated.

Although certain biochemical characteristics of red algae may be more characteristic of one class than the other, the main distinctions that separate Bangiophyceae from Florideophyceae relate to general morphology, reproduction, and life histories. Hence, they are treated separately in the following discussion.

Class Bangiophyceae

Classification and Morphological Diversity

Bangiophyceae is a small class with about 125 species in 29 genera. The group is chiefly marine in distribution, although almost 25% of the species (in 14 genera) occur in fresh water. They are the simplest of all red algae, both vegetatively and reproductively, and they exhibit a wide range of morphology. Most occur on rocks or as epiphytes.

The simplest type of thallus is unicellular (Fig. 11-8A). A number of genera are filamentous. These can be erect, uniseriate or multiseriate, branched (Fig. 11-8D) or unbranched (Fig. 11-8C), or prostrate and **monostromatic** (unistratose) (Fig. 11-8H) or **polystromatic** (multistratose) (Fig. 11-8I). In others, there can be a few regular divisions in the erect, simple branches, indicative of parenchyma (Fig. 11-9F). The most advanced parenchymatous growth attained in the group is in the diffuse, **foliose** type; here cells divide in two planes to form a diffuse, monostromatic (Fig. 11-9A, C) plant body, or in three planes to form a distromatic (Fig. 11-9D, E) thallus. The postulated lines along which these various types may have evolved are diagrammatically illustrated in Figure 11-10. It is possible to derive the upright, parenchymatous thallus from a unicellular form through either the up-

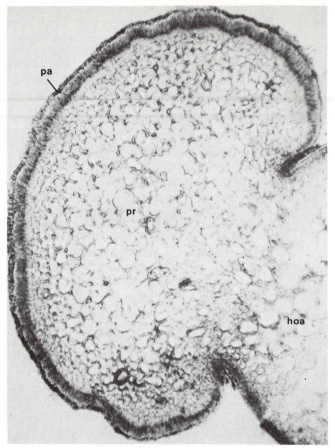

A

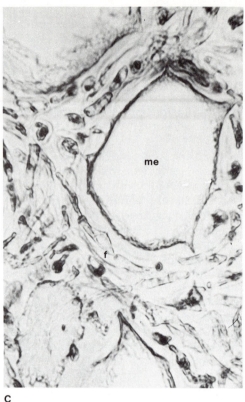

C

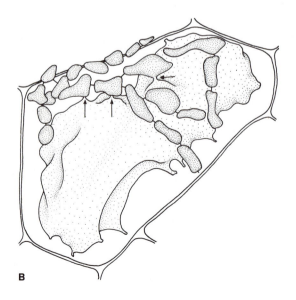

B

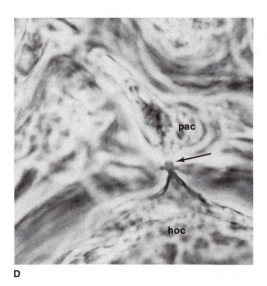

D

right, filamentous forms (Fig. 11-10A-E) or the prostrate discoid type (Fig. 11-10H, I). Apical growth is known only in the uniseriate, branched *Rhodochaete*, which is also unusual in apparently having open (unplugged) primary pit connections between its cells.

The erect thallus can be attached by a differentiated holdfast cell (Fig. 11-8D), by a few basal, **rhizoidal**, multicellular filaments (Fig. 11-8C), or by a number of rhizoidal filaments emanating from cells at the thallus base (Fig. 11-8E). These filaments anchor the thallus to the substrate. There can be a very extensive production of non-septate, rhizoidal processes (Fig. 11-9F) aggregated to form a massive cushionlike disc-shaped holdfast structure.

Although many genera occurring as epiphytes are not selective as to a substrate alga, in some instances they are highly specific, being confined almost entirely to a single algal species (e.g., *Porphyra nereocystis* on *Nereocystis*) or marine seagrass (e.g., *Smithora* on *Zostera* and *Phyllospadix*). *Smithora* has an obligate attachment, by means of its cushionlike holdfast, to the leaves of *Phyllospadix* and *Zostera*. Although it has a full pigment complement and photosynthetic apparatus, an exchange of cellular products in both directions (between host and epiphyte) has been demonstrated. Curiously, despite its obligate occurrence on a specific host in nature, *Smithora* has been grown successfully in culture and on plastic strips in the sea.

Figure 11-7 A red algal parasite, *Harveyella* (Cryptonemiales, Florideophyceae). **A,** transverse section through infected region of host (*Odonthalia*), showing massive proliferation (*pr*) of host tissue, parasite filaments to left of host axis (*hoa*), and outer reproductive region of parasite (*pa*), ×125. **B,** diagrammatic representation showing intracellular penetration of rhizoidal parasite filaments into host cell and secondary pit connections (*arrows*) between cell of host and parasite, ×350. **C,** transverse section through infected region of host, showing intercellular rhizoidal filaments (*f*) of parasite among medullary cells (*me*) of host, ×800. **D,** transverse section through infected region of host, showing secondary pit connection (*arrow*) between host cell (*hoc*) and parasite cell (*pac*), ×1800. (Photographs courtesy of L. J. Goff.)

The class Bangiophyceae is generally divided into three to five orders, depending on the author. Although representatives of only a few of these are considered in detail in this text, the four orders recognized are Porphyridiales, Erythropeltidales, Bangiales, and Rhodochaetales (Table 11-1). The disposition of the taxa in these orders is based primarily on chloroplast features and vegetative organization of the thallus: number and shape of chloroplasts; presence or absence of pyrenoid; and whether the form is unicellular or multicellular, has a branched or unbranched uniseriate or multiseriate filament, has a prostrate disc or cushion of cells, is **saccate**, or has an erect, parenchymatous (diffuse), monostromatic or distromatic, foliose thallus.

Life Histories and Reproduction

Vegetative and Asexual Reproduction Vegetative reproduction by **fragmentation** is common. Various types of spores are also formed. A single spore, produced by simple metamorphosis of a vegetative cell, is called a **monospore**, and the cell a **monosporangium**. In other instances, metamorphosis and division of the protoplast of a vegetative cell result in several spores, called **aplanospores**, and the cell is an **aplanosporangium**. The occurrence of sexual reproduction has been reported in certain genera; however, precise cytological evidence in support of nuclear fusion and the details of meiosis have not yet been fully demonstrated. In some instances, it would appear that some of the cells previously described as gametes in the Bangiophyceae may behave more like spores. Perhaps sexual reproduction has been lost in some members of the group, resulting in **parthenogenetic** development of gametes. However, despite the somewhat incomplete nuclear and chromosomal evidence for the sexual cycle in the Bangiophyceae, there is a morphological alternation or sequence of stages in the life history of certain genera (*Porphyra* and *Bangia*), which have very different morphological phases and several types of spores.

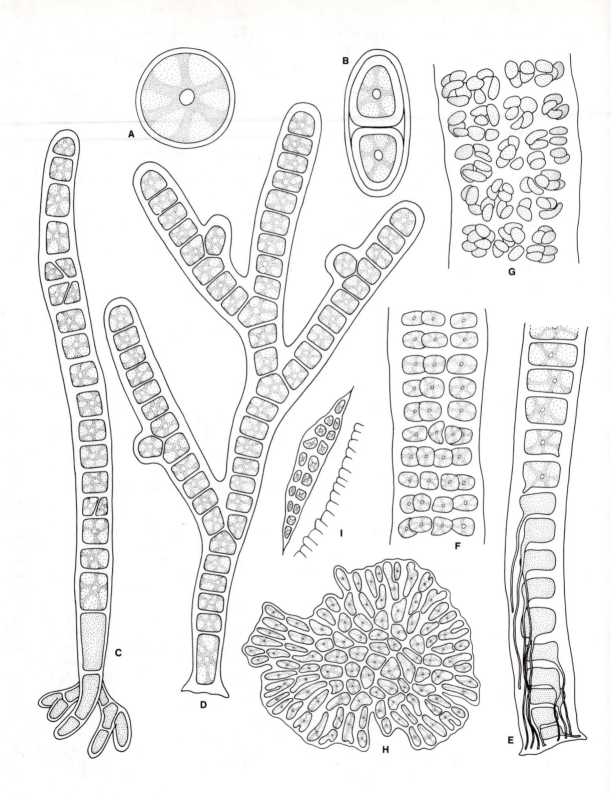

Types of Life History in *Porphyra* In *Porphyra* three different types of life histories (Figs. 11-11–11-13) have been established; these seem to vary with the species and, to some extent, with environmental conditions. In the simplest type (Fig. 11-11), the foliose stage produces monospores (Figs. 11-11B, C and 11-14D). These germinate and develop directly into a similar foliose thallus. In another simple type (Fig. 11-12), the foliose thallus produces aplanospores (8 or 16 per cell) and each aplanospore (Figs. 11-12E and 11-15C) germinates and develops directly into a similar foliose thallus. A more complex type of life history occurs in still other instances, in which a filamentous, shell-inhabiting, conchocelis stage alternates in the life history (Fig. 11-13). This filamentous stage (Figs. 11-13F and 14A–E) was known for a long time as a separate genus, *Conchocelis*, until its affinities were recognized. The same characteristics of reproduction involving a conchocelis stage also occur in *Bangia*.

Sexual Reproduction and the Conchocelis Stage The type of life history in which a conchocelis stage (Figs. 11-13F and 11-14A–E) occurs has generated much controversy about the sexual nature of certain cells. The foliose *Porphyra* thallus (Fig. 11-13A) produces two types (sizes) of reproductive cells that may be borne on separate plants or on the same plant. The larger of these (4 to 64 per cell), are highly pigmented and were referred to classically as carpospores (Figs. 11-13D and 11-15B). They were thought to arise following a zygotic meiosis (in the absence of a confirmed sexual cycle, they were also referred to as α-spores). The smaller ones (16 to 128 per cell), which are almost colorless, were referred to classically as spermatia (Figs. 11-13B and 11-15A). These were thought to be effective in fertilization (in the absence of a confirmed sexual cycle, they were also referred to as β-spores). However, it is now known that carpospores develop following the fertilization of a carpogonial cell, which is a slightly modified vegetative cell, usually with a very indistinct trichogyne. Carpospores germinate to produce the filamentous uniseriate, branched, conchocelis stage (Figs. 11-13F and 11-14A–E). Spermatia frequently appear on the mature thallus as marginal, whitish patches (Fig. 11-13A), and mature carpospores occur as highly pigmented patches. Foliose plants can be bisexual, with both types of patches occurring intermingled or in clearly separated areas; they can also be unisexual, with male and female reproductive structures borne on separate individuals.

At one time it was thought that the absence of primary pit connections was a diagnostic feature of Bangiophyceae. However, although the foliose *Porphyra* thallus lacks pit connections, primary pit connections are present between cells in the conchocelis stage of this alga. Similarly, the filamentous conchocelis stage has apical growth, a feature that is consistent in Florideophyceae but has in the past been regarded as rare in Bangiophyceae.

Conchocelis filaments (Figs. 11-13F and 11-14A–E) become branched and can produce two types of monospores. One of these, the typical monospore (Fig. 11-14D), which arises in a swollen cell (monosporangium) on a conchocelis filament, germinates and develops into a similar conchocelis filament (Fig. 11-13G). The other is produced on a specialized branch (conchospore branch), each cell of which gives rise to a special type of monospore, known as a **conchospore** (Figs. 11-13H–J and 11-14E). The conchospore germinates (Fig. 11-13J) and produces a juvenile thallus (Fig. 11-13K), which develops into a new foliose plant (Fig. 11-13A).

Traditionally, it was assumed that a sexual cycle (with synagamy and meiosis) occurred in *Porphyra*, and this has now been confirmed

Figure 11-8 Morphological diversity in Bangiophyceae. **A, B,** unicellular forms. **A,** *Porphyridium*, ×2500; **B,** *Asterocystis*, after division, before separating into two new cells, ×1200. **C, D,** unbranched and branched filamentous forms. **C,** *Erythrotrichia*, with multicellular rhizoidal holdfast, ×825; **D,** *Goniotrichum*, with undifferentiated holdfast cell, ×800. **E–I,** parenchymatous types. **E–G,** *Bangia*. **E,** basal region of thallus with rhizoidal cells, ×500; **F,** distal vegetative portion of parenchymatous region of thallus, ×500; **G,** fertile portion of thallus, ×500. **H, I,** *Erythrocladia*. **H,** unistratose species, viewed from above ×1000; **I,** multistratose species, in transverse section, ×1000).

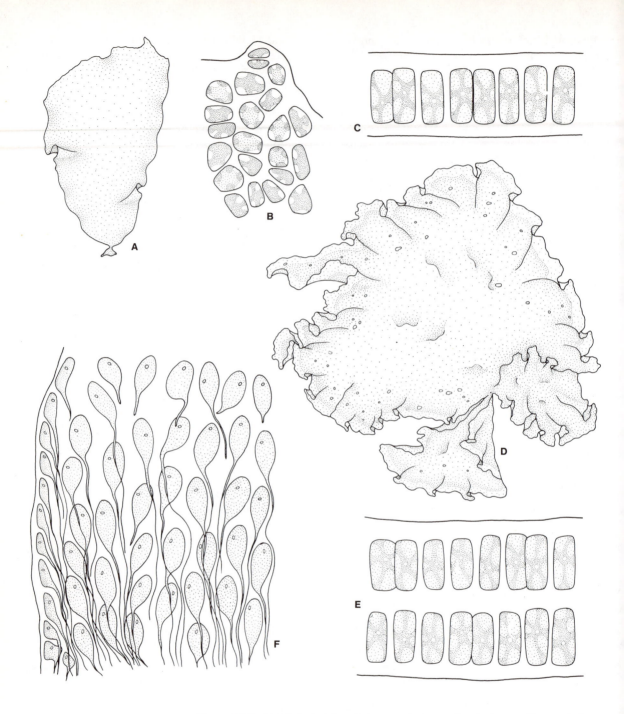

Figure 11-9 Morphological diversity in parenchymatous Bangiophyceae (*Porphyra*). **A–C,** monostromatic species. **A,** habit, ×1.5; **B,** surface view of cells, ×500; **C,** transverse section through portion of thallus, ×500. **D, E,** distromatic species. **D,** habit ×0.5; **E,** transverse section through portion of thallus, ×500. **F,** portion of basal attachment region of thallus as seen in longitudinal section showing rhizoidal cells, ×500.

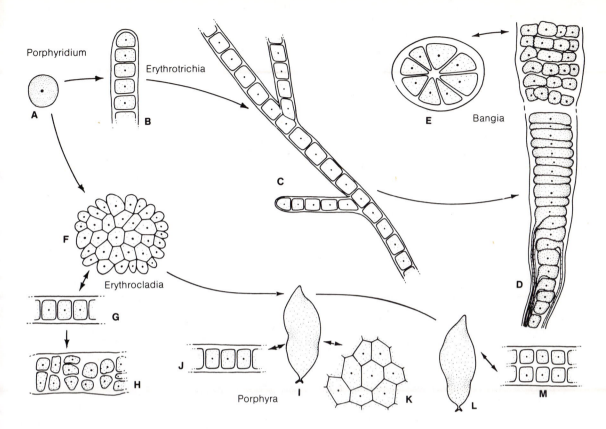

Figure 11-10 Possible evolution of growth types in Bangiophyceae, from unicellular (**A**) to parenchymatous through two possible lines: (1) filamentous (**B–E**) or (2) prostrate disc (**F–M**) types.

cytologically. However, it was also assumed that the carpospores were haploid and resulted from meiosis in the zygote nucleus immediately after fertilization; this would mean that the conchocelis stage is haploid. However, there are reports that the conchocelis stage is diploid. This would suggest that the carpospores result from mitotic divisions of a zygote nucleus; therefore, the conchocelis stage would be diploid, and meiosis must precede or occur during conchospore formation. In all instances where a sexual cycle has been indicated, chromosome counts indicate that the foliose thallus is the haploid stage, and the conchocelis stage is diploid.

Class Florideophyceae

Classification and Morphological Diversity

The Florideophyceae is a large class with nearly 5100 species in about 790 genera. The group is predominantly marine, with less than 3% of the species (about 145 species in 22 genera) occurring in freshwater habitats. They grow on rocks or other usually solid substrates. Florideophyceae are also vegetatively simple, although somewhat more complex than Bangiophyceae.

Florideophyceae consistently have primary pit connections (Figs. 11-3 and 11-16A–C) between adjacent cells in contrast to the Bangiophyceae, in which primary pit connections are uncommon. In more complex thalli, **secondary pit connections** (Fig. 11-16D–H) can occur between adjacent cells in the same filament and between cells in adjacent filaments; in parasitic genera (e.g., *Harveyella*, Fig. 11-7B–D), they can occur between cells of parasite and host. The formation of secondary pit connections is more complex (Fig. 11-16D–H). When a nucleus in a cell (uninucleate or multinucleate) divides, one daughter nucleus can migrate to the periphery of the cell. A protuberance in the cell wall is formed, the daughter nucleus migrates into it, and, in some instances, it is separated as a very small cell. As this protuberance or small cell elongates, it comes in contact with an adjacent cell, and the daughter nucleus passes into it. The resulting connection, the secondary pit connection, is a permanent one. As in the formation of a primary pit connection, there is initially protoplasmic continuity between adjacent cells, but this aperture soon disappears. The secondary type of pit connection is a unique feature in Florideophyceae.

The function of pit connections has not been clearly established. At one time it was thought that there was cytoplasmic continuity through the mature pit connection and that its function was to facilitate the transfer of materials from cell to cell. This is not substantiated by ultrastructural research, which reveals that a plug actually fills the aperture soon after it is formed (Fig. 11-3). However, at certain stages (especially in early **postfertilization** development), these plugs apparently break down and permit subsequent transfer of materials, including nuclei, from one cell to the next. Although primary and secondary pit connections differ in their methods of formation, they are structurally similar when mature. In parasitic red algae, such as *Harveyella* (Fig. 11-7), secondary pit connections between host and parasite are apparently initiated by the host.

There are no unicellular members in the Florideophyceae. The simplest are uniseriate, branched forms (Fig. 11-17). The more complex forms are basically filamentous, but branches are

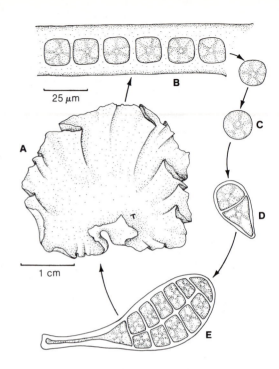

Figure 11-11 Direct type of life history in a monospore-producing, monostromatic species of *Porphyra* (Bangiophyceae). The foliose thallus (**A**) produces monospores (**B, C**) that germinate and develop directly (**D**) into a juvenile thallus (**E**), which produces a new foliose plant (**A**).

laterally fused to varying degrees, so that a **pseudoparenchymatous** thallus results (Figs. 11-18–11-20). There is little tissue differentiation, although aggregation of filaments often results in a compact, small-celled **cortex** and a **medulla** that can be loose and filamentous (Fig. 11-18B, D), or compacted large cells (Fig. 11-20B). The growth of each filamentous branch is by means of a simple, apical cell. In rare instances (Corallinaceae, order Cryptonemiales; Nitophylloideae, order Ceramiales) intercalary cell divisions occur. In *Nitophyllum* (Ceramiales), this results in a type of marginal growth.

Pseudoparenchymatous and loosely branched forms can have either **uniaxial** or **multiaxial** types of growth. In the former, the main axis consists of a single row of large cells (Fig. 11-17E). In multiaxial types, the main axis is

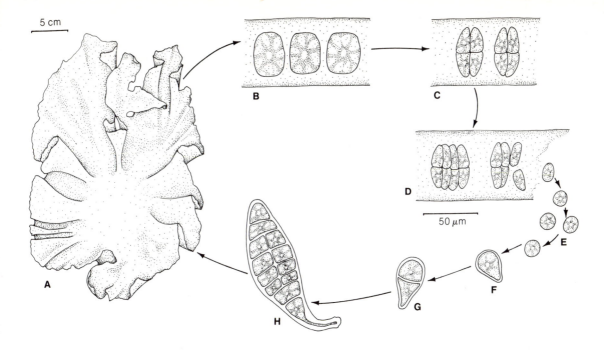

Figure 11-12 Direct type of life history in an aplanospore-producing, monostromatic species of *Porphyra* (Bangiophyceae). The foliose thallus (**A**) produces aplanospores (**B–E**), which germinate and develop directly (**F–G**) into a juvenile thallus (**H**), which produces a new foliose plant.

composed of a number of parallel or almost parallel filaments (Figs. 11-18B and 11-19B, C). In each instance there are principal, main axes of **unlimited growth** accompanied by secondary axes of **limited growth**, with the latter often surrounding the former. In some instances, as in the foliose *Weeksia* (Cryptonemiales), the juvenile thallus has a uniaxial type of growth at first and becomes multiaxial as it matures. The lines along which the uniaxial and multiaxial types may have evolved are illustrated in Figures 11-21 and 11-22.

Considering the relatively simple vegetative growth that occurs in Florideophyceae, the great variety in gross morphology is remarkable. They vary from filamentous microscopic plants to large, profusely branched, or foliose plants. They are usually less than a meter in length, but some (e.g., *Gracilaria*) are as long as four meters.

The class Florideophyceae is generally divided into five to eight orders. The more conservative number recognized in recent years has been six. In this text seven orders (Table 11-1) are recognized: Acrochaetiales, Nemaliales, Cryptonemiales, Gigartinales, Rhodymeniales, Palmariales and Ceramiales. The disposition of genera in these seven orders is based primarily on the female reproductive system, especially the postfertilization stages and embryogeny. Two of these orders, the Rhodymeniales (one of the smallest orders) and the Ceramiales (the largest order), appear to be especially clear-cut and relatively uniform in their reproductive features. On the other hand, Cryptonemiales and Gigartinales are much less uniform in their reproductive features, and as more genera are studied in greater detail, the erection of further orders from these two orders will most likely be justified.

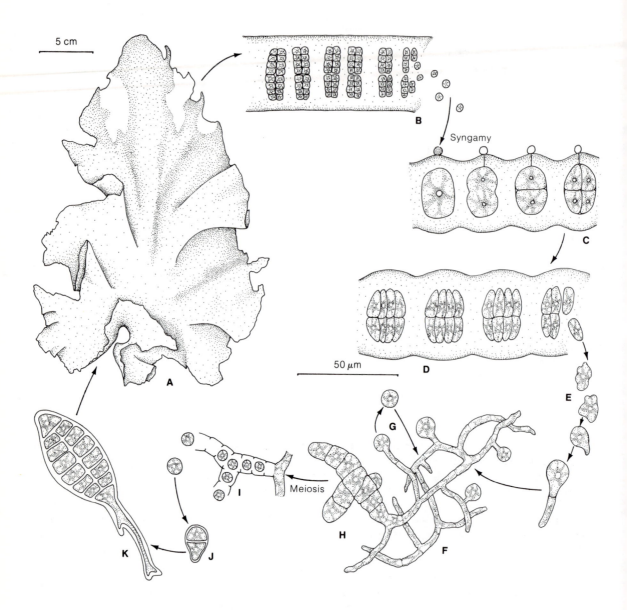

Figure 11-13 Conchocelis-producing type of life history of monostromatic species of *Porphyra* (Bangiophyceae). **A,** foliose, macroscopic thallus with fertile margin; **B–D,** transverse sections of foliose plant. **B,** showing spermatium formation and release; **C,** showing carpogonia, fertilization, and early stages in carpospore formation; **D,** showing carpospore formation and release. **E,** germination of carpospores. **F,** conchocelis stage. **G,** formation and release of monospore from monosporangium on conchocelis stage. **H,** formation of conchospore branch on conchocelis stage. **I,** release of conchospores. **J,** germination of conchospore into juvenile thallus (**K**), which develops into a new foliose plant.

Life Histories and Reproduction

Vegetative and Asexual Reproduction With some more complex thalli, vegetative reproduction is common from a perennial base. Erect branches die back to the base, and the new branches arise later from a basal, undifferentiated region. A variety of spores is produced, especially in some simpler forms. As in the Bangiophyceae, monospores can be formed, or highly differentiated sporangia can divide to produce two or more spores; in most instances, however, the cytological nature and fate of these spores are not well established.

Sexual Reproduction On a morphological basis, sexual reproduction appears to be well established in Florideophyceae; there is usually an alternation or sequence of free-living generations. However, on a cytological level, there is still a great lack of information concerning sexual reproduction and alternation of generations. When a free-living tetrasporophyte occurs, it is often vegetatively indistinguishable from the haploid gametophytes, although many examples are now known where the gametophyte and free-living tetrasporophyte phases are distinctly heteromorphic. Each spermatangium gives rise to a single spermatium. Similarly, each carposporangium gives rise to only one carpospore in Florideophyceae.

The female sexual apparatus and fertilization processes are extremely complex and variable in Florideophyceae. It is impossible in this text to summarize in this respect the characteristic variations that occur in the orders of this class. The cytological evidence for alternating phases in red algae is still far short of that desirable to justify any generalizations. Genotypic sex determination has been demonstrated in a few Florideophyceae (as in *Gracilaria* in the Gigartinales); that is, from a single tetrasporangium that produces four tetraspores by meiosis, two spores produce male gametophytes and two produce female ones. However, there are anomalous situations in which male, female, and tetraspore structures have been found on the same thallus. This situation has not yet been fully clarified at a cytological level. Hybridization has been demon-

strated in species of *Polysiphonia* (Ceramiales), and polyploidy has been reported in *Gracilaria* (Gigartinales) and *Spermothamnion* (Ceramiales). However, genetic studies of red algae in general are very few.

Life Histories The most commonly encountered and best-known type of life history in the Florideophyceae is that occurring in the more complex orders. It comes close to an alternation of isomorphic generations, except for the introduction of an additional diploid generation, the carposporophyte, or gonimoblast.[2] The free-living, gamete-producing generations (male and female) "alternate" with a free-living, tetraspore-producing generation, and the intermediate, diminutive, diploid carposporophyte develops and remains *in situ* on the female gametophyte. In some instances, species of the same genus can have unisexual or bisexual gametophytes. The additional diploid generation (carposporophyte) evidently occurs as a result of delay in the time and place of meiosis. In a few species, one or another of these phases (gametophytes or tetrasporophyte) appears to have been lost (or perhaps is not yet known and may even be so morphologically different as not to have been associated in the life history).

The characteristics of the female reproductive system and the events following fertilization vary markedly in specialized Florideophyceae. *Polysiphonia* (Ceramiales) is the only example of this type of life history that is discussed in detail.

The Polysiphonia-type Life History. In *Polysiphonia* (Fig. 11-23), there is an alternation of free-living, vegetatively similar, haploid male (Fig. 11-23B) and female (Fig. 11-23A) gametophytes with a morphologically identical diploid **tetrasporophyte** (Fig. 11-23S). A diminutive diploid carposporophyte, which remains attached to the female gametophyte (enclosed in a **pericarp**, Fig.

---

[2]Recent ultrastructural studies comparing cytological features of vegetative cells with those of gonimoblast filaments and with cells of parasitic red algae suggest that diploid cells of postfertilization stages making up the carposporophyte have cytological features that might make it more appropriate to interpret the carposporophyte as a multicellular "zygote," or to abandon the term *carposporophyte* and revert to the classical term *gonimoblast* to refer to this diploid phase.

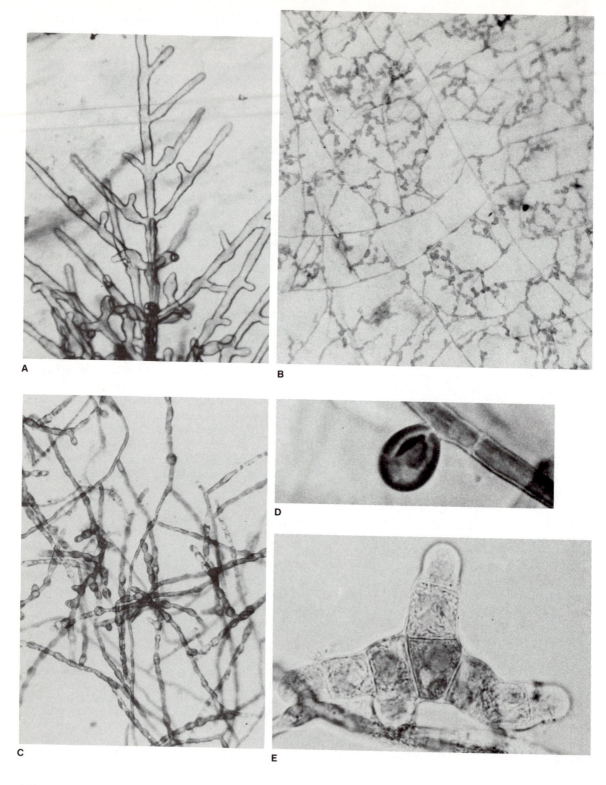

A

B

C

D

E

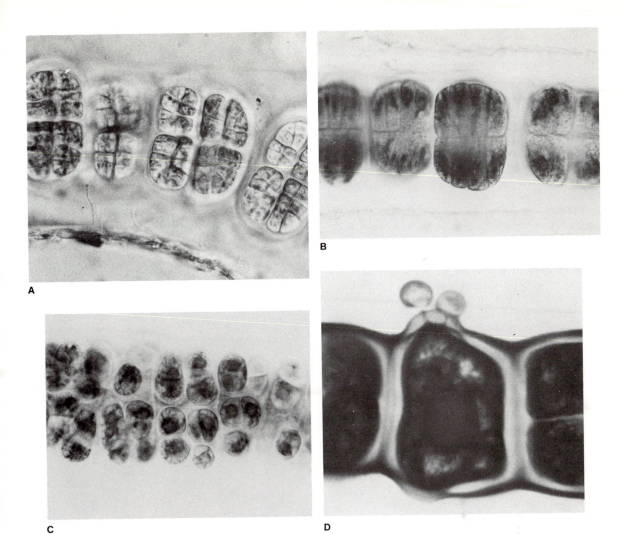

Figure 11-15 Transverse sections of foliose thallus of *Porphyra*, showing reproductive stages. **A, B,** P. *kanakaensis.* **A,** formation of spermatia, ×400; **B,** formation of carpospores, ×450. **C,** aplanospore formation in *P. sanjuanensis,* ×600. **D,** fertilization in *P. gardneri,* showing two spermatia and penetration through wall of trichogyne, ×1500 (**A–C,** courtesy of T. F. Mumford; **D,** courtesy of M. W. Hawkes.)

Figure 11-14 Vegetative and reproductive stages of conchocelis development in Bangiophyceae. **A–C,** vegetative stages in *Porphyra.* **A, B,** conchocelis filaments of *P. nereocystis* in shell; **A,** ×800: **B,** ×175. **C,** conchocelis filaments of *P. nereocystis* in culture, ×400. **D,** monosporangium on conchocelis filament of *Bangia,* ×2000. **E,** conchospore branch on conchocelis filament of *Porphyra,* ×1200. (**A, E,** courtesy of T. F. Mumford; **B, C,** courtesy of M. W. Hawkes; **D,** courtesy of K. Cole.)

11-23L) and produces diploid carpospores, precedes initiation of the free-living tetrasporophyte. Tetraspores (meiospores) produced on the free-living tetrasporophyte (Figs. 11-23T,U) germinate and develop into the free-living gametophytes (Figs. 11-23A,B). Genotypic sex determination occurs, and two of the tetraspores from each tetrasporangium give rise to male gametophytes; the other two give rise to female gametophytes. The cytological alternation, as outlined above, accompanies the morphological sequence of generations.

The mature plant of *Polysiphonia* is a branched thallus with cells in regular tiers (Fig. 11-24B). Each mature tier or segment consists of a **central cell** surrounded by a number of **pericentral cells** of equal length. This is often referred to as a **polysiphonous** type of construction. The number of pericentral cells in the vegetative axes is usually constant in any one species, although it varies from species to species; four is a common number (but 5, 6, or up to 25 can occur). Even where the number of pericentral cells in the vegetative axis is four or more, the reproductive axes usually revert to the basic five.

Growth in *Polysiphonia* is strictly apical, and the apical cell cuts off cells from its posterior face (Fig. 11-25). If the apical cell segment is to produce only a central cell and pericentral cells, daughter nuclei in the dividing apical cell line up along the longitudinal axis (Fig. 11-25B–D). The cell wall that follows this nuclear division develops at right angles to the longitudinal axis. The primary cell undergoes longitudinal division in a regularly alternating sequence within a few segments of the apical cell—that is, from left to right of the first cell cut off (Fig. 11-25J–0). This results in the typical arrangement of a central cell surrounded by four or more pericentral cells.

If the apical cell segment is to produce a branch, daughter nuclei in the dividing apical cell line up diagonally, and the cell wall following is oblique to the longitudinal axis (Fig. 11-25E–H). From the high side of the resulting, wedge-shaped cell, a protuberance arises that initiates the branch (Fig. 11-25H, I). The primary cell thus first forms a branch initial. Subsequently the primary cell divides in typical fashion, producing a central cell surrounded by four or more pericentral cells. Branch initials can be **indeterminate** and repeat the development of

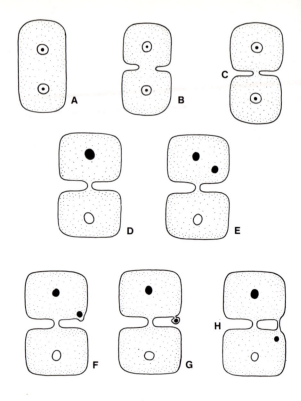

Figure 11-16 Diagrammatic representation of formation of pit connections in Rhodophyta, **A–E,** formation of primary pit connection; **F–H,** formation of secondary pit connection.

Figure 11-17 Morphology and reproduction in uniaxial Florideophyceae. **A, B,** *Antithamnion.* **A,** habit epiphytic on kelp stipe, ×0.5; **B,** terminal portion of thallus, showing uniaxial, loose, filamentous construction, and immature tetrasporangia (*tspn*) (lower one immature), ×240. **C–K,** *Batrachospermum.* **C,** habit, ×1; **D,** apical region of thallus, showing apical cell (*ac*) and filamentous branches, ×475; **E,** portion of main axis, showing uniaxial construction and whorls of filamentous branches, ×165; **F,** more mature portion, showing development of corticating filaments that cover the axial row of cells, ×165; **G,** unfertilized carpogonium (*cp*), ×915; **H,** postfertilization stage, showing initiation of carposporophyte (gonimoblast) filaments (*gb*), ×915; **I,** branch bearing several mature spermatangia (*sp*), ×960; **J,** later stage of carposporophyte development, showing chains of carposporangia (*cspn*), ×790; **K,** portion of thallus, showing mature and empty monosporangia (*mspn*), ×790.

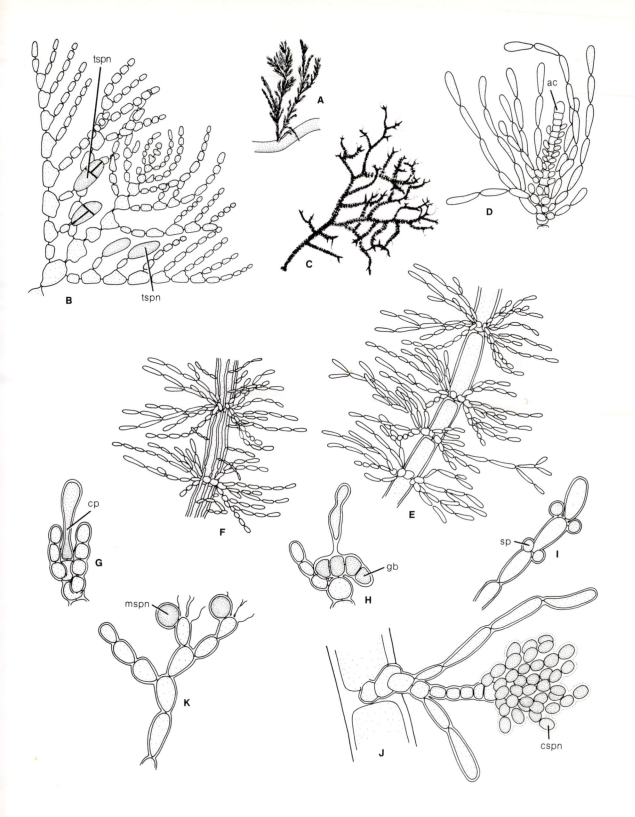

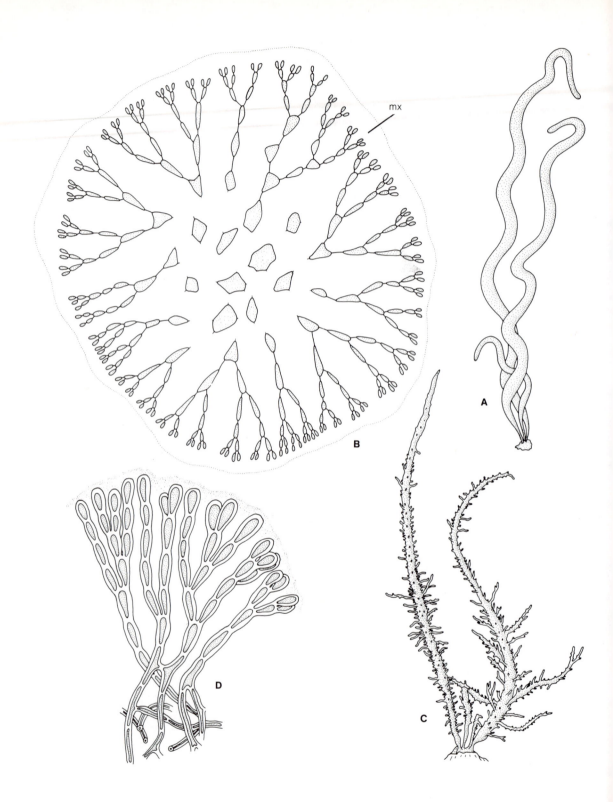

the main axis; they can also be **determinate** and of limited development, remaining partly or wholly **monosiphonous**. These monosiphonous determinate branches are referred to as **trichoblasts** and can be further branched. Trichoblasts can be sterile or fertile. Sexual structures are formed on this determinate type of branch, resulting in distinctive female and male trichoblasts.

On the female trichoblast (Fig. 11-23C), usually only the basal three segments next to the supporting axis become polysiphonous (Fig. 11-23D); the distal portion, which can be simple or branched, remains monosiphonous and is generally rather **evanescent**. It is the first segment above the basal one that becomes fertile, and the last-formed pericentral cell in the fertile segment that produces the carpogonial branch. Before considering the ontogeny of the female reproductive system, there is vegetative development of some adjacent cells that must be mentioned (Figs. 11-23E–H and 11-24E). The last-formed pericentral cells of both the basal segment and the segment above the fertile one cut off a series of vegetative cells. Similarly, the pericentral cells on each side of the fertile one also cut off a series of vegetative cells. The filamentous series of cells resulting from these four vegetative pericentral cells fuse laterally to form an urn-shaped pericarp (Figs. 11-23L and 11-24E, F), which encloses the developing female reproductive system. These filaments do not fuse completely; at the apex, a small pore, or **ostiole** (Fig. 11-24F), remains. The trichogyne extends through it into the water. Development of the sterile pericarp starts before fertilization, and is accelerated after fertilization, keeping pace with the developing carposporophyte within it. Intimately associated cells of the gonimoblast ($2n$) and surrounding pericarp (n) are frequently referred to collectively as the **cystocarp**.

Figure 11-18 Morphology of multiaxial Florideophyceae. **A, B,** *Nemalion*. **A,** habit of several plants, ×0.7; **B,** transverse section through thallus, showing simple multiaxial construction and organization of branched filaments embedded in mucilaginous matrix *mx*, ×415. **C, D,** *Cumagloia*. **C,** habit of a cluster of plants, ×0.5; **D,** transverse section through portion of thallus, showing more complex multiaxial construction and filamentous organization, ×350.

Prior to fertilization, the fertile pericentral cell enclosed within the pericarp undergoes three primary divisions (Fig. 11-23E–G). The first cuts off a cell to one side and results in a lateral sterile cell (Fig. 11-25E); the second cuts off toward the top the **carpogonial branch** initial (Fig. 11-23F); and the third cuts off a basal sterile cell toward the base (Fig. 11-23G). The remaining portion of the fertile pericentral cell, which is attached to the central cell of the fertile segment, is referred to as the **supporting cell** (Fig. 11-23G). Thus, there are three initials attached to the supporting cell at this stage. The lateral and basal sterile cells each divide a few times to form short filaments of sterile cells. These may later serve as a source of nutriment for the developing carposporophyte. The carpogonial branch initial divides to produce a four-celled carpogonial branch (Figs. 11-23H and 11-24E), the terminal cell of which develops into the carpogonium. It has a long, narrow trichogyne that emerges through the pericarp ostiole and protrudes into the water, forming a receptive structure for the spermatium (Fig. 11-23H).

The male reproductive system matures at about the same time as the female. In contrast to the female trichoblast, the male trichoblast (Fig. 11-23N) forms a large number of spermatia. The basal segment of the fertile trichoblast generally remains monosiphonous, as does the often-branched distal portion (Figs. 11-23Q and 11-24C). However, between these two monosiphonous regions, several segments divide to form a central cell and five pericentral cells. All five pericentral cells in each of these fertile, polysiphonous segments become fertile.

Each fertile pericentral cell undergoes a number of transverse divisions to produce three or more spermatangial parent cells (Fig. 11-24D). In turn, each of these divides obliquely, cutting off two or more spermatangia (Fig. 11-24D), which (at maturity) liberate a single, colorless, naked spermatium (Fig. 11-23R). Due to the number of divisions resulting in successively smaller cells, the mature male trichoblast appears as a stalked, club-shaped structure with a large number of small, almost colorless, refractive cells (spermatangia) on its surface (Fig. 11-24C). Spermatia are carried passively by water currents, coming to rest on the trichogyne.

When a spermatium comes in contact with

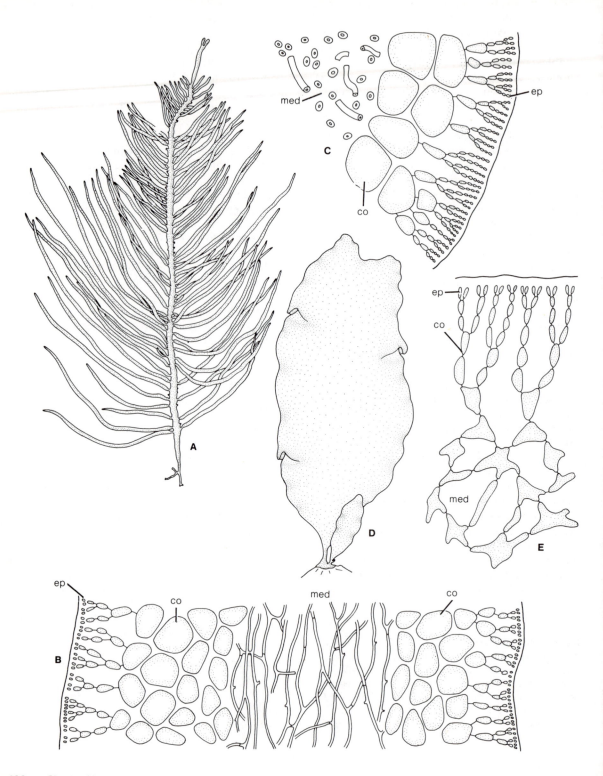

the trichogyne, fusion occurs, and the male nucleus passes down the trichogyne to the base of the carpogonium, where it fuses with the female nucleus. As soon as fertilization occurs, the trichogyne degenerates, and a cell is cut off from the top of the supporting cell. This is the auxiliary cell (Fig. 11-23I). A protuberance very much like that produced during the formation of a secondary pit connection arises from near the base of the fertilized carpogonium. The diploid zygote nucleus migrates into it, eventually being transferred to the auxiliary cell. The haploid nucleus within the auxiliary cell apparently degenerates, and the diploid (zygote) nucleus now present divides mitotically. The first true diploid cell of the carposporophyte generation is formed when a diploid daughter nucleus is cut off in the gonimoblast initial (Fig. 11-23K); this initial cell then divides repeatedly by mitosis to produce a multicellular gonimoblast, or carposporophyte.

As the gonimoblast develops, pit connections between cells of the adjacent female gametophyte (auxiliary cell, carpogonial branch, sterile cells, supporting cell) tend to break down and open up, resulting in extensive cytoplasmic continuity between cells in this area. This probably facilitates the transfer of food reserves from the female gametophyte to the developing gonimoblast. The terminal cells of the filamentous gonimoblast become much enlarged and develop into carposporangia (Fig. 11-23M). At maturity, each carposporangium liberates a single, elongate, densely pigmented, naked carpospore, which escapes through the ostiole. The carpospore is then carried passively in the water and, on contacting a suitable substrate, germinates and develops into a free-living tetrasporophyte generation (Fig. 11-23S) that is morphologically identical to the gametophyte generations (Fig. 11-25A, B).

Figure 11-19 Morphology of multiaxial Florideophyceae. **A–C,** *Neoagardhiella*. **A,** habit, ×0.5; **B,** longitudinal section through portion of thallus, showing filamentous medulla (*med*) and pseudoparenchymatous cortex (*co*), ×60; **C,** transverse section through outer portion of thallus, showing filamentous medulla and pseudoparenchymatous cortex, ×70; **D, E,** *Iridaea*. **D,** habit of plant, ×0.5; **E,** transverse section of outer part of thallus, showing loose, filamentous, inner medulla region (*med*), and more compact, outer, small-celled cortex (*co*), ×700. *ep*, epidermis.

Reproduction in the free-living tetrasporophyte phase is less complicated than on the gametophyte phase. Tetrasporangia (meiosporangia) are produced in ordinary, vegetative, polysiphonous axes of the plant (Figs. 11-23T and 11-26H). However, when the plant becomes fertile, a large number of consecutive segments toward the apices of the branches become fertile. Each segment generally produces one tetrasporangium, which arises indirectly from one of the pericentral cells. The fertile pericentral cell undergoes a number of divisions (Fig. 11-24G, H), somewhat comparable to the primary divisions that occur in the development of the female reproductive system. Three cells, two usually larger than the third, are cut off from the outer face of the fertile pericentral cell. Each of these is a sterile **cover cell** (Fig. 11-24G, H). Then, from the upper portion of the fertile pericentral cell, a single cell, the tetrasporangium, is cut off and protected by the external cover cells. The small cell to which the tetrasporangium and the three cover cells remain attached is the **stalk cell** (Fig. 11-24G); it is comparable to the supporting cell of the carpogonial branch. The tetrasporangium enlarges considerably, and its diploid nucleus divides meiotically to form four haploid nuclei. The cytoplasm then divides into four uninucleate portions. At maturity, the wall of the tetrasporangium breaks down, the intercellular material between the cover cells breaks down, and naked tetraspores (meiospores) are liberated from between the cover cells. These meiospores (Fig. 11-25U) are carried passively in the water, and they germinate to develop into free-living gametophyte generations.

Antithamnion (Ceramiales), which also has a *Polysiphonia*-type life history, is simpler in its vegetative organization (Fig. 11-17A, B). Male branches are simpler in organization, although freely exposed as in *Polysiphonia*. However, the female reproductive system, mature carposporangia, and tetrasporangia are also all freely exposed. A number of genera having a *Polysiphonia*-type life history produce a variety of additional types of spores, including **monospores, bispores, paraspores,** and **polyspores.** Bisporangia (producing two spores) and polysporangia (producing several spores) are thought to be meiosporangia, homologous to tetrasporangia. On the other hand, monosporangia and parasporangia (producing several spores) are thought to pro-

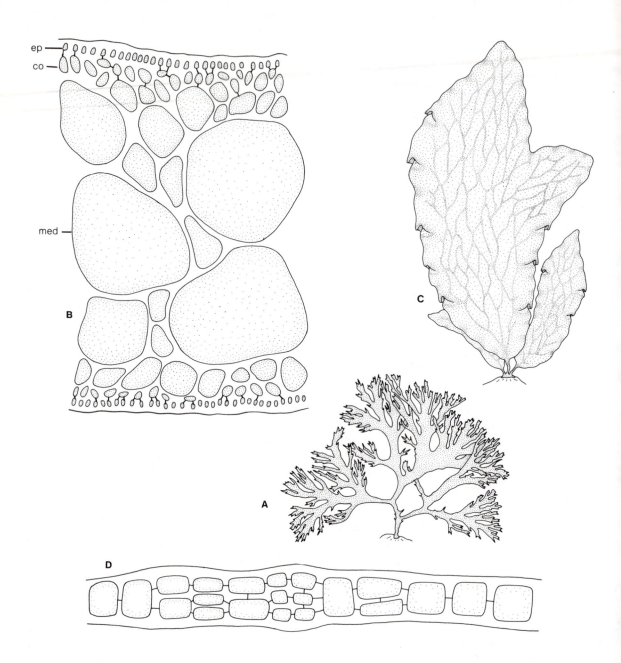

Figure 11-20 Morphological diversity of complex multiaxial Florideophyceae. **A, B,** *Callophyllis.* **A,** habit, ×0.5; **B,** transverse section through portion of thallus, showing pseudoparenchymatous organization (obscuring multiaxial nature of thallus) with compact, large-celled medulla (*med*), small-celled cortex (*co*), and epidermis (*ep*), ×150. **C, D,** *Polyneura,* a membranous form. **C,** habit of veined thallus, ×0.5; **D,** transverse section through portion of thallus in vicinity of vein: *ep,* epidermis.

duce mitospores and apparently function in the same way as the monospores of Bangiophyceae in producing the same type of plant on which they were formed.

Many Florideophyceae having the *Polysiphonia*-type life history have some of their reproductive structures, especially the female system, deeply immersed in cortical tissues of a pseudoparenchymatous thallus (Fig. 11-26). In these pseudoparenchymatous forms, the spermatangia are usually modified (usually smaller and colorless) epidermal cells, although in some instances they are produced in conceptaclelike depressions, as in *Gracilaria* (Fig. 11-26B). The tetrasporangia are also usually modified (usually larger and more densely pigmented) epidermal cells, sometimes scattered (Fig. 11-26J) or sometimes densely aggregated. However, in some instances tetrasporangia are located deep within the cortical tissue, sometimes scattered and sometimes densely aggregated. In these pseudoparenchymatous types, although the trichogyne emerges through the surface cells and is in a receptive position to receive the spermatia, the carpogonium is usually deeply immersed in the cortical tissue of the female gametophyte, where fertilization and subsequent development of the carposporophyte occur. In such instances, the deeply embedded carposporophyte can be completely obscured or appear only as an erumpent swelling. The eventual release of carpospores can occur through a definite ostiole or by a breakdown of the outer tissues of the thallus.

The Nemalion-type Life History. It is in the simpler forms, especially in Nemaliales, that the greatest variation in life history occurs. The longstanding interpretation that *Nemalion* (Fig. 11-27) and other members of the Nemaliales have meiosis immediately following syngamy, resulting in haploid carpospores, is no longer valid.

Just as in Phaeophyta, where culture studies have linked simple filamentous and crustose *Myrionema*like and *Ralfsia*like stages as alternate phases in the life histories of a number of brown algae, recent culture studies of Florideophyceae have yielded similar results. In Florideophyceae there can be diminutive, filamentous or crustose tetrasporophytes produced in some genera. In Nemaliales, *Nemalion* (Fig. 11-27) and

Pseudogloiophloea have an **acrochaetioid** tetrasporophyte phase, and *Bonnemaisonia* has a *Trailliella*like tetrasporophyte stage (Fig. 11-30). As a result of recent culture studies, a number of red algae in other orders have been shown to have life histories differing from the *Polysiphonia* type (Ceramiales). In the Cryptonemiales, *Gloiosiphonia* has been shown to have a diminutive, *Cruoriopsis*like, crustose tetrasporophyte stage, and *Thuretellopsis* to have an *Erythrodermis*like crustose tetrasporophyte stage. In the Gigartinales, *Turnerella* and *Halarachnion* have a diminutive, *Cruoria*like, crustose tetrasporophyte stage; *Ahnfeltia* has a *Porphyrodiscus*like tetrasporophyte stage; and some species of *Gigartina* have a *Petrocelis*like, crustose tetrasporophyte stage. On the other hand, *Gracilaria*, also in the Gigartinales, has the *Polysiphonia*-type life history (Fig. 11-26).

Cytological confirmation of the life history details in Florideophyceae is still very limited. However, it is now very clear that many of these have a sequence of heteromorphic phases or generations, where there is a dominant or conspicuous, free-living gametophyte stage (or stages, in which the plant is bisexual) "alternating" with a diminutive, free-living, filamentous or crustose tetrasporophyte. And again, as in the *Polysiphonia*-type life history, a diminutive, intervening, diploid carposporophyte develops *in situ* on the gametophyte and produces diploid carpospores. *Nemalion* is the only example of this heteromorphic type of life history that will be discussed in detail.

In *Nemalion*, the gametophyte is the conspicuous generation. The plant is an erect, branched or unbranched multiaxial thallus (Figs. 11-18A and 11-27A, B) of densely compacted filaments, embedded in a mucilaginous matrix (Fig. 11-18B). The medullary region has an axial core of intertwined longitudinal filaments, and lateral branches from this axis form a progressively looser filamentous cortex toward the surface. Colorless spermatia are produced superficially in spermatangia borne on the tips of filamentous branches (Fig. 11-27C). Carpogonia, also freely exposed at the tips of filamentous branches, have a well-defined, diminutive trichogyne (Fig. 11-27D). The spermatium rests on the trichogyne and fuses with it; the nucleus passes to the enlarged basal part of the carpogonium (Fig.

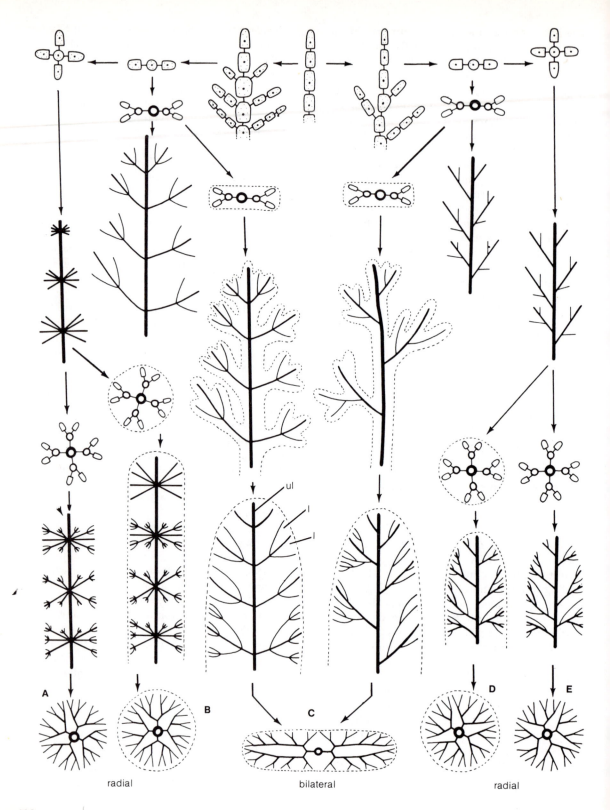

radial   bilateral   radial

11-27G), where it fuses with the female nucleus. The zygote nucleus divides (Fig. 11-27I) mitotically; daughter nuclei are cut off in cells, which then divide repeatedly, forming chains of diploid cells in a tight cluster. Collectively, these diploid cells make up the gonimoblast filaments, or carposporophyte (Fig. 11-27K). The carposporophyte is freely exposed, and terminal cells mature directly into carposporangia, each of which liberates a single, diploid carpospore (Fig. 11-27L). The carpospore germinates and develops into a free-living, acrochaetioid filament (Fig. 11-27M–P). In other members of the Nemaliales, the zygote nucleus is transferred to a cell beneath the carpogonium, and from this lower (hypogynous) cell, the gonimoblast filaments emanate. The filamentous acrochaetioid stage produces tetrasporangia, and meiosis occurs during tetraspore production (Fig. 11-27P, Q). The liberated tetraspores germinate and develop first into a prostrate filamentous stage, from which the typical *Nemalion* gametophyte plants eventually arise (Fig. 11-27A, B).

Pseudogloiophloea (Nemaliales), which can be unisexual or bisexual, is very similar to *Nemalion* in its life history, alternation of generations, and postfertilization development of the female reproductive system. However, in addition it produces monospores (in monosporangia) on both the haploid gametophyte and the diploid tetrasporophyte. Monospores reproduce the type of plant on which they were formed, as an accessory but asexual method of reproduction.

The florideophycean examples discussed in detail so far in this chapter have a type of post-fertilization development that is relatively easy to follow. Postfertilization complexities in the representatives of other orders, especially in the Cryptonemiales and Gigartinales, are beyond the scope of this text to consider in detail. However, a brief discussion of *Dudresnaya* (Cryptonemiales) follows to give some idea of the type of postfertilization development that can occur, and especially to illustrate how more than one carposporophyte can be produced as a result of a single sexual fusion.

Dudresnaya. *Dudresnaya* (Fig. 11-29) is uniaxial, with profusely branched whorls of branched filaments compacted loosely into a mucilaginous, pseudoparenchymatous thallus. Near the surface of the thallus, filamentous axes are loosely arranged; distal portions of the branches are free from each other. All reproductive structures are borne superficially on these terminal branches, much as in *Nemalion*. The carpogonium is borne at the end of a carpogonial branch (Fig. 11-29A) of 5 or more (up to 19 in some species) cells. Auxiliary cells are produced about midway in remote **auxiliary cell branches** (Fig. 11-29B), 9–14 cells in length. Following fertilization, the diploid nucleus passes into a nearby **nutritive cell** (Fig. 11-29C) of the carpogonial branch. A connecting filament (Fig. 11-29D) grows out from the nutritive cell. The diploid nucleus or a derivative of it remains near the apex of the connecting filament as it continues to grow. Eventually, the connecting filament contacts and fuses with the auxiliary cell "diploidizing" the auxiliary cell. The diploid nucleus then enters the auxiliary cell and divides there; one of the resulting derivatives initiates the formation of gonimoblast filaments (Fig. 11-29E) at this point; the other passes through another connecting filament that grows out from this auxiliary cell and fuses with another remote auxiliary cell (Fig. 11-29E). The same phenomenon can occur successively, resulting in several "diploidized" auxiliary cells, from each of which can develop gonimoblast filaments. The compact aggregation of gonimoblast filaments that emanate from each "diploidized" auxiliary cell results in a mature carposporophyte. The mature carposporophyte is thus produced on superficial filamentous branches; there is no vegetative development of a surrounding pericarp. At ma-

Figure 11-21 Growth types in uniaxial Florideophyceae, showing possible evolutionary lines leading to freely branched (**A, E**) and pseudoparenchymatous (**B–D**) types with varied amounts of fusion of filaments of unlimited and limited growth, and with radial (**A, B, D, E**) and bilateral (**C**) symmetry. Diagrammatic cellular details of filaments are shown across top of figure (and above transverse section of each type). Throughout diagrams, filaments of axes are of unlimited (*ul*) growth shown by heaviest lines, or limited (*l*) growth shown by lighter lines; in transverse sectional views, filaments of unlimited growth are shown as heavy circles. Dotted lines indicate fusion of axes and laterals in a common mucilaginous matrix.

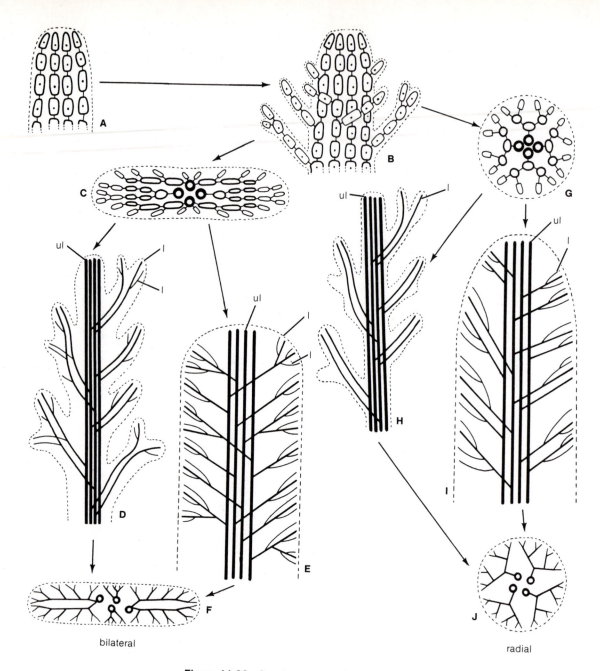

bilateral

radial

Figure 11-22 Growth types of multiaxial Florideophyceae, showing possible evolutionary lines leading to pseudoparenchymatous forms with varied amounts of fusion of filaments of unlimited and limited growth, and with bilateral (*left*) and radial (*right*) symmetry. Filaments of axes are of unlimited (*ul*) growth, shown by heaviest lines, and limited (*l*) growth, shown by lighter lines. **A, B,** diagrammatic cellular detail of filaments. **C, F, G, J,** diagrammatic transverse sectional views. **D, E, H, I,** diagrammatic longitudinal sectional views. In transverse sections (**C, F, G, J**) positions of filaments of unlimited growth are shown as heavy circles. Dotted lines indicate fusion of axes and laterals in a common mucilaginous matrix.

turity, all cells of the carposporophyte become carposporangia.

In related forms, such as *Cryptosiphonia* (Cryptonemiales), the same type of sequential "diploidization" of auxiliary cells can occur; the vegetative filaments near the surface are densely compacted, and the auxiliary cell branches are deeply immersed in the pseudoparenchymatous thallus. Hence, at maturity the carposporophytes are also deeply immersed; at most, they are detectable at the surface from only a slight swelling. In this type of development, the release of carpospores depends on the breakdown of the overlying mass of branched filaments of the female gametophyte.

Distribution and Ecology of Rhodophyta

The red algae are predominantly marine and are more widely distributed than brown algae. Some, such as *Porphyridium* (Fig. 11-8A), can be terrestrial, forming reddish scums on damp surfaces in greenhouses or around the margins of drying garden pools. Many freshwater species are confined to cold, fast-running streams. They are seldom characteristically red; instead, they are a dull green, blue, bluegreen, or sometimes brown to black.

The marine Rhodophyta are most abundant and have attained great diversity in form and color. They are also the most widely distributed geographically and vertically of all the larger algae. Some, such as *Porphyra* and *Gloiopeltis*, occur high in the intertidal region. Others occur at great depths, especially in the tropics, where they have been recorded below 120 meters. Although more species of red algae occur in tropical waters, they are well represented in colder waters. However, they are seldom as conspicuous as the larger brown algae.

The extreme range in color of marine red algae follows vertical variations in the habitat with respect to light conditions. Intertidal forms achieve the greatest diversity in color. Depending on the relative proportion of the various pigments present, they range from a dull green or black high in the intertidal zone to purplish-red, brown, or rosy-red lower in the intertidal zone.

Those from greater depths are generally a bright rosy-red, with phycoerythrin masking the chlorophyll. The presence of the accessory phycobilin pigments permits photosynthesis at great depths. Light of shorter wave lengths (blue light) penetrates deepest in water. Phycobilin pigments absorb most of this light. Light energy absorbed by the accessory pigments is then transferred to chlorophyll *a* in a manner not clearly understood, thus permitting photosynthesis to occur.

Although coralline algae of both the encrusting and articulated types are widespread in cold as well as in warm waters, they are particularly significant in coral reef areas, where the cementing action of the encrusting coralline algae is thought to be primarily responsible for the maintenance of reef structure.

Importance and Uses

Red algae provide, directly and indirectly, a source of detritus and food for marine animals. They are certainly less important on the whole in this respect in cold-water regions than brown algae, but they can be significant. Reproductive cells of the red algae, although nonmotile, are liberated in profusion and form part of the phytoplankton. A single plant of the foliose *Rhodymenia pertusa*, which can reach a length of a meter or more, can produce about 12 million carpospores; a single tetrasporophyte of the same species can produce about 100 million tetraspores.

Red algae are also used directly by humans and are important for industrial and domestic uses, including food. **Dulse**, prepared in several countries from *Palmaria palmata* (= *Rhodymenia palmata*), is used in many ways. It is eaten like candy, used as a relish with potatoes, or cooked in soups. *Porphyra*, known as **purple laver**, or **nori** in Japan, is also used widely as food. The Japanese have artificially cultivated this red alga for many years (Fig. 11-30A–C). Nori is prepared as a dry, flavored product. It is also used in the preparation of Japanese macaroni and in soups and sauces.

Another important use of red algae is as a source of certain valuable phycocolloids. There

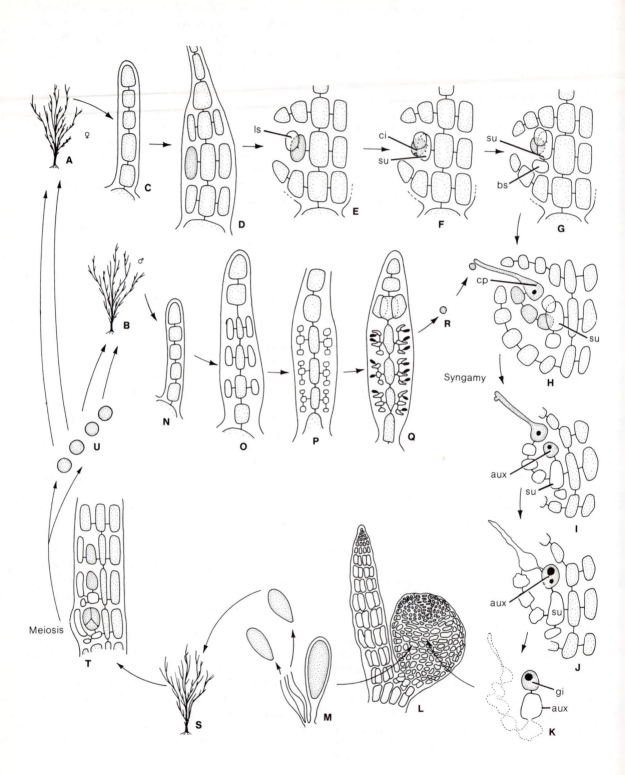

are a number of these substances, the best known of which are *funoran* (funori), *carrageenan*, and agar. Funoran, which is obtained from *Gloiopeltis*, is used as a water-soluble sizing. It is also used in the preparation of certain water-base paints, as an adhesive in hair dressings, and as a starch substitute in laundering.

Carrageenan is a mucilaginous extract obtained chiefly from *Chondrus* (Irish Moss) of the North American Atlantic Coast and from *Eucheuma* of southeast Asia. It was first used by the people of Northern Ireland, who simply wrapped the seaweed (*Chondrus*) in cloth and boiled it in water. The extract obtained was flavored and cooled; when the mixture was firmly set, it was used as a dessert like blanc mange. Carrageenan extracted commercially is still used as food in much the same way. It is used in the food industry as a stabilizing or thickening agent in chocolate milk, cheese, ice cream, and jellied foods; it is also used in cosmetics, insect sprays, and water-base paints.

Agar is probably the most widely used and most valuable product obtained from red algae. It has been known since 1760 in Japan and, prior to World War II, was produced almost exclusively from *Gelidium*. Although Japan is still the largest producer, agar is now available in commercial quantities from a number of genera from several countries, including South Africa, Australia, New Zealand, United States, Chile, and U.S.S.R. Agar has many of the same uses as carrageenan, and many additional uses. It is widely used in microbiology in the preparation of media for culture work. It is a useful therapeutic agent in intestinal disorders, since it is a nonirritant bulk producer that can absorb and hold water and at the same time act as a mild laxative. It is used also to make capsules to enclose antibiotics, sulfa compounds, vitamins, and other substances when a slow release of the medicant is desired at a point beyond the stomach. It is used as a dental impression material. In certain special breads for diabetics, agar replaces starch; however, agar and carrageenan are not good as a source of nutrition for human consumption because they are relatively indigestible.

Phylogeny

Although some red algae have left excellent fossil records, the origin of the group as a whole is obscure. Most fossil records are of the lime-encrusting forms, such as coralline algae. It is apparent that encrusting coralline algae, such as *Lithothamnium*, were as important and active in reef building in past periods (especially during the Jurassic) as they are now. Some phylogeneticists believe that red algae probably had their origin during the Archeozoic Era from some unknown, probably uncalcified, Precambrian ancestor. During the Paleozoic Era, calcareous red algae were represented by the Solenoporaceae, and closely related forms, which appear in the fossil record during the late Cambrian, but were rare until late in the Ordovician. By the mid-Silurian, two structurally distinct types had differentiated in the Solenoporaceae: *Solenopora* and *Parachaetetes*. During the Pennsylvanian period, *Archaeolithophyllum*, *Cuneiphycus*, and several closely related forms appeared. Before the end of the Paleozoic, forms that clearly forecast the appearance of true coralline algae, including recent encrusting and articulated types, are recognizable. However, many recent genera were well established by the end of the Mesozoic Era, from the Cretaceous onward, and, especially in the

Figure 11-23 Life history of *Polysiphonia* (Ceramiales, Florideophyceae). **A, B,** female and male gametophytes. **C–H,** stages in development of female branch (trichoblast), showing formation of lateral sterile cell (*ls*), carpogonial branch initial (*ci*), and basal sterile cell (*bs*). (Note: details of surrounding pericarp are omitted in **H–K.**) The four-celled carpogonial branch with carpogonium (*cp*) in **H** is fully developed and attached to supporting cell (*su*). **N–R,** stages in development of male branch (trichoblast), showing formation of spermatangia and release of spermatia. **I,** fertilization is complete and auxiliary cell (*aux*) is formed. **J,** movement of zygote nucleus into auxiliary cell. **K,** formation of carposporophyte (gonimoblast) initial (*gi*). **L,** mature carposporophyte enclosed in pericarp. **M,** carposporangia, liberating carpospores. **S,** tetrasporophyte. **T,** fertile branch of tetrasporophyte with tetrasporangia forming (lowest tetrasporangium mature). **U,** tetraspores (meiospores). (For structural details and representative sizes of structures refer to Fig. 11-24.)

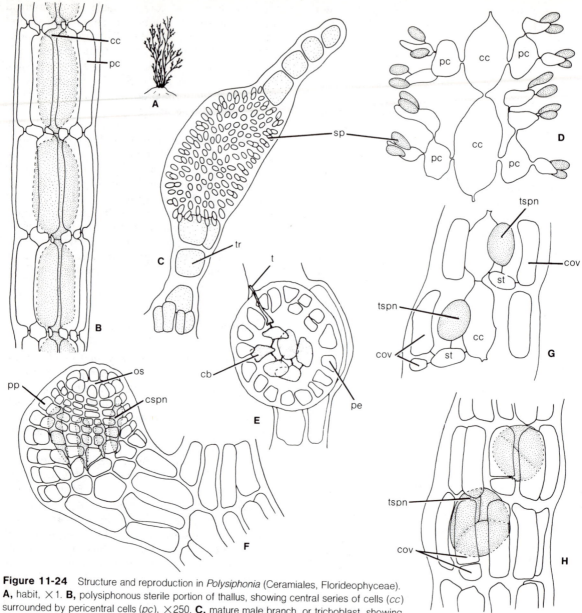

Figure 11-24 Structure and reproduction in *Polysiphonia* (Ceramiales, Florideophyceae). **A,** habit, ×1. **B,** polysiphonous sterile portion of thallus, showing central series of cells (*cc*) surrounded by pericentral cells (*pc*), ×250. **C,** mature male branch, or trichoblast, showing dense mass of spermatangia (*sp*), ×200. **D,** longitudinal section through two fertile segments of male trichoblast, showing derivation of spermatangia from pericentral cells, and relationships to central cells, ×500. **E,** carpogonial branch (*cb*) with terminal carpogonium and emergent trichogyne (*t*) surrounded by pericarp (sectional view), ×200. **F,** cystocarp, showing ostiole (*os*) and outer sterile pericarp wall (*pp*), through which some of fertile terminal carposporangia (*cspn*) of carposoporophyte are shown, ×150. **G,** longitudinal section through two fertile segments of tetrasporophyte, showing attachment of immature tetrasporangia (*tspn*) with stalk (*st*) and cover cells (*cov*), ×250. **H,** surface view of fertile portion of thallus, showing mature tetrasporangia as seen through protecting cover cells (*cov*), ×250.

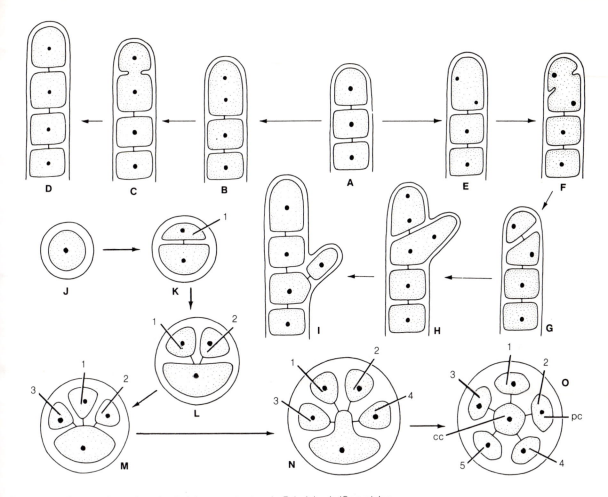

Figure 11-25 Apical growth and polysiphonous structure in *Polysiphonia* (Ceramiales, Florideophyceae). **A–D,** stages in division of apical cell to produce nonbranching segments. **E–I,** stages in division of apical cell (note position of nuclei in apical cell) to produce a branch-forming segment. **J–O,** divisions in segments (as seen in transverse section) cut off from apical cell, resulting in polysiphonous structure of axis with central (axial) cell and several (in this example, five) pericentral cells. Sequence in division is indicated by number: *cc*, central cell; *pc*, pericentral cell.

Tertiary, genera such as *Lithothamnium* and *Lithophyllum* are clearly recognizable.

It is thought by some that the *Solenopora* type was the evolutionary line that gave rise to *Archaeolithothamnium* and the extant *Lithothamnium*, and that the *Parachaetetes* type was the evolutionary line that gave rise to two separate series—one through *Archaeolithophyllum* to the extant *Lithophyllum* and *Mesophyllum*, and the other through

Cuneiphycus to the modern, articulated coralline algae. However, on the basis of recent morphological and cultural studies, especially concerning patterns of spore germination, it is suggested that a systematic arrangement of modern coralline algae into articulated and nonarticulated groups is unnatural. It now appears that some of the articulated taxa are more closely related to certain nonarticulated genera than to

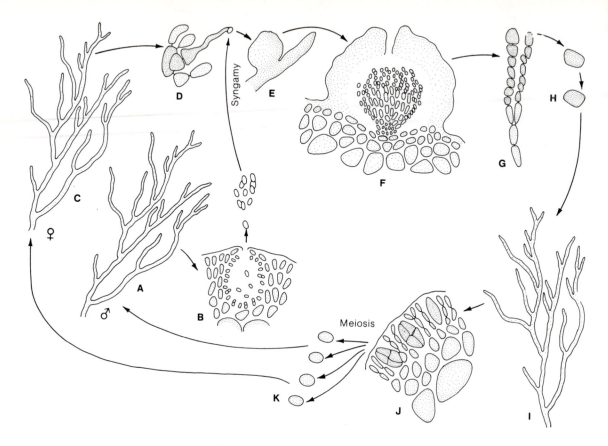

Figure 11-26 Life history of *Gracilaria* (Gigartinales, Florideophyceae). **A,** male gametophyte. **B,** conceptacle of spermatangia liberating spermatia. **C,** female gametophyte. **D,** spermatium attached to trichogyne of carpogonium before fertilization. **E,** immature cystocarp. **F,** transverse section through mature cystocarp, showing carposporophyte enclosed within ostiolate, sterile pericarp wall of gametophytic tissue. **G,** portion of carposporophyte, showing chains of carposporangia. **H,** carpospore. **I,** tetrasporophyte. **J,** tetrasporangia (meiosporangia) in surface region of mature thallus; **K,** tetraspores released from tetrasporangium; these germinate to produce new male and female gametophytes.

other articulated genera. These phenetic data suggest that articulated genera may have evolved from two distinct groups of early, nonarticulated ancestors.

It has been suggested that the red algae and bluegreen algae are closely related, because of a number of similarities in their biochemical and certain ultrastructural characteristics. The presence of phycobilin pigments, forms of starch and oil stored as reserves, and unstacked thylakoids have been given greatest attention, but the ab-

sence of motile cells in each group and the presence of primary pit connections (in some of each group) have also been emphasized. The major obstacle in this suggested relationship is the difference in organization of the genetic material. In the eukaryotic red algae the genetic material is clearly organized in a nucleus, whereas in the prokaryotic bluegreen algae, it is present as a nucleoid.

However, since the fossil history of the recent Rhodophyta is known with certainty only as

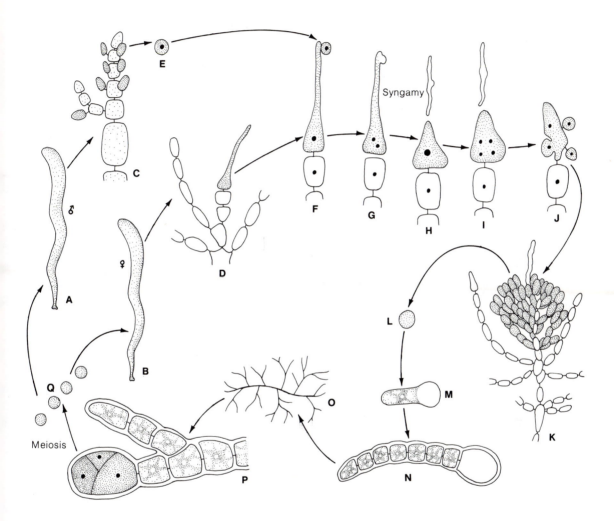

Figure 11-27 Life history of *Nemalion* (Nemaliales, Florideophyceae). **A, B,** gametophytes. **C,** branch of male gametophyte with several spermatangia and one released spermatium (**E**); **D,** branch of female gametophyte with carpogonium. **F,** spermatium adjacent to trichogyne (on carpogonium). **G,** spermatium nucleus in base of carpogonium. **H,** syngamy completed. **I,** after meiosis, trichogyne degenerates. **J,** initiation of gonimoblast filaments from fertilized carpogonium. **K,** mature carposporophyte with numerous carposporangia. **L,** carpospore. **M, N,** germinating carpospore. **O,** filamentous tetrasporophyte (acrochaetioid stage). **P,** portion of branch of tetrasporophyte, showing tetrasporangium with three of four tetraspores visible. **Q,** tetraspores (meiospores). (For details and representative sizes of comparable structures, as in *Batrachospermum,* refer to Fig. 11-17C–K.)

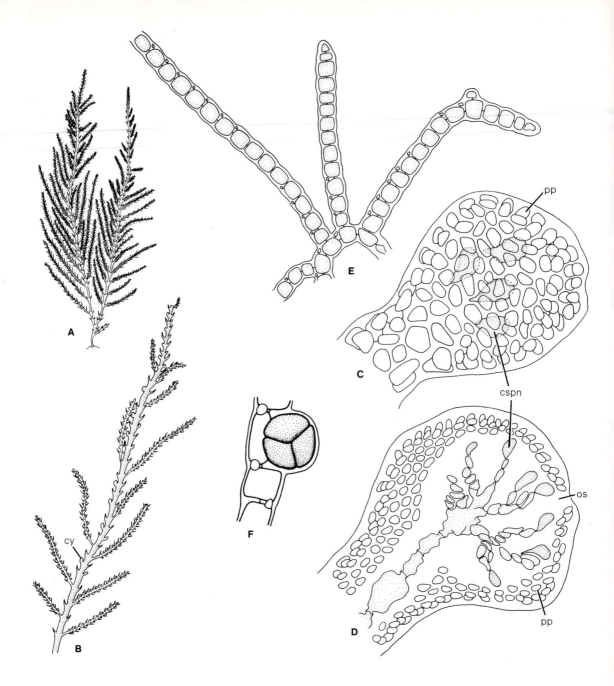

Figure 11-28 *Bonnemaisonia* (Nemaliales, Florideophyceae). **A–D,** female gametophyte. **A,** habit ×0.3; **B,** fertile portion of branch, showing position of cystocarps (*cy*), ×1; **C,** surface view of cystocarp, showing sterile pericarp wall (*pp*) through which some terminal carposporangia (*cspn*) of carposporophyte are seen, ×300; **D,** transverse section of cystocarp, showing outer sterile pericarp wall, ostiole (*os*), and enclosed carposporophyte, ×250. **E, F,** filamentous tetrasporophyte (*Trailliella*-stage). **F,** portion of filamentous thallus showing tetrasporangium with three of four tetraspores (meiospores) visible, ×400.

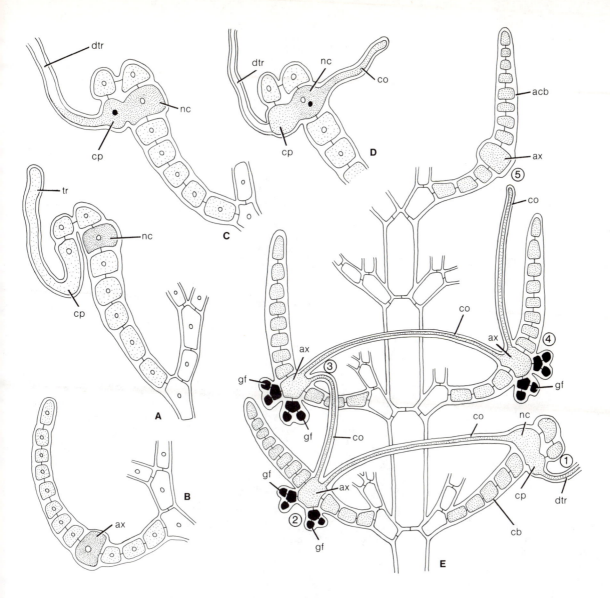

Figure 11-29 Diagrammatic representation of carpogonial and auxiliary cell branches and postfertilization stages of development in *Dudresnaya* (Cryptonemiales, Florideophyceae). **A,** carpogonial branch, showing trichogyne (*tr*), carpogonium (*cp*), and adjacent nutritive cell (*nc*) prior to fertilization. **B,** auxiliary cell branch, showing intercalary position of auxiliary cell prior to "diploidization." **C,** carpogonial branch, showing fusion of base of carpogonium (*cp*) with adjacent nutritive cell (*nc*) following fertilization, with diploid nucleus in base of carpogonium. **D,** distal portion of carpogonial branch, showing connecting filament (*co*) growing out from nutritive cell (*nc*), which contains diploid nucleus. **E,** fertile region of thallus, showing relationships of carpogonial branch (*cb*) to auxiliary cell branches (*acb*), position of auxiliary cells (*ax*), connecting filaments (*co*) proceeding sequentially (in order indicated by circled numbers, 1–5) from one "diploidized" auxiliary cell to the next, and initiation of gonimoblast filaments (*gf*) from "diploidized" auxiliary cells: *dtr*, degenerating trichogyne.

A

B

Figure 11-30 **A–C,** nori culture in Japan. **A,** early Japanese drawing showing ancient method of cultivating *Porphyra* using bundles of bamboo brush forced into the bottom to catch naturally liberated spores in the water. **B, C,** cultivation in Tokyo Bay. **B,** net partly submerged and suspended from fixed bamboo poles with meshes covered by growth of *Porphyra* plants. **C,** Japanese "fisherman" at low tide tending emergent nets covered with growth of *Porphyra* plants. (**A,** from Smith, 1905; **B, C,** photographs by A. Miura, courtesy of S. Ueda.)

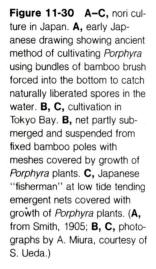

C

far back as the Cretaceous, there is little phylogenetic evidence for an origin of the red algae from bluegreen algae, which are well represented in the Precambrian. Perhaps the best one can suggest is that red algae probably arose from a nonflagellated but nucleated organism after the bluegreen algae had evolved. The absence of centrioles, as far as is presently known, in any stage of the life history of a red alga would argue against a derivation from a flagellated ancestor. Rhodophyta are obviously taxonomically remote from other groups of algae. The similarity of the sexual reproductive processes of some Florideophyceae to certain ascomycetous fungi has led some botanists to suggest that fungi may have descended from a red algal prototype, but this speculation has few adherents.

The relationship between the two classes of red algae is also largely a matter of speculation. Similarities in their biochemical and ultrastructural characteristics, both with respect to pigments and storage products, provide a strong basis for keeping these otherwise morphologically different groups together. The absence of a sexual process in many Bangiophyceae provides a striking difference in the two classes. However, it may well be that sexual reproduction has been lost in most Bangiophyceae. The classical concept of the absence of pit connections in the Bangiophyceae was, at one time, a strong point of difference in the two classes. But the discovery of primary pit connections in several members of Bangiophyceae, and especially in the conchocelis stage of *Bangia* and *Porphyra*, has now eliminated this barrier. It may well be that the filamentous conchocelis stage of *Porphyra* represents a prototype in the red algal line from which Florideophyceae have arisen and that Bangiophyceae represent a dead end culminating in the parenchymatous type of plant exemplified by *Porphyra*. The presence of a girdle thylakoid in the chloroplast of the conchocelis stage (and its absence in the foliose *Porphyra* phase) suggests that a conchocelislike alga may be considered as a prototype of Florideophyceae.

The filamentous line is obviously the one that has been most highly developed in red algae. Although one can suggest relationships among the Florideophyceae, in the present state of uncertainty concerning the orders, especially in the Nemaliales, this too becomes highly spec-ulative. Some attempts at deriving relationships within the orders have been made, but these relate primarily to the differences in postfertilization stages and embryogeny—too complicated a subject to undertake in this text.

However, it would appear on the basis of present information that the basic type of red algal life history is a modification of an alternation of isomorphic generations. One can assume that a delay in meiosis, with retention of the zygote on the female gametophyte, has led to development of the diploid, "parasitic" carposporophyte in the life history of Florideophyceae and that a further delay has resulted in the production of diploid carpospores. This is the *Polysiphonia*-type life history. A modification of this basic type, in which an alternation of heteromorphic generations occurs, would be a second step. This is the *Nemalion*-type life history. This would suggest that the branched filament (which in *Porphyra* is believed to be diploid) and the acrochaetioid stage (which has been shown to be diploid) in some Florideophyceae are homologous.

Certainly among Florideophyceae, if one accepts the significance placed on the nature of the female reproductive structure, postfertilization stages, and embryogeny, it appears that Acrochaetiales is the most primitive order and that Nemaliales, Cryptonemiales, Gigartinales, Palmariales, Rhodymeniales, and Ceramiales are progressively advanced groups. However, the vegetative theme throughout the Florideophyceae is one of remarkable simplicity, despite the great external variety that is expressed.

References

Abbott, I.A., and Dawson, E. Y. 1978. *How to know the seaweeds*. Wm. C. Brown Co. Pubs., Dubuque, Iowa.

Abbott, I. A., and Hollenberg, G. J. 1976. *Marine algae of California*. Stanford University Press, Stanford, Calif.

Bold, H. C., and Wynne, M. J. 1978. *Introduction to the algae*. Prentice-Hall, Englewood Cliffs, N.J.

Chapman, V. J., and Chapman, D. J. 1973. *The algae*. 2nd ed. Macmillan & Co., London.

Chen, L. C.-M.; Edelstein, T.; and McLachlan, J. 1969. *Bonnemaisonia hamifera* Hariot in nature and in culture. *J. Phycol.* 5:211–20.

Chiang, Y.-M. 1970. Morphological studies of red algae of the family Cryptonemiaceae. *Univ. Calif. Publ. Bot.* 48:vi + 1-98.

Conway, E.; Mumford, T. F., Jr.; and Scagel, R. F. 1975. The genus *Porphyra* in British Columbia and Washington. *Syesis* 8:185-244.

Dixon, P. S. 1963. The Rhodophyta: some aspects of their biology. *Oceanogr. Mar. Biol. Ann. Rev.* 1:177-96.

——. 1970. The Rhodophyta: some aspects of their biology. *Oceonogr. Mar. Biol. Ann. Rev.* 8:307-52.

——.1973. *Biology of the Rhodophyta.* Oliver & Boyd, Edinburgh.

Dixon, P. S., and Irvine, L. M. 1977. *Seaweeds of the British Isles.* Vol. 1. *Rhodophyta.* Part 1. *Introduction, Nemaliales, Gigartinales.* British Museum (Natural History); The George Press, Kettering, Northhamptonshire.

Drew, K. M. 1949. *Conchocelis*-phase in the life-history of *Porphyra umbilicalis* (L.) Kütz. *Nature* 164:748.

——. 1951. *Rhodophyta.* In Smith, G. M., ed. *Manual of phycology.* Chronica Botanica, Waltham, Mass.

——. 1956. Reproduction in the Bangiophycidae. *Bot. Rev.* 22:553-611.

Edelstein, T. 1970. The life history of *Gloiosiphonia capillaris* (Hudson) Carmichael. *Phycologia* 9:55-59.

Fan, K. 1961. Morphological studies of the Gelidiales. *Univ. Calif. Publ. Bot.* 32:315-68.

Fritsch, F. E. 1945. *The structure and reproduction of the algae.* Vol. 2. Cambridge Univ. Press, Cambridge.

Garbary, D. J.; Hansen, G. I.; and Scagel, R. F. 1980. A revived classification of Bangiophyceae (Rhodophyta). *Nova Hedwigia* 33:145-66.

——. 1981. The marine algae of British Columbia and northern Washington: Division Rhodophyta (Red Algae), Class Bangiophyceae. *Syesis* 13:137-95.

Goff, L. J., and Cole, K. 1976. The biology of *Harveyella mirabilis* (Crytonemiales, Rhodophyceae). IV. Life history and phenology. *Can. J. Bot.* 54:281-92.

Hawkes, M. W. 1978. Sexual reproduction in *Porphyra gardneri* (Smith *et* Hollenberg) Hawkes (Bangiales, Rhodophyta). *Phycologia* 17:329-53.

Hoek, van den, C., and Jahns, H. M. 1978. *Algen.* George Thieme Verlag, Stuttgart.

Hommersand, M. H. 1963. The morphology and classification of some Ceramiaceae and Rhodomelaceae. *Univ. Calif. Publ. Bot.* 35:165-366.

Hommersand, M. H., and Searles, R. B. 1971. Bibliography on Rhodophyta. In Rosowski, J. R., and Parker, B. C. eds. *Selected papers in phycology.* Dept. of Botany, Univ. of Nebraska, Lincoln, Neb.

Irvine, D. E. G., and Price, J. H., eds. 1978. *Modern approaches to the taxonomy of red and brown algae.* Academic Press, New York.

Johansen, H. W. 1969. Morphology and systematics of coralline algae with special reference to *Calliarthron. Univ. Calif. Publ. Bot.* 49:viii + 1-98.

——. 1976. Current status of generic concepts in coralline algae (Rhodophyta). *Phycologia* 15:221-44.

Johnson, J. H. 1959. Studies of Silurian (Gotlandian) algae. *Quart. Col. School Mines* 54(1):vi + 1-173.

——. 1960. Paleozoic Solenoporaceae and related red algae. *Quart. Col. School Mines* 55(3):vi + 1-77.

——. 1962. The algal genus *Lithothamnion* and its fossil representatives. *Quart. Col. School Mines* 57(1):1-111.

Jones, W. E., and Smith, R. M. 1970. The occurrence of tetraspores in the life history of *Naccaria wiggii* (Turn.) Endl. *Brit. Phycol. J.* 5:91-95.

Kylin, H. 1956. *Die Gattungen der Rhodophyceen.* Gleerup Folog, Lund.

Lee, R. E. 1980. *Phycology.* Cambridge Univ. Press, Cambridge.

Littler, M. M. 1972. The crustose Corallinaceae. *Oceanogr. Mar. Biol. Ann. Rev.* 10:311-47.

Martin, M. T. 1969. A review of life histories in the Nemalionales and some allied genera. *Brit. Phycol. J.* 4:145-58.

Meer, van der, J. P., and Todd, E. R. 1980. The life history of *Palmaria palmata* in culture. A new type for the Rhodophyta. *Can. J. Bot.* 58:1250-56.

Norris, R. E. 1957. Morphological studies on the Kallymeniaceae. *Univ. Calif. Publ. Bot.* 28:251-334.

Ó hEocha, C. 1971. Pigments of the red algae. *Oceanogr. Mar. Biol. Ann. Rev.* 9:61-82.

Papenfuss, G. F. 1966. A review of the present system of classification of the Florideophycidae. *Phycologia* 5:247-55.

Percival, E., and McDowell, R. H. 1967. *Chemistry and enzymology of marine algal polysaccharides.* Academic Press, New York.

Pueschel, C. M. 1980. A reappraisal of the cytochemical properties of rhodophycean pit plugs. *Phycologia* 19:210-17.

Ramus, J. 1967. The developmental sequence of the marine red alga *Pseudogloiophloea* in culture. Univ. Calif. Publ. Bot. 52:ix + 1–42.

Scagel, R. F. 1953. A morphological study of some dorsiventral Rhodomelaceae. *Univ. Calif. Publ. Bot.* 27:1–108.

Searles, R. B. 1968. Morphological studies of red algae of the order Gigartinales. *Univ. Calif. Publ. Bot.* 43:viii + 1–100.

Smith, G. M. 1955. *Cryptogamic botany.* Vol. 1. *Algae and Fungi.* 2nd ed. McGraw-Hill Book Co., New York.

South, G. R.; Hooper, R. G.; and Irvine, L. M. 1972. The life history of *Turnerella pennyi* (Harv.) Schmitz. *Brit. Phycol. J.* 7:221–33.

Sparling, S. R. 1957. The structure and reproduction of some members of the Rhodymeniaceae. *Univ. Calif. Publ. Bot.* 29:319–96.

Taylor, W. R. 1957. *Marine algae of the northeastern coast of North America.* 2nd ed. Univ. of Michigan Press, Ann Arbor.

Tokida, J., and Hirose, H. 1975. *Advance of phycology in Japan.* Junk, The Hague.

von Stosch, H. A. 1965. The sporophyte of *Liagora farinosa* Lamour. *Brit. Phycol. Bull.* 2:486–96.

Wagner, F. S. 1954. Contributions to the morphology of the Delesseriaceae. *Univ. Calif. Publ. Bot.* 27:279–346.

West, J. A. 1969. The life histories of *Rhodochorton purpureum* and *R. tenue* in culture. *J. Phycol.* 5:12–21.

———. 1972. The life history of *Petrocelis franciscana.* *Brit. Phycol. J.* 7:299–308.

12

BRYOPHYTA (MOSSES AND LIVERWORTS)

Mosses and moss allies belong to a single division, Bryophyta, which includes land plants that are extremely simple in morphology and anatomy. Both sporophyte and gametophyte are generally conspicuous, with the sporophyte entirely different in structure from the gametophyte.

The sporophyte of bryophytes is generally dependent upon the gametophyte and is attached to it throughout its life span; it is structurally simple, never consisting of more than one sporangium surmounting a stalk (**seta**), and always annual. The gametophyte is generally the conspicuous part of the life history; it is autotrophic, often "leafy," usually perennial, and may have a simple "vascular" system. The growth of both sporophyte and gametophyte is usually by an apical cell.

Often bryophytes are called archegoniates because, like ferns and fern allies, they have a peculiar, multicellular, female sex organ, the **archegonium,** in which a single egg is enclosed. The archegonium is flask-shaped (Fig. 12-1), whereas in "higher" vascular plants the egg is enclosed in a much reduced structure of more complex cytological origin and development.

In this text, the division Bryophyta includes liverworts (Class Hepaticae), hornworts (Class Anthocerotae), and mosses (Class Musci). Bryophytes show no close relationships to any living plant group, nor do they appear to have been ancestral to any other plant group. Of all plant groups, the subdivision Rhyniophytina (Division Tracheophyta), shows the greatest similarities; on this basis, it is suggested to have shared a common ancestor with bryophytes. Another theory is that bryophytes and vascular plants arose independently from green algal ancestors.

All bryophytes are relatively small, the smallest being almost microscopic. The largest erect forms are up to 80 cm tall; some aquatic species are more than a meter long, and some epiphytic mosses reach lengths of more than 60 cm.

Most bryophytes absorb their moisture and dissolved nutrients directly into the cells in which they are used. The conduction of water in most is over the surface of the plant, often aided by capillary spaces among leaves or rhizoids. A few possess a well-differentiated, internal conducting system. Most mineral nutrients come from atmospheric moisture, rainwash from vascular plants that shade the bryophytes, or, to a lesser extent, from water near the soil surface. Most bryophytes absorb and lose water rapidly.

There are about 320 genera of hepatics with approximately 8000 species, 5 genera of hornworts with approximately 100 species, and approximately 700 genera of mosses with more than 14,300 species.

Gametophyte

In bryophytes, the gametophyte reaches its greatest diversification in the plant kingdom. It is usually the most conspicuous generation, and its most structurally complex part is the **gametophore,** which produces the sex organs. The spore of a bryophyte usually germinates to produce a **protonema,** which is always relatively simple in organization and without tissue differentiation. The protonema ultimately produces a specialized apical cell, from which the gametophore arises. The gametophore is a flattened thallus in many liverworts and most hornworts, but in mosses and most liverworts it is leafy. Thus, the gametophyte consists of two distinctive phases, protonema and gametophore.

Bryophytes lack true leaves and stems, but since they bear morphologically similar structures that perform functions similar to those of true leaves and stems, the terms *leaf* and *stem* are used here for convenience. More correctly, stems should be referred to as **caulids** and leaves as **phyllids.**

In leafy bryophytes, leaves are always sessile and usually a single cell thick (**unistratose**). The stem is usually structurally simple, with an epidermis, a layer of thick-walled cortical cells, and sometimes a central cylinder. In some, there is remarkable tissue organization in the stem of leafy forms and in the flattened thallus of others. The gametophore is usually affixed to the substratum by **rhizoids.** In most hepatics and hornworts, these are unicellular, but in mosses they are multicellular and generally multibranched, uniseriate strands. Most gametophores are perennial.

In bryophytes, the flask-shaped, multicellular archegonium is barely visible to the unaided eye (Figs. 12-1A–C and 12-2A). The unistratose

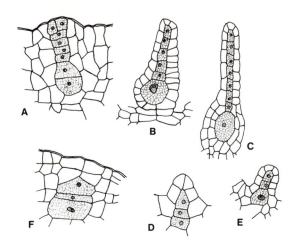

Figure 12-1 Archegonia of various archegoniate plants. **A,** *Anthoceros* sp., ✕325. **B,** *Marchantia* sp., ✕325. **C,** *Sphagnum* sp., ✕325. **D,** *Gleichenia* sp., ✕325. **E,** *Equisetum* sp., ✕210. **F,** *Isoetes* sp., ✕215.

neck is generally five or six cells in circumference and surrounds a single row of central neck-canal cells. The lower, swollen portion of the archegonium is the **venter;** its walls are often more than one cell thick (**multistratose**) and enclose a single egg. The archegonium is usually attached to the gametophore by a short, multicellular stalk. At maturity, a cap cell at the apex of the neck folds back, exposing the tip of the neck. Meanwhile the neck-canal cells have dissolved; their semifluid remains flow into the surrounding water that initiated the rupture of the apex of the archegonium. Generally, each gametophore produces several archegonia, often surrounded by a protective envelope of tissue or leaves.

The multicellular male sex organ, the **antheridium,** is elongate or spherical and is generally attached to the gametophore by a stalk (Fig. 12-2B). Its outer, sterile jacket is unistratose, but there are numerous inner cells, each of which produces a motile sperm. The sperm is somewhat coiled and has two anterior whiplash flagella, ultrastructurally similar to those of some green algae (Fig. 12-2C).

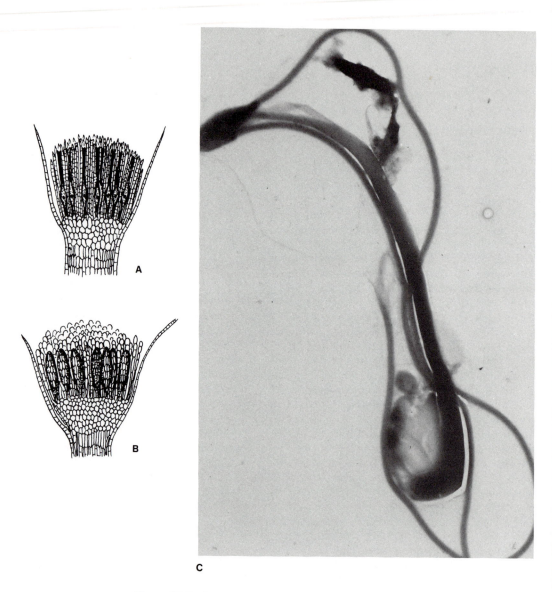

Figure 12-2 Sex organs of bryophytes. **A,** archegonia of a moss with intermixed paraphyses, × 10. **B,** antheridia of a moss with intermixed paraphyses, × 10; **C,** sperm of the liverwort *Blasia pusilla,* showing flagella, ×9500. (**C,** after Z. B. Carothers, ''Comparative studies on spermatogenisis in bryophytes,'' *Biological Journal of the Linnaean Society* 7 (Supplement):71–84, 1975, with permission of Academic Press.)

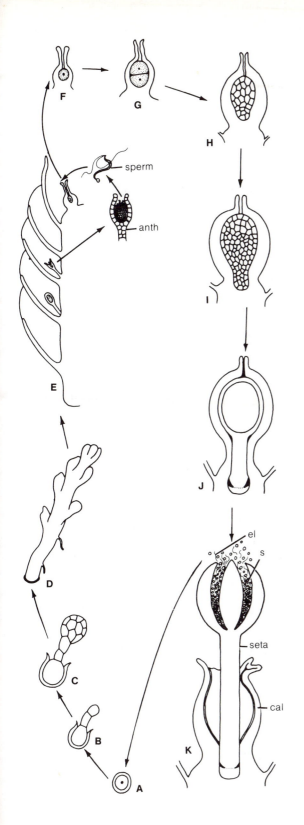

Sporophyte

Although a few bryophytes are not known to produce sporophytes, most develop one by mitotic division of the zygote. Often only a single egg is fertilized on each sexual branch of the gametophore, but sometimes several sporophytes can emerge from a single sexual branch, the result of the fertilization of several eggs in as many archegonia.

Each sporophyte produces a single sporangium at the apex of an unbranched stalk termed the *seta*. The base of the sporophyte, the **foot,** penetrates the apex of the sexual branch and occasionally extends short, haustoriumlike swellings into the tissue of the gametophore. As the sporophyte develops, the cells below the venter and stalk of the archegonium undergo many further divisions, enlarging to become a protective cap **(calyptra)** which encloses the young, chlorophyll-rich embryo.

Life History

The life history of a bryophyte begins with the germination of a spore. Light and moisture are necessary, and in most instances the spore germinates after it is shed from the sporangium. When the spore coat is ruptured, a germ tube usually emerges to begin the structurally simple protonema (Figs. 12-3 and 12-4) that precedes production of the more complex gametophore. The protonema is very reduced in most hepatics and hornworts, but in mosses it often forms an extensive, branched, system of uniseriate filaments. Ultimately, it forms a bud of undifferentiated cells, one of which achieves apical dominance and cuts off cells that initiate the structurally more complex gametophore.

Figure 12-3 Life cycle of typical leafy hepatic. **A–D,** germination of spore and development of gametophyte. **E,** detail of apex of fertile shoot, showing antheridia, archegonium, and fertilization. **F–J,** development of zygote and differentiation of sporophyte. **K,** mature sporangium bursting to shed spores: *an,* antheridium; *cal,* calyptra; *el,* elater; *seta,* seta; *s,* spore; *sperm,* sperm. (After Schuster.)

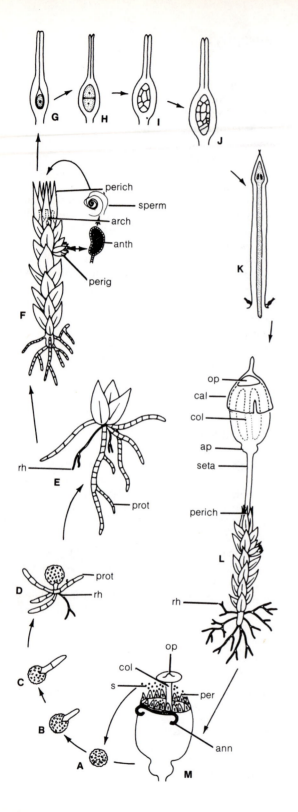

The gametophore bears the sex organs. Expelled in a mass from the mature antheridium, sperms float to the surface of the water film that initiated antheridial rupture. They spread quickly and rotate randomly near the water film surface, moving horizontally largely by movement of the water film. As sperms enter the vicinity of the fluid that has diffused from the neck-canal cell, their movement is guided into the archegonium, which has the greatest concentration of this fluid. The fusion of sperm and egg appears to set up a hormonal response in the surrounding archegonia of the same sexual branch, and fertilization of adjacent archegonia usually does not occur.

The zygote begins to divide immediately after fertilization, producing an embryonic sporophyte. As the embryo develops, cell division and enlargement are also initiated in the venter and stalk of the archegonium and the cells below it, producing a sheathing calyptra that protects it. In hepatics and in a few mosses, including *Andreaea* and *Sphagnum*, the calyptra protects the entire sporophyte until sporangium differentiation occurs and meiospores are produced. But in most mosses, the calyptra is torn away from the apex of the **perichaetial shoot** (female sexual branch) as the seta elongates, carrying the calyptra upward on its tip, where it protects the developing sporangium. In the hepatics, the seta generally elongates rapidly after differentiation of all sporophytic tissues; the calyptra is ruptured, and the sporangium emerges, leaving the calyptra sheathing only the base of the seta. In hornworts, there is no seta, and the calyptra is generally lost early. In mosses and liverworts, sporophyte growth is determinate, and all spores mature at approximately the same time. In hornworts, the

Figure 12-4 Life cycle of typical moss. **A–E,** germination of spore and development of gametophyte. **F,** mature gametophore, showing antheridium, archegonia, and fertilization. **G–K,** development of zygote and differentiation of sporophyte. **L,** gametophore bearing mature sporophyte. **M,** detail of sporangium, showing spore release: *ann,* annulus; *an,* antheridium; *ap,* apophysis; *arch,* archegonium; *cal,* calyptra; *col,* columella; *op,* operculum; *per,* peristome tooth; *perich,* perichaetium; *perig,* perigonium; *prot,* protonema; *rh,* rhizoid; *seta,* seta; *s,* spore; *sperm,* sperm.

intercalary meristem near the base of the sporangium differentiates new sporogenous tissue throughout the favorable season.

In bryophytes, the sporangium usually has a multistratose, sterile jacket; enclosed by it, spore mother cells are ultimately differentiated, undergoing meiosis and producing tetrads of meiospores.

Wind dispersal of spores is important in most bryophytes, and many mechanisms have evolved that enchance it. **Hygroscopic** structures represented by **peristome teeth** in mosses (Bryidae), in the elaters of hepatics, and in the twisting seta of Bryidae and "valves" of some hornworts are all such mechanisms. The production of an elongate seta by many mosses and hepatics also improves spore dissemination by expanding the sporangium well above the gametophore.

Spore viability is brief in some bryophytes, but spores occasionally remain viable for ten or more years. Moisture and light are prerequisites to spore germination. In most bryophyte spores, a germ tube emerges, rupturing the spore wall, and the protonemal phase is produced. Soon after the gametophore appears, the protonema usually disappears.

In most bryophytes, the gametophytes reproduce asexually (Figs. 12-5 and 12-6). Indeed, some mosses and liverworts are known only as gametophytes, and therefore reproduce only asexually. **Gemmae** are undifferentiated masses of cells produced on various parts of the gametophyte. Gemmae produce the same type of protonemal structures as spores, and they give rise to gametophores. **Propagula** are also produced

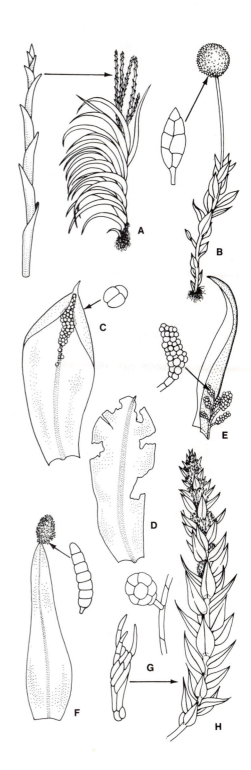

Figure 12-5 Asexual reproduction in mosses. **A,** *Dicranum flagellare,* with deciduous flagelliferous shoots, ✕4. **B,** *Aulacomnium androgynum,* with gemmiferous shoot, ✕3. **C,** *Tortula papillosa,* leaf with gemmae on costa, ✕5. **D,** *Tortula fragilifolia,* showing gaps, fragments from which serve as propagants, ✕5. **E,** *Grimmia torquata,* with gemmiferous clusters at costa base, ✕4. **F,** *Ulota phyllantha,* with gemmae at costa apex, ✕5. **G,** rhizoidal gemma of *Bryum violaceum,* ✕25. **H,** gemma clusters of *Pohlia annotina,* ✕4.

by some bryophytes. These are diminutive gametophores with an apical cell already differentiated; thus, when they fall from the parent gametophore, they produce a gametophore directly without an intervening protonemal phase. Fragmentation of the gametophore also serves as an effective means of vegetative propagation. Brittle leaves, deciduous leaves, and brittle short branches are all means of vegetative reproduction that many bryophytes exhibit. Asexual reproduction is so frequent in many bryophytes that it doubtless diminishes variability in the population and may slow down evolution. On the other hand, if a random mutant does produce a gametophore, its chances of persisting are greatly enhanced by vegetative propagation.

Class Hepaticae (Liverworts)

The Hepaticae form a highly distinctive evolutionary line clearly isolated from other bryophytes. Although hepatics reach their greatest diversity in the tropics, they are most luxuriant in humid, temperate climates, where they grow chiefly in shaded sites and occasionally in open, dry sites. Some are highly tolerant of desiccation, a few can survive in extremely saline sites in arid regions, and some are aquatic.

Most liverworts are relatively small, but the gametophores vary from tiny strands less than

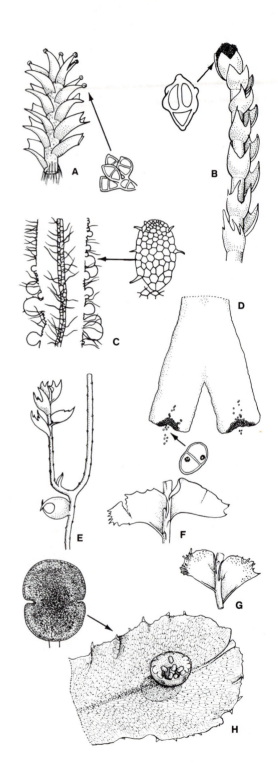

Figure 12-6 Asexual reproduction in hepatics. **A,** *Lophozia ascendens,* with gemmae on apical leaf lobes, ×15. **B,** *Anastrophyllum minutum,* with gemmiferous mass at apex of shoot, ×45. **C,** *Metzgeria* sp., with marginal gemmae on thallus, ×15. **D,** fragment of gemmiferous section of *Riccardia palmata* thallus, showing terminally produced gemmae, ×15. **E,** *Plagiochila tridenticulata* shoot, showing barren stem from which propagating deciduous leaves have fallen, ×14. **F,** *Plagiochila yokogurensis fragilifolia* shoot sector, showing partially deciduous leaves that can propagate new gametophytes, ×9. **G,** *Plagiochila virginica* shoot sector with propagula-bearing leaves, ×6. **H,** *Marchantia polymorpha* thallus with gemma cup containing gemmae, ×2. (**A–G,** after Schuster.)

0.5 mm in diameter, to coarse, broad, flattened thalli up to 5 cm broad and more than 20 cm long. In hepatics, the protonemal stage is normally reduced and ephemeral. This contrasts with the rather extensive protonemata of most mosses. The gametophore is sometimes thallose, but is generally leafy. In some instances, the thallus has an elaborate system of pores and air chambers, a feature unknown in other bryophytes.

In **chlorophyllose** cells, there are numerous chloroplasts, much as in mosses, but differing from many hornworts. Characteristic compound oil bodies are also frequent in hepatics, but are absent in other bryophytes. In most instances, rhizoids are unicellular and unbranched. The calyptra is always ruptured by the sporophyte as the seta elongates and is left as a sleeve at the base of the seta.

The sporophyte is even more distinctive than the gametophyte. In contrast to mosses, an **operculum** is rare and a **peristome** is never present; the seta is of thin-walled cells and is ephemeral, held rigid by turgidity and usually persisting briefly. In many instances, the sporangium has outer cells of the jacket with annular or spiral thickenings. Spirally thickened, hygroscopic **elaters** are usually mixed among the spores, and aid in scattering them to be carried away by air currents. Stomata are never present in sporangial walls. The sporangium generally dehisces by four longitudinal lines. The sporangium and its contents are fully differentiated before elongation of the seta, a feature in marked contrast to mosses.

The hepatics appear to be an extremely ancient group. Fossil material interpreted to be hepatic is known from the Devonian. Material obviously of modern genera is known from Baltic amber deposits of Tertiary age. It is impossible to state whether the "leafy" condition or thallose condition is more primitive. The fossil record suggests that the simple, thallose condition is primitive and complex thallose and leafy conditions are more advanced. Morphological details of extant hepatics strongly suggest that the leafy condition is generalized, and the elaborate thallus is specialized.

The hepatics show several distinctive evolutionary lines, none of which can be convincingly considered as ancestral to any others. These are treated conveniently as various orders. They are dealt with here in linear sequence, but in each order there are some genera that appear at the same level of elaboration as some in other orders.

Order Calobryales

Two families make up the order Calobryales (Fig. 12-7). The family Takakiaceae is structurally simple, and is therefore considered more generalized; it contains *Takakia*, with two species. The family Calobryaceae contains *Haplomitrium*, with ten species. *Takakia* occurs in alpine and subalpine regions of Japan, North Borneo, India, and the North Pacific area of North America (Alaska, British Columbia) (Fig. 12-7C). *Haplomitrium* is predominantly tropical and subtropical; like *Takakia*, it shows a very interrupted distributional pattern. *Haplomitrium hookeri* (Fig. 12-7A, B) shows a very erratic distribution in the Northern Hemisphere, whereas other species are endemic to various tropical and subtropical areas in both the Southern and Northern Hemispheres, especially Australia and southeast Asia.

The small gametophore of the Calobryales is a leafy, erect, chlorophyllose shoot arising from a generally subterranean, colorless root system; it lacks rhizoids. In both *Haplomitrium* and *Takakia*, the gametophores are usually less than 2 cm high.

The leaves are in three ranks, and in most representatives all leaves are structurally alike. In *Takakia* there are **mucilage hairs** of two types on erect shoots and on roots. In *Haplomitrium*, roots lack such mucilage hairs, but the whole system can be invested with mucilage. In those species in which roots occur, they appear to be absorptive in function; no other bryophytes appear to have evolved such a system. The shoots have a cuticle, and thus water uptake from the atmosphere is low or slight; the same appears to be true for leaves.

Haplomitrium has flattened, somewhat fleshy leaves that are usually unistratose and roughly elliptical to circular in outline. Leaf cells are uniformly thin-walled; each cell has many oil bodies and several chloroplasts. The shoot is internally differentiated; cortical cells have a cuticle, and the innermost cells form a central strand. The

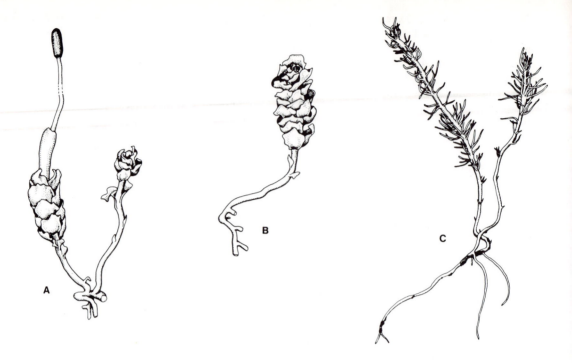

Figure 12-7 Morphology of Calobryales. **A, B,** *Haplomitrium hookeri.* **A,** habit, showing sporophyte with elongated stem calyptra sheathing base of seta, ×4; **B,** habit of gametophore, showing terminal clusters of antheridia, ×4. **C,** *Takakia lepidozioides*, habit, ×4.

central strand conducts fluid from the root system to the leafy gametophore.

The leaves of *Takakia* are generally less than 0.5 mm long and consist of one to four tapered segments, each an attenuated cone. As in *Haplomitrium*, each cell is uniformly thin-walled and has several chloroplasts and many oil bodies.

In *Haplomitrium*, sex organs are **anacrogynous;** although archegonia often appear as a terminal group, the apical cell persists after their production. If fertilization does not take place, further growth of the shoot can occur. The gametophores are unisexual (**dioicous**). In *Takakia*, antheridia have never been observed. Archegonia are few, anacrogynous near the apex, and lateral on the shoot. Each archegonium has a rather massive stalk, a several-layered venter wall, and a neck of six cell rows. In *Haplomitrium*, antheridia are either in a terminal cluster surrounded by several enlarged leaves or are lateral as well as terminal. Antheridia are numerous,

stalked, and spherical to ovoid. Archegonia are similarly located and numerous; each tends to have a thick stalk, a massive venter, and a neck of four cell rows.

The sporophyte of *Haplomitrium* is terminal on the shoot. In early development of the sporophyte, both calyptra and shoot apex grow and enlarge, forming a massive, enclosing, protective **stem-calyptra.** After the sporangium matures, the pale green seta elongates rapidly. The cylindrical sporangium emerges through the apex of the stem-calyptra. The sporangium opens by one or four longitudinal lines. The elaters are long and slender, numerous, and scattered among the spores. The sporangium wall is unistratose, and each cell has vertically arranged, thickened bands. Sporophytes are unknown for *Takakia*.

In *Haplomitrium*, gemmae are not produced in nature, although they are sometimes produced in laboratory cultures. In *Takakia*, the brittle leaves are readily deciduous and probably serve as vegetative propagules. The rhizomatous extensions from gametophores expand the colonies of all species.

The order Calobryales represents an independent evolutionary line that retains a number of structures considered generalized, including massive archegonia, anacrogyny, massive seta, absence of rhizoids, and radial symmetry of the stems. On the other hand, a number of derived features appear, including the unistratose sporangium wall, elaborate mucilage cells in *Takakia*, root system, and central strand in the stem.

Order Jungermanniales

The order Jungermanniales (scale mosses) contains over two-thirds of all known hepatics. Approximately 230 genera and more than 7000 species in nearly 40 families have been recognized. These are distributed from the Arctic to the Antarctic, reaching their greatest diversity in humid tropical forests.

Epiphytes often form attenuated festoons on branches, or sheathe the trunks of trees. Other species form dense carpets and cushions over rocks and forest floors. A few species are aquatic; others grow in tropical and subtropical humid forests on living evergreen leaves.

The color of gametophores varies through a wide range of greens through yellows, browns to nearly black, and red to red-purple. Some species have a distinctly bluish appearance, and others are nearly white. Color results from pigments in cell walls, in oil bodies and in chloroplasts.

The gametophore is generally dorsiventrally flattened, with two lateral rows of leaves. On the under surface of the (normally) reclining stem is usually a row of ventral leaves, termed **amphigastria** (Fig. 12-8F). Leaf arrangement in the Jungermanniales is of three general types (Fig. 12-9). Elaboration of the leaf outline shows much variety in this order (Fig. 12-10) and is more complex than in any other group of bryophytes. Leaves are generally unistratose, and in most genera, cells are **isodiametric** and of the same general shape throughout the leaf. Cell walls are of uniform thickness or are markedly thickened at the corners, exhibiting conspicuous **trigones** (Fig. 12-11A–C, F–H).

Most genera have numerous oil bodies (Fig. 12-11) in each leaf cell. These tend to be larger than the numerous chloroplasts, and are of a consistent morphology in each species. Oil bodies are usually colorless, and are simple and homogeneous, or made up of aggregated spherules that form rounded or segmented clusters.

Rhizoids are unicellular in most genera. They arise from the superficial cells of the stem cortex and sometimes from leaf bases. In the most generalized Jungermanniales, rhizoids are few or absent; in less generalized genera, they are scattered on the ventral surface of the stem; in more derived or specialized genera, they are restricted to specific parts of the stem or leaves. In some genera that are considered specialized in other respects, rhizoids are absent.

The stem is structurally simple. Outer cortical cells are generally chlorophyllose and have oil bodies. Their outermost wall is generally smooth and somewhat thickened. The inner cell generally lacks chloroplasts and is parenchymatous. The stem is usually circular in transverse section, but in some instances is flattened. Gametophores with a flattened stem that bears no leaves are essentially thallose. Fungi often invade some of the ventral cells of reclining stems.

In the Jungermanniales, branching follows several regular patterns and has been used as a feature to suggest interrelationships among genera. The modes of branch origin are entirely different from those in other bryophytes, although the gross appearance is similar to that in some mosses. Strictly dichotomous branching is rare in the order. **Pinnate** and irregular branching are frequent. All of these types of branches are lateral in position. In some instances, branches arise from the ventral surface, as in *Bazzania*; these branches are termed **postical**.

Most species of Jungermanniales are unisexual, but there are various manifestations of the bisexual **(monoicous)** condition. Sex organs are always discrete, stalked, and on the surface rather than immersed. Antheridia are spherical or ovoid, and each is borne on a multicellular, slender stalk. They are solitary or clustered and ensheathed in the axil of a lateral leaf that differs from other lateral leaves of the gametophore. The ensheathing leaves are termed **perigonia.**

The archegonia always terminate a branch of the main shoot, using the apical cell in their pro-

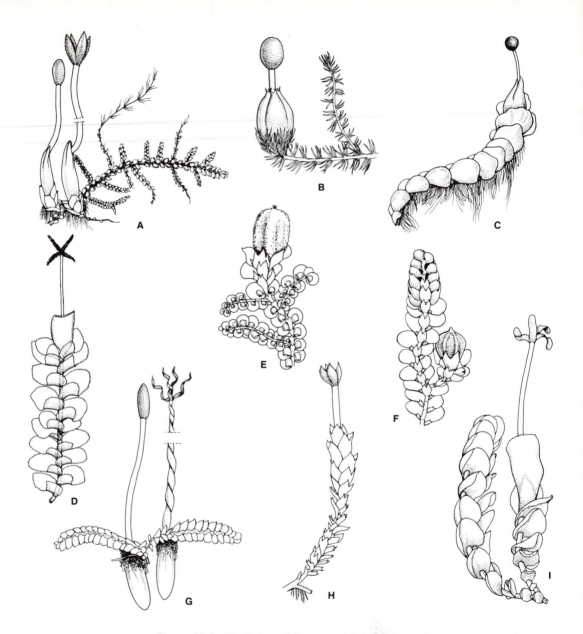

Figure 12-8 Morphology of Jungermanniales, showing various sporophytes and gametophores. **A,** *Lepidozia reptans* shoot with ripe sporangia, ×5. **B,** *Blepharostoma trichophyllum* shoot with sporophyte, ×12. **C,** *Solenostoma hyalinum* shoot with sporophyte, ×8. **D,** *Scapania nemorosa* shoot with open sporangium, ×5. **E,** *Frullania dilatata* shoot with perianth (note also helmet-shaped lobules and notched amphigastria), ×9. **F,** *Lejeunea flava* shoot, showing terminal perigonial branch and lateral perianth-bearing branch (note also notched amphigastria and lobules of lateral leaves), ×15. **G,** *Calypogeja fissa* shoot, showing fleshy subterranean perigynium and emergent sporophytes, ×4. **H,** *Marsupella sprucei* shoot with mature sporophyte, ×12. **I,** *Plagiochila asplenioides* shoot with flattened perianth and emergent sporophyte, ×4. (**A, C–E, G, H,** after Muller, with permission of Akademische Verlagsgesellschaft, Geest and Portig, K.-G. Leipzig: **F,** after Schuster.)

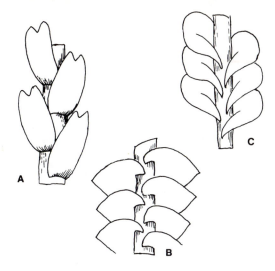

Figure 12-9 Leaf arrangement in Jungermanniales (diagrammatic) showing leaf attachment, as seen from upper surface of shoot. **A,** transverse. **B,** incubous. **C,** succubous.

duction and thus inhibiting further growth of that shoot or branch. This condition is termed **acrogynous** and is restricted to the Jungermanniales. There are several archegonia per shoot, and they are frequently surrounded by the **perianth** (Fig. 12-8A–F, I), a protective sleeve formed of fused leaves. In genera interpreted as generalized, a perianth is absent, but enlarged perichaetial leaves protect the archegonia. In some instances, archegonia are protected by the growth of stem tissue that surrounds them, so that they are sunken in a sac made up of stem tissue.

When the sporophyte is mature, the seta elongates rapidly, pushing the sporangium well above the gametophore and its protective envelopes. The seta varies from slender (only four cells in circumference and less than 0.1 mm in diameter), to massive (numerous cells in circumference and nearly 1 mm in diameter). The sporangium varies from cylindrical to spherical. The jacket is two to eight cells thick, and its outer cells are generally thickened by brown to black, nodular to semiannular thickenings.

Numerous elaters are among the spores. These are unicellular, cylindrical structures, tapered at both ends, which vary in length and in

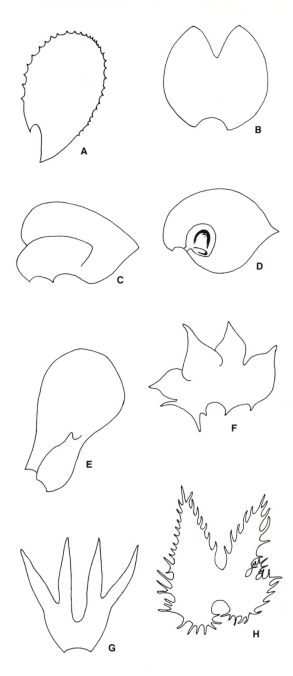

Figure 12-10 Variation in leaf shape in Jungermanniales. **A,** *Plagiochila,* unlobed. **B,** *Marsupella,* equally bilobed. **C,** *Scapania,* complicate bilobed, showing simple lobule. **D,** *Frullania,* complicate bilobed, showing elaborate, helmet-shaped lobule. **E,** *Cololejeunea,* showing pocket. **F,** *Lophozia,* showing many lobes. **G,** *Pseudolepicolea,* showing bifid leaf. **H,** *Ptilidium,* ciliate, many-lobed leaf.

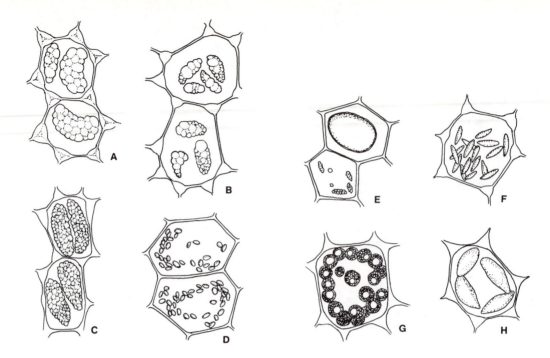

Figure 12-11 Oil bodies of Hepatics (magnification approx. ×600). **A,** median cells of *Cheilolejeunea clausa.* **B,** *Cryptocolea imbricata,* also showing trigones. **C,** basal cells of *Frullania kunzei,* also showing trigones. **D,** *Lejeunea minutilobula.* **E,** *Diplasiolejeunea rudolphiana,* showing differing oil bodies in each cell. **F,** *Brachiolejeunea bahamensis.* **G,** *Lophozia silvicola.* **H,** *Acrobolbus ciliatus* (After Schuster.)

number of helical thickenings. In most instances, elaters are highly hygroscopic and uncoil violently. The sporangium opens by four longitudinal or (rarely) helical lines that run the full length of the sporangium. Upon drying, the "valves" fold back or uncoil rapidly, exposing the sporangium contents (Fig. 12-12A–C). Elaters immediately respond by springing into the air, loosening the spores and discharging them to be carried away by air currents. In *Frullania,* elaters are attached to both the top and base of the interior of the sporangium. As the "valves" bend outward, they stretch the elaters until the tension is so great that the lower ends of the elaters break loose, and snap the "valves" outward and violently throwing spores into the air (Fig. 12-12D–G).

Spores are derived from a four-lobed spore-mother cell and are usually unicellular when dispersed. Spores are generally small, 6–30 μm in diameter, and numerous. Spore sculpturing, of

unknown functional significance, shows numerous patterns.

Spore germination patterns are highly diverse. The spore can produce a uniseriate or multiseriate filament, which soon initiates an apical cell that produces the leafy gametophore; in some instances the spore produces an endosporic or exosporic cell mass that differentiates an apical cell. There are also various elaborations on these three generalized types. In a few instances, the multiseriate protonema is persistent and bears reduced, leafy, sexual branches directly.

Asexual reproduction is frequent in the Jungermanniales. The fragmentation of living shoots of the gametophore can isolate fragments that can each initiate an apical cell and produce a new gametophore. Many species have brittle leaves; these fragments or even entire leaves can produce new gametophores (Fig. 12-6E, F). Some leaf cells of a number of species are able to initiate small buds that produce miniature, leafy

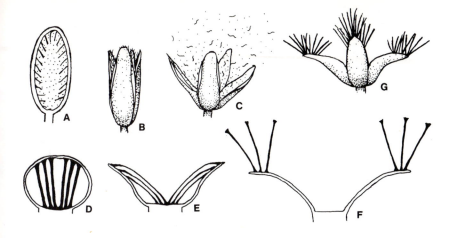

Figure 12-12 Spore discharge in Jungermanniales. **A–C,** *Cephalozia bicuspidata.* **A,** longitudinal section through sporangium, showing elaters and spores (diagrammatic). **B,** sporangium, showing longitudinal splitting of sporangium wall, exposing the elaters to air. **C,** sporangium, showing rapidly uncoiling elaters leaping from sporangium, dislodging spores, and throwing them into the air. **D–G,** *Frullania dilatata.* **D,** longitudinal section through sporangium, showing elaters attached to walls and base of sporangium (diagrammatic); **E,** sporangium opening, showing elaters strongly stretched by opened sporangium; **F,** sporangium, showing that elaters have become rapidly detached from base of sporangium, thus flicking sporangial walls outward quickly, and throwing spores into the air; **G,** view of ruptured sporangium, shown diagrammatically in **F.** (Modified after C. T. Ingold, *Spore discharge in land plants,* 1939, with permission of Oxford University Press.)

gametophores; these fall off and grow into gametophores of normal proportions (Fig. 12-6G). Deciduous masses of cells, called **gemmae,** are often produced on leaf margins or at shoot apices (Fig. 12-6A, B).

The Jungermanniales do not appear in the fossil record until the Tertiary. Some beautifully preserved material is in Baltic amber deposits, and many of these specimens are referred to extant genera, implying that the order originated before the Tertiary. The relationships of the Jungermanniales appear to be more with the Metzgeriales and Calobryales than to any other orders.

Order Metzgeriales

The order Metzgeriales contains 8 families with approximately 25 genera, the largest of which are *Riccardia* and *Metzgeria.* The order is worldwide but reaches its greatest diversity in the humid tropics. Most species grow in damp, shaded habitats. *Cryptothallus* is a nonchlorophyllose saprobe and grows under the moist, acid litter of mosses and leaves, generally in forested sites. It is restricted to northern Europe and Greenland.

The gametophore is usually thallose but is sometimes leafy (Fig. 12-13). Thalli are highly diverse, from a simple, thickened thallus (Fig. 12-14A, B, D, G), to a double-winged thallus on a central stem (Fig. 12-14F), to a lamellate condition (Fig. 12-14C). In *Hymenophyton,* a creeping rhizomatous portion produces erect, flattened, dichotomizing axes (Fig. 12-14E) that resemble some genera of the filmy ferns. A striking feature of this genus is its possession of a nonlignified "vascular" system composed of tracheidlike pitted cells that conduct water through the thallus. Nests of pitted cells are in the thickened portions

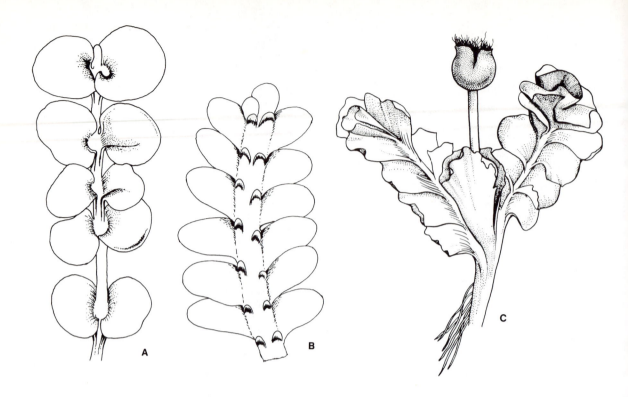

Figure 12-13 Metzgeriales; habits of a variety of morphological types. **A,** *Phyllothallia nivicola* thallus with leaflike lobes, ×4. **B,** *Treubia insignis* thallus, showing leaflike lobes and scales on upper surface of shoot, ×3. **C,** *Fossombronia cristata* habit, showing sporophyte on gametophore, ×25. (**B,** after Schuster and Scott; **C,** after Schiffner.)

of the thalli of other genera of this order, but they do not form a continuous system. The thallus branching pattern in most instances is irregular, although more regular branching patterns occur (Fig. 12-14D). The thallus is generally prostrate on the substratum.

Except for pitted cells in several genera, the cells of the thallus are mainly thin-walled. The upper cells are chlorophyllose, whereas those deeper in the thallus lack chlorophyll. Besides many chloroplasts in each cell, there are numerous oil bodies in the cells of most genera. However, in scattered cells in *Treubia*, there are single, large, complex oil bodies. In the thallus of *Blasia*, there are endogenous colonies of the bluegreen alga *Nostoc* scattered in the thallus (Fig. 12-14G); and, the thallus superficially resembles that of the hornworts. In most genera of Metzgeriales,

the rhizoids are all smooth and confined mainly to the thicker portion of the thallus.

The thalli are always anacrogynous and are generally unisexual. Antheridia are variously arranged: embedded in pits in the upper surface of the thallus (Fig. 12-14B), exposed on the upper surface of the thallus, or surrounded by a flap of tissue (Fig. 12-14F). The archegonia are generally enclosed within a chamber made up of an extended flap of tissue, the **perigynium,** which forms a conspicuous structure in some genera (Fig. 12-14A) but in others is relatively reduced.

In a number of genera (e.g., *Aneura*), an extensive stem-calyptra surrounds the developing sporophyte; in others the perigynium serves the same function.

In most members of the order, the seta is long, raising the sporangium well above the surface of the gametophore. The sporangium wall is two to five cells thick, and outer cells have semi-annular thickenings. The sporangium opens somewhat irregularly in some genera or by one, two, four, or even six longitudinal lines. Within the sporangium there is sometimes a column of sterile tissue, the **elaterophore,** to which some elaters are attached. Elaters are also scattered among the spores in all genera. Spores are derived from a conspicuously four-lobed spore mother cell. Spores are generally unicellular when shed, but in some instances they are multicellular. The germinating spore usually produces a short filament, but an apical cell is differentiated early, initiating the production of the more complex gametophore. In some genera, the gametophore is so similar among the species that they must be identified on the basis of spore sculpturing, which shows great diversity throughout the order.

Hepaticites devonicus, of Devonian time, is the oldest material attributed to the Hepaticae. It has been suggested to resemble strongly the genus *Pallavicinia* of the Metzgeriales, but only fragments of the gametophore are known (Fig. 12-15A, B). Carboniferous fossils presumed to be hepatics also resemble several metzgerialean genera. All are placed conservatively in the form genus *Hepaticites,* and different taxa resemble the genera *Treubia* (compare Figs. 12-13B and 12-15C), *Metzgeria, Aneura,* and *Fossombronia.* If these fossils are correctly attributed, this group of plants existed as early as the late Deuonian.

Order Sphaerocarpales

The order Sphaerocarpales contains 3 genera and approximately 30 species in 2 extant families. The fossil family Naiaditaceae contains only *Naiadita,* with 1 species (Fig. 12-16F–I). The extant genera are found primarily in warm temperate to subtropical countries, where they are locally abundant. Both *Sphaerocarpos* and *Geothallus* grow on somewhat clayey soil, whereas *Riella* and the fossil *Naiadita* are submerged aquatics, growing on mud and in quiet pools. Some species of *Riella* are **halophytes.** All genera appear to be winter annuals, occurring primarily in regions of wet winters and dry summers. *Geothallus* is endemic

to a very restricted area in southern California.

The most distinctive feature of the order is the unistratose "bottle" that surrounds each sex organ and thus the sporophyte (Fig. 12-16C–E). The sporophyte is also distinctive, since a seta is essentially absent, and the unistratose sporangium jacket lacks any ornamentation on the cell walls. Elaters are absent (but **nurse cells** are present), and spores are dispersed upon decay of the sporangial jacket, rather than by a specialized dehiscence.

Riella has several features unique to the order. The gametophore has a well-differentiated, erect stem, and the rhizoids are confined mainly to the base of this stem (Fig. 12-16A, B). On one side of the stem there is a unistratose ruffled "wing." If antheridia are present, they are always in pockets at the edge of this wing. On the opposite side of the stem there are two rows of leaflike scales; discoid gemmae sometimes replace some of the scales. Early in its development, the *Riella* gametophore differentiates an intercalary meristem, a feature not shared by any other hepatic. This intercalary meristem continues to produce a new stem with its associated appendages. Thus, new sex organs are produced throughout the life of the thallus. The female sex organs are produced in "bottles" on the stem at various intervals opposite the ruffle. The stem is 0.5–2 cm tall and sometimes forks.

Naiadita (Fig. 12-16F–I) is the most completely understood of all fossil bryophytes, thanks largely to the study of T. M. Harris (1939), who possessed well-preserved material from the Upper Triassic. Since the plant is exceedingly small (1–3 cm tall), its preservation and detailed study are the more remarkable.

Naiadita presumably grew as submerged dense colonies on fine-particled muds of shallow coastal pools. Gametophores were erect, 1–3 cm tall, and generally unbranched, except for those that bore gemmiferous branches. Rhizoids are borne mainly near the base of the shoot. Lanceolate, unistratose leaves are arranged in a loose spiral. Archegonia are lateral and lack a surrounding "bottle," which, as the sporophyte matures, is replaced by a stem calyptra and a stalk that develops beneath it. Antheridia are unknown. Gemma cups terminate the leafy shoot of a short, lateral branch (Fig. 12-16F). Within the cup is a single, oval gemma. The sporangium is spherical,

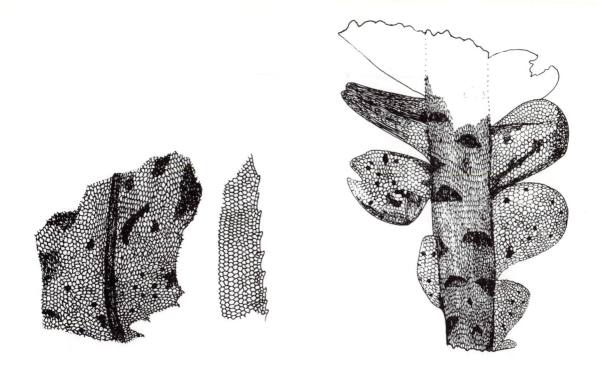

Figure 12-15 Fossil Metzgeriales. **A,** fragment of the Devonian hepatic *Hepaticites devonicus,* showing similarity to the modern genus *Pallavicinia,* ×12. **B,** margin of thallus of *H. devonicus,* showing teeth, ×15. **C,** the Carboniferous hepatic *Hepaticites kidstoni,* showing similarity to the modern genus *Treubia,* ×35. (**A, B,** modified from Hueber; **C,** modified from Walton.)

Figure 12-14 Metzgeriales; habits of a variety of morphological types. **A,** *Pellia neesiana* sporophyte-bearing thallus lobe, ×6. **B,** *P. neesiana* antheridial thallus, ×6. **C,** *Petalophyllum ralfsii* thallus with lamellae on upper surface (note calyptra surrounded by campanulate perigynium from which sporophyte emerges), ×6. **D,** *Riccardia multifida,* portion of thallus with sporophyte and antheridial lobes, ×5. **E,** *Hymenophytum flabellatum* thallus with sporophyte, ×3. **F,** *Apometzgeria pubescens* thallus with antheridial branches on under surface, ×8. **G,** *Blasia pusilla* thallus with gemma "bottles," ×5: *nc, Nostoc* colonies.

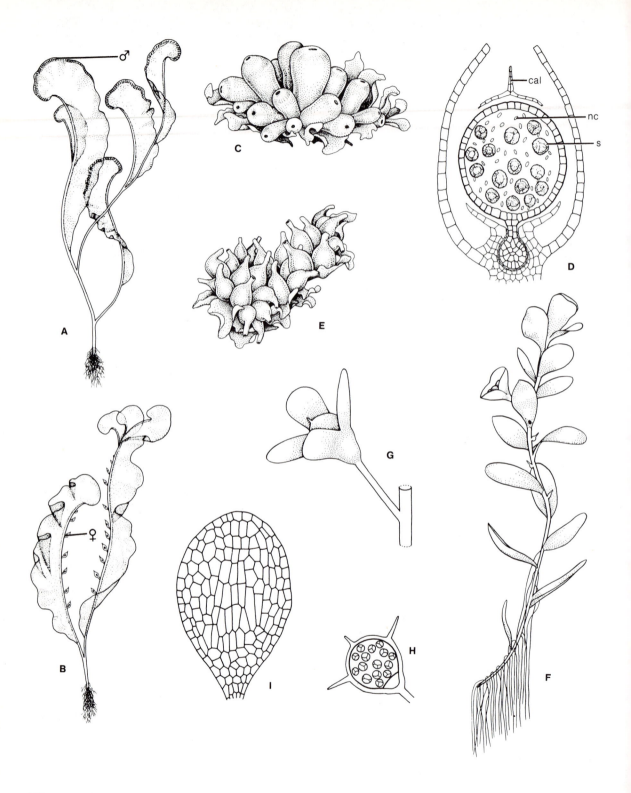

with a small, hemispherical foot; it is partially embedded in a stem-calyptra and fringed by leaves at its equator. Only spores are within the sporangium.

Sphaerocarpos and *Geothallus* resemble each other in many ways. Both are consistently unisexual, and the thallus has a thickened, stemlike central portion with leaflike lobes (Fig. 12-16C, E). In nature, the lobes are apparent but the "stem" is obscure. Mature thalli are usually 0.2–1.5 cm in diameter and are pallid green. The "bottles" cover much of the upper surface of the thallus. Simple, smooth rhizoids attach the thicker portion of the thallus to the substratum. Sporangia have nurse cells among the spores, and the spores of most species remain united in tetrads when shed. This insures that both male and female gametophytes are adjacent to each other, but reduces variability within the populations, since mating can be between sibling thalli. In *Geothallus*, **tubers** are formed within the thallus behind the growing points. These serve as vegetative propagules, and are rich in food storage products; they tolerate desiccation and germinate upon the return of favorable conditions.

Largely because of the detailed studies of C. E. Allen (1930), the genetics of *Sphaerocarpos* is better understood than that of any other bryophyte. In *Sphaerocarpos*, sex-correlated chromosomes were first discovered in plants. In the archegonium-bearing plants, one chromosome is much larger than the seven others; in the antheridium-bearing plant, one chromosome is much smaller than the others. In each tetrad of spores, two spores produce male gametophores, and the others produce female gametophores. Since the sex of thalli can be determined readily on the basis of the shape of the "bottles," it is an extremely convenient organism for genetical studies. The maternal thallus determines the spore sculpturing, thus demonstrating a type of somatic inheritance.

It is impossible to suggest any close relationships of the Sphaerocarpales with any other hepatics. As will be noted in the discussion of the Marchantiales, there are genera that share some sporophytic similarities to the Sphaerocarpales. These appear to be examples of convergent evolution in response to similar environmental selective pressures.

Order Monocleales

The order Monocleales has one genus, *Monoclea*, and two very similar species (Fig. 12-17) that are widely distributed in Central and South America, the subantarctic islands, and New Zealand. The thalli are somewhat leathery (Fig. 12-17A, B) and form extensive, olive green to dark green mats in humid to wet sites.

The thallus is unisexual, dichotomously branched, up to 20 cm long and up to 5 cm wide. It is multistratose and shows no tissue differentiation. The upper cells are somewhat smaller than the rest of the thallus cells, and contain chloroplasts; deeper in the thallus, chloroplasts are absent. Some scattered cells contain a single, brown oil body. Rhizoids are confined to the undersurface of the thallus. Most are perpendicular to the thallus surface and attach the thallus to the substratum, and others form a tufted, central line along the thallus length and are oriented parallel to the length.

The antheridium-bearing thallus produces small, padlike receptacles, generally behind the growing point of each dichotomy of the thallus (Fig. 12-17A). Many antheridia are embedded in the antheridial pads, each antheridium occupying a small chamber that opens to the upper surface by a pore. Mucilage hairs line the inner wall of these antheridial chambers and secrete slime into the chamber as the antheridium matures. Elongate, sleevelike chambers occur in the same

Figure 12-16 Sphaerocarpales. **A,** habit of *Riella americana* thallus with antheridia on margin, ✕3. **B,** *R. americana* thallus with sporophytes on stem, ✕3. **C,** *Sphaerocarpos texanus* archegoniate plants, showing "bottles" containing sporangia, ✕15. **D,** longitudinal section through *Sphaerocarpos* sporangium and "bottle," ✕35. **E,** *S. texanus* antheridial plant, showing antheridium-containing "bottles" ✕50. **F,** reconstruction of the Triassic hepatic *Naiadita lanceolata*, ✕6. **G,** sporangium of *N. lanceolata* with subtending bracts, ✕10. **H,** longitudinal section through sporangium of *N. lanceolata*, showing tetrads of spores, ✕8. **I,** leaf of *N. lanceolata*, ✕50: *col,* columella; *nc,* nurse cells; *s,* spores. (**A, B,** after Studhalter; **F–I,** after Harris, with permission of the British Museum, Natural History.)

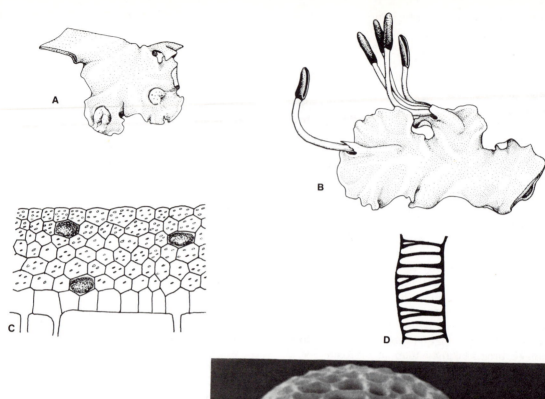

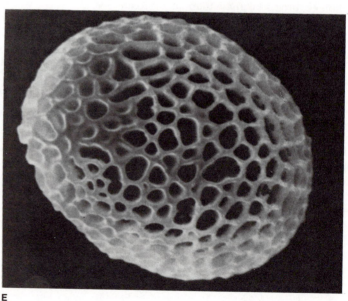

Figure 12-17 Monocleales: *Monoclea forsteri*. **A,** portion of thallus with antheridial pads. ×0.75. **B,** portion of thallus with sporophytes, showing longitudinal dehiscence of the sporangia, ×0.75. **C,** transverse section of thallus, showing oil bodies and ventral rhizoids, ×120. **D,** cell of wall from longitudinal section of sporangium, showing thickenings of walls, ×100. **E,** scanning electron micrograph of a spore, ×5600. (**E,** after Taylor, et al., with permission of *Bryologist*.)

general position of the archegonium-bearing thallus as the antheridial pads on the antheridium-bearing thallus. Each chamber contains several archegonia and their surrounding mucilage hairs.

The sex organs most distant from the growing tip of the thallus mature first; successively, other sex organs mature in the same receptacle or chamber. In consequence, fertilization can occur successively as the archegonia mature, and multiple sporophytes frequently emerge from each archegonial chamber (Fig. 12-17B). The seta is massive and colorless and elongates rapidly, carrying the cylindrical sporangium high above the thallus surface. The sporangial jacket is unistratose, and the cells have thickenings on all walls except the outer one (Fig. 12-17D). The sporangium opens by a single, longitudinal line, exposing elongate elaters that spring from the sporangial cavity and throw the spores into the air.

This order shares many features with entirely unrelated orders. It seems most reasonable to treat Monocleales as an independent evolutionary line, probably closely allied to the Marchantiales.

Order Marchantiales

The most specialized of the hepatics are in the order Marchantiales (Fig. 12-18). The order contains 11 families with approximately 30 genera and 450 species, nearly half of which are in the genus *Riccia*. The order is world-wide and in all climatic regions.

Most Marchantiales grow on moist earth or mud; a few are commonly found in water. Most species are perennial and annually produce sex organs and sporophytes.

The prostrate gametophore is always thallose, generally dichotomously branched, and varies from a few millimeters to more than 2 cm in diameter. It is generally internally differentiated into several distinct tissues. The upper surface forms a unistratose epidermis, the outer walls of which are often cutinized. The epidermal cells have little or no chlorophyll. In most genera, the epidermis is punctured by special pores that open to intercellular air chambers beneath. The pores are surrounded by several cells and are simple or compound. The compound pores are surrounded by a barrel-shaped structure formed of several superimposed tiers of cells. In rare instances, as in *Preissia*, the lowermost rank of cells in the "barrel" collapses with water loss, partially closing the pore and preventing further water loss from the interior tissues of the thallus. As turgor is regained in these cells, the pore opens again. In most genera the pore is continually open.

The layer of air chambers is apparent from a surface view of the thallus. Each pore is surrounded by a polygonal depression that outlines the boundaries of the air chamber beneath. The air chambers represent the photosynthetic tissue, comparable to chlorenchyma in the leaves of vascular plants. Chloroplasts are located in the cells forming the walls of the chamber or are in uniseriate filaments that often grow from the floor of the chamber. Scattered cells of both epidermis and other layers contain no chloroplasts, but a single, large, complex oil body. Cells below the chlorophyllose region are predominantly parenchymatous; some cells are filled with mucilage, and occasional ones have pits in the walls that do not form a continuous conducting system. This parenchymatous tissue is often invaded by fungi. A ventral epidermis encloses this layer.

From some of the lower epidermal cells, particularly those of a thickened band of tissue of the middle of the thallus, emerge colorless rhizoids of several types: smooth-walled (either broad or slender), those with peglike thickenings on the inner walls, and those with alternate constrictions in the walls, giving them a corkscrewlike appearance (Fig. 12-19E). Rhizoids are either perpendicular to the thallus surface and attach it to the substratum, or they diverge backward away from the growing point and are longitudinal to the thallus. These latter rhizoids, plus unistratose scales in two or more rows on the undersurface of the thallus, serve to conduct water along the length of the thallus.

There are various other types of tissue organization represented in the order. In some genera the entire thallus is constructed of air chambers and their chlorophyllose, associated cells. Others have a restricted ventral area of parenchymatous cells, and the rest of the thallus is made up of several layers of air chambers. In *Riccia*, the chlorophyllose tissue is composed of vertical columns of cells interspersed with elongate air chambers that open to the upper surface of the thallus (Fig. 12-19A).

Thalli may be bisexual or unisexual. Antheridia are always enclosed within a chamber that opens to the upper surface of the thallus. Archegonia are embedded in the thallus, as in *Riccia*, or are on the thallus surface and surrounded by a protective envelope, as in most genera.

Figure 12-18 Marchantiales, showing variation in morphology. **A,** *Asterella gracilis,* habit of portion of thallus bearing carpocephala with sporangia, ×3. **B,** *Ricciocarpus natans,* habit of aquatic thallus, ×6. **C,** *Carrpos monocarpus* thallus surrounding carpocephalum containing sporangium, ×15. **D,** *Mannia rupestris,* portion of thallus with carpocephala (note slits forming openings into air chambers of thallus), ×6. **E,** *M. sibirica,* unisexual thallus with single carpocephalum (note operculum on sporangium), ×5. **F,** *Conocephalum conicum* thallus fragment with carpocephala bearing sporangia, ×2. **G,** *Targionia hypophylla,* portion of thallus with ventral sporangia, ×3. **H,** *Neohodgsonia mirabilis,* unisexual thallus with carpocephalum (*left*) and antheridiophore (*right*), ×4. (**A,** after Hattori and Schimizu; **D, E,** after Schuster.)

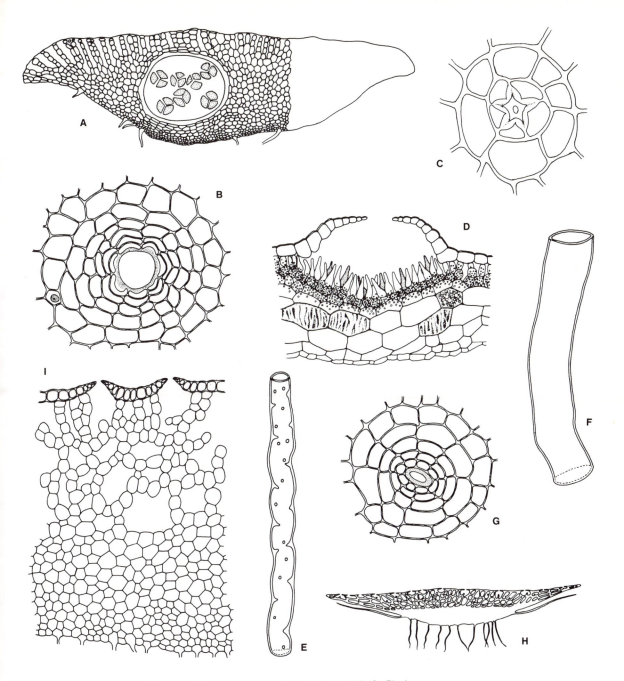

Figure 12-19 Anatomy of Marchantiales (transverse sections through thalli). **A,** *Riccia beyrichiana,* showing sporangium containing spores, ×20. **B,** surface view of air-chamber pore of *Conocephalum conicum,* ×180. **C,** surface view of compound pore of *Athalamia hyalina,* ×200. **D,** air chamber of *C. conicum,* showing chlorophyllose filaments, ×180. **E, F,** rhizoids of *C. conicum,* ×360. **G,** surface view of air-chamber pore of *Reboulia hemisphaerica,* ×180. **H,** thallus of *R. hemisphaerica,* ×7.5. **I,** detail of thallus and air chambers of *R. hemisphaerica,* ×180. (**A, B, G–I,** after Schuster.)

In many genera, the antheridia are embedded in a **peltate** receptacle that terminates a specialized extension of the gametophore. This branch is termed the **antheridiophore** (Fig. 12-18H). The stalk of the branch is perpendicular to the main thallus; it is very short when sperms are shed but elongates afterward. In other genera, antheridia are in a sessile cushion that disintegrates soon after sperms are shed (Fig. 12-18E). Sperms sometimes are dispersed widely by splashing raindrops. In still other genera, all sex organs are embedded deep in the thallus.

Archegonia are also often borne on a similar specialized extension, the **archegoniophore.** Differing from antheridia, archegonia are exposed on the upper surface and near the margins of the receptacle. After fertilization, there is much growth of the upper surface of the receptacle, and sheathing envelopes partially enclose developing sporophytes, which become oriented so that the sporangium faces downward or outward in many instances. When the archegoniophore bears sporangia it is termed the **carpocephalum.** The carpocephalum is morphologically diverse among genera (Fig. 12-18A, D–F, H).

The sporangium varies from spherical to elongate; its jacket is unistratose, and the walls sometimes have spiral thickenings or no ornamentation whatsoever. Within the sporangium, spores are produced from an unlobed spore mother cell and are generally mixed with elaters. The elaters tend to be shorter and stouter than those of other hepatics and, though hygroscopic, do not uncoil abruptly. In some genera, including *Riccia*, elaters are absent but nurse cells, similar to those of the Sphaerocarpales, are present. The sporangium opens in different ways: by four longitudinal lines, by an operculum, by irregular rupturing, or even by decomposition of the sporangial jacket.

Spore ornamentation is variable and often elaborate (Fig. 12-20). It is used as a taxonomic characteristic to assist in distinguishing species of *Riccia*. Spore size varies from 10 μm to about 40 μm. Spores are unicellular, but in *Conocephalum* they germinate endosporously before being shed and are thus multicellular. Spore germination patterns are highly diverse among genera but constant for a given species. A short, filamentous protonema usually precedes the production of a terminal plate of cells. This plate initiates an apical cell that gives rise to the thallus. In *Conocephalum*, no protonemal filament is produced; the plate emerges directly from the multicellular spore.

Asexual reproduction in the order is rarely by gemmae. When present, these gemmae are discoid, on slender stalks, and are formed in gemmae cups on the surface of the thallus (Fig. 12-6H). Among the gemmae are mucilage cells. The swelling of the mucilaginous material ruptures the stalks of the gemmae, and raindrops splash the gemmae away from the cup. Asexual reproduction also results from fragmentation of the thallus and, in rare instances, from the formation of endogenous gemmae.

The fossil record of the Marchantiales is rather scanty. The oldest fossil (of Carboniferous age) suggestive of this order is *Thallites lichenoides.* Other fossil material of Mesozoic age closely resembles the genera of Marchantiales. *Hepaticites cyathoides* of the Triassic resembles *Cyathodium,* whereas *Marchantites hallii* of the Lower Cretaceous closely resembles marchantiaceous genera. *Ricciopsis florinii* of the Triassic is strongly reminiscent of *Riccia.* Thus the fossil record suggests the early presence of all main evolutionary lines in the Marchantiales.

The order Marchantiales, then, represents the peak of gametophytic specialization in the hepatics, having not only a great array of morphological types but also showing numerous patterns of spore germination and sporophyte types. Its relationships are closest to the Monocleales, but it represents a widely divergent evolutionary line within the hepatics, set apart in both thallus morphology and in details of gametophyte development.

Class Anthocerotae (Hornworts)

Hornworts superficially resemble thallose liverworts, particularly some of the Metzgeriales. They have both sporophytic and gametophytic features that separate them as an independent evolutionary line in the bryophytes:

1. A single chloroplast is in each cell in most genera.

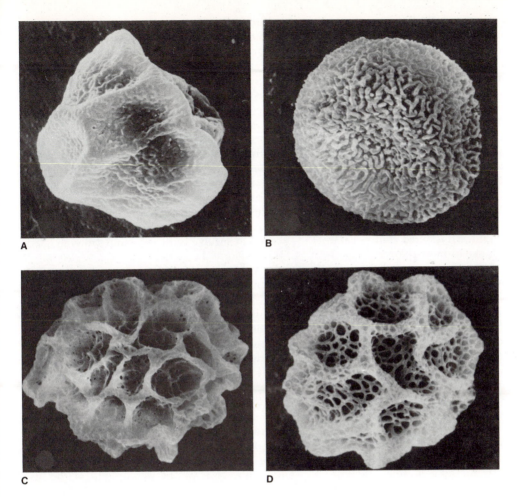

Figure 12-20 Scanning electron micrographs of spores of Marchantiales. **A,** *Neohodgsonia mirabilis,* ×3000. **B,** *Marchantia berteroana,* ×4000. **C,** *Targionia hypophylla,* ×1000. **D,** *Asterella tenera,* ×1200. (After Taylor et al., with permission of *Bryologist.*)

2. A pyrenoid is in the chloroplast in most genera.

3. The archegonium is embedded in the thallus and is not a discrete organ.

4. Sex organs are derived from a superficial cell, but one that never bulges above the thallus surface.

5. Occasionally, stomatelike structures are present in the thallus.

6. The sporophyte is of indeterminate growth, thus differentiating new sporogenous tissue throughout its life.

7. An intercalary meristem is present at the base of the sporophyte.

8. Elaters are generally multicellular.

9. Opening of the sporangium is generally by two longitudinal lines.

Other features are shared by either the Musci or Hepaticae.

The class contains at least 100 species in a single order, the Anthocerotales. A single family with five genera makes up the order. *Notothylas, Anthoceros,* and *Phaeoceros* are rather widely distributed throughout the world, particularly in milder climates, but *Dendroceros* and *Megaceros* are essentially tropical and subtropical. Most species grow on shaded, humid soil, but a few are epiphytic, and others grow in relatively exposed sites on somewhat clayey substrata.

All genera have a flattened thallose gametophore, normally a rosette composed of radiating, overlapping, dichotomous branches. In some genera, the thallus is 1–4 cm wide, but in others it can exceed 5 cm in width. Occasionally, the thallus has a midriblike, thickened central band (as in *Dendroceros*), but in most the central portion is multistratose and thins toward the margin. In some species of *Dendroceros*, the midrib resembles a stem with two rows of irregular, lateral, unistratose leaves. The thallus is either annual or perennial.

The upper cells of the thallus are chlorophyllose, but deeper in the thallus chloroplasts are absent. In most genera there is a single, large, lens-shaped chloroplast in each cell, and a pyrenoid is associated with the plastid. However, in *Megaceros* there are numerous chloroplasts in each cell, and pyrenoids are absent in some species. Most cells of the thallus are thin-walled, but in *Dendroceros* the central cells of the stem have pitted walls similar to those of some thallose hepatics. Within the thallus there are frequently mucilage-filled intercellular cavities that sometimes open to the ventral surface by stomatelike pores (Fig. 12-21B). Often these cavities are invaded by colonies of the bluegreen alga *Nostoc*. The hornwort and alga appear to assume a symbiotic partnership in which the bryophyte receives a source of nitrogen and the alga receives moisture and is protected from environmental extremes. The thallus is attached to the substratum by thin-walled, unicellular rhizoids.

Most species are bisexual, with the sex organs embedded in the upper surface of the thallus. Their mode of origin is very unlike that in other bryophytes: the antheridial initial is derived from the inner cell resulting from transverse division of a cell on the upper surface of the thallus; thus it is endogenous (Fig. 12-21G). This initial can give rise to several antheridia

within a chamber. The archegonial initial is also a superficial cell, and various divisions produce cover cells, neck-canal cells, and egg (Fig. 12-21H). The cells surrounding the contents of the archegonium do not differ conspicuously from others in the thallus; thus the archegonium is not a discrete organ.

Marginal gemmae are produced by a few hornworts, and several species produce **perennating** tubers similar to those of the genus *Geothallus* in the Sphaerocarpales. The tubers can tolerate some desiccation. In still other species, all cells of the thallus die except the growing point with its apical cell and adjacent tissue. All of these vegetative devices permit the gametophore to survive until favorable growing conditions occur.

The sporophyte is always cylindrical and tapers to the apex (Fig. 12-21A). It pierces the upper surface of the thallus and is generally perpendicular to it. The enlarged foot is embedded in the thallus (Fig. 12-21E). The sporophyte varies from a few millimeters up to 16 cm tall. In all species it is green until it begins to shed spores. Early in its development, an intercalary meristem is formed near the base of the sporophyte, above the foot. This meristem continues to differentiate new sporophytic tissue throughout the life of the sporophyte and gives it indeterminate growth—a feature common in vascular plants but otherwise unknown in bryophytes. Except for the foot, the bulk of the sporophyte is the spo-

Figure 12-21 Morphology and anatomy of Anthocerotales. **A,** habit of *Phaeoceros laevis,* with mature sporangia, ×5. **B,** pores of gametophytic thallus of *Anthoceros,* ×225. **C,** stoma with guard cells of *Anthoceros* sporangium wall, ×150. **D,** longitudinal section through mature sporangium (note pseudoelaters among sport tetrads), ×100. **E,** base of *Anthoceros* sporangium, ×160. **F,** pseudoelaters of Anthocerotales (*upper, A. punctatus; lower, Megaceros endivaefolius*), ×500. **G,** young antheridium filling antheridial chamber (note single lenticular chloroplast in each cell), ×200: **H,** transverse section through fertile thallus, showing mucilage chambers (*black*) ×80: *arch,* archegonium; *col,* columella; *ft,* foot; *spl,* sporogenous layer. (**B,** after Goebel.)

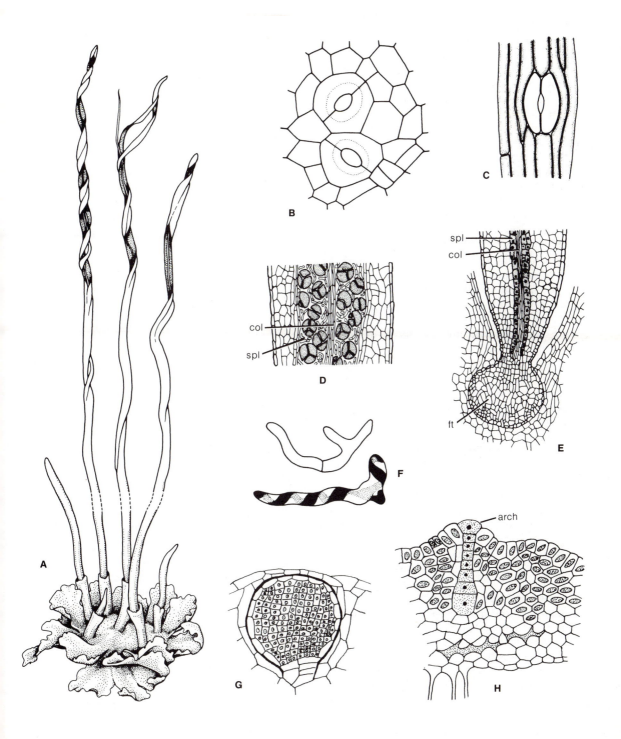

rangium. The jacket is four to five cells thick, and its outer cells are chlorophyllose. In the epidermis of most species, there are characteristic stomata (Fig. 12-21C). The sporogenous layer generally forms a dome-shaped region that overtops a central columella, a feature shared by the mosses *Andreaea* and *Sphagnum*. In *Notothylas*, a columella is missing. Among the spores of Anthocerotae, there are pseudoelaters (Fig. 12-21F). These are often multicellular and have irregularly thickened or spirally thickened walls.

In *Dendroceros*, very elongate pseudoelaters superficially resemble tracheids of vascular plants. Spiral thickenings are on the outer cells of the columella and on the inner cells of the jacket in the same genus. However, in the genus *Notothylas*, pseudoelaters are perpendicular to the length of the sporophyte. Spores are often retained in tetrads until they are shed.

The sporangium usually opens along two opposing longitudinal lines that meet at the apex of the sporangium. Sometimes the two "valves" coil as they dry, which helps to scatter the spores (Fig. 12-21A). The dehiscence lines extend downward toward the sporangium base as sporogenous tissue matures; thus spores are shed over a relatively extended period.

In many hornworts, spores are viable for only a brief period; in others they remain viable after more than ten years. The spore may germinate endosporously and rupture the spore coat irregularly, forming an apical cell soon after emerging. In most, however, a germ tube emerges from the **trilete** scar and forms a short filament that soon initiates a lobate, conical cell mass. In this mass, an apical cell is soon differentiated, which gives rise to a characteristic thallus.

Fossil spores attributed to the Anthocerotae are reported from Tertiary deposits in Hungary. Some features in rhyniophyte fossils suggest that the Anthocerotae may have been derived from an ancestor common to this group. The rhyniophyte *Horneophyton* has a columella; the spore ornamentation resembles that of some species of *Anthoceros*; and the sporophyte is a naked axis, albeit one that is dichotomously branched.

Hornworts and hepatics have so many features in common that it is reasonable to assume a relationship. In both gametophyte and sporophyte, there are several common features: growth and structure of the thallus, rhizoid structure, the single sporangium produced by each sporophyte, and the nature of the sterile cells among the spores. The coincidence that all of these features were evolved independently is not impossible, but without any connecting "links" in either the fossil or extant record, no absolute decision can be made.

The few features shared with mosses also point to a very distant relationship. The gametophytic structure is entirely unlike that of any moss. The single sporangium produced by the sporophyte is common to all bryophytes, but in hornworts and mosses a columella is present in the sporangium. The jacket of the sporangium also has stomata in both of these bryophyte classes. It is possible that these few characteristics could have been derived independently by parallel and convergent evolution in response to similar selective factors in the environment.

Class Musci (Mosses)

Musci represent the most diverse assemblage of bryophytes, with more than 680 genera and somewhat fewer than 15,000 species. These can be separated readily into six distinctive evolutionary lines, corresponding to the subclasses.

In many respects, the mosses are similar to the "leafy" liverworts (order Jungermanniales). In gross terms, the gametophore and sporophyte are structurally similar. There are several fundamental differences:

1. The protonema is normally extensive and usually forms a uniseriate, branched system that precedes the appearance of the gametophore.

2. The gametophore is always leafy, even when extremely reduced, and in most instances, leaves are borne in more than three ranks.

3. Complex oil bodies are never present.

4. Rhizoids are always multicellular.

5. Leaves commonly have a multistratose midrib (costa).

6. Leaf cells are commonly elongate and rarely have trigonous corners.

7. Leaves are rarely lobed.

8. The sporangial wall frequently has stomata and always lacks spiral or nodular ornamentation of the outer cells.

9. The sporangium usually opens by an apical lid, beneath which hygroscopic peristome teeth are frequently present.

10. A columella is present in the interior of the sporangium.

11. The seta is usually long and wiry, and the sporophyte may persist for a long period so that the spores can be shed over an extended interval.

12. The seta elongates before the sporangium is fully differentiated, and the calyptra is torn from the gametophore to form a protective cap over the apex of the elongating sporophyte.

Compared to those of liverworts and hornworts, the moss sporophyte shows a great array of diversity, and gametophores show somewhat less. However, in mosses there is greater tissue elaboration than in other bryophytes, although no mosses have such an elaborate system of chlorenchyma and associated gas-exchange pores as the order Marchantiales of the liverworts. On the other hand, in the moss gametophore an elaborate "vascular" system is sometimes present—a system never so markedly developed in other bryophytes. Also, in some mosses there is a clear development of a "vascular" system in the seta of the sporophyte, a feature paralleled, but not matched, in the columella of some hornworts.

Research in moss taxonomy is still in very early stages; many taxa have been described, but few have received the critical study of a monographer. It is probable that when mosses are studied carefully, the number of valid species may not exceed 7000.

Gametophyte

Spore germination patterns are essentially exosporic, but initial germination is rarely endo-sporic. The protonema is chlorophyllose and is attached to the substrate by means of multicellular, uniseriate, nonchlorophyllose rhizoids that have oblique end walls to the cells. The protonema usually has a uniseriate, branched, chlorophyllose, creeping portion, the cells of which also have oblique walls.

Gametophore structure is highly variable among mosses. There are unbranched shoots in many genera of the subclass Polytrichidae, a complex system of branches in some genera of the subclass Bryidae, and a unique system of clusters of lateral branches in the subclass Sphagnidae. The branches are generally lateral on a main shoot and arise from cortical cells; occasionally, they form a condensed terminal cluster.

The position of the archegonia shows two main patterns: (1) terminating the main shoot, the apical cell of which is used to produce archegonia and their associated structures; in these mosses the sporophyte also terminates the main shoot and is described as **acrocarpous;** (2) appearing on shortened lateral branches, so that the apical cell of the main shoot is not used up in producing archegonia but continues to differentiate new growth from year to year; in these mosses the sporophytes are on budlike, reduced, lateral branches on the main shoot and are termed **pleurocarpous.**

Leaves are always sessile and are generally arranged spirally on the stem. Leaves are generally unistratose except for the costa (or midrib), where they are multistratose. Leaf cells are variously ornamented by thickenings or bulging of the exposed surfaces of the cells, and the walls of adjacent cells are frequently irregularly thickened or pitted. Leaf shapes vary considerably among mosses; details of the arrangement of differently shaped or ornamented cells within the leaf are remarkably constant within each moss species.

Besides the leaves on the stem, there are sometimes small, unistratose, branched, chlorophyllose structures termed **paraphyllia** (Fig. 12-31). These can thickly cover the stem among the leaves, as in the genera *Thuidium* and *Hylocomium*. It is possible that they are important in external water conduction along the stem.

Moss gametophores are frequently unisexual; in bisexual mosses, archegonia are generally on different branches than are antheridia. Arch-

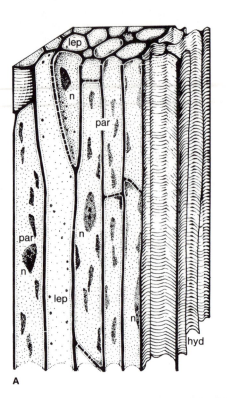

A

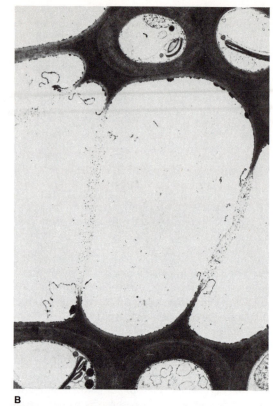

B

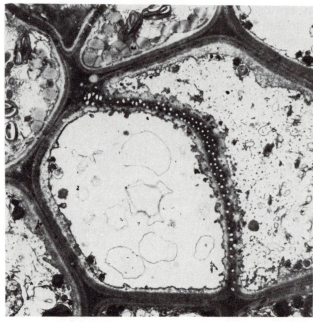

C

Figure 12-22 Conducting systems in mosses. **A,** diagrammatic representation of the organization of conductive tissue in a leafy stem of *Polytrichum commune,* showing hydroids (*hyd*), and leptoids (*lep*), arranged much as are xylem and phloem tissues in vascular plants: *n,* nucleus; *par,* parenchyma. **B,** transmission electron micrograph of hydroids of *P. commune,* ×2000: *st,* stereid. **C,** transmission electron micrograph of leptoids of *P. commune,* ×3500. (Courtesy of C. Hébant.)

egonia are generally surrounded by specialized leaves termed perichaetial leaves, and among the archegonia are scattered filaments termed paraphyses. Antheridia also usually have paraphyses intermixed among them, and these are enclosed by specialized leaves termed **perigonial leaves.**

The stems of mosses are usually structurally more complex than those of hepatics. The epidermal cells, usually termed the *outer cortical cells,* are often structurally different from those of the center of the stem. They usually have thicker walls, but in some instances they are very thin-walled. In many instances, the center of the stem consists of a central strand of smaller cells. The stem cells are mainly longitudinally elongate and taper at both ends. In some mosses, especially in the subclass Polytrichidae, some of these cells resemble sieve-tubes in both structure and function and are termed **leptoids.** In other mosses, especially in the central strand, there are also clusters of tracheidlike cells, the **hydroids** (Fig. 12-22). In some instances, these form a simple "vascular system", with the hydroids and leptoids uniting with a similar system in the leaf costa. The stem is densely clothed in rhizoids, which may be important in external capillary conduction of water.

Vegetative propagation of the gametophore is by the production of rhizoidal gemmae (Fig. 12-5G), by gemmae on leaves (Fig. 12-5C, E, F) and at the tips of stems and branches (Fig. 12-5B), and by small propagula in leaf axils (Fig. 12-5H). Other vegetative means of reproduction are by the fragmentation of rhizoids, of the whole gametophore, or of leaves (Fig. 12-5D). Some mosses appear to depend entirely on such vegetative propagation. It is probable that most expansion of local populations of gametophores is by vegetative propagation.

Sporophyte

Compared to other bryophytes, the mosses show the greatest diversity in sporophyte structure. In most, the sporophyte consists of a sporangium at the apex of a usually rigid seta which is composed of longitudinally elongate cells. Outer cells of the seta tend to be thick-walled and smooth, and inner cells have thinner walls. In some instances, there is a central strand of somewhat smaller cells. In some mosses, the seta has both hydroids and leptoids, whereas in others only leptoids are present, and in still others both are lacking.

While the sporophyte is growing it is rich in chlorophyll. Growth tends to be comparatively slow, sometimes taking several months from the time the egg is fertilized until the time spores are shed from the mature sporangium. Elongation of the seta is also generally very slow, in contrast to its rapid elongation in liverworts.

The sporangium wall is usually several cells thick, and outer cells usually have thicker walls than inner cells. Stomata are often present in the sporangium, particularly near its base. The stomata are generally exposed but are sometimes immersed in a chamberlike depression.

The opening of the sporangium usually follows a concentric line of weakness that circles the sporangium near the apex; the lid so formed is called the **operculum.** In a number of mosses, spores are shed only after the sporangium decomposes or is broken by chance. Normally, the sporangium is partially ensheathed by the calyptra until the operculum is shed.

In most mosses there are teeth around the mouth of the sporangium. These are the **peristome teeth,** which are exposed when the operculum is shed. Various devices have evolved that assist in releasing spores from the sporangium. In most instances, moving air is the main vector for spore dispersal.

In most mosses, the sporangium has central, sterile tissue, the columella, which is surrounded by a cylinder of spores.

Although it is part of the gametophore, the calyptra is so intimately associated with the sporangium that it is treated here. As the sporangium enlarges, the calyptra is usually ruptured so that most of it caps the sporangium, is torn away from the gametophore by the elongating seta, and protects the tip of the seta. As the sporangium enlarges, the calyptra usually remains at the apex of the sporangium, enveloping the operculum. The shape and size of the calyptra appear to exert some control over the differential growth of the sporangium beneath it, thus influencing the orientation of the sporangial mouth.

Subclass Sphagnidae

The genus *Sphagnum* (peat moss) is the sole representative of the subclass Sphagnidae. About 150 species can be distinguished. The genus is distributed worldwide, but forms extensive peatlands only in the Northern Hemisphere. All species grow in wet places, either in water or in sites where free water is available.

The subclass is so markedly different from other mosses that some authors would place it in its own class, Sphagnopsida, parallel to other classes of bryophytes. Features unique to the subclass include the distinctive branching of the gametophore, details of leaf and stem anatomy, dehiscence of sporangium, its unique ecology, and other more technical features.

The Gametophore The gametophore is often large, particularly in aquatic species, where it can exceed 50 cm in length. The living portion of the gametophore rarely exceeds 10 cm, and it is usually much less. The gametophore usually consists of a main axis on which leaves, when mature, are inconspicuous, nonchlorophyllose, widely spaced, and spirally arranged. Generally, at every fourth leaf of the spiral a **fascicle** of branches emerges. Three to eight branches emerge from each point on the stem; two or three of these generally diverge, and the remainder droop downward against the stem. The branch leaves are closely **imbricate** and contain both chlorophyllose and nonchlorophyllose cells (Fig. 12-23A).

The leaves are unistratose, lack costae, and are made up of a network of elongate chlorophyllose cells that encircle larger, swollen, dead **hyaline** cells. Generally, five to six chlorophyllose cells surround each hyaline cell (Fig. 12-23D). The outer walls of the hyaline cells are generally perforated by several pores, and the walls are reinforced by **fibril** thickenings (Fig. 12-24C).

The apex of the shoot of the gametophore is generally composed of a condensed mass of branches, which, as the stem elongates, become fascicles of branches (Fig. 12-23A). The mature gametophore is perennial and lacks rhizoids.

The main stem of the gametophore is somewhat brittle and aggregations of gametophores support each other or the aquatic medium in which the plants grow give them support. The stem is differentiated into two tissues: cortex and central cylinder. The cortical cells are often enlarged, hyaline, porose, and have fibril wall thickenings (Fig. 12-23I). In others, they are hyaline and porose, but lack fibril thickenings. The central cylinder of the stem is formed of an outer region of thick-walled cells, often strongly pigmented, whereas the innermost cells are thin-walled.

The gametophores are bisexual or unisexual. Antheridial branches replace the divergent branches of a fascicle and resemble such branches. The antheridia are spherical, and each is borne on a slender stalk—one at the base of each leaf of a perigonial branch. Perigonial branches are rather swollen near their upper portion and are often more brightly colored than the rest of the gametophore. Perichaetial branches are reduced to a few leaves and also re-

Figure 12-23 Morphology and anatomy of subclass Sphagnidae. **A,** habit of sporophyte-bearing shoot of *Sphagnum palustre,* ×1. **B,** branch leaf of *S. palustre,* ×6. **C,** stem-leaf of *S. palustre,* ×6. **D,** leaf cells of *S. palustre* (note network of chlorophyllose cells surrounding porose hyaline cells; also fibril thickenings of walls of hyaline cells), ×165. **E,** transverse section of *S. palustre* leaf, showing relationships of hyaline and chlorophyllose cells, ×165. **F,** same for *S. squarrosum,* ×215. **G.** same for *S. magellanicum,* ×235. **H,** transverse section of branch-stem of *S. palustre* (note hyaline outer cells and dark-walled cells of central axis), ×30. **I,** external view of stem of *S. palustre,* showing hyaline porose outer cells with their fibril thickenings, ×25. **J,** thallose protonema of *Sphagnum* bearing young, leafy shoot, ×75. **K,** spores of *S. papillosum,* showing sporewall sculpturing and trilete face, ×1100. **L,** longitudinal section through *Sphagnum* sporangium, showing anatomy, ×95. **M,** transverse section of branch-stem of *S. tenellum,* ×120. **N,** external view of branch-stem of *S. tenellum,* showing hyaline retort cells, ×120: *col,* columella; *ft,* foot; *ps,* pseudopodium; *spl,* sporogenous layer. (**H, I,** after Breen, with permission of University of Florida Press; **L,** after Flowers, with permission of Brigham Young University Press, from Flowers, Seville, *Mosses: Utah and the West.* Provo, Utah: Brigham Young University Press, 1973.)

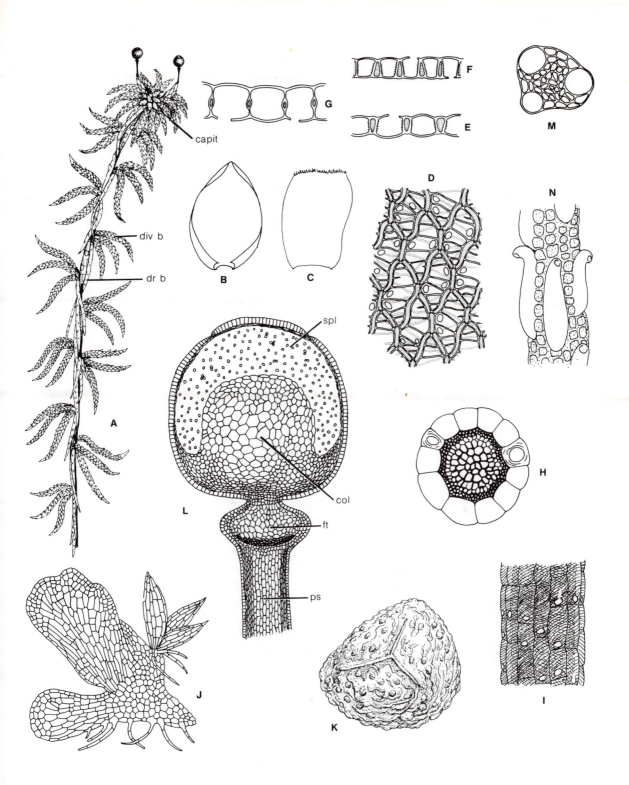

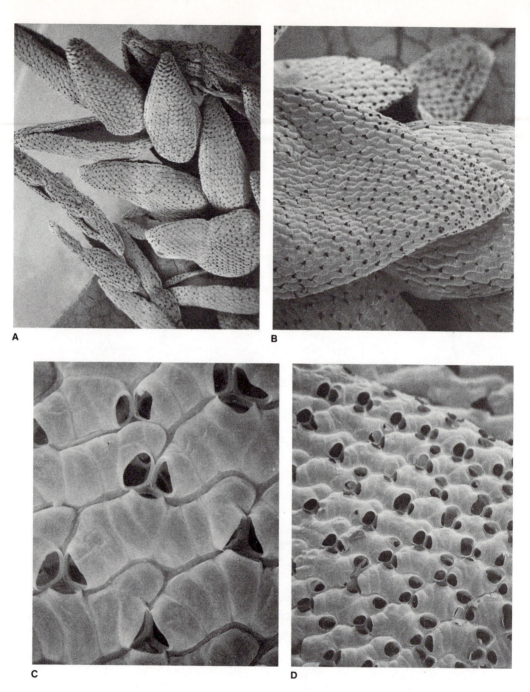

Figure 12-24 Scanning electron micrographs of *Sphagnum* gametophore. **A,** divergent and pendent branch, showing leaf orientation, ×70. **B,** view of one leaf, showing arrangement of pores (*black*) and hyaline cells, ×65. **C,** portion of the abaxial surface of leaf, showing pores, fibril thickenings, and outlines of hyaline cells of leaf, ×260. **D,** hyaline cells and pores with intervening chlorophyllose cells, showing fibril thickenings across hyaline cells, ×700. (After Mozingo et al., with permission of *Bryologist*.)

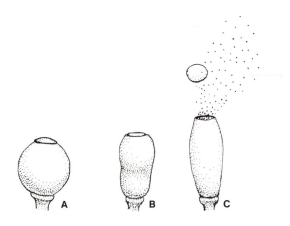

Figure 12-25 Spore dispersal in the genus *Sphagnum* (diagrammatic), showing "air gun" mechanism. **A,** mature sporangium on pseudopodium. **B,** as sporangium dries its diameter decreases, placing sporangium contents under high pressure. **C,** when pressure is sufficient, operculum is thrown off and spores are jetted into the air (see text). (Modified after C. T. Ingold, *Spore discharge in land plants,* 1939, with permission of Oxford University Press.)

place the divergent branches near the apex of the main stem. Perichaetial leaves tend to be larger than normal vegetative leaves and enclose three to four archegonia.

The Sporophyte   After fertilization of one egg, the apex of the perichaetial shoot, just beneath the archegonia, begins to elongate somewhat. As the sporophyte matures, there is rapid elongation of the gametophytic tissue beneath it, extending the sporophyte on a leafless **pseudopodium** (Fig. 12-23L) well beyond the perichaetium. The sporophyte is made up of a subspherical sporangium and an enlarged foot. The seta is virtually absent, and its role is supplanted by the production of the pseudopodium.

As it matures, the sporangium is completely enclosed within a calyptra; the capsule enlarges, and the calyptra is torn. The sporophyte is chlorophyllose until it matures, when the walls become brownish and chlorophyll disappears. The epidermis of the sporangium has many rudimentary stomata. The sporangium opens by means of an operculum. Within the sporangium, the sporogenous layer forms an inverted cup over the

central, dome-shaped columella. The rupture of the operculum is remarkably specialized (Fig. 12-25). As the sporangium matures, the cells of the columella collapse and are replaced by gaseous material. As the sporangium dries, it shrinks in circumference and compresses the gas within and beneath the spores. This internal pressure can build up to five atmospheres and becomes so great that the operculum is thrown off explosively. With this release of pressure, the spores are ejected into the air and carried away.

The spore germinates to produce a short filament terminated by a fragile, multicellular, unistratose, chlorophyllose thallus with short rhizoids extending to the substratum (Fig. 12-23J). This thallus ultimately produces a bud near its base, which initiates the leafy gametophore. The expansion of colonies of *Sphagnum* occurs mainly by vegetative reproduction. When broken off, young branches can ultimately produce an entire gametophore.

Ecology and Uses   Because of the abundance of dead, porose hyaline cells, the gametophores of *Sphagnum* absorb great quantities of fluid, sometimes up to 20 times their own weight. This characteristic is significant both in their commercial use and in their role in vegetation.

Floating mat species of *Sphagnum* that grow on lake and pool margins add to the acidity of these water bodies. The living or dead *Sphagnum* is able to adsorb cations from the water, and thus leave it more acidic. In the highly acidic conditions of such water bodies, the aquatic flora is often very limited. Frequently the water is also darkly stained by organic acids leached from the adjacent vegetation, making colonization by algae and othe plants restricted. This results in a turbid, highly acidic, poorly oxygenated water body, which inhibits the decomposition of organic material entering it, whether formed by the resident *Sphagnum* or adjacent terrestrial vegetation.

The floating mat, first restricted to the perimeter of the lake, expands toward the center. The accumulation of organic material in the marginal shallows makes these areas sufficiently solid that other plants can invade them. The decrease in free water and increase in acidity first affect the species of *Sphagnum* that have produced these conditions. Although they cannot persist, other *Sphagnum* species can. These then move in, with an array of vascular plants and other mosses. The ultimate result of such succession and persistent organic sedimentation is the filling in of the water body with **peat.** Peat is composed of a matrix of partially decomposed *Sphagnum* and the remains of plants that grew with the *Sphagnum* as well as the remains of material blown or washed into the lake.

Such lake peat deposits are important in preserving a record through time of the sequence of vegetational change adjacent to the lake. Some pollen and spores that fall annually upon the surface of such a lake ultimately sink to the bottom and are preserved from decay. Year after year, new deposits are added, each layer forming a record of the spores of many plants from the surrounding vegetation of that year. If water circulation does not disturb these deposits, careful analysis of the record of a sequence of strata, beginning with those deposited earliest, reveals the succession of vegetation through time.

Most peatland is not a result of lake-deposited peat but results from the establishment of *Sphagnum* and other plants tolerant to waterlogged habitats. Because of its highly absorptive capacity, the growth of *Sphagnum* tends to increase waterlogging. This, in turn, decreases air circulation in the soil and impedes decomposition of accumulating organic material, so that peat accumulates.

There are extensive areas of peatland in many parts of the boreal zone of the Northern Hemisphere. In North America, the Hudson Bay Lowlands of Canada have a continuous area of peatland nearly as large as the British Isles. In the Soviet Union, there is an even more extensive area in the west Siberian Lowland.

The boreal terrain contains many areas of impeded drainage as well as many natural waterbodies occupying depressions produced by continental glaciations. The combination of short, cool summers, waterlogged conditions, and low nutrient availability favors the growth of *Sphagnum*, and thus vast expanses of peatland develop.

Several species of *Sphagnum* are brightly colored. The reddish pigment of some species is caused by anthocyanin and is confined to cell walls. The pigment tends to increase coincident with high daytime and low nighttime temperatures. It is reduced in poorly illuminated areas and is correlated with high sugar content of the living cells.

Most human use of *Sphagnum* is determined by the remarkable absorptive capacity of the gametophores and by the extremely light weight of dried plants. *Sphagnum* peat has been used for many years as insulating material for homes. It has also been used as fuel for fires. Its replacement of cotton as a surgical dressing was of particular importance during much of the first World War when cotton was unavailable.

Fossils   Permian fossil materials of the genera *Vorcutannularia*, *Jungagia*, and *Protosphagnum* show cell arrangement similar to that of immature *Sphagnum* leaves (Fig. 12-26). However, in these fossil genera the leaves are costate. In *Vorcutannularia*, leaf arrangement is also more like that of Bryidae than of Sphagnidae. Undoubted *Sphagnum* spores and leaves are known from as early as the lower Jurassic, whereas modern species can be determined from Quaternary material.

Subclass Andreaeidae

The Andreaeidae include a single order, Andreaeales; a single family, Andreaeaceae; and two genera: *Andreaea*, with probably fewer than 50 distinct species, and *Andreaeobryum*, with one. The order is widespread in temperate to frigid climates throughout the world.

Andreaea grows on exposed siliceous rock, where it usually forms blackish to red-brown cushions. Most species tolerate some desiccation; some are confined to very wet sites or areas of long snow persistence. *Andreaeobryum* grows on limestone in northwestern North America.

Andreaea shares with *Sphagnum* the fact that the sporangium is raised above the perichaetium on a pseudopodium and also in the position of the sporogenous tissue that overarches the columella (Fig. 12-27H). In sporangium dehiscence, it is unique among mosses; there are longitudinal lines of weakness; thus the sporangium wall gapes open, often by four slits, as in many of the hepatics. However, in *Andreaea*, the slits usually do not extend to the apex (Fig. 12-27A, B). *Andreaeobryum* has a short seta.

The protonema is generally multiseriate and

has some biseriate as well as uniseriate rhizoids. Although very simple in morphology and anatomy, the gametophore most resembles that of some Bryidae. The brownish to black gametophore is irregularly branched, and usually tightly affixed to the substratum by rhizoids at the base of the shoot (Fig. 12-27A, C). The leaves are spirally arranged, crowded, and generally

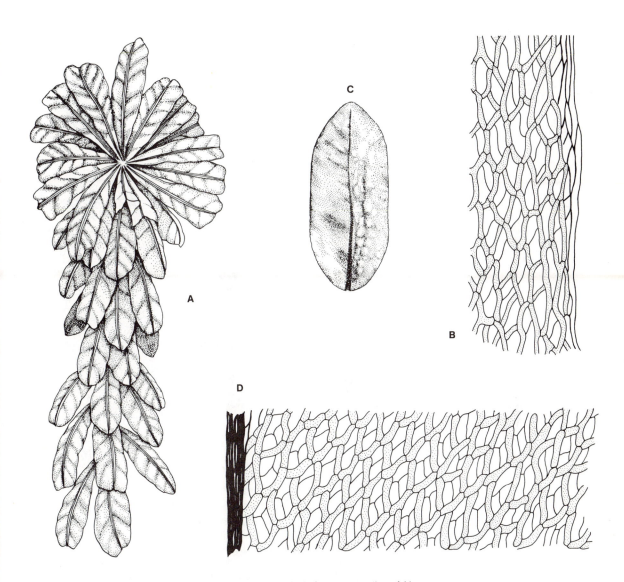

Figure 12-26   Fossil mosses (possibly related to Sphagnidae). **A,** reconstruction of *Vorcutannularia plicata*, moss of the Permian, ×2. **B,** cellular detail of leaf of *V. plicata*, showing hyaline and chlorophyllose cells as well as leaf margin, ×100. **C,** leaf of *Protosphagnum nervatum*, a Permian moss, ×8. **D,** cellular detail of leaf of *P. nervatum*, showing chlorophyllose and hyaline cells, ×100. (After Neuberg.)

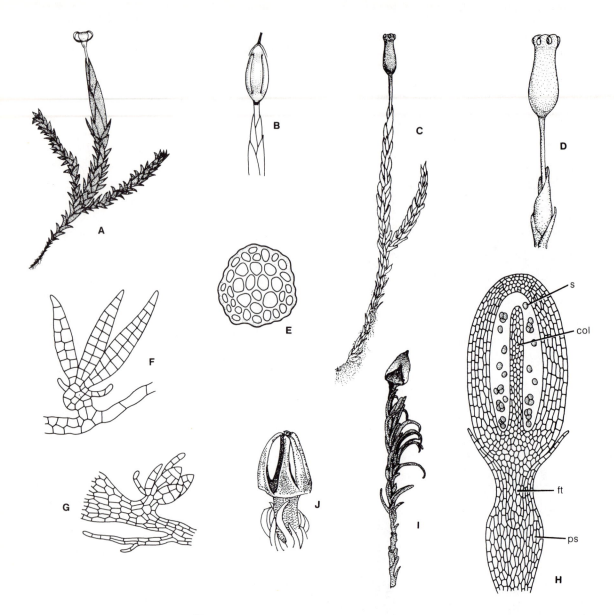

Figure 12-27 Morphology and anatomy of subclass Andreaeidae. **A,** habit sketch of *Andreaea rupestris* bearing dehiscing sporangium, ×8. **B,** sporangium of *A. rupestris* when moist, ×15. **C,** habit of *A. wilsoni* bearing dehiscing sporangium, ×5. **D,** detail of sporangium of *A. wilsoni,* showing peristome-toothlike splits around capsule mouth. ×8. **E,** transverse section of stem of *A. rupestris,* showing undifferentiated cell structure, ×135. **F,** portion of straplike protonema of *Andreaea* with frondiform flaps, ×130. **G,** portion of protonema of *Andreaea* bearing a single, young, leafy shoot, ×130. **H,** longitudinal section through sporangium of *Andreaea,* ×130. **I,** *Andreaeobryum macrosporum* habit, showing sporophyte with seta and calyptra, ×3. **J,** *A. macrosporum,* detail of opening sporangium, X5: *col,* columella; *ft,* foot; *ps,* pseudopodium; *s,* spore. (**F, G,** after Ruhland.)

ovate to lanceolate. Except for those around the sex organs, the leaves are all similar. They are costate in some species, with a multistratose midrib in the center of the leaf. In others a costa is absent. Most leaves are unistratose except at the costa. Leaf cell walls are smooth, or each cell may bear a conspicuous thickened wart or papilla on its exposed surfaces. All leaf cells are small and chlorophyllose, although this is obscured by the darkly pigmented cell walls. The stem is of thick-walled cells throughout, and walls are pigmented (Fig. 12-27E). The mature gametophore varies from 1–20 cm in length, but is usually much smaller and bears leaves 0.5–3 mm long. The gametophores are bisexual or unisexual. Perichaetial leaves terminate the main shoot; thus the gametophore is acrocarpous. The perichaetial leaves are usually more than twice as large as the vegetative leaves of the shoot. There are several archegonia in each perichaetium, and they are intermixed with a few short paraphyses. Antheridia occur on reduced, bulbiform, lateral branches, and perigonial leaves envelop the several subspherical to elongate antheridia; a few **filiform** paraphyses are usually present.

The sporophyte consists of a small, elliptical sporangium that, when mature, is raised above the perichaetium by a pseudopodium in *Andreaea*, and by a short seta in *Andreaeobryum*. The opening of the sporangium in *Andreaea* is by four to six longitudinal lines of weakness. As the sporangium dries, it shrinks in length, the openings gape apart, and the spores fall out. Rewetting closes the sporangium, which again gapes open upon drying. In *Andreaea*, the calyptra forms a small apical cap on the sporangium and is ephemeral. In *Andreaeobryum*, it is large and persistent (Fig. 12-27I), enveloping the sporangium until it opens.

Spores of *Andreaea* are usually multicellular when shed, a feature shared by a few other bryophytes. When the spore coat is ruptured, the spore contents produce a multiseriate filament, or thallus. This branches irregularly, and biseriate branches form rhizoids that affix the thallose primary protonema to the hard rock substratum. Produced on the upper surface of this thallus are clusters of biseriate, determinate flaps of tissue (Fig. 12-27F). These can serve as vegetative propagules or as an extension of the photosynthetic surface. These protonemal flaps are similar to

those in the Tetraphidae. Ultimately, one or several buds are produced on the thallus, from which a leafy stem arises. The thallus is microscopic and disappears as soon as the leafy gametophores are produced.

Despite some superficial similarities to *Sphagnum*, *Andreaea* is probably more closely related to other subclasses of mosses, but its mode of sporangium dehiscence, and the nature of its protonema and rhizoids are not shared by any other mosses. It cannot be considered as derived from or very closely related to any other subclass of mosses.

Subclass Tetraphidae

The subclass Tetraphidae consists of the single order Tetraphidales, with one family, Tetraphidaceae, containing the genera *Tetraphis* and *Tetrodontium*. *Tetraphis* includes two species and is restricted to the Northern Hemisphere, largely in temperate forests, where it is especially common on shaded, rotting logs and stumps or occasionally on sandstone. *Tetrodontium* has two (or three) species and is confined to siliceous rock. Like *Tetraphis*, it grows in poorly illuminated sites, in this case in humid, shaded crevices and overhangs of cliffs. *Tetrodontium* is widely scattered, but local, in the Northern Hemisphere, and very rare in New Zealand.

The Tetraphidae have a number of features that strongly contrast with those of the Sphagnidae and Andreaeidae: the sporangium, which is raised high above the perichaetium by a long stalk, or seta, has an operculum; there is a peristome of four multicellular teeth (Fig. 12-28E); the sporogenous layer forms a cylinder around the columella instead of overarching it; persistent rhizoids are uniseriate and branched, and the gametophore produces apical masses of gemmae.

The gametophores of *Tetraphis* form extensive, short turfs of erect, unbranched shoots (Fig. 12-28A). The shoots that first arise from the protonema produce apical clusters of gemmae (Fig. 12-28A) that are surrounded by a cuplike group of blunt leaves. Each disclike gemma surmounts a slender stalk, which is easily broken. Upon germination, the gemma produces a primary pro-

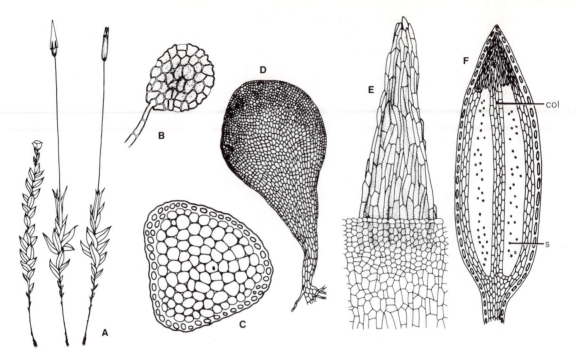

Figure 12-28 Morphology and anatomy of subclass Tetraphidae: *Tetraphis pellucida.* **A,** habit (gemma-bearing plant at *far left*), ×4. **B,** lenticular gemma, ×225. **C,** transverse section of stem, showing slight differentiation in cells, ×35. **D,** frondiform protonemal flap, ×35. **E,** multicellular peristome tooth, ×150. **F,** longitudinal section through sporangium, ×10: *col,* columella; *s,* spore. (**D,** after Ruhland.)

tonema of uniseriate filaments, from which emerges a secondary protonema of perpendicular, unistratose frondiform flaps (Fig. 12-28D). These flaps increase the photosynthetic surface of the protonemal stage. Later, from the bases of these flaps, both gemmiferous shoots and archegonium-bearing shoots arise, in which the leaves are in three to eight ranks. At the apex of the shoot is the perichaetium, surrounding eight to ten archegonia. No paraphyses are present. If fertilization does not occur, the archegonium-bearing shoot gives rise to one or two lateral branches that emerge just beneath the perichaetium. Each of these branches is terminated by a cluster of larger and blunter leaves surrounding the antheridia and their intermixed filamentous paraphyses. Superficially, these perigonial branches resemble gemmiferous shoots.

The leaves of the gametophores are costate but are otherwise unistratose. The cells are hex-agonal and filled with numerous chloroplasts. The stems are relatively undifferentiated, with thicker-walled outer cells and a central cylinder of parenchyma cells; usually a small central strand of thicker-walled, small hydroid cells is also present. The entire gametophore is rarely taller than 3 cm.

The sporophyte is always formed of an elongate seta surmounted by a cylindrical sporangium. The sporangium is almost completely enclosed by a pleated, conical calyptra. Beneath the conical operculum are four multicellular peristome teeth. These are formed from four equal fractions of all cells enclosed by the operculum, a feature unique to this subclass. In *Tetraphis,* the sporangium has no stomata, although in *Tetrodontium* a few are present. Within the sporangium, the sporogenous layer forms a cylinder around the central, cylindrical columella. The spores are small and are shed when the operculum falls off. The peristome teeth gape open somewhat when dry, releasing spores slowly. When wet, the teeth tend to close the mouth of the sporangium. Germinating spores form a filamentous protonema that gives rise to protonemal flaps and later to gemmiferous shoots and archegonium-bearing gametophores.

In *Tetraphis*, the **frondiform** protonemal flaps disappear soon after the leafy gametophores appear, but in *Tetrodontium* the flaps are long-persistent, and the gametophores are usually reduced to a few leaves. Unlike those of *Tetraphis*, the leaves in *Tetrodontium* are nearly ecostate, and the stem is less than 1 mm long. Gemmae are absent in *Tetrodontium*, and the protonemal flaps tend to be lanceolate to tongue-shaped.

The Tetraphidae are often treated as belonging to the Bryidae because of the general similarity of the gametophore to that of many Bryidae. Because of their multicellular peristome teeth, as well as other features, the Tetraphidae appear to be an isolated evolutionary line, probably more closely related to the Polytrichidae than to the Bryidae, but not very closely related to any extant subclass of mosses.

Subclass Polytrichidae

Some of the most conspicuous of mosses, the hair-cap mosses, are included among the Polytrichidae, which are distributed throughout the world.

The subclass includes a single order, Polytrichales, with a single family, Polytrichaceae, containing 20 genera with somewhat fewer than 400 species. Most species grow on somewhat acidic to neutral soil. Some are highly tolerant of desiccation, and grow in brightly illuminated sites. Others are restricted to humid, shaded sites, and some are common in boggy or wet habitats.

In most genera, the gametophores are conspicuous. *Dawsonia* can produce erect, unbranched shoots up to 80 cm tall (Fig. 12-29I); in *Rhacelopus* and some species of *Pogonatum*, the gametophore is reduced to a few leaves and is less than 3 mm tall. In the latter instances, the protonema is perennial and forms an algalike film over the substratum.

Most species are unisexual, and gameto-phores are usually perennial. The antheridium-bearing gametophore is remarkable in its annual production, in many species, of an apical rosette of brownish, enlarged perigonial leaves surrounding the large mass of antheridia and their intermingled paraphyses. The center of the rosette remains meristematic and thus produces an extension of leafy stem, ultimately surmounted by another perigonial rosette that matures the following year (Fig. 12-29C). On the other hand, the archegonium-bearing shoot produces a perichaetium at the apex of the main shoot. The three perichaetial leaves sheathe the archegonia and their associated paraphyses. Further growth of the archegonium-bearing shoot follows the intercalary production of a new apical cell from beneath the perichaetium and from the stem cortex.

In many Polytrichidae, the aerial shoot is continuous with a subterranean organ, often termed a **rhizome.** This is heavily sheathed with rhizoids and is sometimes branched, producing aerial shoots at intervals along its length. The rhizome functions as both an anchoring and an absorptive organ, but its main function is probably in the vegetative expansion of a clone.

The erect shoot is usually unbranched, but in some genera it is branched. The stem is flexible, and aggregated gametophores frequently form tall turfs.

The leaves are spirally arranged. The leaf base is often colorless and sheathes the stem whereas the lamina diverges outward exposing the upper surface. The leaf usually has a broad costa, and the adaxial surface of the costa is often covered by numerous perpendicular, longitudinal, unistratose plates of tissue, the **lamellae** (Fig. 12-29F). These are chlorophyllose and greatly expand the photosynthetic surface of the leaf. In some instances, the lamina curves over the lamellae, partially enclosing them in a broadly cylindric chamber. The lamellae thus become a chlorenchymalike layer, protected from desiccation by the unistratose, colorless lamina, which acts as an epidermis. In most species, the leaves are toothed. The apical cells of the lamellae are often markedly thickened or ornamented; the abaxial surface of the leaf is frequently cutinized.

The stems of some Polytrichidae are differentiated into conspicuous tissues comparable to the tissues in shoots of vascular plants (Figs.

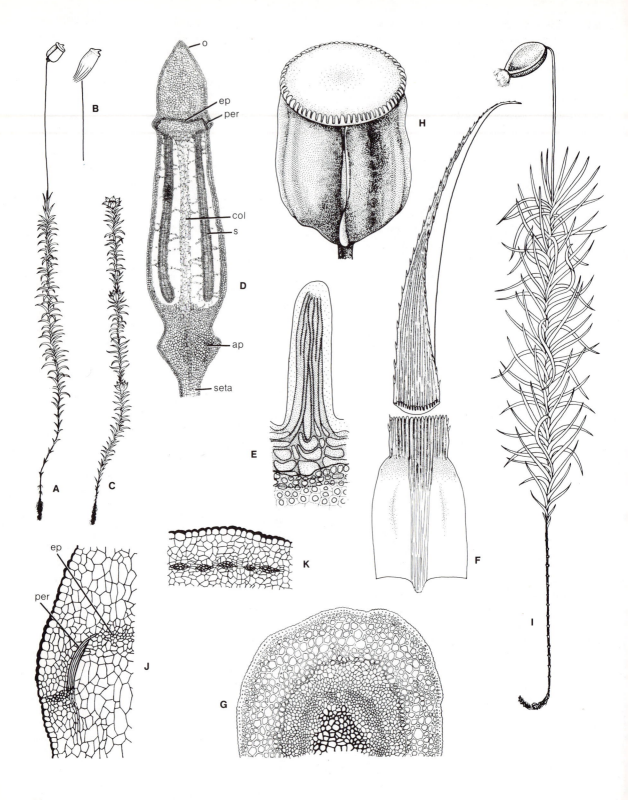

12-29G and 12-22). The outer layer of the stem constitutes the epidermis, generally composed of much-thickened, darkly pigmented, elongate cells. The cortex forms a broad, multistratose layer, in which the outer cells are thick-walled, and the inner cells are parenchymatous and predominantly chlorophyllose. Occasionally leaf traces are in this cortical region, and these are sometimes continuous with the axial cylinder. The axial cylinder is composed of a central cylinder of conducting cells, the hydroids, comparable to the tracheary elements of vascular plants. Surrounding the hydroids is an interrupted cylinder of leptoids, comparable to sieve cells. The hydroids are water-conducting, whereas the leptoids conduct metabolites. Leaf traces also contain hydroids and leptoids; thus there is a vascular system in the gametophore of the Polytrichidae comparable to that in the sporophyte of Tracheophyta.

The rhizomatous portion also has hydroids and leptoids continuous with those of the aerial stem, but their arrangement is somewhat different. The central strand is three-lobed, with hydroid cells scattered among smaller, thick-walled cells, the **stereids**. Between the lobes of the strand are groups of leptoids. Rhizoids that emerge from the epidermal cells are uniseriate and branched.

The sporophyte in the Polytrichidae always has a wiry seta bearing an erect or suberect, generally cylindrical sporangium. In *Polytrichum*, however, the sporangium is distinctly four-angled (Fig. 12-22H). Stomata are sometimes present in the epidermis. In many instances, they are confined to the groove between the body of the sporangium and the swollen upper end of the seta, the **apophysis**, which forms a neck to the sporangium.

The axial strand of the seta is of hydroids, outside of which there are sometimes leptoids. These are surrounded by several rows of thick-walled cells and bounded on the outside by the epidermis. The foot pierces the upper end of the gametophore, thus forming a conducting system leading from the gametophore upward to the sporangium. When young, the seta is chlorophyllose, but as it ages, the cells become reddish to yellow. The mature seta is usually brownish.

The sporangium is sometimes of complex anatomy, with a cylinder of air spaces between the sporangial jacket and the sporogenous layer, and another cylinder of air spaces between the columella and the sporogenous layer. The sporangium always opens by an operculum, beneath which is a membrane composed of the expanded apex of the columella, the **epiphragm**. Overarching the edge of the epiphragm, and attached to it, are the blunt, multicellular peristome teeth. There are 16, 32, or 64 in number, depending on the genus, and are not hygroscopic. The spores are minute and must pass through the small openings between the peristome teeth and the epiphragm. Any slight movement of the mature sporophyte causes some spores to be released; thus they are released gradually rather than all at once.

In several genera, there are no peristome teeth, but the epiphragm is still present. In such genera, as the sporangium dries, it tends to shrink in length, leaving an opening between the epiphragm and sporangium rim through which spores escape. In the genus *Dawsonia*, the peristome is made of numerous rows of multicellular, filamentous, intertwined teeth, and an epiphragm is absent. In other Polytrichidae, each tooth is relatively short and stout. In most instances, each tooth is derived from four concentric layers of cells, but in *Dawsonia* eight concentric layers are involved, and each tooth can be derived from a single vertical cell row.

Figure 12-29 Morphology and anatomy of subclass Polytrichidae. **A–H,** *Polytrichum commune.* **A,** sporophyte-bearing plant after calyptra has fallen, ×1; **B,** sporangium with calyptra in place, ×1; **C,** antheridial plant, showing three successive years' crops of perigonia, ×1; **D,** longitudinal section through sporangium (idealized), ×25; **E,** multicellular peristome tooth, ×75; **F,** leaf, showing lamellae on upper surface, ×12; **G,** transverse section of stem, showing differentiation into tissuelike areas (hydroids in center), ×100; **H,** sporangium, showing peristome teeth and epiphragm, ×8. **I,** habit of *Dawsonia superba*, ×1. **J,** longitudinal section of *Polytrichum commune* sporangium near apex, ×50. **K,** transverse section of sporangium near apex, showing multicellular teeth (*darkened bands*), ×50: *ap*, apophysis; *col*, columella; *ep*, epiphragm; *o*, operculum; *per*, peristome; *seta*, seta; *s*, sporogenous layer. (**D,** after Gibbs; **G, J, K,** after Flowers, with permission of Brigham Young University Press, from Flowers, Seville, *Mosses: Utah and the West.* Provo, Utah: Brigham Young University Press, 1973.)

The calyptra sheathes the immature sporangium. It is usually completely invested with a mass of tawny hairs that originate from near its apex.

Generally less than 10 μm in diameter, the minute spores germinate to produce an extensive, branched heterotrichous protonema. Erect branches (the secondary protonema) are strongly chlorophyllose, and transverse walls are at right angles, whereas the creeping portion and rhizoids (the primary protonema) have oblique end walls. In most instances, the protonema disappears with the appearance of leafy gametophores. In others, in which the gametophores are greatly reduced, the protonema is perennial.

Relationships of the Polytrichidae are far from clear. In peristomial structure, the subclass bears some affinity to the Tetraphidae. However, Polytrichidae are far more specialized. In gametophytic structure, they closely approach the acrocarpous Bryidae in fundamental structure, but there are numerous features almost restricted to the Polytrichidae, including the nature of rhizomes, lamellar structure of leaves, and the complexity of stem anatomy. No other mosses have such peristome teeth or an epiphragm. Fossils attributed to this subclass are known from the Eocene. Earlier fossils are highly suggestive of the Polytrichidae, but their imperfect preservation does not permit confident determination.

Subclass Buxbaumiidae

Although small, the subclass Buxbaumiidae shows surprising gametophytic and sporophytic diversity. The four genera are generally treated in two families: the Buxbaumiaceae, with *Buxbaumia* (10 species); and the Diphysciaceae with *Diphyscium* (19 species), *Theriotia* (probably a single species) and *Muscoflorschuetzia*, with one species. All are restricted to acidic or neutral sites—some on earth or rotten wood and others on rock. They are rather widespread in northern temperate to subtropical climates and are uncommon in the tropics and Southern Hemisphere.

The gametophore of *Buxbaumia* is the most reduced one known for the mosses. It consists of a branched, uniseriate protonema that gives rise to minute perichaetia composed of six to ten

nonchlorophyllose leaves surrounding one to five archegonia (Fig. 12-30B, C). A stem is barely evident. The antheridial shoot is borne on the same protonema and is even more reduced: it consists of a simple bud made up of a unistratose flap of tissue folded around a single, spherical antheridium (Fig. 12-30E).

In the Diphysciaceae, the gametophore is more conspicuous (Fig. 12-30F). The stem is short and bears several ranks of spirally arranged, costate leaves. In *Diphyscium* and *Muscoflorschuetzia*, most of the leaf cells are chlorophyllose, but in *Theriotia* there is specialization of the leaf cells of the multistratose leaf. There is a single, central stratum of chlorophyllose cells, and the rest of the leaf is composed of hyaline cells lacking chlorophyll. As in *Buxbaumia*, the gametophores are unisexual, but with several archegonia intermixed with reduced paraphyses. Much like those in the Polytrichidae, the perichaetial leaves are often brownish and lack chlorophyll, but like those in some species of *Buxbaumia*, they are ciliated. The antheridial plants arise as dwarf shoots beneath the perichaetial shoot. Filiform paraphyses are intermixed among the antheridia.

In the Diphysciaceae, the uniseriate, branched protonema produces protonemal flaps in much the same way as do the Tetraphidae. However, in the Diphysciaceae the flap is shaped rather like a golf tee (Fig. 12-30G). As in *Tetraphis*, these flaps disappear as the leafy gametophores are formed.

In *Buxbaumia*, the sporophyte always has a conspicuous, rough seta that is terminated by an obliquely oriented sporangium, the upper diagonal face of which is somewhat flattened (Fig. 12-30A). The small operculum is generally erect, and the calyptra barely covers it. When immature, the sporophyte is green and must synthesize most metabolites for its growth. The sporangium has a similar, elaborate, internal anatomy common to some Polytrichidae and a few Bryidae. Between the jacket of the sporangium and the sporogenous layer, there is an extensive cylinder of air space. This is supported by filaments that traverse the space between the jacket and the outer boundary of the sporogenous layer.

Like that of many Bryidae, the peristome of *Buxbaumia* consists of an **endostome** (inner peristome) and an **exostome** (outer peristome) (Fig.

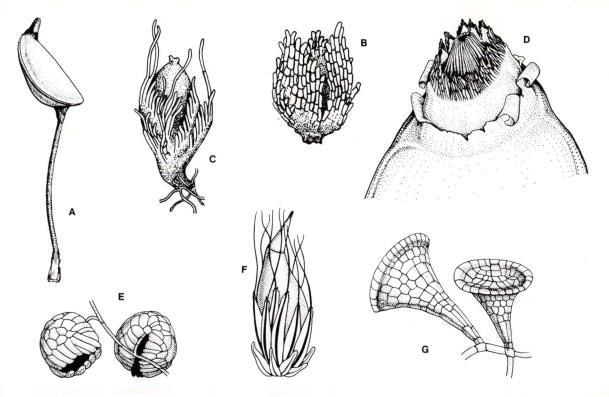

Figure 12-30 Morphology and anatomy of subclass Buxbaumiidae. **A–E,** *Buxbaumia aphylla.* **A,** habit, ×7; **B,** gametophore with single archegonium, ×120; **C,** young gametophore with developing sporophyte, ×120; **D,** apex of sporangium, showing peristome (note outer multicellular peristome and inner pleated cone), ×13.5: **E,** antheridial "branches," external view, ×225. **F, G,** *Diphyscium foliosum.* **F,** habit of sporangium-bearing gametophore, ×8. **G,** protonemal flaps, ×25. (**E,** after Goebel; **G,** after Ruhland.)

12-30D). The endostome is a colorless, truncate, pleated cone with 32 pleats and is composed of thickened remnants of the walls of two adjacent layers of cells. Because the endostome is somewhat twisted, spores must escape through a narrow, apical opening. Any external pressure, even a gentle touch of the jacket of the sporangium, causes spores to be puffed out into the air. The outer rows of peristome teeth alternate with the grooves of the pleats. There can be up to four concentric rows of these exostomial teeth. These teeth are also formed of thickened cell wall remnants derived from two adjacent concentric layers of cells. Their thickening is much greater than in the endostome, and they are opaque and separate. They are not hygroscopic. Outside the exostomial teeth, there is often a multicellular

ring of cells termed a **false annulus.** It is often extensive and is hygroscopic. Thus, in *Buxbaumia* the peristome is the most elaborate known in mosses.

Near the base of the sporangium, where it joins the seta, there is a conspicuous swelling, the apophysis. In the epidermis of the apophysis are often a few stomata.

Sporangia are similar in the Diphysciaceae. Each sporangium is partially sheathed by perichaetial leaves and has a very short seta, evident only after careful dissection. In shape and anatomy, the sporangium is identical to that of *Buxbaumia,* but the endostome is of 16 pleats rather than 32, and exostomial teeth are absent or rudimentary. In *Muscoflorschuetzia,* no teeth are present.

Within this single subclass, then, there is a remarkable reduction series in sporophytes, with the most reduced gametophore associated with the most elaborate sporophyte, and the most elaborate gametophore, on the other hand, associated with the most reduced sporophyte. The Buxbaumiidae seem most closely related to the Bryidae. The leaves in the Diphysciaceae are more like those in some genera of Bryidae than in most other subclasses of mosses.

Subclass Bryidae

Most mosses are contained in the subclass Bryidae. There are at least 75 families, more than 650 genera, and approximately 14,000 species. Most families are poorly studied, and few genera are well understood. Even general evolutionary lines are unclear, and inadequate information has produced a certain lack of uniformity in classification.

Species of Bryidae are found nearly everywhere a plant can grow. Some tolerate extended periods of desiccation, and others are aquatic. Most flourish in environments of high humidity where competition with other plants is slight. Many have strict substratum requirements.

The Bryidae share with other subclasses a diversity of features, many of which appear to have been selected through parallel or convergent evolution. Since these various subclasses are all limited to similar environments, it is inevitable that there should be a certain degree of similarity in gross structure. Other, more detailed features indicate a common ancestor. The meager fossil material supplies no assistance in clarifying these problems.

In the Bryidae, there appear many features that are absent in all other subclasses—a situation inevitable in such a vast assemblage of genera and species. Not all Bryidae share all the features given here. Indeed there are two prominent and independent evolutionary lines within the Bryidae: those in which the gametophore is acrocarpous, and the peristome is of articulate, opaque teeth in a single series; and those in which the gametophore is pleurocarpous, and the peristome is of an exostome of opaque, articulate teeth and an endostome of hyaline, articu-

late teeth. Unfortunately, not all Bryidae fit neatly into these two lines; some share features of both lines.

Features essentially restricted to the Bryidae include, in the gametophore, the pleurocarpous condition, paraphyllia (Fig. 12-31) on the stem, leaf gemmae, and strong differentiation of the alar cells in leaves (Fig. 12-32F). Features that unite the Bryidae are largely sporophytic, particularly the details of peristomial structure and development. The peristome teeth are never derived from more than three concentric rows of cells; this is in strong contrast to all other subclasses of peristomate mosses, in which more than three concentric rows are involved in tooth formation. The peristome teeth are always articulate and consist, in large part, of only cell wall fragments, not of complete cells. The elaboration of cell wall ornamentation and gross morphology of the peristome teeth reaches its peak of complexity in this subclass. The peristome teeth are generally hygroscopic and are important in spore dissemination.

The Gametophore The gametophore of the Bryidae is always leafy, although in a few instances it is reduced to only sexual branches, and photosynthetic activity is taken over by a persistent protonema—a phenomenon noted also in a few Polytrichidae. The gametophore varies from a height of less than 2 mm to more than 5 cm. In reclining forms or in species festooning branches or clothing tree trunks, lengths can exceed 40 cm. In aquatic forms, as in some species of *Brachelyma*, the gametophore can approach a meter in length. In those with a more elaborate gametophore, leaf arrangement is generally spiral and in more than three ranks. Leaves are strongly divergent or imbricate and frequently alter their position in response to changes in moisture.

There is much diversity in the growth forms of the Bryidae. Most acrocarpous genera form turfs or cushions (Fig. 12-33E–G, I, J), whereas pleurocarpous genera also form mats, wefts, or loose colonies of dendroid gametophores (Fig. 12-33A, C, D, H). The growth form is strongly affected by the microclimate, particularly relative humidity and shade.

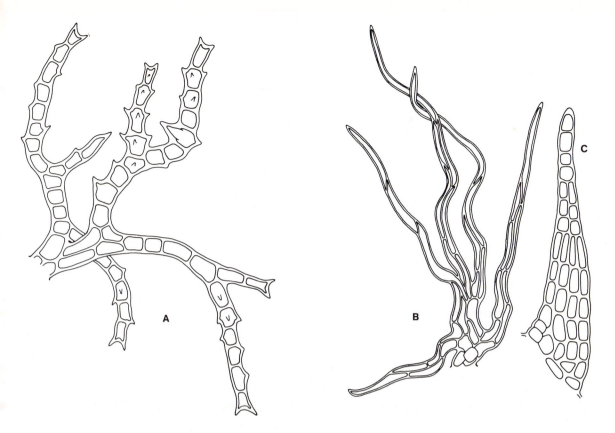

Figure 12-31 Variety in paraphyllia of subclass Bryidae. **A,** *Thuidium delicatulum*, ×685. **B,** *Hylocomium splendens*, ×685. **C,** *Lescuraea incurvata*, ×685.

In a few instances, as in *Fissidens,* the leaves are in two ranks. This appears to be a derived condition, in which the apical cell has two cutting faces instead of the normal three. In other instances, although the apical cell has three cutting faces, only two of the three initials cut off from it produce leaves. In still other instances, the lateral segments produce large leaves, whereas the other segment produces a smaller leaf, similar to the amphigastria of the Jungermanniales. This is another of the many examples of convergent evolution in bryophytes.

Branches generally arise below the axil of a leaf, although in many instances they can arise anywhere on the stem. If growth of the apical cell of the stem is inhibited, a new apical cell arises behind the old stem apex. Although the branches appear to arise in the leaf axil, they ac-

tually are derived from the stem segment that has given rise to the leaf above the branch.

Branching is of several main types. In most acrocarpous genera, the erect shoots form **innovations** near the apex of the shoot that has remained dormant during the preceding unfavorable season. This leads to a rather irregular branching pattern. In most other instances, branching in acrocarpous Bryidae is irregular, and branches can arise in any plane. In pleurocarpous Bryidae, branching also can be irregular, but the gametophore frequently reclines. In still others, branching is loosely or strongly pinnate. In most instances, branches arise in a single plane on opposite sides of the stem. They are occasionally several times pinnate, resulting in a highly complex branch system, as in *Hylocomium splendens* (Fig. 12-33C).

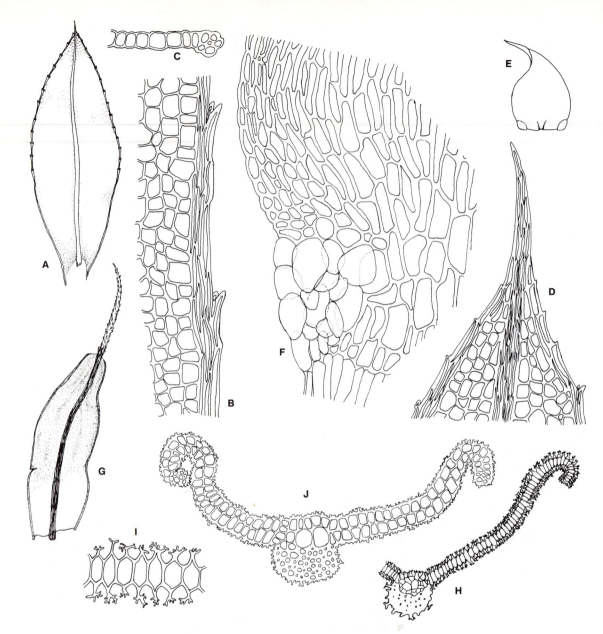

Figure 12-32 Morphology of leaves of subclass Bryidae. **A–D,** *Mnium marginatum.* **A,** leaf outline, showing costa and differentiated margin, ×8; **B,** differentiated leaf margin, showing cellular detail, ×120; **C,** transverse section of leaf margin, ×120; **D,** leaf apex, ×120. **E–F,** *Hypnum lindbergii.* **E,** leaf outline, showing alar region and double costa, ×20; **F,** swollen alar cells (detail), ×385. **G–I,** *Tortula papillosissima.* **G,** leaf ×8; **H,** transverse section of leaf, showing remarkable papillae and stereid cells making up much of costa, ×24; **I,** detail of leaf cells in transverse section, ×120. **J,** *Tortula bistratosa,* transverse section of bistratose papillose leaf, ×80. (**A–D, G–J,** after Flowers, with permission of Brigham Young University Press, from Flowers, Seville, *Mosses: Utah and the West.* Provo, Utah: Brigham Young University Press, 1973.)

Leaf shape is more variable than in other mosses (Fig. 12-34) and may differ from the main stem to the branches. The first-formed leaves on the main stem and on branches, especially of pleurocarpous taxa, are generally smaller than those of the rest of the shoot. The apex is generally acute, but can be blunt or obtuse; sometimes it extends as a long awn (Fig. 12-34G). The margins are toothed or entire, and the marginal cells can be of markedly different shape than those of most of the leaf (Fig. 12-32A–D).

The leaf is generally unistratose, but margins can be thickened and multistratose, and the costa is generally multistratose (Fig. 12-32). The leaf is costate in most acrocarpous mosses as well as in many pleurocarpous genera. In some pleurocarpous genera, and less frequently in acrocarpous genera, there can be more than one costa, or the costa is absent. The anatomy of the costa often shows a distinctive arrangement of stereid cells and cells of much greater diameter (Fig. 12-32H, J).

The arrangement of cells within the leaf is termed its **areolation** (Fig. 12-32). The "shoulders" of the leaf are often made up of **alar** cells distinctly different in shape and color from those of the rest of the leaf. These alar cells, are a diagnostic feature of several genera and species (Fig. 12-32F). The leaf cells are smooth, or their exposed walls are variously thickened with complex papillae (Fig. 12-32H–J) or are swollen outward **(mammillose).** The walls bounding adjacent cells are often perforated by pits. Irregular thickening of the cell walls often gives the leaves a striking areolation. Leaf cell shape is highly diverse.

Besides leaves, some genera of pleurocarpous Bryidae also have paraphyllia on the stems. The paraphyllia are chlorophyllose, unistratose, determinate outgrowths from the stem epidermis. They are multicellular and sometimes much-forked, as in *Thuidium* and *Hylocomium* (Fig. 12-31). Their function is uncertain; they may form a capillary network for external water conduction. Resembling the paraphyllia are the **pseudoparaphyllia.** These structures appear to be aberrant, reduced leaves that surround the rudiments of incipient branches, and sometimes the base of an emergent branch. Pseudoparaphyllia are confined to pleurocarpous Bryidae.

Most Bryidae also have rhizoids on the stems. They are of two types: large, profusely branching systems that arise from cells near buds of the stem, in much the same position from which branches originate; and short, thin, sparsely branched systems arising from scattered, single cells of the stem epidermis. In some Bryidae, rhizoids are confined to the leaf base and costa. Rhizoids are generally brownish to reddish brown, and the end walls of the cells are generally oblique. The rhizoids of Bryidae are always uniseriate, and their cell surfaces are usually smooth. Besides attaching the gametophore to the substratum, rhizoids also assist in capillary conduction of water over the surface of the stem. In some instances, they produce rhizoidal gemmae, or "tubers," that are important in asexual reproduction.

The stem structure in the Bryidae appears to be less complex than in either Polytrichidae or Sphagnidae. The epidermal cells are either extremely thick-walled or swollen and thin-walled. These cells generally have smooth outer walls, but they are occasionally papillose. Cortical cells tend to be thick-walled stereids. The central cylinder of the stem is generally of parenchymatous cells, but the very center of the stem sometimes has a strand of smaller cells, usually of hydroids. In some genera all of the stem cells are parenchymatous. In most Bryidae, water conduction is largely external, with lesser conduction in the central cylinder. The differentiation of leptoids, known in the Polytrichidae, is not apparent in gametophores of the Bryidae.

The gametophores are unisexual in most Bryidae. In acrocarpous Bryidae, the perigonia are generally terminal on a main branch, but they are sometimes on reduced lateral branches. The perichaetia are always terminal in acrocarpous Bryidae, although later growth of lateral shoots from just beneath the perichaetium makes the perichaetium appear lateral. In pleurocarpous Bryidae, both perigonia and perichaetia are on reduced lateral branches, and many are produced by each gametophore. In bisexual Bryidae, archegonia and antheridia are rarely intermixed, as in *Leptobryum*; they are often on separate branches of the same shoot.

Archegonia in the pleurocarpous Bryidae are always surrounded by a well-differentiated, bulbiform perichaetium of closely sheathing leaves. This is also the case in many acrocarpous Bryidae,

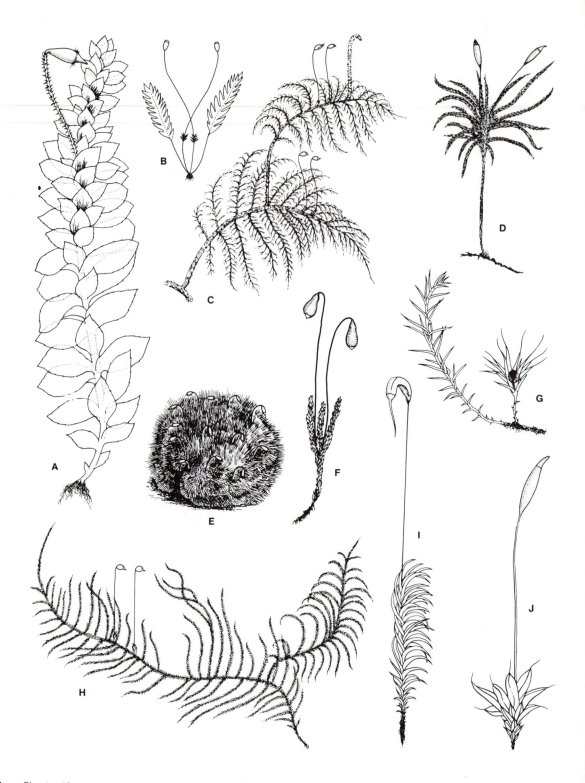

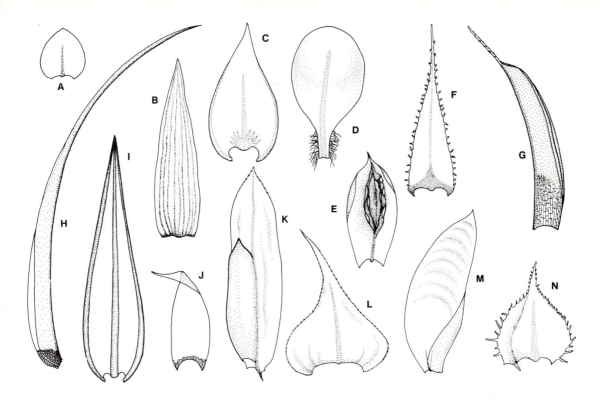

Figure 12-34 Variety in leaves of subclass Bryidae. **A,** *Hygrohypnum smithii*, ×15. **B,** *Orthothecium rufescens*, ×25. **C,** *Antitrichia curtipendula*, ×20. **D,** *Oedipodium griffithianum*, ×15. **E,** *Pterigoneurum ovatum*, ×30. **F,** *Leucolepis menziesii*, stem-leaf, ×25. **G,** *Tortula ruralis*, ×13. **H,** *Dicranum scoparium*, ×15. **I,** *Sciaromium tricostatum*, ×20. **J,** *Brotherella roellii*, ×25. **K,** *Fissidens adianthoides*, ×15. **L,** *Stokesiella oregana*, ×15. **M,** *Neckera douglasii*, ×25. **N,** *Thelia hirtella*, ×30.

Figure 12-33 Variety in gametophytes and sporophytes of subclass Bryidae. **A,** *Eriopus remotifolius* (note amphigastrialike underleaves), ×4. **B,** *Schistostega pennata*, ×6. **C,** *Hylocomium splendens*, ×1. **D,** *Climacium dendroides*, ×1. **E,** *Grimmia pulvinata*, ×1. **F,** *Bryum* sp., ×1. **G,** *Archidium alternifolium*, ×6. **H,** *Stokesiella oregana*, ×2. **I,** *Dicranum scoparium*, ×1. **J,** *Tortula muralis*, ×2. (**A,** after Goebel.)

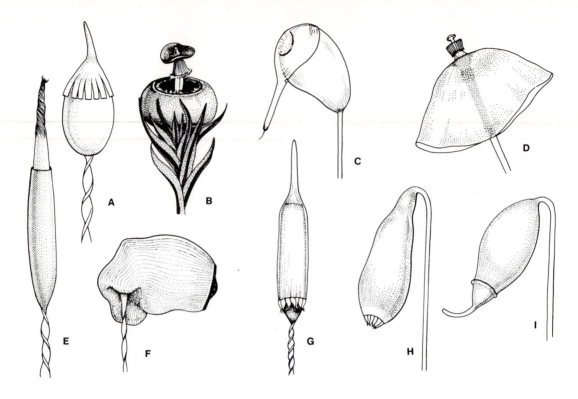

Figure 12-35 Variety in sporangia and calyptras of subclass Bryidae. **A,** *Rhacomitrium lanuginosum*, with calyptra, ×20. **B,** *Scouleria aquatica*, dehiscing, ×10. **C,** *Funaria hygrometrica*, with calyptra, ×15. **D,** *Splachnum luteum*, showing extensive hypophysis (sporangium dehiscing), ×6. **E,** *Tortula princeps*, dehiscing, ×7. **F,** *Philonotis fontana*, with operculum in place, ×8. **G,** *Encalypta ciliata*, with calyptra, ×10. **H,** *Leucolepis menziesii*, dehiscing, ×6. **I,** *Stokesiella oregana*, with operculum in place, ×10.

although in some genera the perichaetial leaves are indistinguishable from vegetative leaves. Generally there is a central mass of archegonia with associated, intermixed paraphyses. In some instances, there are axillary archegonia and paraphyses in perichaetial leaves nearest the center of the perichaetium (as in *Aulacomnium*). The paraphyses are generally chlorophyllose uniseriate filaments, in which the terminal cells, depending on the species, are swollen, elongate, or pointed. Occasionally perichaetial leaves are highly elaborated, as in the case of *Hedwigia*, which has complex, antlerlike marginal hairs.

In a few instances, the archegonium-producing gametophore is of normal size and form, but the antheridium-producing gametophore is re-duced to little more than the perigonium. This "dwarf male" gametophore is epiphytic on the archegonium-producing gametophore. However, if the antheridium-bearing gametophore grows on soil, away from the influence of the archegonium-producing plant, it sometimes is of normal morphology, similar to the gametophores of the opposite sex. This is the case in *Leucobryum* and *Trachybryum*.

The Sporophyte In most Bryidae, the sporophyte consists of an elongate seta terminated by a sporangium. Sporangium orientation and ornamentation vary among genera (Fig. 12-35). If present, stomata are usually restricted to the base

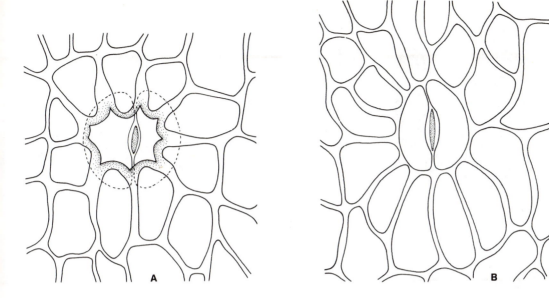

Figure 12-36 Stomata of sporangia in subclass Bryidae. **A,** immersed stoma of *Orthotrichum ohioense,* ×540. **B,** superficial stoma of *Funaria hygrometrica,* ×675.

of the sporangium, near the apophysis if one is present. The stomata are usually exposed on the surface, but in some instances they are deeply immersed, with adjacent epidermal cells bulging over the guard cells (Fig. 12-36). In the genus *Splachnum,* the apophysis is remarkably expanded. In some species, it produces a broad, flaring skirt that curves outward, umbrellalike, over the upper part of the seta (Fig. 12-35D). In this same genus, the sporangium is brightly colored, sometimes vinous red, and exudes a fetid substance that attracts flies.

The sporangium generally has an operculum, although a few Bryidae lack one. The operculum is released as the sporangium dries out and thus shrinks in diameter. In many Bryidae, a ring of specialized cells separates the operculum from the rest of the sporangium. These cells have somewhat thicker walls than the other sporangial cells and are elastic; upon drying, the cells of this ring, the **annulus,** curl back, thus breaking the operculum loose.

In many acrocarpous Bryidae, the peristome consists of a single row of 16 peristome teeth

(Fig. 12-37B, E, F, G). Each of these teeth can be forked, and they are opaque, with brownish or yellowish pigment and varied ornamentation. In such mosses, the teeth are articulate and formed of cell fragments from two concentric rows of adjacent cells. Only the inner wall of the outer concentric row and the outer wall of the inner concentric cell row remain to form the teeth; the remainder of the cell row disintegrates and does not participate directly in peristome structure.

In other acrocarpous Bryidae, and most pleurocarpous Bryidae, the peristome is of two concentric rows of peristome teeth, 16 in each row (Fig. 12-37A, C, D, H). The outer row is generally of opaque teeth, whereas the inner tends to be fragile and transparent. Such "double" peristomes are formed from three concentric rows of cells. The outer peristome (exostome) is formed in the same general way as in the Bryidae that have a single peristome, whereas the inner peristome (endostome) is formed of the inner wall of the middle concentric row of cells and the outer wall of the innermost concentric row. The teeth of the exostome, or of the single peristome

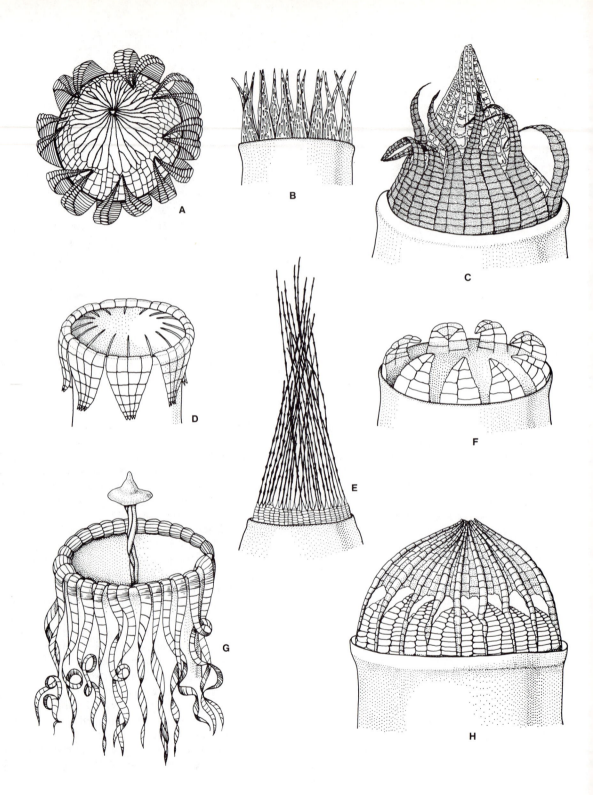

in many Bryidae, are hygroscopic. In many species, the teeth curl inward when humid, catch spores on their jagged inner faces, and flick outward when dry, throwing spores into the air. In other Bryidae, the teeth curve inward when dry and outward when humid. In still others, the teeth, though hygroscopic, do not assist in the extraction of spores from the sporangium by wind or raindrops. In most sporophytes, the sporangium is somewhat inclined when mature, which aids the gravitational movement of spores from deep within the sporangium toward the mouth and into the path of the hygroscopic teeth. Spores are thus shed gradually from the sporangium, improving the chances that some will be released when conditions are favorable for germination.

The seta is smooth and wiry in most species, since both epidermis and cortex are of stereid cells. The central portion of the seta is often of somewhat thinner-walled cells and occasionally has a central strand. This strand is of hydroids, and leptoids are sometimes associated with it. The seta often twists as it dries and untwists with moistening, a feature that may aid in spore dispersal. The foot of the seta penetrates into the apex of the gametophore. In some instances, the seta is very short, and the sporangium is immersed among perichaetial leaves.

In most Bryidae, there is a cylindrical columella, although in the genus *Archidium* there is no columella.

The calyptra is well developed in all Bryidae, always covering the operculum and often sheathing the entire capsule, as in the genus *Encalypta*. In most instances, the calyptra is smooth, but it can be papillose, with erect bristles, or pleated. The lower margin is occasionally lacerate, but is usually entire.

Spores of Bryidae are variable in number, size, and shape (Fig. 12-38). In some genera, spore morphology is very distinctive and provides a valuable taxonomic characteristic for separating species. In most genera, the spores are unicellular when shed; in a few genera, including all members of the family Dicnemonaceae, the spores are multicellular (Fig. 12-38I). In these genera, the filamentous protonemal stage is bypassed, the apical cell being derived directly from the multicellular spore. The largest unicellular spores are those of *Archidium*, occasionally 200 μm in diameter (Fig. 12-38A), but most are less than 20 μm in diameter. In a few genera, including some species of *Macromitrium*, each sporangium contains spores of two distinct sizes. The smaller spores generally produce only antheridium-bearing gametophores, whereas the larger spores produce archegonium-bearing gametophores. This is a type of **heterospory (or anisospory)**, in which two cells of a single tetrad give rise to spores of different size than the other two.

Moisture and light are required for the germination of spores in nature. Most germinating spores produce a protonema of a uniseriate, branched filament, similar to that in many Polytrichidae. The erect branches of this heterotrichous protonema are strongly chlorophyllose and sometimes rather brittle, serving vegetatively to propagate another protonema elsewhere. The protonema ultimately produces gametophores in response to a substance that inhibits further growth of a protonemal branch, but results in a "bud" of undifferentiated cells, in which an apical cell is differentiated. Many buds can be produced on a single protonema; thus a population of gametophores is produced. In most instances, the protonema disappears as the leafy gametophores grow larger.

Uses Bryidae have very few reported uses. In the Bronze Age, however, some Bryidae were used as caulking material in boat-building. Bryidae were also frequently used for plugging drafty openings in dwellings and for mattress material. Even now, large Bryidae are sometimes used as packing material for living nursery plants and fragile china.

Figure 12-37 Variety in peristomes of subclass Bryidae. **A,** *Timmia bavarica,* showing endostome and exostome, ×35. **B,** *Coscinodon cribrosus,* ×50. **C,** *Fontinalis antipyretica,* showing conic endostome and exostome, ×30. **D,** *Ulota megalospora,* showing endostome and exostome, ×50. **E,** *Rhacomitrium canescens,* ×35; **F,** *Octoblepharum albidum,* ×50. **G,** *Tayloria splachnoides,* showing conic operculum and exostome, ×50. **H,** *Cinclidium stygium,* showing domelike endostome and exostome, ×50. (**A,** after A. S. Lazarenko, with permission of *Bryologist;* **G, H,** after Schimper.)

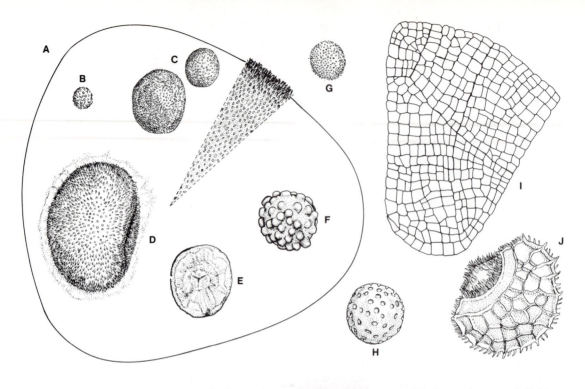

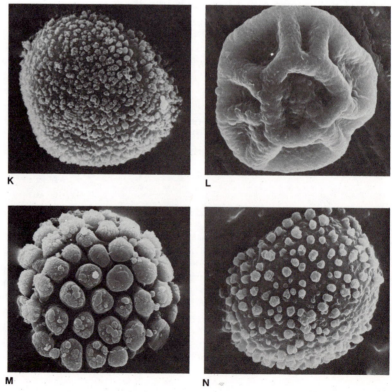

Figure 12-38 Variation in size and morphology of spores of subclass Bryidae, ×350. **A,** *Archidium alternifolium,* showing outline of spore to indicate size and wedge to show sculpturing. **B,** *Ulota megalospora.* **C,** *Macromitrium comatum,* showing heterospory. **D,** *Ephemerum minutissimum;* **E,** *Encalypta ciliata.* **F,** *Encalypta rhabdocarpa.* **G.** *Octoblepharum albidum.* **H,** *Bruchia brevifolia.* **I,** *Dicnemon calycinum,* multicellular spore. **J,** *Bruchia drummondii.* **K–N,** scanning electron micrographs of spores of *Encalypta.* **K,** *E. alpina,* ×1500; **L,** *E. ciliata,* ×1400; **M,** *E. rhaptocarpa,* ×1400; **N,** *E. affinis,* ×1900. (**K–N,** courtesy D. H. Vitt.)

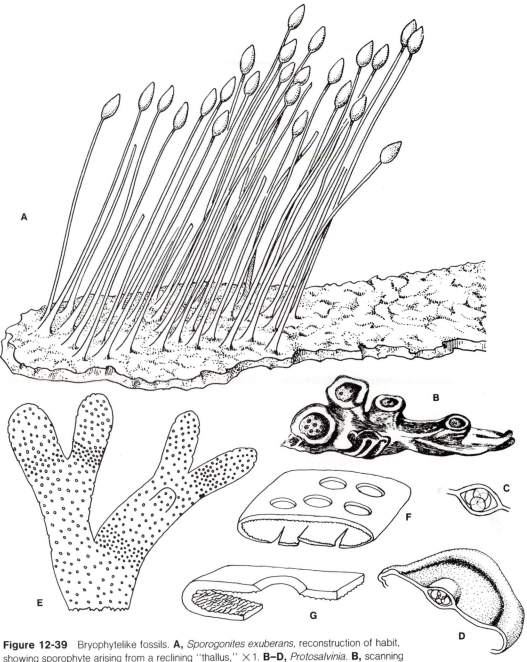

Figure 12-39 Bryophytelike fossils. **A,** *Sporogonites exuberans,* reconstruction of habit, showing sporophyte arising from a reclining "thallus," ×1. **B–D,** *Protosalvinia.* **B,** scanning electron micrograph of thallus, showing cellular pattern, ×15; **B,** reconstruction of thallus, showing cuplike structures containis spore tetrad chambers, ×1; **C, D,** reconstruction of spore tetrad chambers, ×7. **E–G,** *Spongiophytum nanum.* **E,** reconstruction of thallus, showing pores, ×3; **F,** reconstructed segment of thallus, showing pores on thicker, supposedly upper, face, ×20; **G,** section of pore, showing beveled edges and cellular reticulum on inner face of cuticle, ×40. (**A,** after Andrews; **B–D,** after Niklas and Phillips; **E–G,** after Chaloner et al.)

Bryophytelike Fossils

There are a number of fossils of Devonian age that have a structure strongly resembling that of bryophytes. If their interpretation as plants is correct, they supply evidence of a bryophytic structure adapted to a littoral or more or less amphibious existence. They show features that strongly imply that they were adapted, at least in part, to tolerate some desiccation or to depend on air for spore dispersal.

Sporogonites exuberans, of Lower Devonian age (Fig. 12-39A), consists of unbranched, suberect stalks up to 10 cm tall. The plant grew in dense masses, extending from a thalluslike base. Each stalk (about 0.5 mm in diameter) was surmounted by an erect, elliptic sporangium that attained a length up to 9 mm and a diameter of 2–4 mm. The stalk is somewhat enlarged at the base where it emerges from the "thallus." The only cellular structures available are spores. These were homosporous and weakly papillose.

The structure of *Sporogonites* is strongly suggestive of a sporophyte of a bryophyte. Without any knowledge of the gametophore, it is difficult to be certain. In any event, it does provide evidence that this type of simple sporophyte did exist at that time. The fact that many long-stalked sporophytes emerge from a flattened surface implies that this organism might be a bryophyte. If it is, then it shares features of both mosses and hepatics.

Protosalvinia (Fig. 12-39B–D) is better known, thanks to the careful study and analysis of its structure by Niklas and Phillips (1976). This organism has an irregularly lobed thallus up to 6 cm long, with lobes to 2 cm wide. It may have formed floating colonies, much like the modern liverwort *Ricciocarpus*. The thallus contains cuticlelike material, suggesting a tolerance to some desiccation. On the upper surface of the thallus are cuplike structures, and embedded in them are tetrads of spores, each tetrad in its own chamber. These spores appear to have been adapted for air dispersal, since they show evidence of a cuticlelike material. This organism shows features unlike any living plant group. It gives evidence of the presence of the bryophytic lifeform in the earliest land plants, but the nature of its sporangial chambers is unlike anything in the bryophytes.

Spongiophytum nanum (Fig. 12-39E–G) is an even more baffling organism. Indeed, whether it is truly a plant remains uncertain. The form is strongly suggestive of a porose, lobate thallus, and cellular detail is present. The presumably flattened thallus is hollow, and very regular pores are confined mainly to its upper surface. Again, there is evidence of cuticlelike material, suggestive of tolerance to desiccation. This growth form resembles that of the marchantialean liverworts, but there is insufficient detail available to suggest that *Spongiophytum* is a liverwort. If it is a thallus, the pores suggest a gas exchange adaptation.

These organisms, and others that doubtless await discovery, all contribute evidence of the significance of bryophytic lifeforms among the earliest terrestrial organisms.

References

Allen, C. E. 1930. Inheritance in a hepatic. *Science* 71:197–204.

Anderson, L. E. 1974. Bryology: 1947–72. *Ann. Missouri Bot. Gard.* 61:56–85.

Andrews, H. N. 1960. Notes on Belgian specimens of *Sporogonites. The Palaeobotanist* 7:85–89.

Bauer, L. 1962. On the physiology of sporogonium differentiation in mosses. *Bot. J. Linn. Soc.* 58:1–21.

Benson-Evans, K. 1964. Physiology of the reproduction of bryophytes. *Bryologist* 67:431–45.

Bonnot, E. J., ed. 1972. *Les problèmes modernes de la bryologie.* Soc. Bot. France. Colloque, 1972, Paris.

Bopp, M. 1962. Development of the protonema and bud formation in mosses. *Bot. J. Linn. Soc.* 58:305–9.

———. 1968. Control of differentiation in fern allies and bryophytes. *Ann. Rev. Plant Physiol.* 19:361–80.

Bower, F. O. 1935. *Primitive land plants.* Macmillan & Co., London.

Brotherus, V. F. 1924–25. *Musci (Laubmoose).* In Engler, A., and Prantl, K. *Die Natürlichen Pflanzenfamilien.* Vols. 10, 11. 2nd ed. Wilhelm Engler Verlag, Leipzig.

Campbell, D. H. 1918. *The structure and development of mosses and ferns.* Macmillan & Co., London.

Campbell, E. O. 1971. Problems in the origin and classification of bryophytes with particular reference to liverworts. *N. Z. J. Bot.* 9:678–88.

Chaloner, W. G.; Mensah, M. K.; and Crane, M. D. 1974. Nonvascular land plants from the Devonian of Ghana. *Palaeontology* 17:925–47.

Clarke, G. C. S.; Greene, S. W.; and Greene, D. M. 1971. Productivity of bryophytes in polar regions. *Ann. Bot.* 35:99–108.

Clarke, G. C. S., and Duckett, J. G., eds. 1979. *Bryophyte systematics.* Academic Press, New York.

Clausen, E. 1964. The tolerance of hepatics to desiccation and temperature. *Bryologist* 67:411–17.

Conard, H. S., and Redfearn, P. L. 1979. *How to know the mosses and liverworts.* 2nd ed. Wm. C. Brown Co. Pubs., Dubuque, Iowa.

Crum, H. A. 1972. The geographic origins of the mosses of North America's eastern deciduous forest. *J. Hattori Bot. Lab.* 35:269–98.

Crum, H. A., and Anderson, L. E. 1981. *Mosses of Eastern North America.* Columbia Univ. Press, New York.

Dixon, H. N., and Jameson, H. G. 1924. *The student's handbook of British mosses.* 35th ed. V. V. Sumfield, Eastbourne.

Doyle, W. T. 1970. *The biology of higher cryptogams.* Macmillan & Co., London.

Flowers, S. 1974. *Mosses: Utah and the West.* Brigham Young Univ. Press, Provo, Utah.

Fulford, M. 1964. Contemporary thought in plant morphology: Hepaticae and Anthocerotae. *Phytomorphology* 14:103–19.

———. 1965. Evolutionary trends and convergence in the Hepaticae. *Bryologist* 68:1–31.

Gimingham, C. H., and Robinson, E. T. 1950. Preliminary investigations on the structure of bryophytic communities. *Trans. Brit. Bryol. Soc.* 1:330–34.

Goebel, K. 1930. *Organographie der Pflanzen.* Teil 2. Bryophten-Pteridophyten. G. Fischer Verlag, Jena.

Gorham, E. 1957. The development of peat lands. *Quart. Rev. Biol.* 32:145–66.

Greig-Smith, P. 1950. Evidence from hepatics on the history of the British flora. *J. Ecol.* 38:320–44.

Grout, A. J. 1903–1908. *Mosses with a hand-lens and microscope.* Mt. Pleasant Press, Harrisburg, Pa.

———. 1928–40. *Moss flora of North America, north of Mexico.* 3 vols. Publ. by author, Newfane, Vt.

Grubb, P. J. 1970. Observations on the structure and biology of *Haplomitrium* and *Takakia*, hepatics with roots. *New Phytol.* 69:303–26.

Grubb, P. J., Flint, O. P., and Gregory, S. C. 1968. Preliminary observations on the mineral nutrition of epiphytic mosses. *Trans. Brit. Bryol. Soc.* 5:802–17.

Harris, T. M. 1939. *Naiadita*, a fossil bryophyte with reproductive organs. *Ann. Bryol.* 12:57–70.

Hébant, C. 1970. A new look at the conducting tissue of mosses (Bryopsida): their structure, distribution and significance. *Phytomorphology* 20:390–410.

———. 1973. Diversity of structure of the waterconducting elements in liverworts and mosses. *J. Hattori Bot. Lab.* 37:229–34.

———. 1978. *The conducting tissues of bryophytes.* J. Cramer, Vaduz.

Herzog, T. 1925. Anatomie der Lebermoose. In Linsbauer, K., ed. *Handbuch der pflanzenanatomie.* Vol. 2. Abt. 2, Teil Bryophyten. Gebrüder Borntraeger, Berlin.

———. 1926. *Geographie der Moose.* G. Fischer Verlag, Jena.

Huneck, S. 1974. Neue Ergebnisse zur Chemie der Moose, eine Ubersicht. Teil 4. *J. Hattori Bot. Lab.* 38:609–31.

Ingold, C. T. 1959. Peristome teeth and spore discharge in mosses. *Trans. Bot. Soc. Edinburgh* 38:76–88.

Inoue, H. 1960. Studies in spore germination and the earlier stages of gametophyte development in the Marchantiales. *J. Hattori Bot. Lab.* 23:148–91.

Jovet-Ast, S. 1967. Bryophyta. In Boureau, E., ed. *Traité de paléobotanique.* Vol. 2. Masson et Cie, Paris.

Keever, C. 1957. Establishment of *Grimmia laevigata* on granite and its method of establishment. *Ecology* 38:422–29.

Keever, C.; Oosting, H. J.; and Anderson, L. E. 1951. Plant succession on exposed granite of Rocky Face Mountain, Alexander County, North Carolina. *Bull. Torrey Bot. Club* 78:402–21.

Lacey, W. S. 1969. Fossil bryophytes. *Biol. Rev.* 44:189–205.

Lewis, K. R. 1961. The genetics of bryophytes. *Trans. Brit. Bryol. Soc.* 4:111–30.

Longton, R. E. 1974. Genecological differentiation in bryophytes. *J. Hattori Bot. Lab.* 38:49–65.

Lorch, W. 1931. Anatomie der Laubmoose. In Linsbauer, H., ed. *Handbuch der Pflanzenanatomie.* Vol.7. Abt. 2, Teil Bryophyten. Gebrüder Borntraeger, Berlin.

McLure, J. W., and Miller, H. A. 1967. Moss chemotaxonomy. *Nova Hedwigia* 14:111–25.

Marchal, E., and Marchal, E. 1911. Aposporie et sexualité chez les mousses. *Bull. Acad. Roy. Belgique Cl. Sci.* 1911:750–76.

Martensson, O., and Nilsson, E. 1974. On the morphological colour of bryophytes. *Lindbergia* 2:145–59.

Miller, H. A. 1974. Rhyniophytina, alternation of generations, and the evolution of bryophytes. *J. Hattori Bot. Lab.* 38:161–68.

Moutschen, J. 1962. Quelques tendances de la génétique des mousses. *Bull. Soc. Roy. Bot. Belgique* 95:61–71.

Nehira, K. 1974. Phylogenetic significance of the sporeling pattern in Jungermanniales. *J. Hattori Bot. Lab.* 38:151–60.

Neuberg, M. F. 1958. Permian true mosses of Angaraland. *J. Palaeontol. Soc. India* 3:22–29.

Niklas, K. J., and Phillips, T. L. 1976. Morphology of *Protosalvinia* from the Upper Devonian of Ohio and Kentucky. *Am. J. Bot.* 63:9–29.

Parihar, N. S. 1965. *An introduction to Embryophyta.* Vol. 1. *Bryophyta.* 5th ed. Central Book Depot, Allahabad, India.

Paton, J. A., and Pearce, J. V. 1957. The occurrence, structure, and function of the stomata in British bryophytes. *Trans. Brit. Bryol. Soc.* 3:228–59.

Proskauer, J. 1960. Studies in the Anthocerotales. VI. *Phytomorphology* 10:1–19.

Ramsay, H. P. 1966. Sex chromosomes in *Macromitrium. Bryologist* 69:295–311.

Rastorfer, J. R. 1962. Photosynthesis and respiration in moss sporophytes and gametophytes. *Phyton* 19:169–77.

Ratcliffe, D. A. 1968. An ecological account of Atlantic bryophytes in the British Isles. *New Phytol.* 67:365–439.

Reimers, H. 1954. *Bryophyta: Moose.* In Engler, A. *Syllabus der Pflanzenfamilien.* Vol. 1. 12th ed. Gebrüder Borntraeger, Berlin.

Richards, P. W. 1950a. *A book of mosses.* Penguin Books, London.

———. 1950b. *Bryophyta.* In Turrill, W. B. ed. *Vistas in botany.* Vol. 1. Pergamon Press, London.

Sarafis, V. 1971. Modern methods in the investigation of bryophyte morphology. *N. Z. J. Bot.* 9:725–38.

Savicz-Ljubitzkaja, L. I., and Abramov, I. I. 1959. The geological annals of the Bryophyta. *Rev. Bryol. Lichenol.* 28:330–42.

Schofield, W. B. 1969. *Some common mosses of British Columbia.* British Columbia Provincial Museum Handbook, No. 28.

Schofield, W. B., and Crum, H. A. 1972. Disjunctions in bryophytes. *Ann. Missouri Bot. Gard.* 59:174–202.

Schuster, R. M. 1966. Studies on Hepaticae. XV. Calobryales. *Nova Hedwigia* 13:1–63.

———. 1966, 1969, 1974, 1980. *The Hepaticae and Anthocerotae of North America east of the hundredth meridian.* Vols. 1–4. Columbia Univ. Press, New York.

———. 1969. Problems of antipodal distribution in lower land plants. *Taxon* 18:46–91.

Shacklette, H. S. 1965. Bryophytes associated with mineral deposits and solutions in Alaska. *U. S. Geol. Survey Bull.* 1198-C.

Sjörs, H. 1961. Surface patterns in boreal peatland. *Endeavour,* 20:217–24.

Smith, A. J. E. 1978. *The mosses of Great Britain and Ireland.* Cambridge Univ. Press, Cambridge.

Smith, G. M. 1955. *Cryptogamic botany.* Vol. 2. *Bryophytes and Pteridophytes.* 2nd ed. McGraw-Hill Book Co., New York.

Smith, J. C. 1966. The liverworts *Pallavicinia* and *Symphyogyna* and their conducting systems. *Univ. Calif. Publ. Bot.* 39:1–46.

Steere, W. C. 1947. A consideration of the concept of genus in Musci. *Bryologist* 50:247–58.

———. 1972. Chromosome numbers in bryophytes. *J. Hattori Bot. Lab.* 35:99–125.

Stotler, B. C. 1972. Morphogenetic patterns of branch formation in the leafy Hepaticae. A resumé. *Bryologist* 75:381–403.

Streeter, D. T. 1970. Bryophyte ecology. *Sci. Prog. Oxf.* 58:419–34.

Suire, C., ed. 1978. *Congrès International de Bryologie.* Bryophytorum Bibliotheca. Vol. 13. J. Cramer, Vaduz.

Tamm, C. O. 1964. Growth of *Hylocomium splendens* in relation to tree canopy. *Bryologist* 67:423–26.

Thieret, J. W. 1955. Bryophytes as economic plants. *Econ. Bot.* 10:75–91.

Verdoorn, F., ed. 1932. *Manual of bryology*. Martinus Nijhoff, The Hague.

Vitt, D. A. 1968. Sex determination in mosses. *Michigan Botan.* 7:195–203.

Watson, E. V. 1971. *The structure and life of bryophytes.* Hutchinson Univ. Library, London.

Willis, A. J. 1964. Investigations on the physiological ecology of *Tortula ruraliformis. Trans. Brit. Bryol. Soc.* 4:668–83.

The various plant groups described in the preceding chapters illustrate the diversity of nonvascular plants. The range extends to all main features: size, morphology, anatomy, cytology, genetics, physiology, biochemistry, ecology, and reproductive processes. As stressed throughout, variations in the structure and function of modern plants are the end results of genetics and natural selection acting through at least 3 **aeons,** or **Gyr** ($= 10^9$ years). During this time, some of the main events in botanical evolution occurred: the formation of prokaryotes, the evolution of eukaryotes, syngamy and meiosis, alternation of generations, and adaptations to terrestrial habitats.

13

EVOLUTION AND PHYLOGENY

Early Evolution

The origin and evolution of prokaryotes and nonvascular eukaryotes are linked intimately with events that occurred during the earth's geologic history. Recent estimates of the earth's age range from 4 to 6 Gyr. Relative to this, the earliest fossils, resembling bacillar bacteria (Fig. 13-1) and coccoid bluegreen algae, have been found in rocks in Western Australia dated at approximately 3.5 Gyr. By contrast, the first apparent eukaryotes have been found in rocks of the late Precambrian, about 1.3 Gyr old. Thus the fossil record demonstrates clearly that prokaryotes were the earliest organisms to evolve on earth, and supports the belief that eukaryotes evolved from prokaryotes prior to the late Precambrian interval.

An interesting and informative model showing the main events in the evolution of primitive earth and early life forms is presented in Figure 13-2. Drafted by Preston Cloud (1976), a leading student of the Precambrian, it shows some of the main events between 4.6 and 0.7 Gyr. One significant event proposed by Cloud was a period of gravitational heating after the origin of the earth. This could have resulted from frictional heat generated by tides produced when the moon was captured by the earth's gravitational field, meteorite showers, or by the natural evolution of heat from radioactive decay, or a combination of these factors.

In the interval between approximately 4.5

and 3.5 Gyr, the earth's atmosphere probably contained gases such as H_2, CO_2, CO, N_2, SO_2 and HCl, but little or no molecular oxygen. This was the period of biogenesis, leading to the evolution of the first prokaryotes, including the first autotrophs. We do not know whether the first prokaryotes were more like bacteria or bluegreen algae, as both occur together. Presumably the early prokaryotic forms were anaerobic and heterotrophic and derived their energy from fermentation. Once the photosynthetic process had become established in the first photoautotrophs, at around 3.5 Gyr, oxygen would have been produced. As indicated in Figure 13-2, oxygen would have been dispersed first into the hydrosphere; it probably would have taken a long time to establish atmospheric oxygen and the prerequisite ozone layer in the outer atmosphere.

During this same time interval, beds of banded iron formation (*BIF* in Fig. 13-2) are found extensively in sediments. Banded iron consists of alternating layers of iron-rich and iron-poor silica. It has been suggested that this alternation resulted from periods when ferrous iron (unoxidized) was deposited in sites of bluegreen algal growth, where it was oxidized to ferric iron by the oxygen released by the algae. This process lasted until about 2.0 Gyr, at which time there was a marked change in deposition from banded ironstones to red beds. This appears to have marked the time when oxygen had been released and accumulated in the hydrosphere to the extent that it began to escape to the atmosphere. Here it would have produced a high rate of oxidation in detrital terrestrial and marginal marine sediments, resulting in the extensive red beds of the later Precambrian. The buildup of an oxygenated atmosphere and an ozone layer appear to coincide with the origin of eukaryotes; many believe that these are two of the main factors in eukaryotic evolution.

A prominent feature in the Precambrian is the presence of **stromatolites**—domed or columnar formations in carbonate rocks and cherts. They are particularly well developed in the Pongola and Bulawayan rock series of South Africa (Fig. 13-2), as well as in western Australia. Bluegreen algal cells have been found in some stromatolites. From the study of modern stromatolitic analogs, such as those from beaches in Australia (Fig. 13-3), it is generally believed that most

Figure 13-1 Electron micrographs of bacillar bacteria from the Precambrian of South Africa. (From E. S. Barghoorn and J. W. Schopf, *Science* 152:758–63, 6 May 1966, with permission of the authors and the American Association for the Advancement of Science. Copyright 1966 by the American Association for the Advancement of Science.)

Precambrian stromatolites were formed by the growth of carbonate-secreting bluegreen algae.

It appears that most morphological and (by inference) physiological features of bluegreen algae had evolved by 2.0 Gyr. Any additional evolution in later times appears to have been at or below family level, although the latest group to evolve, the Stigonematales, did not appear until the Devonian Period, about 370 **Ma** (= years $\times$ 10^6).

The relative conservatism in the evolution of bluegreen algae may be explained by their physiological flexibility, as evidenced by their wide ecological tolerance, and their genetic stability. Being photosynthetic and requiring few nutrients, they can exist in relatively impoverished

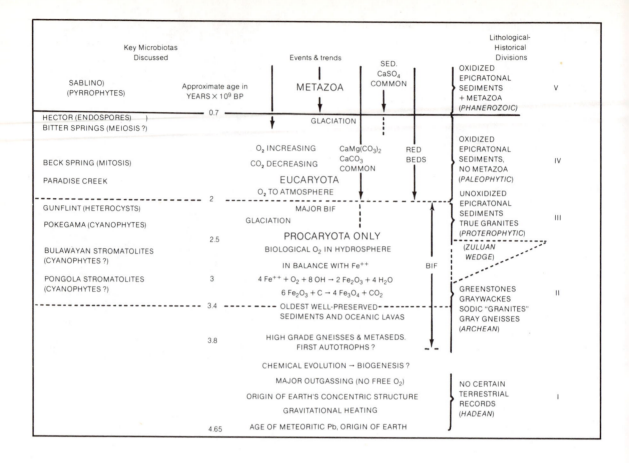

Figure 13-2 Proposed scheme for early evolution of the biosphere, lithosphere, and atmosphere: *BIF*, banded iron formation. (From P. Cloud, 1976, with permission of the author and the Paleontological Society, publishers of *Paleobiology*.)

ecological niches. Their genetic stability may be the result of the presence of several sets of genetic material, which provide protection against mutational changes that would otherwise be expressed in individuals with a single set of genes. Furthermore, being primarily asexual, they lack the means of increasing variability that results from meiosis and syngamy in eukaryotes. The net result of these factors has been a relatively slow rate of evolutionary change in bluegreen algae.

The record of the bacteria is so meager that no evolutionary trends can be identified. The presence of bacillar forms at about 3.5 Gyr, together with bacterialike filaments and rodlike structures in the mid-Precambrian, suggest a diversity of form in early periods. Bacteria have also been credited with the formation of some of the extensive iron ore deposits around the western Great Lakes in the late Precambrian.

Evolution of Eukaryotes

As mentioned in Chapter 3, there is a sharp evolutionary break between prokaryotes and eukaryotes that first evolved some time in the mid- to late Precambrian. The essential features that set eukaryotes apart so distinctly are the double

Figure 13-3 Stromatolitic formation on a beach at Shark Bay, Western Australia. (Courtesy of Australian Information Service.)

membrane that limits organelles, including the nucleus, mitochondria, and photosynthetic plastids, together with differentiated endoplasmic reticulum, golgi bodies, lysosomes, microtubules, microbodies, and flagella. This difference in structure is so sharp that many evolutionists and taxonomists believe division of living organisms into these two groups is more fundamental and essential than distinguishing green plants, animals, fungi, and protists into separate kingdoms!

Although there is no direct proof from either fossils or living members, the general belief is that eukaryotes evolved from prokaryotic ancestors. Without direct evidence as to how this came about, two main schools of thought have

developed to account for the likely evolutionary pathways. Until the 1960s, most researchers adhered to the **autogenous** school. According to several autogenous hypotheses, the various membrane-bounded organelles of eukaryotes arose from step-by-step changes in the non-bounded photosynthetic, respiratory, and genetic units of the prokaryotic cell. In this progression, there was gradual compartmentalization, condensation, and specialization of the functional units, until the discrete, membrane-bounded organelles of eukaryotes were evolved (Fig. 13-4A). This involved the engulfing of the functional units by a process known as **cytosis.** In such a scheme, mitochondria would have

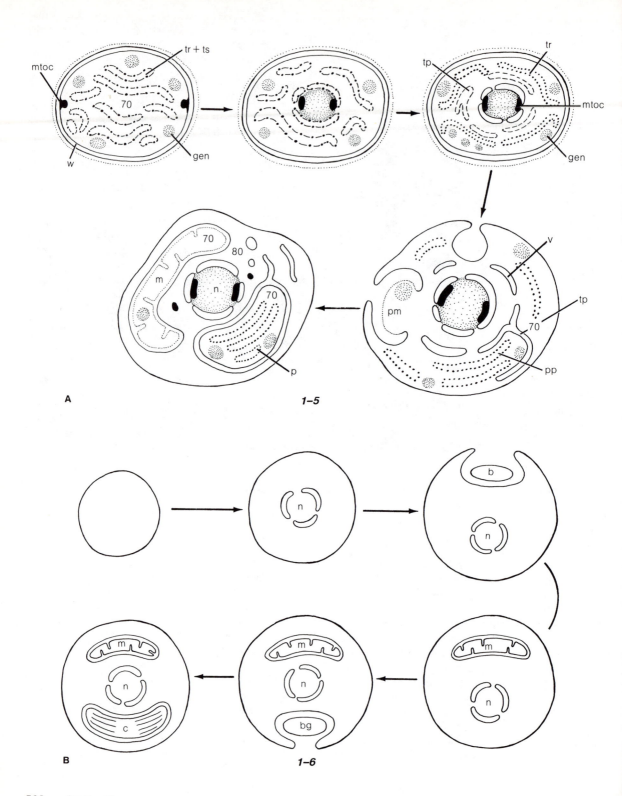

A

1–5

B

1–6

evolved by invaginations of the plasma membrane encircling cytoplasmic DNA particles. Similarly, chloroplasts of eukaryotic cells would have evolved by a gradual envelopment of both photosynthetic pigments and cytoplasmic DNA particles, possibly by budding and branching thylakoid membranes. Additional details can be found in the papers of Cavalier-Smith (1975, 1978) and Taylor (1976, 1978, 1979).

The other main view on the origin of the eukaryotes is that they obtained their organelles, particularly the DNA-containing mitochondria and chloroplasts, by a series of **endosymbioses** of prokaryotic cells (Fig. 13-4B). In contrast, this has been called a **xenogenous** hypothesis. Although not a new idea, the general xenogenous hypothesis and its detailed steps have been amplified in recent years, notably by Margulis. In general, Margulis and others postulate that a series of endosymbioses occurred in the middle part of the Precambrian. They believe that the first step was the symbiosis of **protomitochondria,** most likely in the form of an early bacterium that had already differentiated protomitochondria. Likewise, the second main step

was the endosymbiotic incorporation of blue-green algal cells that then acted as **protoplastids;** these later evolved step-by-step into the full-fledged chloroplasts of eukaryotic algae (Fig. 13-4B, step 5). Margulis also derived the 9 + 2 flagellum by endosymbiosis of flagellated cells. This last step has not gained the general acceptance that mitochondrial and plastid symbioses have, mainly because DNA has not been found associated with either centrioles or basal bodies. However, there is some evidence favoring the derivation of eukaryotes by endosymbiosis, including numerous examples of endosymbioses of algae in many invertebrates. Although these and other supporting features cannot be detailed here, this exciting concept can be followed up in the excellent papers by Margulis, Cavalier-Smith, Pickett-Heaps, and Taylor.

The different pattern of chromosome structure in the dinoflagellates has been suggested as representing an evolutionary stage intermediate between the prokaryotic nucleoid and the eukaryotic nucleus. Lacking histones, and having complex whorls of DNA fibrils distinct from the nucleoplasm, the dinoflagellates have been

Figure 13-4 A *1–5,* Hypothetical model of autogenous evolution of a eukaryotic cell from a prokaryotic ancestor by condensation, compartmentalization, and specialization. *1,* a prokaryotic cell, showing a wall (*w*), thylakoid lamellae performing both photosynthesis and respiration (*tr* + *ts*), several areas of genome material (*gen*), and early stages of microtubule organizing centers (*mtoc*). *2,* a later stage, showing some of the genome material concentrated into a nucleoidlike mass surrounded loosely by membranes, and containing the microtubule organizing centers. *3,* still later stage, showing a nuclear envelope around the pronucleus, and the specialization of membranes into photosynthetic thylakoids. (*tp*) and respiratory thylakoids (*tr*); note the continued presence of cytoplasmic genome. *4,* a stage in which the cell wall has been lost; the membrane is undergoing a series of endocytoses, forming vacuoles (*v*), and surrounding some of the cytoplasmic genomes with membranes to form protoplastids (*pp*) and protomitochondria (*pm*). *5,* compartmentalization of organelles is essentially as in the eukaryotic cell, with nucleus (*n*), mitochondria (*m*) and plastids (*p*) delimited by membranes. (After F. J. R. Taylor, 1976, with permission of the author and the International Association of Plant Taxonomists.)

Figure 13-4 B *1–6,* Hypothetical model of the xenogenous formation of a eukaryotic cell by a series of endosymbioses. *1,* ancestral, nonphotosynthetic prokaryotes. *2,* the formation of a nucleus (*n*) by concentration of genome material surrounded by a nuclear envelope. *3,* endosymbiosis of a bacterium (*b*) to provide mitochondrial component (*m*) shown in *4. 5,* endosymbiosis of the bluegreen algal cell (*bg*) to provide the chloroplast component. *6,* eukaryotic plant cell with the three essential membrane-delimited organelles. (After F. J. R. Taylor, 1974, with permission of the author and the International Association of Plant Taxonomists.)

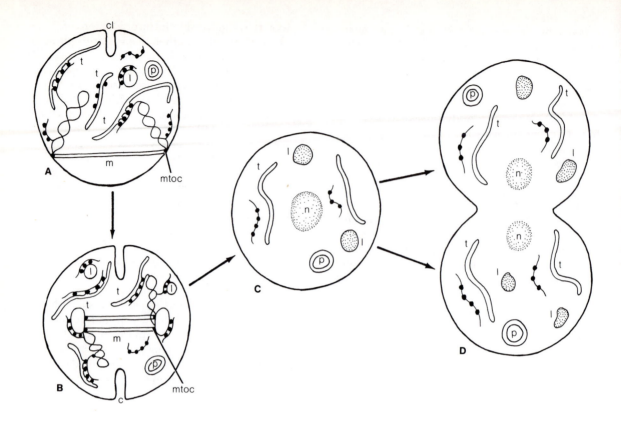

Figure 13-5 Probable evolutionary stages in the development of mitosis. **A,** an amoeboid "prealga," showing the evolution of spindle microtubules (*m*) as a device for separating the two daughter chromosomes into separate cells following cleavage (*cl*) along the furrow. **B,** chromosome attachment sites, initially at the cell membrane, have moved to more central regions by endocytosis; this would permit the two daughter nuclei to segregate into two daughter cells following cleavage at *cl*. **C,** origin of sexual process by fusion of wall-free cells to double the chromosome complement. (After Cavalier-Smith, 1975, with permission of the author and *Nature*.)

Figure 13-6 A scheme showing possible diversification of eukaryotes following evolution of the ancestral phagocytic amoeboid "protoalga" from an amoeboid phagocytic "prealga." (After Cavalier-Smith, 1975, with permission of the author and *Nature*.)

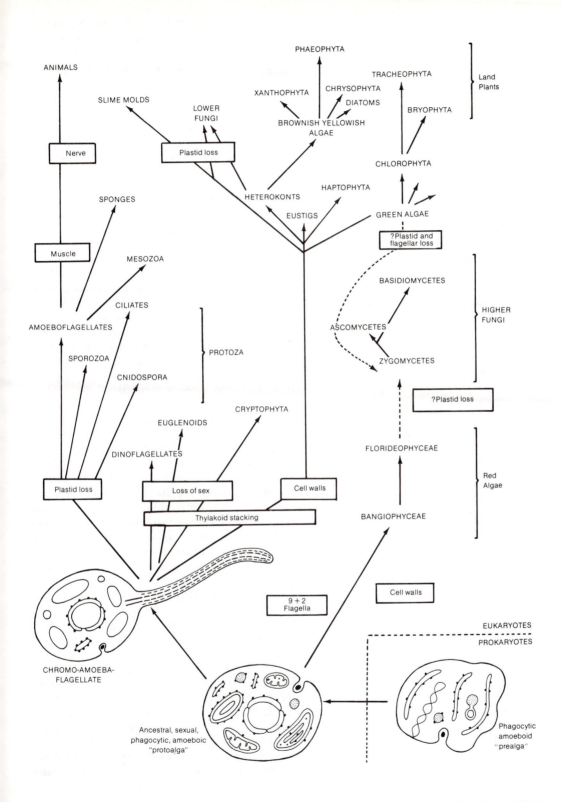

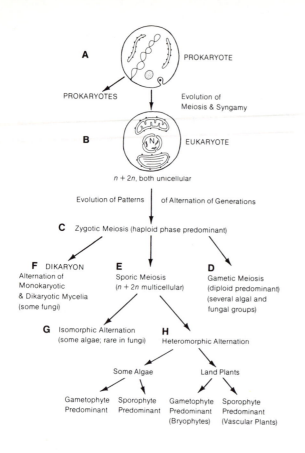

Figure 13-7 Probable evolutionary developments (**A–F**) of the main types of alternation of generations in plants.

called **mesokaryotes.** Although this opinion may have some merit, the general cell organization is typically eukaryotic, and the nucleus is regarded as a variant of the basic eukaryotic type.

Mitosis, Sex, and Alternation of Generations

The evolution of eukaryotes in an oxygenated atmosphere in the mid- to late Precambrian was almost certainly accompanied by the development of mitotic cell division, meiosis and syngamy, and alternation of generations. These processes are of utmost significance to plants.

Their development provided for the duplication of cells (mitosis), which results in relatively large and complex plant bodies; the introduction of more variability through the sexual process (meiosis and syngamy); and the elaboration of different kinds of life histories in plants (alternation of generations).

The development of mitosis apparently originated with the evolution of eukaryotes. The formation of DNA and histones (except in dinoflagellates) into discrete, linear chromosomes would permit replication of duplicate sets of genetic material for each daughter cell. Although fossil evidence is lacking, it has been speculated by Cavalier-Smith that the evolution of the mitotic process was accompanied by the development of spindle microtubules, nuclear envelopes, and the cleavage furrow for cell division (Fig. 13-5). It has also been suggested that the sexual process started very early in the evolution of the eukaryotic cells, along with the mitotic process, and before development of the nuclear envelope. Cavalier-Smith postulated that the early amoeboid **prealga** would have had one circular chromosome. This could have divided by a primitive mitosis to provide segregation. Recombination could have resulted from cell fusion by cytosis, a process that would be selectively advantageous by providing twice the internal food supply of unfused cells, particularly during times of starvation (Fig. 13-5).

Regardless of the details of their evolutionary development, the inception of meiosis and syngamy was a very significant step in plant evolution. The variability introduced by segregation and recombination followed by natural selection acting on new combinations resulted in the evolution of diverse eukaryotes. One general scheme showing the presumed evolutionary developments following the establishment of eukaryotes is shown in Figure 13-6.

The inception of the diploid phase or generation as part of the sexual cycle was an important step in the evolution of eukaryotes. In addition to increasing the reservoir of genetic variability, it would also provide protection against detrimental mutations. This in turn could account for the fact that diploid-dominant organisms have managed to evolve relatively complex tissues, organs, and physiological processes for adaptation to rigorous ecological conditions. Notable examples are the vascular plants and vertebrates on land, and the larger algae of exposed intertidal zones, such as the Laminariales and Fucales.

The alternation of the haploid and diploid generations in the meiotic-syngamy cycle has been expressed in four main patterns, based on the timing of meiosis in the life history (Figs. 2-8, 2-9, 13-7). We do not have a record of the actual evolution of the four patterns. However, from our knowledge of life histories in modern plants relative to the fossil records, it seems probable that the pattern of **zygotic** meiosis was the earliest to evolve (Fig. 13-7C). It also seems likely that the **gametic** (Fig. 13-7D) and **sporic** (Fig. 13-7E) types are two separate patterns that evolved from the zygotic pattern by delay in meiosis until the multicellular diploid generation had developed, later producing either gametangia or sporangia. Although we do not know precisely when these two patterns developed, it was probably in the interval of late Precambrian to early Cambrian, from which we have the first records of plants whose modern descendants have the same pattern of alternation. The fourth pattern of alternation, the **dikaryon** (Fig. 13-7F), is different from the others and is limited to some fungi. Of obscure origin, it consists basically of a fusion of gametic cytoplasm without nuclear fusion, resulting in binucleate cells.

Evolutionary Patterns

In attempting to recognize true evolutionary relationships among groups of plants, botanists are hampered basically by a lack of good fossil series to show various stages in the evolution of taxa. In a few instances among nonvascular plants, we can catch the main stage of evolution because the fossil record is reasonably good. Examples of these are more common among the carbonate- or silica-depositing algae, such as the Dasycladales (Chlorophyta), the Corallinaceae (Rhodophyta), and diatoms (Chrysophyta). Such a phylogenetic series is shown in Figure 13-8 for some of the main dasyclad genera from the Ordovician Period to the present. However, even with such series, we do not know from which group or groups the Dasycladaceae (Dasycladales) arose, or which other algae (if any) arose from that family. Hence the phylogenetic relationships of the Dasycladaceae to other algal groups are not known for sure. Because of this, botanists working with nonvascular plants use various characteristics possessed by modern members in an attempt to establish relationships as closely as possible. For this, they use biochemical, physiological, reproductive, and structural criteria. A good example has already been noted in the use of pigments, storage products, and flagellation in classifying and relating algal groups (Table 2-4). The basic assumption is that the more characteristics that taxa share, the closer they are related; hence they have diverged less with evolution. If the taxa have become separated to a greater extent by evolution, they will share fewer characteristics. Such series showing the degree of relationship among living taxa are termed **phenetic series.**

Recent developments in the field of cytogenetics have given a new approach to understanding relationships. Particularly valuable are the number, size, and shape of chromosomes. In general, the results of chromosome studies have supported conclusions previously made on structural grounds.

Comparative phytochemistry is another field of study that has revealed some interesting series of compounds within plant groups, and promises to play an increasingly important role in our understanding of relationships, phylogeny, and classification. Recently, electrophoresis has been applied successfully to some algal groups. Among other things, it has shown a clear-cut distinction in the ribosomal RNA of prokaryotes and eukaryotes.

Fungi

Probable fungal hyphae called *Eomycetopsis* have been found from the late Precambrian (about 1 Gyr). Also, spores that are probably fungal have been found in beds of similar age. However, spores and hyphal remains have never been found connected; hence the fungal relationship has not been proved beyond doubt. Other reports have funguslike filaments occurring in earlier parts of the Precambrian, but they have not been shown unequivocally to be those of fungi. The earliest undoubted fungi are found in Paleozoic rocks at about 400 Ma. Many of these Pal-

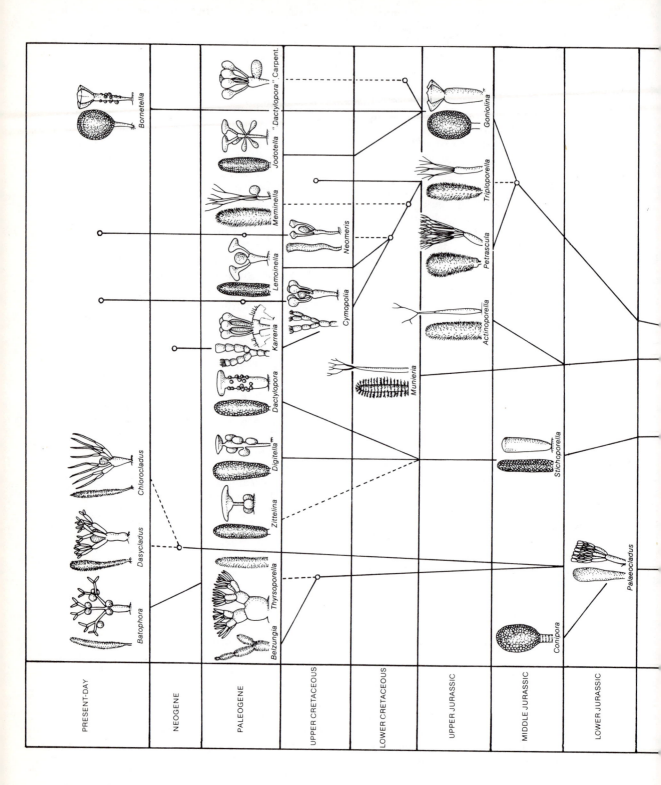

PRESENT-DAY

NEOGENE

PALEOGENE

UPPER CRETACEOUS

LOWER CRETACEOUS

UPPER JURASSIC

MIDDLE JURASSIC

LOWER JURASSIC

Bornetella

Jodotella "*Dactylopora*" Carpent.

Meminella

Goniolina

Lemoinella

Neomeris

Triploporella

Karreria

Cymopolia

Petrascula

Dactylopora

Actinoporella

Chlorocladus

Digitella

Muneria

Zittelina

Stichoporella

Dasycladus

Palaeocladus

Batophora

Thyrsoporella

Belzungia

Conipora

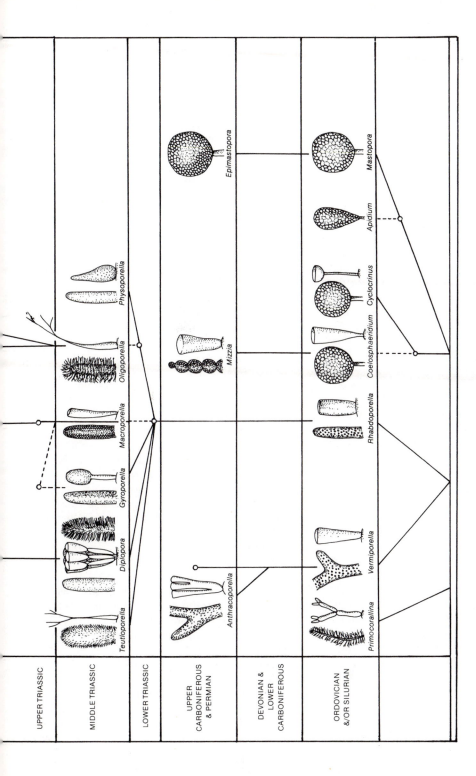

Figure 13-8 Phylogeny of genera of dasyclad algae. Circles within or at the end of solid lines indicate the occurrence of the genus within the geological time zone. Broken lines indicate a measure of uncertainty in the evolutionary link between the genera. (Modified after Pia, 1922.)

Evolution and Phylogeny **515**

eozoic fungi are very similar to modern zygomycetes. Examples are hyphal vesicles and cysts within the tissues of several vascular plants in the early Devonian Rhynie chert from eastern Scotland.

Although several possible examples of ascomycetes have been reported from the Carboniferous, there are no undoubted remains of ascomycetes until the Cretaceous. The earliest basidiomycete is from the latter half of the Carboniferous, with a well-developed clamp connection, suggesting that this major fungal group evolved before that period of geologic time. Given these records, it is reasonable to assume that the zygomycetes are the ancestral group. Although it is often assumed that ascomycetes evolved from zygomycetes, and basidiomycetes from ascomycetes, the fossil record provides no direct evidence other than the earlier occurrence of zygomycetes. However, comparative structural and biochemical criteria suggest to some that the most likely evolutionary progression was zygomycetes-ascomycetes-basidiomycetes. There is also a suggestion that this evolutionary pattern corresponds to the colonizing of more terrestrial habitats by ascomycetes and particularly by basidiomycetes. A theory that fungi were derived from certain red and/or green algae has never gained a large following and is probably an example of parallel evolution.

A phenetic scheme outlining possible relationships of various fungal groups is shown in Figure 13-9.

Algae

The algal divisions recognized in this text have a varied fossil record, and our knowledge of their relationships is equally varied. Known fossil records for algal divisions are shown in Figure 13-10, together with those of the two prokaryotic divisions and microfossils of unknown affiliations that are probably plant cells, called **acritarchs.** It is noteworthy that next to acritarchs, the oldest remains are members of the Chlorophyta from the late Precambrian, followed by the Rhodophyta of the Cambrian interval.

Some caution is required in interpreting the fossil history of algal divisions. There is heavy bias on the calcareous and siliceous members simply because they have been better preserved. In noncalcareous and siliceous algae, many of the fossil records comprise one or two often fragmentary occurrences. Furthermore, what we find is only a smattering of the total number of individuals, species, and even genera that lived throughout the vast period represented in the geological record. Despite all these shortcomings, the discovery of fossils such as Precambrian cyanophytes has done much to further our knowledge of early evolution, and the same applies to discoveries of later fossils such as the calcareous rhodophytes and dasyclad chlorophytes (Fig. 13-8).

The general relationships of the divisions are shown in Figure 13-12. As noted frequently throughout previous chapters, these relationships are based largely on similarities in pigments, flagellation, major cell coverings, and main storage products. On this basis, all of the divisions except the Chrysophyta show a fairly uniform set of characteristics throughout the various classes and orders. Hence, each appears to represent a distinct evolutionary line.

Based on the presence of chlorophylls, there appear to have been three main groups in the algae (Fig. 13-11). The first group with chlorophyll *a* comprises only the Rhodophyta and the prokaryotic cyanophytes. It appears as though the Rhodophyta may have evolved directly from cyanophytan prokaryotes, possibly through intermediates such as *Cyanidium caldarium*, a blue rhodophyte (Porphyridiales) that is both **thermophilic** and **acidophilic.** *Cyanidium* has been variously assigned to the Cyanophyta, the Chlorophyceae, and the Cryptophyceae, but recent work has suggested that it is a rhodophyte, and may represent a transitional form.

The second main grouping includes those classes of Chlorophyta with chlorophylls *a* and *b* and starch-type storage polysaccharides. This is the group that is generally considered in part to be ancestral to the main terrestrial groups, bryophytes and vascular plants. The third line comprises those divisions with chlorophylls *a* and *c* and laminaranlike storage polysaccharides; it has been called by some authors the Chromophyta in contradistinction to the Chlorophyta.

Many attempts have been made to draw up

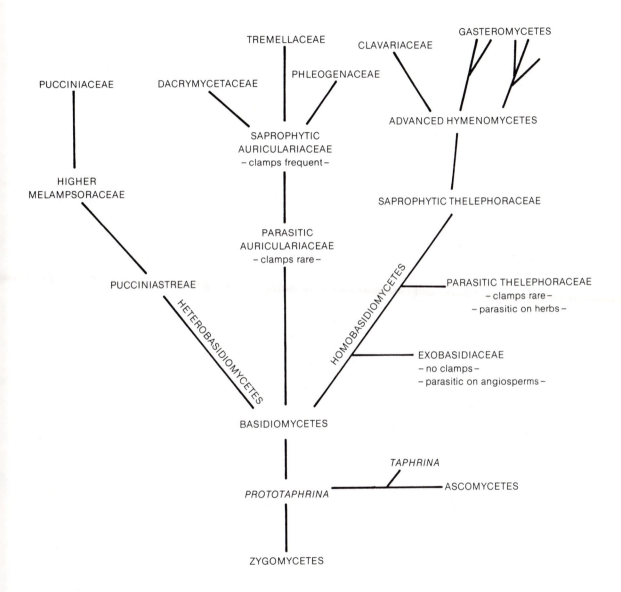

Figure 13-9 Phenetic series showing the probable relationships of the main fungal groups. (After D. B. O. Savile, with permission of the author. Reproduced by permission of the National Research Council of Canada from the *Canadian Journal of Botany* 33:60–104, 1955 and from Ainsworth and Sussman, eds., *The Fungi,* vol. 3, 1968, with permission of Academic Press.)

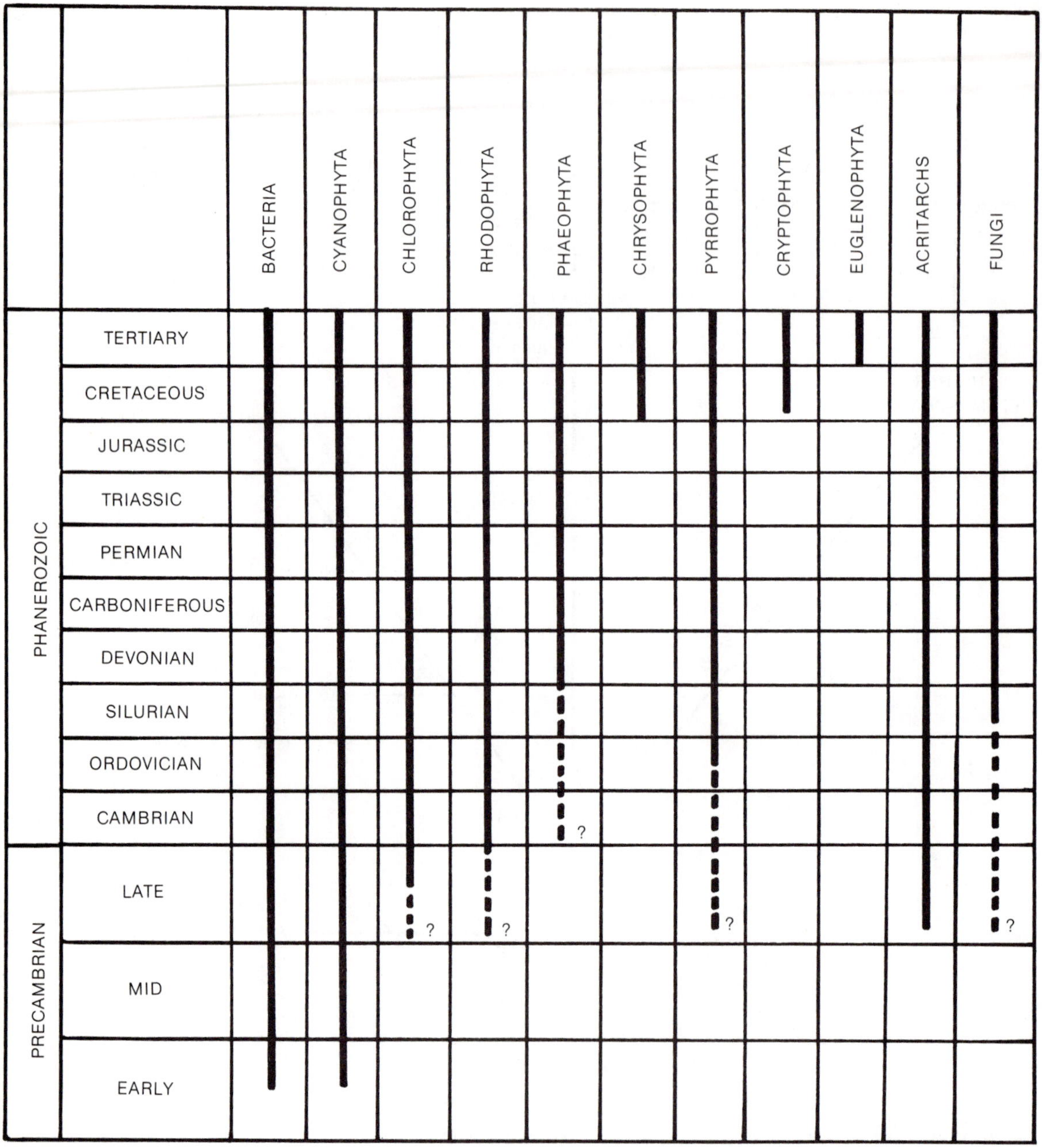

Figure 13-10 Geologic ranges of the main prokaryotic and eukaryotic divisions.

		BACTERIA	CYANOPHYTA	CHLOROPHYTA	RHODOPHYTA	PHAEOPHYTA	CHRYSOPHYTA	PYRROPHYTA	CRYPTOPHYTA	EUGLENOPHYTA	ACRITARCHS	FUNGI
PHANEROZOIC	TERTIARY											
	CRETACEOUS											
	JURASSIC											
	TRIASSIC											
	PERMIAN											
	CARBONIFEROUS											
	DEVONIAN											
	SILURIAN											
	ORDOVICIAN											
	CAMBRIAN					?						
PRECAMBRIAN	LATE			?	?			?				?
	MID											
	EARLY											

Figure 13-10 Geologic ranges of the main prokaryotic and eukaryotic divisions. Broken lines connect questionable or unverified occurrences with established ranges.

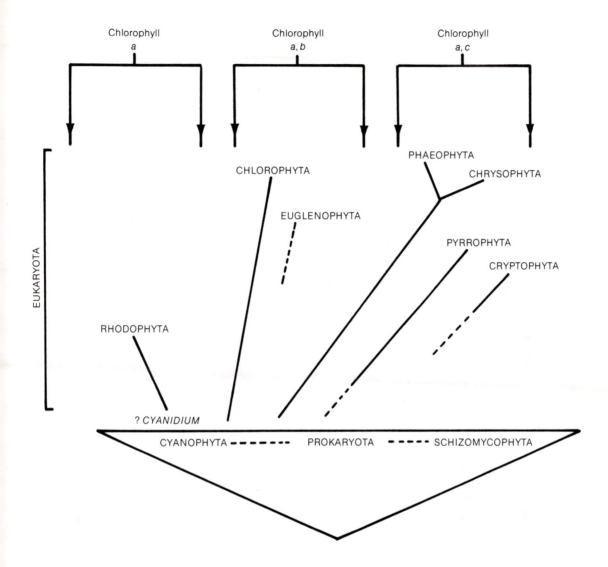

Figure 13-11 Scheme showing possible derivations of eukaryotic algae from pro-karyotes, and distinctions of the groups based on chlorophylls. (Based on Klein and Cronquist, 1967; R. M. Klein, *New York Academy of Science,* annals 175:623–33, 1970, with permission of New York Academy of Science; and V. J. Chapman and D. J. Chapman, *The algae,* 2d ed., 1973, by permission of MacMillan London and Basingstoke.)

schemes within the algal divisions to show evolutionary relationships and trends among classes and orders. Although such schemes clearly reflect the degree of relationship among the taxa, they cannot be taken as proof of how the various taxa actually evolved through time. However, such phenetic schemes do tell us how much evolutionary divergence has taken place among the taxa, and sometimes in which direction. One of the more current schemes is presented here to il-

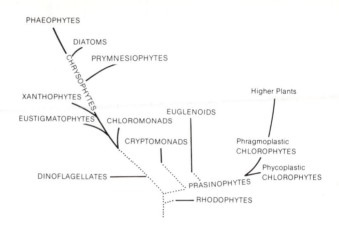

Figure 13-12 Contemporary scheme showing relationships of algal groups based on ultrastructural, morphological, and biochemical criteria. (After F. J. R. Taylor, reprinted with permission of the Society of Protozoologists from the *Journal of Protozoology* 23:28–40, 1976.)

lustrate the relationships of divisions, based on as many ultrastructural, morphological, and biochemical criteria as are available (Fig. 13-12).

Rhodophyta The evolutionary record based on fossil members of the family Corallinaceae was discussed in detail in Chapter 11. The relationships of other families and orders of extant Rhodophyta are rather obscure and need much more investigation. The general consensus is that the two classes Bangiophyceae and Florideophyceae are reasonably related by similarities in biochemical and ultrastructural characteristics. The main difference is the apparent lack of a sexual process in many of the Bangiophyceae. It has been postulated that the filamentous conchocelis stage of *Porphyra* with pit connections may represent a link with the Florideophyceae. Also, the Bangiophyceae, apparently lacking a sexual cycle in many instances, may be an evolutionary dead end. Although it is purely provisional, the scheme outlined in Figure 13-13 gives an approximation of the phenetic relationships of some of the main orders of the group.

Phaeophyta It was pointed out in Chapter 10 that the fossil record is spotty, although remains ascribed to brown algae extend from the Paleozoic to the present. Also, ancestors of the group are unknown. In assessing the relationships of orders, it is generally agreed that the Ectocarpales represent a group to which all others are closely related except the Fucales. Hence, a phenetic scheme such as that shown in Figure 13-14 represents a reasonable picture of what are believed to be the relationships in the division. The Fucales are considered to represent a group only distantly related to the others, mainly because of marked differences in the form, flagellation of the sperm, and life history.

There are two distinct morphological forms in the phaeophytes, excluding the Fucales. One consists of filamentous forms, ranging from simple filaments through filament clusters to complex pseudoparenchymatous forms. The other is parenchymatous, ranging up to the largest and most complex aquatic plants with meristems and differentiated tissues. Within both of these, the reproductive pattern ranges from isogamy through anisogamy to oogamy.

Chlorophyta The geologic record of the Chlorophyta is the oldest and one of the best-documented records of the algal groups. Certain chlorophytan fossils extend from the late Precambrian to the present. The geologic ranges of the fossil green algal representatives are shown in Figure 13-15. It is interesting, and perhaps significant, that the oldest known chlorophytes are thalloid and lime-encrusting, and that they antedate the earliest undoubted filamentous chlorophytes from the later Devonian Period by about 230 million years. We do not know whether this is a true indication that the filamentous habit evolved later, or whether filamentous types have not been preserved and discovered.

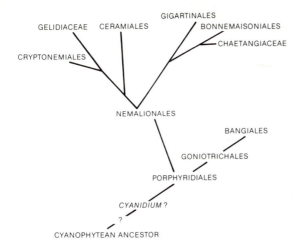

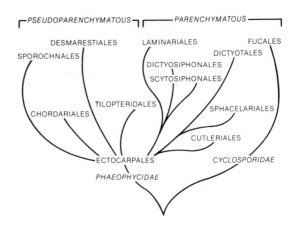

Figure 13-13 Phenetic series showing the relationships of the main groups of the Rhodophyta and their possible connection with the cyanophytes through *Cyanidium*. (After Fan, 1961, with permission of University of California Press.)

Figure 13-14 Phenetic relationships of the orders of the Phaeophyceae. (After Scagel, R. F., "The Phaeophyceae in perspective." *Oceanogr. Mar. Biol. Annual Rev.* 4:123–194, 1966, with permission of George Allen & Unwin Ltd., London.)

Significantly, the first fossils of the siphonous Bryopsidophyceae occur in the late Precambrian. These are genera of the carbonate-encrusting Dasycladales that have a reasonably good phylogenetic history (Fig. 13-8). This indicates that the multinucleate condition characteristic of the Bryopsidophyceae had probably become established relatively early in the overall history of the Chlorophyta. Together with the record of another bryopsidophycean order, the Codiales, from the earliest Cambrian, it indicates that the Bryopsidophyceae have formed a discrete evolutionary line since late Precambrian times.

The record of the Charophyceae extends from the Devonian to the present. Our knowledge of evolution in this group is limited to carbonate-encrusted oogonia and their encircling sheath cells, collectively called gyrogonites. These are often well preserved and occur in sufficient numbers to be very useful as geologic time and stratigraphic indicators. The type of arrangement of sheath cells in the gyrogonites appears to reflect the evolution of several main lines of charophytes (Fig. 13-16).

Because the charophytes have several features similar to those of land plants, it has been suggested they might be related. The similarities include a large central vacuole, numerous discoid chloroplasts, sterile cells surrounding both male and female gametangia, and sperms with banded flagella. The general consensus is that charophytes are probably an offshoot of a chlorophytan line that led to land plants, rather than being directly ancestral.

The record of the Prasinophyceae extends from the Ordovician, about 500 Ma to the present. The fossils are cystlike structures called *Tasmanites* and appear virtually identical to cysts of the modern planktonic genus *Pterosperma* (Fig. 13-17). The cystlike bodies are thick-walled and have **foveae** extending into the wall to varied depths. From the long record of *Tasmanites*, it appears very probable that the Prasinophyceae has remained as a separate and stable genetic line since its beginning. Based on pigmentation, it is likely that the Prasinophyceae are more closely related to the Chlorophyceae than to other classes.

Although the fossil record of some orders of

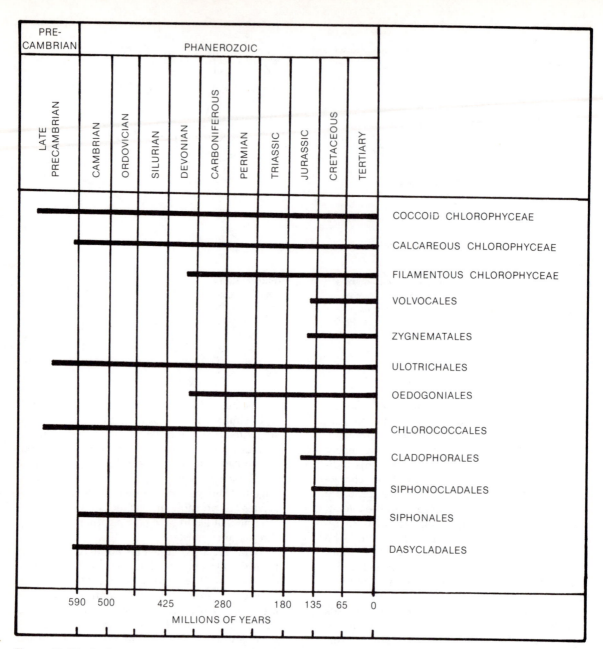

Figure 13-15 Geologic ranges of the main groups of the Chlorophyta.

Figure 13-16 Distribution of charophyte gyrogonites from Devonian to recent times. Note the relatively short ranges of the Devonian families and the Clavatoraceae in the late Jurassic-Cretaceous interval, and the dextral (right hand) spiral in the Devonian Trocholiscaceae. (After L. J. Grambast, 1974, with permission of International Association of Plant Taxonomy.)

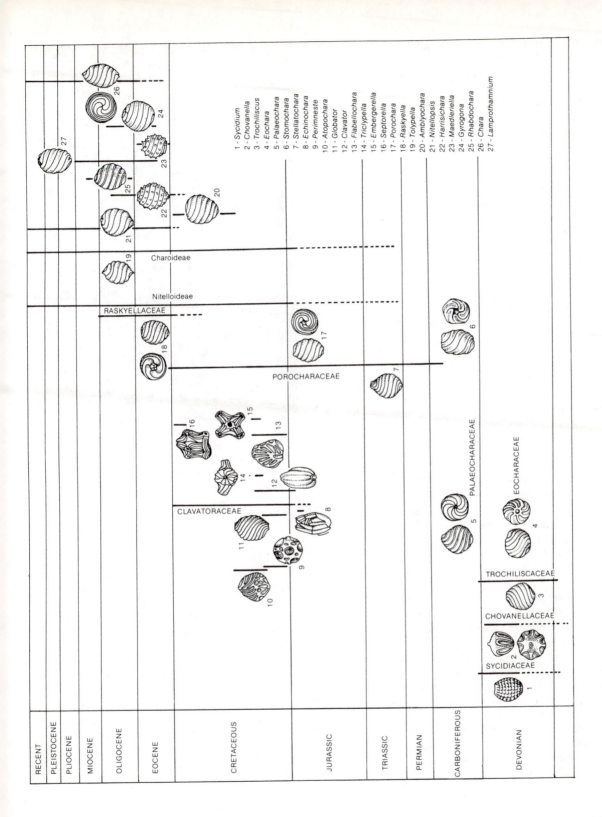

1 - Sycidium
2 - Chovanella
3 - Trochiliscus
4 - Eochara
5 - Palaeochara
6 - Stomochara
7 - Stellatochara
8 - Echinochara
9 - Perimneste
10 - Atopochara
11 - Globator
12 - Clavator
13 - Flabellochara
14 - Triclypella
15 - Embergerella
16 - Septorella
17 - Porochara
18 - Raskyella
19 - Tolypella
20 - Amblyochara
21 - Nitellopsis
22 - Harrisichara
23 - Maedleriella
24 - Gyrogona
25 - Rhabdochara
26 - Chara
27 - Lamprothamnium

Charoideae

Nitelloideae

RASKYELLACEAE

POROCHARACEAE

CLAVATORACEAE

PALAEOCHARACEAE

EOCHARACEAE

TROCHILISCACEAE

CHOVANELLACEAE

SYCIDIACEAE

RECENT
PLEISTOCENE
PLIOCENE
MIOCENE
OLIGOCENE
EOCENE
CRETACEOUS
JURASSIC
TRIASSIC
PERMIAN
CARBONIFEROUS
DEVONIAN

the Chlorophyta is relatively good, it cannot shed much light on the degree of relationships among other orders, particularly those with a short fossil history. For this it is necessary to group the classes and orders in phenetic series that reflects our present knowledge of similarities. Such a scheme is presented in Figure 13-18.

Chrysophyta   As outlined in Chapter 8, this division includes a heterogeneous group of algal classes that share similar pigments, chloroplasts, laminaranlike storage products stored outside the chloroplast, and (except in the Prymnesiophyceae) one tinsel plus one smooth, often reduced, flagellum. Many of the species are small and nonmotile and inhabit freshwater habitats. They have a very short fossil history; the coccolithophores (Prymnesiophyceae) range from the Triassic to the present, whereas the record of some other groups extends from the Cretaceous to the present. With the exception of the diatoms, chrysophytes are not very widespread or numerous in the sedimentary record. Thus the fossil record provides little evidence of relationships among the classes. Based mainly on similarities of pigmentation and storage products, the chrysophytan classes—particularly the diatoms, chrysophytes and xanthophytes—are believed to be related to the Phaeophyta (Fig. 13-19). The remaining three classes appear to have distant relationships at best and are questionably related to the first three. Both the fossil distributions and probable phylogeny of diatoms are shown in Figure 13-20.

Algal Potpourri   As noted in Chapter 9, the divisions Pyrrhophyta, Cryptophyta, and Euglenophyta appear to be very distantly, if at all, related. The fact that the members of all three are flagellated and have a groove is almost certainly the result of parallel evolution. Also, differences in flagellation and pigmentation are major, suggesting that if they ever had close ancestry, a great deal of evolutionary divergence has occurred to the present time. Some suggest that euglenoids have been derived from prasinophytes, but the origins and relationships of the dinoflagellates and cryptophytes are clearly unknown

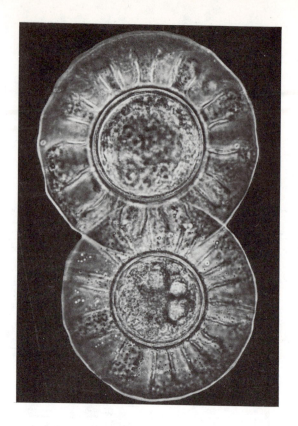

Figure 13-17   Cyst of the plankton prasinophyte genus *Pterosperma*, ×350. (Photograph courtesy F. J. R. Taylor.)

(Fig. 13-10). As mentioned on p. 353, all three divisions may have originated by endosymbioses of photosynthetic and colorless flagellated cells.

In recent years, an interesting field of study on fossil dinoflagellates has blossomed. The cyst stages of dinoflagellates are resistant. They are often abundant in Mesozoic and Tertiary rocks and are recovered in palynological preparations (Fig. 13-21). These display a great range in form and structure, and a classification has been growing in recent years for easy reference. Although a few cyst types can be related to modern dinoflagellates, the modern affiliations of many others are unknown. Relating cysts to extant forms promises to be very valuable in developing our knowledge of evolution in the Pyrrhophyta.

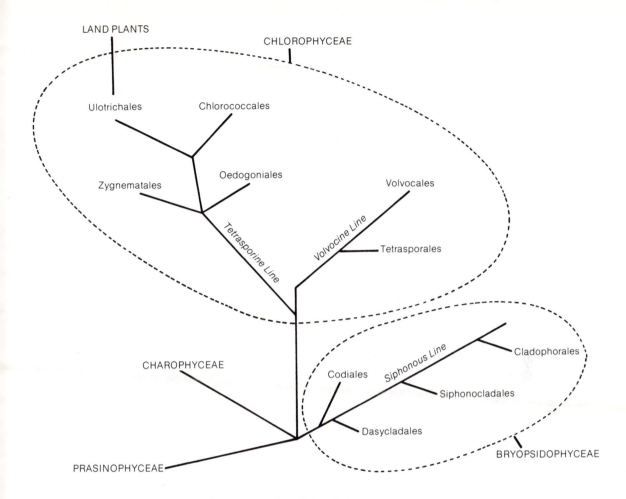

Figure 13-18 Phenetic relationships of main groups of the Chlorophyta.

Evolution of Land Plants

The origin of land plants—bryophytes and vascular plants—is very problematical and has stimulated much interest among botanists for many decades. As mentioned in Chapter 7, the general consensus is that land plants evolved from algae; the Chlorophyta are usually chosen because of similarities in pigmentation, cell wall formation, ultrastructure of motile cells, food manufacture, and storage products. Although several groups have been cited as probable ancestors, the most likely are among the Ulotrichales.

The transition from an aquatic to a terrestrial mode was one of the most important evolutionary events in the history of plants, comparable to the evolution of terrestrial from aquatic vertebrates in the animal kingdom. In both instances, derivatives of the evolutionary succession had to be adapted for the relatively hostile environment on land. In order to survive, it was absolutely essential that features be evolved for obtaining, distributing, and protecting against the loss of water. These features are found in sporophytes of vascular plants and some bryophytes. They include a surface cuticle and stomata to regulate water loss and conducting cells for distributing water.

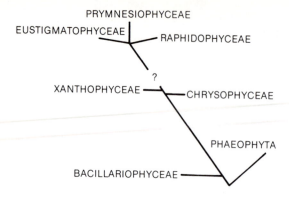

Figure 13-19 Phenetic series of chrysophyte families, and relations to the Phaeophyta.

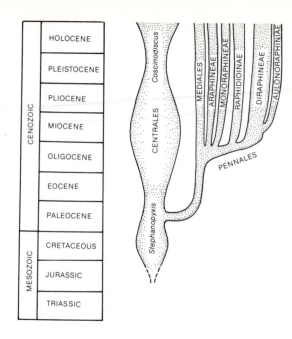

Figure 13-20 Diagram of geological range and probable phylogeny of diatoms. (After A. P. Zhuse, "Bacillariophyta," *Osnovi Paleontologii*, vol. 14, 1963, with permission of the Academy of Sciences of the U.S.S.R.)

The search for evolutionary links between algal ancestors and land plants usually focuses on selecting algae with the most likely ancestral characteristics. In a hypothetical model, such algae would be chlorophytan, filamentous, and heterotrichous and would have two multicellular generations. Fertilization would occur *within* the female gametangium, and the zygote would germinate directly into a diploid sporophyte that would eventually produce meiospores in sporangia. Some of these features are found in the modern Ulotrichales, especially in the genus *Coleochaete*.

The steps in the hypothetical model of evolution onto land are illustrated in Figure 13-22A–G, based largely on Jeffrey (1962). The model postulates that the ancestral algae were growing in the shallow waters of sites such as embayments, ponds, or lakes. They were essentially similar to freshwater chlorophytes such as *Coleochaete*, with a haploid, heterotrichous plant body (Fig. 13-22A). Sexual reproduction was likely oogamous, with fertilization occurring in the oogonium. This would result in a 2*n* zygote, which acted as a resting stage. On resumption of favorable conditions, the zygote underwent meiosis to form zoospores that developed into haploid plants.

It seems likely that one of the earliest evolutionary developments would have been the retention of the zygote on the haploid plant body, essentially as happens in *Coleochaete*, where sterile cells form around the oogonium after fertil-ization (Fig. 13-22B). This would have provided some protection for the zygote against periods of desiccation when water levels dropped. Once this was accomplished, the stage would be set for a multicellular, diploid generation to develop and remain attached to the gametophyte for nourishment. Although there are no known instances of this in *Coleochaete* or other chlorophytes, there are red algae (e.g., *Griffithsia* and *Callithamnion*) that show analagous attachment of the (carpo)sporophyte to the gametophyte. With further evolution, the sporophyte would have become larger and assumed its main function of spore dispersal (Fig. 13-22C). As habitats became more emergent, some sporangium-bearing branches would have become exposed for prolonged periods to air and the effects of drying. The evolution of a cuticle on such emergent sporophytes would protect against desiccation and provide a selective advantage for aerial spore dispersal.

In this evolutionary progression, there

Figure 13-21 A fossil dinoflagellate cyst, *Wetzeliella*, from the Eocene of Arctic North America, ×800.

would have been different selection pressures exerted on the two generations. Selection pressures would result in some plants remaining adapted for an aquatic existence to fulfill the main function of producing gametes for fertilization. At the same time, selection pressure on others would result in progressive adaptation of the sporophyte for its main function of dispersing spores. In order for spores to be disseminated on land, selection pressures would favor emergent sporophytic filaments (Fig. 13-22D, E).

Continuation of the two sets of selection pressures through long periods of time would result in increased emergence of the sporophyte, cutinization of emergent parts, the development of conducting tissues, and the formation of water control structures, the stomata (Fig. 13-22F, G). Thus a sporophyte adapted for existence in the relatively hostile environment on land and carrying out its prime function of spore dispersal for increasing the population could have become established. At the same time, the gametophyte remained tied to water, or at least to habitats with sufficient water to fulfill the prime function of gamete production and fertilization.

In this process of evolution onto land, there was probably a major split into two main evolutionary pathways (Fig. 13-22F). One of these, leading to the bryophytes, would result in the gametophyte maintaining its characteristics and independence (Fig. 13-22F, G). This would be a natural response of plants living in water, in very wet or humid sites, or in shade provided by topographic features. In these sites, there would be sufficient water to supply the gametophyte and also to insure the prime functions of gamete production and union.

The other evolutionary pathway leading to vascular plants expressed the opposite trend in generation development (Fig. 13-22F, G). There was increased selection for larger sporophytes that were able to obtain and conduct water and nutrients, to photosynthesize, and to disperse spores over drier sites. In this evolutionary pathway, the trend was toward increased independence of the sporophyte, accompanied by better conduction and protection of tissues against water loss. At the same time, selection worked against large, flimsy gametophytes requiring water and favored smaller and fast-maturing gametophytes that could produce gametes quickly during periods of available water, such as during or after rains. The gametophytes of the earliest vascular plants would probably have been submerged. Those of later vascular plants on terrestrial sites may well have been wholly or partially subterranean, like those of some modern species of *Lycopodium*. Some may eveny have been infected with mycorrhizal fungi and saprobic. Whatever the case, the trend in the evolution of vascular plants in terrestrial habitats would have been toward decreasing size and complexity in the gametophyte, and increasing size and complexity in the sporophyte.

Bryophyta

The bryophytes, with a fossil record from the Devonian, have three main evolutionary lines: Hepaticae, Anthocerotae, and Musci. All three of these are recorded before the end of the Paleozoic era, indicating that the main phyletic lines became differentiated early in the history of bryophyte evolution. The fossil record of the Hepaticae is the oldest and also contains the most fossil representatives, but there is no evidence to document that it was ancestral to the other lines (Fig. 13-23). Although the Anthocerotae are often

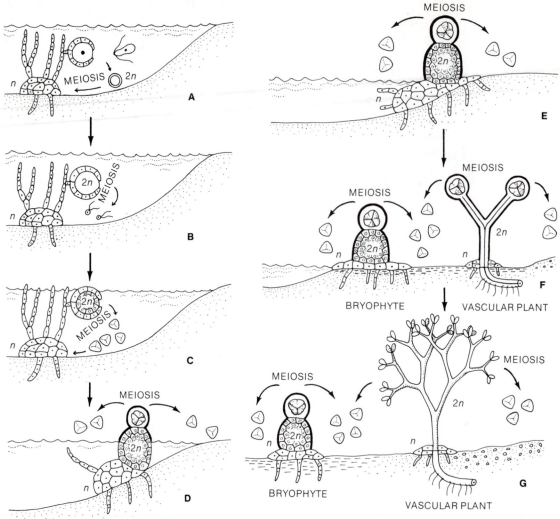

Figure 13-22 A–G, stages in a hypothetical evolutionary series from algae to early land plants (based in part on Jeffrey 1962). The first stage (**A**) represents a haploid, heterotrichous green alga (such as *Coleochaete,* Ulotrichales), with fertilization occurring in the oogonium, and the 2*n* zygote dispersing before undergoing meiosis. In **B,** the 2*n* zygote is retained within the oogonium, and meiosis occurs within the zygote. In stage **C,** the 2*n* generation is still retained on the haploid plant, but has become multicellular; the water is shallower, producing intermittent drying on the upper branches of both the haploid and diploid generations. In **D,** the diploid generation has developed sporangium-bearing branches, which are emergent, cutinized, and adapted for dispersing meiospores through the air. Further emergence and elaboration of the sporophyte (**E**), accompanied by reduction of the gametophyte, increases the efficiency of spore dispersal on the land. **F** indicates selection of these early emergents; some remain haploid dominant (**F1**) and are selected for wet/shaded habitats. Others (**F2**) become diploid dominant, and are selected for drier areas. Continued selection (**G**) results in two distinct land-plant stocks: a gametophyte-dominant bryophytic line (**G1**) adapted for gamete dispersal and damp/wet/shaded regions, and a sporophyte-dominant vascular plant line (**G2**) adapted for spore production in drier, more upland sites.

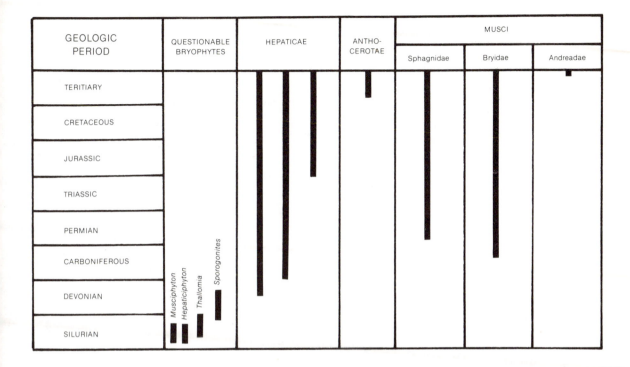

Figure 13-23 Geological record of the main lines of bryophytes. (After Lacey, 1969.)

credited with being an ancestral link between bryophytes and chlorophytan precursors, the fossil record of Anthocerotae is very short and late and hence provides no support for the hypothesis. The fact that the main evolutionary lines show well-established characteristics of their respective groups also suggests that the bryophytes evolved earlier than their first record in the Devonian. The record may be foreshortened because they were not preserved, have not been found, or have not been recognized conclusively as bryophytes (e.g., *Sporogonites*, Fig. 13-23).

Summary

In summary, the 4-Gyr history of the earth shows progressive evolution of plants and fungi from unicellular prokaryotes to complex, multicellular eukaryotes. The general pattern indicates that the evolution of eukaryotes occurred with a buildup in O_2 in the hydrosphere and eventually in the atmosphere (Fig. 13-24). This was accompanied by the development of the new and significant processes of mitosis, the sexual cycle, and alternation of generations. These set the stage for the subsequent evolution of the multicellular and relatively complex plants of several algal groups, with adaptations to diverse ecological niches in the Paleozoic and subsequent times. The final major evolutionary steps were the movement onto land and the development of fungi, bryophytes, and vascular plants as major lines. With such a long history, we can only predict that the evolution of plants and fungi will continue for a long time into the future. The challenge is for us to ensure, as potential manipulators, that we don't prevent it from happening.

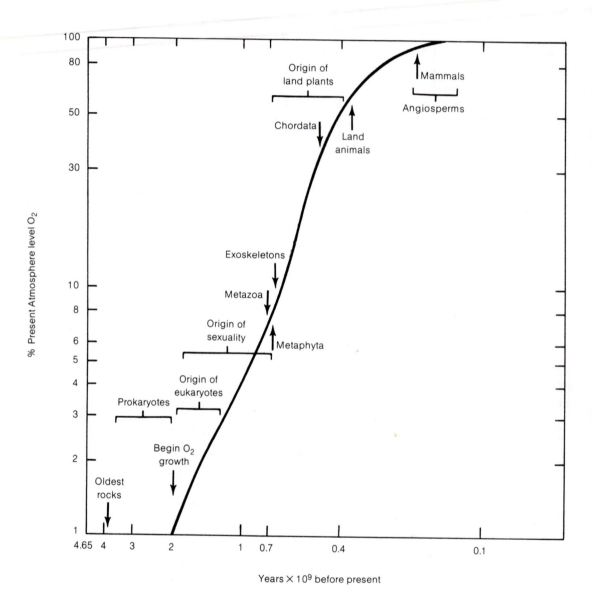

Figure 13-24 Apparent timing of some of the main events in biospheric evolution relative to hypothetical levels of atmospheric oxygen. Both scales logarithmic. (Modified slightly from P. Cloud, 1976, with permission of the author and the Paleontological Society, publishers of *Paleobiology*.)

References

Adey, W. H., and Macintyre, I. G. 1973. Crustose coralline algae; a re-evaluation in the geological sciences. *Geol. Soc. Am. Bull.* 84:883–904.

Balch, W. E.; Magrum, L. J.; Fox, G. E.; Wolfe, R. S.; and Woese, C. R. 1977. An ancient divergence among the bacteria. *J. Mol. Evol.* 9:305–11.

Barghoorn, E. S., and Schopf, J. W. 1965. Microorganisms three billion years old from the Precambrian of South Africa. *Science* 152:758–63.

Brock, T. D. 1973. Evolutionary and ecological aspects of cyanophytes. In Carr, N. G., and Whitton, B. A., eds. *The biology of blue-green algae.* Blackwell, London.

Buetow, D. E. 1976. Phylogenetic origin of the chloroplast. *J. Protozool.* 23:41–47.

Cavalier-Smith, T. 1975. The origin of nuclei and of eukaryotic cells. *Nature* 256:463–68.

———. 1978. The evolutionary origin and phylogeny of microtubules, mitotic spindles, and eukaryote flagella. *Biosystems* 10:93–114.

Cloud, P. 1976. Beginnings of biospheric evolution and their biogeochemical consequences. *Paleobiology* 2:351–87.

Demoulin, V. 1974. The origin of Ascomycetes and Basidiomycetes: the case for a red algal ancestry. *Bot. Rev.* 40:315–45.

Dodge, J. D. 1974. Fine structure and phylogeny in the algae. *Sci. Prog.* (Oxford) 61:257–74.

Doemel, W. N., and Brock, T. D. 1973. Bacterial stromatolites: origin of laminations. *Science* 184:1083–85.

Downie, C. 1967. The geological history of the microplankton. *Rev. Palaeobot. Palynol.* 1:269–81.

Flugel, E., ed. 1977. *Fossil algae: recent results and developments.* Springer-Verlag, Berlin, Heidelberg, New York.

Fulford, M. 1965. Evolutionary trends and convergence in the Hepaticae. *Bryologist* 68:1–31.

Grambast, L. J. 1974. Phylogeny of the Charophyta. *Taxon,* 23:463–81.

Jeffrey, C. 1962. The origin and differentiation of the archegoniate land-plants. *Bot. Not.* 115:446–54.

Jovet-Ast, S. 1967. Bryophyta. In Boureau, E., ed. *Traité de paléobotanique.* Vol. 2. Masson et Cie, Paris.

Klein, R. M. 1970. Relationships between blue-green and red algae. *Ann. New York Acad. Sci.* 175:623–33.

Lacey, W. S. 1969. Fossil bryophytes. *Biol. Rev.* 44:189–205.

Loeblich, A. R., Jr. 1974. Protistan phylogeny as indicated by the fossil record. *Taxon* 23:277–90.

Margulis, L.; To, L.; and Chase, D. 1978. Microtubules in prokaryotes. *Science* 200:1118–24.

Monty, C. 1973. Precambrian background and Phanerozoic history of stromatolitic communities, an overview. *Ann. Soc. Geol. Belg.* 96:585–624.

Pirozynski, K. A., and Malloch, D. W. 1975. The origin of land plants: a matter of mycotrophism. *Biosystems* 6:153–64.

Reijnders, L. 1975. The origin of mitochondria. *J. Molec. Evol.* 5:167–76.

Savile, D. B. O. 1968. Possible interrelationships between fungal groups. In Ainsworth and Sussman, eds. *The fungi.* Vol. 3. *The fungal population.* Academic Press, New York.

Schopf, J. W. 1970. Precambrian micro-organisms and evolutionary events prior to the origin of vascular plants. *Biol. Rev.* 45:319–52.

———. 1977. Biostratigraphic usefulness of stromatolitic Precambrian microbiotas: a preliminary analysis. *Precambrian Res.* 5:143–73.

Schwartz, R. M., and Dayhoff, M. O. 1978. Origins of prokaryotes, eukaryotes, mitochondria and chloroplasts. *Science* 199:395–403.

Stebbins, G. L. 1970. Variation and evolution in plants. In Hecht, M. K., and Steere, W. C., eds. *Essays in evolution and genetics in honor of Theodosius Dobzhansky.* Appleton-Century-Crofts, New York.

Steere, W. C. 1958. Evolution and speciation in mosses. *Amer. Nat.* 92:5–20.

Taylor, F. J. R. 1976. Autogenous theories for the origin of eukaryotes. *Taxon* 25:377–90.

———. 1978. Problems in the development of an explicit phylogeny of the lower eukaryotes. *Biosystems* 10:67–89.

———. 1979. Symbionticism revisited: a discussion of the evolutionary impact of intracellular symbioses. *Proc. Roy. Soc. Lond. B.* 204:267–89.

Woese, C. R. 1977. Endosymbionts and mitochondrial origins. *J. Molec. Evol.* 10:93–96.

Woese, C. R., and Fox. G. E. 1977. The concept of cellular evolution. *J. Molec. Evol.* 10:1–6.

GLOSSARY

abaxial Situated facing away from the axis of the plant.

abstrict To separate and discharge part of the plant (applied to release of basidiomycete spores).

acervulus (plural, **acervuli**) Discoid or pillow-shaped fungal structure in which conidia and conidiophores are formed.

acidophilic The ability to thrive in an acid medium (e.g., the moss *Sphagnum*).

acrasins Substances that are secreted by myxamoebae and that regulate streaming together during the aggregation stage of Dictyosteliales.

acritarch Unicellular organic microfossil with resistant and often highly ornamented outer test. Some considered probably to represent unicellular algae (dinoflagellates); others may represent various stages of plants and animals.

acrocarpous In mosses, a growth form in which the gametophore is erect, and the sporophyte terminates the main axis.

acrochaetioid Having a branched filamentous form similar to *Acrochaetium* (=*Audouinella*) in Rhodophyta.

acrogynous In Jungermanniales, a condition in which the apical cell produces the female sex organs, thus bearing the sporophyte terminally.

adaxial Situated toward the axis of the plant.

adenosine triphosphate See **ATP.**

advanced The state of certain characteristic(s) in a more evolved taxon.

aeciospore Dikaryotic spore produced in an aecium of the Uredinales (Heterobasidiomycetes).

aecium (plural, **aecia**) Structure (often cup-shaped) in which aeciospores are formed.

aeodan Complex phycocolloid substance occurring in cell walls of some red algae.

aeon A period of geologic time of 1×10^9 years; also spelled *eon*.

aerobe An organism requiring the presence of free oxygen.

aerobic The ability to function as an aerobe.

aethalium A sessile, rounded, or pillow-shaped fructification formed by a massing of all or part of the plasmodium in myxomycetes.

agar Complex phycocolloid substance occurring in cell walls of some red algae; also prepared as a commercial product and used to solidify culture media.

aggregation Movement of amoebae in dictyosteliomycetes toward one point prior to grex formation.

ahnfeltan Complex phycocolloid substance occurring in cell walls of some red algae.

akinete Thick-walled resting cell or spore in the algae, generally in-corporating original vegetative cell wall.

alar cells Cells at basal margin of a moss leaf where it joins the stem.

algal layer In stratified lichen thalli, the zone in which algal cells occur (see also **phycobiont**).

algin Phycocolloid substance occurring in cell walls and intercellular spaces of brown algae; commercially marketed.

alginate Generalized name given to a salt of alginic acid.

alginic acid Carboxylated polysaccharide material found in cell walls and intercellular spaces of brown algae.

allele Any one of several alternative forms of a gene at a given chromosome locus.

alloploid Polyploid resulting from an interspecific cross, thus containing two different genomes.

allopolyploidy The formation of a polyploid as a result of chromosome doubling in a hybrid in which there are few (or no) homologous chromosomes.

allotetraploid Tetraploid derived by doubling of the two different genomes from an interspecific cross.

alternation of generations Alternation of sexual gamete-producing phase with a meiospore-producing stage; usually the alternation of a haploid with a diploid generation.

alveolus (plural, **alveoli**) Open elongated chamber on diatom frustule (wall) from central axis to margin.

ameiotic Reduction of nuclear material by cleavage or splitting without visible chromosome formation.

amitotic Division of nuclear material by cleavage or splitting without visible chromosome formation.

amoeboid Resembling an amoeba; amorphous and changing in shape.

amphigastria Ventrally located row of leaves in Jungermanniales (Hepaticae).

amphigenous Growing all around, or over entire surface.

amphipolyploid See alloploid.

amphithecium Outer layer of cells in the early embryonic sporangium of bryophytes.

amyloid Appearing blue-black after addition of iodine solution, giving a starchlike reaction.

amylopectin Storage polysaccharide starch composed of α-1,6 and α-1,4 glucoside linkages; also known as "branching factor" of starch.

amylose Portion of the storage polysaccharide starch, composed of α-1,4 glucoside linkages; presence gives amyloid reaction.

anacrogynous Condition in some hepatics in which the female sex organs are lateral in position, having been formed from subapical cells; thus the sporophyte is borne laterally.

anaerobe Organism able to grow in the absence of free oxygen; oxygen is toxic to many anaerobes.

anaerobic Ability to function as an anaerobe.

analogy Similarity due to the common appearance, or perfor-

mance of a common function, and not due to descent from a common ancestor.

anastomosis (plural, **anastomoses**) Parts joined or coming together, as in fusion of two filaments.

aneuploid Differing from the usual diploid chromosome number by the addition or loss of an extra chromosome.

anisogamous Having anisogamy.

anisogamy Fusion of gametes of similar form, one of which is always larger; generally the larger gamete is considered the female and the smaller the male.

anisospory Condition of spores in bryophytes in which the spores of a tetrad are of two different sizes, each size producing a gametophore of a different sex.

annulus A ring: in basidiomycetes, remnants of the partial veil on the stipe; in bryophytes, a specialized ring of cells on the sporangium indirectly involved in spore release.

antheridiophore Specialized branch bearing antheridia in members of the Marchantiales (Hepaticae).

antheridium (plural, **antheridia**) Sex organ producing male gametes: in fungi and algae, a single cell; in bryophytes, many cells, including sterile jacket cells.

antibody Protein produced in a living body in response to introduction of foreign particles or cells (antigens); the antibody is specific and combines with its inducing antigen.

anticlinal Perpendicular to the circumference or surface.

antigen Substance which, when introduced into the body, stimulates the production of defensive antibodies; bacterial flagella, cap-

sules, cell walls, and many other substances function as antigens.

aplanogamete Nonmotile gamete.

aplanosporangium Sporangium in which nonmotile spores (aplanospores) are produced.

aplanospore Nonflagellated spore.

apophysis Swollen basal area of sporangium in some mosses.

apothecium (plural, **apothecia**) Ascocarp (often cupulate or discoid) in which the hymenium is exposed at maturity of the ascospores.

archegoniophore Specialized branch bearing archegonia in members of the Marchantiales (Hepaticae).

archegonium (plural, **archegonia**) Multicellular sex organ producing single female gamete; generally flask-shaped with elongate neck and swollen portion containing single egg.

archesporium Mass of cells from which sporogenous cells originate.

areola (plural, **areolae**) Markings in diatom wall consisting of thin areas bordered by ridges of siliceous material and often containing many finer pores.

areolation Arrangement of network of cells in the bryophyte leaf.

armored Having cellulose plates under plasma membrane, as in some dinoflagellates.

arthrospore Spore resulting from hyphal fragmentation.

articulated Jointed or segmented.

artificial taxon Same as unnatural taxon.

ascocarp Ascus-containing structure, or "fruiting body," of ascomycetes.

ascogenous hypha In ascomycetes, hypha that develops from the ascogonial surface after plasmogamy and gives rise to asci.

ascogonium (plural, **ascogonia**) In ascomycetes, female cell that receives nuclei from the antheridium.

ascospore Spore formed within an ascus, typically the result of a meiotic division followed by a mitotic division.

ascostroma (plural, **ascostromata**) A stroma within which locules and asci develop.

ascus (plural, **asci**) A saclike cell in which ascospores are produced; in Euascomycetes, the cell in which both karyogamy and meiosis occur.

ascus mother cell Cell which enlarges to become the ascus in ascomycetes.

asexual reproduction Production of more individuals identical (except in some fungi) to the parent without syngamy and meiosis.

assimilative Growing and absorbing food; vegetative or nonreproductive.

ATP Adenosine triphosphate: a substance containing energy in the form of high energy phosphate bonds (this bond energy directly or indirectly drives all energy—requiring processes of life.)

autoecious Term applied to rust fungi that require a single host to complete the life cycle.

autogenous Refers to the origin and differentiation of eukaryotic organelles within the cell, rather than from outside (see **xenogenous**).

autoploid Polyploid resulting from the duplication of a single genome.

autopolyploid A polyploid resulting from chromosome doubling in a fertile plant.

autopolyploidy The formation of a polyploid as a result of chromosome doubling in a plant in which many, or even all, of the chromosomes have homologues present.

autotroph Organism that requires for its growth only inorganic substances and light or chemical energy.

autotrophic The ability to function as an autotroph.

auxiliary cell In some Florideophyceae, a cell to which the diploid zygote nucleus is transferred, where growth of the carposporophyte is initiated.

auxiliary cell branch In Florideophyceae, the specialized filament of cells in which the auxiliary cell occurs.

auxospore In diatoms, large cell often produced as a result of sexual reproduction.

auxotroph Organism that needs an external supply of some organic substances, especially vitamins.

avirulent Condition in which a strain of a known pathogenic species is unable to establish itself and reproduce in the host.

axil Angle formed between axis and attached appendage.

axile Central in position, as in some algae in which the chloroplast is centrally located in the cell.

bacillus (plural, **bacilli**) Straight, rod-shaped bacteria.

bacteriochlorophyll Photosynthetic pigment occurring in purple photosynthetic bacteria.

bacteriophages Viruses occurring in bacteria; also termed *phages*.

basal body Cylindrical body composed of a ring of nine units, each of which is made up of three microtubules, present at the base of each flagellum in motile cells (also called a *kinetosome*).

basal cell In *Polysiphonia*, cell cut off from the supporting cell.

basidiocarp Basidium-bearing structure, or "fruiting body," of basidiomycetes.

basidiospore Spore formed exogenously on a basidium, generally following karyogamy and meiosis.

basidium (plural, **basidia**) Cell in which karyogamy and meiosis occur and upon which basidiospores are borne.

basipetal Development or maturation from the apex toward the base.

β-carotene Accessory carotene pigment present in chloroplasts of land plants and most algal groups.

benthic (benthonic) Living on and generally attached to the bottom of aquatic habitats.

binary fission Reproduction occurring when a single cell divides into two equal parts.

biotypes Organisms with the same genotype.

biseriate Having two linear rows of cells.

bisexual Having both sexual reproductive structures (male and female) produced by any one individual.

bispore In Florideophyceae, spore produced in a bisporangium (two spores are produced in each sporangium, and they are believed to be meiospores).

blade The broadened, flattened portion of the thallus of large marine algae.

blanc mange Dessert made from milk and gelatinous, mucilaginous, or starchy substances.

bloom Dense growth of planktonic algae giving a distinct color to the water body.

bothrosome In Labyrinthulomycetes, organelle responsible for producing the membranes of the ectoplasmic net.

brackish With salinity less than that in the marine environment (generally less than 20 parts per thousand).

bryologist One who studies bryophytes (liverworts and mosses).

budding Type of asexual reproduction in which a small protuberance develops, enlarges, and is separated from the parent cell.

caducous Deciduous; not persistent.

calcareous Limy; consisting of or containing calcite, or calcium or magnesium carbonate.

calcite Precipitate of calcium carbonate.

callose Cell wall constituent in some cells of Laminariales (brown algae).

calyptra The enlarged tissue below archegonium that surrounds and protects the developing sporophyte in bryophytes, forming a sheathing cap over the sporangium in most mosses.

capillitium In some fungi and myxomycetes, threadlike strands (often forming a network) interspersed with spores.

capitate Having a head.

capitulum Small head; in Sphagnidae (Musci), dense tuft of branches at apex of gametophyte.

capsule A case: a nonliving layer of slime material of varying thickness and viscosity, situated immediately outside the cell wall or outer membrane in some prokaryotic cells; may consist of a

single polysaccharide or of such material plus amino acids, proteins, etc.

carotene General name for group of orange, carotenoid pigments.

carotenoid A yellow, orange, or red pigment that often serves as an accessory pigment in photosynthesis (or may serve as a protection against radiation); includes xanthophylls and carotenes; soluble in organic solvents (e.g., ethanol and acetone).

carpocephalum (plural, **carpocephala**) Specialized, erect, sporangium-bearing branch in Marchantiales (Hepaticae).

carpogonium (plural, **carpogonia**) Female gametangium in the red algae.

carpogonial branch In Florideophyceae, specialized filament of cells, at the end of which is borne the carpogonium.

carposporangium (plural, **carposporangia**) Sporangium produced directly or indirectly as a result of division of the zygote nucleus in red algae.

carpospore Spore produced by a carposporangium in red algae.

carposporophyte Collection of carposporangia occurring in chains on the gonimoblast filaments in Florideophyceae; also referred to as the *gonimoblast*.

carrageenan Complex phycocolloid occurring in the cell wall of some red algae (commercially marketed).

caulid Main shoot, or "stem," of the bryophyte gametophore, which supports the photosynthetic and reproductive organs.

cell plate Collection of vesicles forming new membrane and wall in a dividing cell.

cellulose Main cell wall compo-

nent of plants (excluding most fungi); composed of β-1,4, 1,6 glucoside linkages.

central cell Cell that is axile, or central, in position in a polysiphonous thallus (e.g., *Polysiphonia, Rhodophyta*) which is surrounded by a series of pericentral cells.

centric Central; common name for diatoms with markings radially arranged.

centriolar plaque Structure associated with nuclear envelope and lying at the poles of the spindle during nuclear division; believed to be a homologue of the centriole.

centrifugal(ly) Developing from the center toward the outside.

centriole Cell organelle consisting of nine units, each composed of three microtubules, lying adjacent to the nucleus and often concerned with the flagella and spindle formation; generally characteristic of animal cells and motile plant cells.

centripetal(ly) Developing from the outside toward the center.

centroplasm Inner, often colorless, part of protoplast in Cyanophyta.

cephalodium (plural, **cephalodia**) Epiphytic lichen growing as wartlike protuberance on upper surface of host lichen.

chemoautotroph Autotrophic organism that derives energy from chemical reactions, such as oxidation-reduction reaction of inorganic compounds.

chemotactic The attraction of motile cells by certain chemical substances.

chemotroph Organism deriving energy from chemical reaction; see **chemoautotroph**.

chitin Polysaccharide cell wall material composed of glucoside units and nitrogen.

chlamydospore   Thick-walled, nondeciduous spore produced by rounding up of hyphal cell protoplast and secretion of new wall.

chlorobium chlorophyll   Photosynthetic pigment occurring in the green sulfur bacteria.

chlorococcine group   Series of green algae that are nonmotile and nonfilamentous in vegetative condition.

chlorophyll   General name for green photosynthetic pigment, soluble in organic solvents (e.g., ethanol and acetone).

chlorophyllose   Having chlorophyll.

chloroplast   Cell organelle with membranes (thylakoids) containing photosynthetic pigments.

chromatid   One of two identical longitudinal halves of a chromosome.

chromatin   Nucleoprotein complex of chromosomes.

chromatophore   Nonlamellar, pigment-bearing cell organelles occurring in photosynthetic bacteria and animals.

chromoplasm   Outer, generally pigmented part of protoplast in Cyanophyta.

chromosome   Elongate structure in nucleus made of DNA and protein-containing genes in a linear order.

chrysolaminaran   Storage polysaccharide composed of β-1,3, 1,6 glucoside linkages occurring in chrysophytes and xanthophytes.

cilium (plural, **cilia**)   Marginal filamentous hair: in the lichens, hairy extensions on the thallus; in bryophytes, may occur on the leaves or among the inner peristome teeth; also used as motile organelle almost identical to a short flagellum.

circadian rhythm   Natural rhythm: occurs, for example, in the maturation and firing of sporangia in *Pilobolus*, in which patterns are repeated over 24-hour cycles.

cladistics   The erection of a classification based on shared derived traits.

cladist   One who practices cladistics.

clamp connection   In basidiomycetes, looplike structure connecting adjacent hyphal cells; it is produced during formation of new cells by dikaryotic hyphae.

class   A level in the Linnaean hierarchy that includes closely related orders; also, a general term for a group of related objects.

classification   The arrangement of objects into classes; the act or process of classifying.

classify   To arrange objects into groups.

cleistothecium (plural, **cleistothecia**)   Ascocarp in which the asci are completely enclosed at maturity of ascospores.

clonal   With reference to a **clone**. Group of individuals propagated vegetatively by mitosis from a single ancestor; all individuals of a clone are of the same biotype.

coccoid   Spherical cell type and growth form or morphological type.

coccus (plural, **cocci**)   Spherical bacterial cell.

coccolith   Scale composed of $CaCO_3$ on organic base occurring in the prymnesiophytes.

coenobium (plural, **coenobia**)   Colony of algal cells of a definite number, not increasing at maturity; usually has definite arrangement of cells.

coenocyte   Multinucleate cells or thallus in which nuclei are not

separated from one another by septa as in plasmodia of myxomycetes, hyphae of lower fungi, and some green algae (Bryopsidophyceae) and xanthophytes.

coenocytic   Having the condition of a coenocyte.

coenozygote   Multinucleate zygote in some zygomycetes.

colony   In bacteria, yeasts, and algae, a mass of individuals or cells of one species living together; in some instances it consists of a few individuals that can be attached to one another in a definite or regular pattern.

columella   Small column: in myxomycetes, often a continuation of stipe into sporangium that can be capillitial in nature; in zygomycetes, bulbous septum separating sporangiophore and spore case; in Anthocerotae and Musci, central column of sterile cells in the sporangium surrounded by the sporogenous layer.

columellate   Having a columella.

conceptacle   In some brown algae, a cavity in the thallus in which gametangia are formed.

conchospore   In Bangiophyceae, spore produced on conchocelis stage singly in a conchosporangium; conchosporangia are usually produced in a series at the end of a conchospore branch.

conidiophore   Specialized hypha bearing conidia.

conidium (plural, **conidia**)   In bacteria (Actinomycetales), used for chains of endospores formed in hyphal tips; in fungi, essentially a separable portion of a hypha that functions as an asexual propagule.

conjugate division   Simultaneous division of the paired nuclei in dikaryotic cells.

conjugation   Fusion, usually of isogametes or isogametangia: in bacteria, transfer of genetic material from a donor to a recipient; in phycomycetes and some green algae, fusion of gametangial protoplasts; in yeasts and others, fusion of cell protoplasts following development of a conjugation tube.

connecting filament In Florideophyceae, the nonseptate filament that grows out of the base of the carpogonium and transfers the diploid nucleus to the auxiliary cell (or in some instances, from one auxiliary cell to another).

context Sterile inner part of the cap or pileus in basidiomycetes.

continental shelf The gradually sloping, submerged (below sea surface) portion of a continental region that occurs in depths less than about 200 m.

contractile vacuole Small, pulsating organelle of some algae and fungi; generally in motile cells near the flagellar end.

coralline Refers to the lime-encrusted red algae of the Corallinaceae (Florideophyceae).

coralloid Erect, branched, shrublike, or corallike in form.

cortex In some brown algae and some red algae, tissue internal to the epidermis but not in the central position; in lichens, the compact surface layer or layers of the thallus.

corticated Having a cortex; parenchymatous or appearing so.

costa (plural, **costae**) Ridge: in the diatoms, wall marking formed by well-defined siliceous ridges; in bryophytes, the midrib area of leaf or thallus.

cover cells In Florideophyceae, the superficial cells of the thallus that are attached to the supporting cells and cover the tetrasporangium.

crenulate Scalloped.

crossing-over Exchange of parts of two chromatids.

crozier (formation) In ascogenous hyphae, formation of a hook in which conjugate nuclear division occurs, followed by cytokinesis; formation may or may not immediately precede ascus formation.

cruciate Cross-shaped.

crustose Encrusting: lichen growth form in which the thallus adheres tightly to the substrate.

cuticle The external, waxy layer covering the outer walls of epidermal cells: in bryophytes, it consists of a fatty compound called *cutin* that is almost impermeable to water; in algae, it may contain other compounds.

cutin Waxy material covering external cell surfaces of some bryophytes (the layer is referred to as a *cuticle*).

cyanophycin granule Protein storage body occurring in the blue-green algae.

cyanophyte starch Storage polysaccharide composed of α-1,4, 1,6 glucoside linkages; considered to be amylopectin.

cyphella (plural, **cyphellae**) Cup-shaped depression forming regular opening through the lower cortex of some foliose lichens.

cyst Resistant, sporelike body (often thick-walled) that develops by the rounding up of reproductive cells (lower fungi and myxomycetes) or vegetative cells (bacteria and some algae); in algae, forms within vegetative cell.

cystidium (plural, **cystidia**) Sterile structure produced in the hymenium of some basidiomycetes.

cystocarp In Florideophyceae, the structure comprising the gonimoblast (carposporophyte) and

surrounding pericarp (gametophyte tissue).

cytokinesis Cytoplasmic division, usually following nuclear division.

cytological With reference to the structure of the cell.

cytosis The budding, fusion, evagination, or invagination of cell membranes, as in cell cleavage (see also **endocytosis, exocytosis,** and **phagocytosis**).

deciduous Falling off; not persistent.

deoxyribonucleic acid See **DNA**.

determinate (growth) Having a fixed, definite limit.

detritus Particulate organic matter.

dextral Having right-hand helical arrangement: in charophytes, refers to sheath cells around the oogonia.

diatomaceous earth Deposits of the silicified walls of fossil diatoms.

diatomite Diatomaceous earth; fossil deposits of diatoms.

dichotomous Branching in which the two arms are more or less equal.

dictyosome Complex of flattened double lamellae, often with peripheral vesicles; similar to golgi apparatus.

diffuse growth Cell division and elongation occurring throughout the plant.

dikaryon In ascomycetes and basidiomycetes, hyphae in which each compartment contains a pair of compatible nuclei; initiated through plasmogamy of compatible gametangia or hyphae.

dikaryotic Having the condition of a dikaryon.

dimorphic In fungi, condition in which the assimilative structure can exist either as budding cells or as hyphae (mainly Hemiasco-mycetes and Heterobasidio-mycetes).

dioecious Condition in seed plants in which separate pollen-bearing and ovule-bearing cones or flowers are borne on different plants.

dioicous Condition of bryophyte gametophyte in which the gametophore bears only one kind of sex organ; unisexual.

diploid Having a single set of paired chromosomes (twice the number of chromosomes as in the gametes); 2*n*.

diploidization Process in many Florideophyceae in which a specialized haploid cell (auxiliary cell) receives a diploid zygote nucleus or a derivative of it; sometimes used in fungi for *dikaryotization*.

discoid Round and/or flattened with rounded margins; dish-shaped.

distal Remote from place of attachment.

distromatic Having thallus two cells thick.

division A level in the Linnaean hierarchy that includes closely related classes, usually restricted to the plant kingdom; identical to phylum in classification of animals.

DNA (deoxyribonucleic acid) Nucleic acid in cells, primarily in nucleus but also in mitochondria and chloroplasts; main (genetic) component of chromosomes, involved in transmission of heredity.

downgrade evolutionary pathway (downgrade theory) The hypothesis that the leafy habit was the first to evolve in bryophytes, with the thalloid habit derived later (opposite to upgrade theory).

dulse Dried preparation of *Palmaria palmata* (Rhodophyta); commercially marketed.

ecotype A population, or series of populations, of a species that shows adaptation to a local habitat, and in which the adaptations are genetically controlled.

ectoplasmic net Network of tubules within which gliding cells of Labyrinthulomycetes move, and the structurally similar "rhizoids" of *Thraustochytrium* and similar organisms.

effused Referring to the type of basidiocarp that is spread out or flattened.

effused-reflexed Referring to a basidiocarp with a resupinate portion attached to the substrate and an upper portion extending out like a shelf.

ejectosome In some flagellated algae (dinoflagellates, cryptophytes, raphidophyte), the cytoplasmic organelle ejected when the organism is disturbed (see also **trichocyst**).

elater Sterile hygroscopic cell among the spores in the sporangium of many Hepaticae.

elaterophore In liverworts, a specialized, elater-bearing mound of cells arising from either the base or roof within the sporangium.

embryo Immature, incompletely differentiated organism; may or may not be retained by parent plant for part of its development; usually diploid, developing from zygote.

embryogeny Formation of the embryo.

encapsulated Development within a capsule.

encrusted Adherent and completely covered with lime, as in some of the prostrate coralline red algae.

encrusting Having a prostrate habit, in which the thallus is usually in a flattened, thin layer tightly attached to and spreading over the surface of the substrate.

endexine Inner layer of exine of bryophyte spores.

endogenous Development from the inside: in the Jungermanniales (Hepaticae), refers to a branch produced by internal stem cells that pushes through mature cells.

endophyte Growing within another plant.

endoplasmic reticulum A network of membranes in the cytoplasm, connected with the nuclear envelope.

endosporal Referring to the development of a gametophyte within the confines of the spore wall.

endospore Spore formed within parent cell: in the bacteria, a thick-walled resistant spore; in blue-green algae, a thin-walled spore. The term is also used for the inner layer of a spore wall.

endosporic Undergoing cell divisions before the spore coat is ruptured.

endostome Inner peristome teeth of bryophytes.

endosymbiosis The ingestion by one cell of another, which forms an integral part of the new cell.

endothecium Inner layer of cells in the early embryonic sporangium of bryophytes.

enucleate Lacking a nucleus.

epiphragm In most Polytrichidae (Musci), a multicellular, parchmentlike membrane closing the mouth of the sporangium after the operculum has fallen; consists of the expanded apex of the columella.

epiphyte Growing upon another

plant, but not making any nutritional connection.

epitheca (plural, **epithecae**) Outer cell half, or frustule, of diatom cell; often refers only to outer siliceous wall.

erose Irregularly eroded, as though gnawed.

eukaryotic With reference to organisms (**eukaryotes**): containing membrane-bound organelles (e.g., nucleus, mitochondria, chloroplasts) within the cells.

evagination An outgrowth, or unsheathing.

evanescent Of short duration.

evolution Change with time, resulting in biological diversity that can be arranged in hierarchy.

evolutionary divergence The process by which populations become different from each other as a result of evolving in different directions.

exoenzyme An enzyme secreted or formed externally to the protoplast of a cell and functioning outside the cell proper (chemically breaking down organic materials as a source of energy for growth).

exogenous Developing on the outside.

exospore Type of spore formed basipetally and one at a time in some bluegreen algae.

exosporic Undergoing cell divisions after the spore coat is ruptured.

exosporium The thin, loosely fitting outer layer or coating surrounding the endospore wall of some bacteria.

exostome Outer peristome teeth of bryophytes.

extant Living; used in reference to present-day plants.

extranuclear Occurring outside the nuclear envelope.

extranuclear division Nuclear division in which the spindle forms outside the nuclear envelope; the latter structure often disappears during the division.

eyespot Red to orange carotenoid-containing organelle in motile cells of many algae.

facultative parasite An organism with the ability to exist either saprobically or parasitically, or an organism capable of using one of alternate functional pathways; faculative heteromorph (see **mixotroph**).

false annulus A ring of cells outside the peristome, but beneath the operculum (in *Buxbaumia*).

false branching In some bluegreen algae, breaking of a filament, with one or both ends protruding from the sheath and appearing as a branch.

family A level in the Linnaean hierarchy within an order which includes closely related genera.

fascicle A cluster.

fasciculate In clusters.

fertilization tube Branch from male gametangium that transfers male nuclei or gametes to female gametangium in oomycetes.

fibril Submicroscopic threadlike structure: in *Sphagnum*, refers to microscopic fiberlike thickenings on the inner faces of large hyaline cells in the leaf or stem cortex.

fibrillar Made up of fibrils; fibrous; microscopic or submicroscopic.

filamentous Elongate, threadlike, multicellular growth form or morphological type.

filiform Thread-shaped.

fission Splitting in two; characteristic especially of bacteria, some yeasts and myxomycetes and bluegreen algae; also termed *binary fission*.

fixation Term used for preparing biological material for study (involves preservation of material); also, conversion of N_2 to nitrites and nitrates by bacteria.

flagellate Refers to cells that possess organelles for motility; also a growth form or morphological type.

flagelliferous Whiplike.

flagellin A protein like substance that found in muscle, composing the flagella of bacteria.

flagellum (plural, **flagella**) Long, whiplike organelle controlling movement of motile cell: in eukaryotes, external to cell but enclosed within plasma membrane; in bacteria, not surrounded by membrane; distinguished from cilium by length and type of movement.

floridean starch Polysaccharide storage product composed of $\alpha1,4$, 1,6 plus 1,3 glucoside linkages occurring in red algae; somewhat similar to amylopectin.

floridorubin A carmine-purple pigment of unknown function found in certain florideophyte red algae of the family Rhodomelaceae (Ceramiales).

floridoside Storage product of some red algae.

foliaceous Having leaflike shape.

foliose Leaflike: in lichens, a growth form in which the flattened, prostrate thallus may be easily removed from the substrate; in algae, a flattened, usually erect, bladelike thallus.

foot In bryophytes, haustorium-like, basal sporophytic cells.

foot cell In laboulbeniomycetes, the lowermost cell of the thallus, embedded in cuticle of the insect host.

forespore Clear area in cytoplasm of bacterial cell that becomes surrounded by a refractile wall in endospore formation.

form family Family consisting of form genera; that is, genera based upon asexual reproductive structures.

form genus In fungi, genus name based on morphology of asexual structures; in fossil plants, a generic name for those fossils with the same form or morphology.

form order In Fungi Imperfecti, order based on imperfect (i.e., asexual) structure.

fovea (plural, **foveae**) Pit in the wall of palynomorphs, such as spores, pollen, or dinoflagellate cysts (as in *Tasmanites* of the Prasinophyceae).

fragmentation Breaking apart.

free-nuclear division Stage in development in which unwalled nuclei result from repeated division of a primary nucleus.

frondiform Leaflike: in some Musci, flaplike, unistratose, and erect.

fructification Reproductive organ or fruiting structure.

fruiting The production of fruits and, by analogy, the formation of spore-bearing bodies of some nonvascular plants.

frustule Diatom cell wall containing organic material and silicon dioxide.

fruticose Lichen growth form in which the thallus is shrublike and generally branched.

fucoidan Phycocolloid occurring in cell walls and intercellular spaces of brown algae commercially marketed.

fucosan vesicles In brown algae, refractive vesicles, usually aggregated about the nucleus, containing a tanninlike substance; also called *physodes*.

fucosterol A sterol occurring in a number of algal groups (e.g., brown algae and red algae).

fucoxanthin Xanthophyll pigment characteristic of chrysophytes and brown algae.

funoran Phycocolloid of some red algae; commercially marketed.

furcellaran Phycocolloid of some red algae; commercially marketed.

gametangial contact Sexual reproduction in which, following contact of gametangia, nuclei are transferred from the antheridium to the eggs through the fertilization tube (in oomycetes).

gametangial copulation Sexual reproduction of some lower fungi (zygomycetes) in which the entire protoplasts of two gametangia fuse.

gametangium (plural, **gametangia**) Structure producing gametes: in fungi and algae, generally a single cell; in bryophytes, a multicellular structure with an outer, sterile, protective layer.

gamete Sex cell; capable of fusion with another gamete to form a zygote.

gametic meiosis The pattern of alternation of generations in which gametes are either produced directly by meiosis from the multicellular diploid plant, or after several mitotic divisions of individual haploid cells, as in *Fucus* (brown algae).

gametogenesis Formation of gametes.

gametophore The elaborate leafy or thallose structure in bryophytes that bears the sex organs.

gametophyte Gamete-producing plant; generally haploid, and producing gametes by mitosis.

gamone A chemical substance involved in bringing about sexual fusion (of gametes).

gas vacuole Array of cylindrical vesicles, occurring in prokaryotic cells.

gelatinous Composed of proteinacous (or, in some plants, polysaccharide) material: in li-chens, a growth form appearing moist; in fungi and algae, jellylike or slimy; mucilaginous.

gemma (plural, **gemmae**) Specialized cell (oomycetes) or group of cells for vegetative reproduction; in bryophytes, produced on the gametophore.

gemmiferous Bearing gemmae.

gene Unit of inheritance, arranged in a linear sequence on the chromosome; specifically, an arrangement of nucleotides in DNA that will code for a specific protein.

gene mutation Change in structure of a gene; also known as *point mutation.*

generalized The state of a characteristic that is common (general) in a taxon.

general purpose classification A classification of interest to the greatest number of biologists and upon which nomenclature is based.

genetic Pertaining to features resulting from the action of genes.

genome The basic set of chromosomes (n) contributed by each parent via gametes.

genotype Genetic constitution of an organism, determined by the assemblage of genes it has.

genotypic With reference to the genotype.

genotypic sex determination Determination of sex by single allele difference.

genus (plural, **genera**) A level in the Linnaean hierarchy within a family that includes closely related species.

geotropism Growth response of a plant to gravitational pull.

germ tube Hypha produced by a germinating fungal spore.

girdle Encircling, or middle: in dinoflagellates, transverse groove containing flagellum; in diatoms, region where frustules overlap.

girdle view   Side view (of diatom or dinoflagellate).

gleba   Spore-producing zone in basidiocarp of some gasteromycetes.

globose   Nearly spherical.

globule   In charophytes, multicellular male reproductive structure (gametangium), including sterile and fertile cells.

glycogen   Complex polysaccharide similar to starch but probably with more amylopectin; does not give blue-black color with iodine solution.

glycoprotein   Group of enzymes composed of sugar units and amino acids; important in surface reactions of cells (aggregation, clumping, fusion, etc.).

golgi apparatus   Collection of cytoplasmic vesicles involved in storage and secondary products; also called *dictyosome*.

gonidium (plural, **gonidia**) Chain of cells, or single cells, functioning in reproduction of filaments (Beggiatoales) that are released from filament apices and have gliding movement.

gonimoblast   Collection of gonimoblast filaments and carposporangia occurring in the Florideophyceae; also referred to as the *carposporophyte*.

gonimoblast filaments   Cells of carposporophyte (or *gonimoblast*) bearing carposporangia.

granum (plural, **grana**)   A stack of closely appressed (or fused) thylakoids.

grex   In dictyosteliomycetes, the sluglike structure produced by aggregation of myxamoebae; also called a *pseudoplasmodium*.

groove   Longitudinal region present in cryptophytes and euglenophytes.

Gyr   Short form for *gigayear*, or years × 10⁹; equivalent to *aeon*.

gyrogonites   Name for the lime-encrusted, fossilized oogonia and encircling sheath cells (*nucule*) of charophytes.

halophyte   A plant which grows normally in soil impregnated with salts.

haploid   Having a single set of unpaired chromosomes; the chromosome complement present in the gametes; 1*n*.

hapteron (plural, **haptera**)   An attaching structure in some brown algae; usually multicellular, branched, and rootlike.

haptonema   Apical, coiled organelle in some prymnesiophytes, composed of endoplasmic reticulum extension, surrounded by plasma membrane.

haustorium (plural, **haustoria**) An absorptive structure that derives food from its host by penetrating the cell wall (host cell protoplast is not penetrated).

hematochrome   Red pigment granules, probably xanthophyll, occurring in some green algae.

heterocyst   Differentiated cell in some filamentous bluegreen algae involved in nitrogen fixation.

heteroecious   Term applied to rust fungi requiring two hosts to complete their life cycle.

heterogeneous   Differing or unlike; heterozygous.

heterokaryosis   Condition in which mycelium is heterokaryotic.

heterokaryotic   Refers to mycelium in which two or more genetically distinct types of nuclei occupy the same cytoplasm; both haploid and diploid nuclei can be present.

heterokont   Having flagella of unequal length on a motile cell.

heteromerous   Lichen thallus in which algal cells are restricted to a specific layer, creating a stratified appearance.

heteromorphic   Morphologically unlike.

heterospory   Having spores of two different sizes.

heterothallism   Condition in which sexual reproduction requires contact between two morphologically similar but genetically distinct thalli (e.g., Mucorales).

heterotrichous   Having **heterotrichy.**

heterotrichy   Growth form in which erect filaments arise from prostrate portion in some algae and bryophytes.

heterotroph   Organism that requires an external source of one or more organic compounds as an energy and carbon source for growth, etc.

heterozygous   Having two different alleles for a given gene at the same locus of homologous chromosomes.

holdfast   An attaching, discoid or rootlike structure of some algae.

homogeneous   Having the same nature or consistency.

homokaryotic   Refers to a mycelium in which all nuclei have the same genetic makeup.

homologous   Refers to structures that have a similar origin.

homologous chromosomes Chromosomes that contain identical sets of loci and pair during meiosis.

homology   Similarity due to descent from a common ancestor.

homoplasy   A similarity that does not identify natural taxa.

homothallism   In zygomycetes, and basidiomycetes lacking distinguishable male and female gametangia, the condition in which a single thallus is able to reproduce sexually without interaction of two differing thalli; in ascomycetes, the condition of species that are hermaphroditic and self-fertile.

homozygous   Having the same alleles for a given gene at the same locus on homologous chromosomes.

honeydew   Sugary liquid exuded together with spores; presumably functions in attracting insects that disperse the spores.

hormogonium (plural, **hormogonia**)   In the filamentous bluegreen algae, multicellular, filamentous segment capable of gliding motion; can be reproductive unit.

host   Living organism serving as a substrate and/or energy source for another.

hyaline   Clear or colorless; used with reference to dead cells in the leaves and stems of some mosses, especially *Sphagnum*.

hydroids   Specialized, water-conducting cells in bryophytes.

hygroscopic   Readily absorbing and retaining moisture; refers to certain cells or structures that respond to changes in humidity.

hymenium   Aggregation of asci or basidia and related sterile structures in a continuous layer; also termed *fertile* or *fruiting layer*.

hypha (plural, **hyphae**)   One of the tubular filaments composing mycelium.

hypnospore   Spore formed inside parental cell and secreting new wall.

hypogeous   Developing below soil surface.

hypogynous   Originating or developing below: in some Florideophyceae, beneath the carpogonium.

hypothallus   Thin, shiny, membranous, adherent film at base of fructification of myxomycetes.

hypotheca (plural, **hypothecae**)   Inner cell half of diatom cell; often refers only to inner siliceous wall.

imbricate   Closely overlapping.

imperfect stage   In fungi, stages in which asexual reproductive structures (conidiospores) are produced.

incubous   Leaf insertion in Jungermanniales (Hepaticae) in which the upper margin of a leaf lies on top of the lower margin of the leaf directly above it on the same side of the stem.

independent assortment   The chance distribution of alleles in meiosis; the distribution of alleles from one pair of homologous chromosomes having no effect on the distribution of alleles from another pair of homologous chromosomes.

indeterminate (growth)   Growth that occurs indefinitely, or without limit.

information storage and retrieval system   The storage of data in such a fashion that they are arranged into classes of similar objects and the presence in a class can be used to predict unknown traits in an object.

inheritance   Traits in offspring that come from the parental generation.

innovation   New branch arising from an old stem, in bryophytes.

inoperculate   Opening of a sporangium or ascus by an irregular tear or plug to discharge spores.

intercalary   Inserted within or between two cells or tissues.

intercalary meristem   Area of new cell production situated some distance from the apex.

internode   Portion of axis between two nodes.

interphase   Between stages: in mitosis and meiosis, period when there is no chromosomal division, although DNA replication may occur; in dictyostelids, period between active assimilative and aggregation phases in which the myxamoebae cease feeding and undergo internal changes (dictyosteliomycetes); also used with reference to a nondividing nucleus.

intertidal region   Portion of continental shelf periodically exposed between the highest and lowest tide levels.

intranuclear   Occurring or remaining inside the nuclear envelope.

intranuclear division   Nuclear division in which the spindle lies within the nuclear envelope, and the latter structure remains intact throughout the division process.

invagination   An ingrowth, or ensheathing.

inversion   In asexual reproduction in *Volvox*, the process when the new colony completely turns inside out, so that flagellar ends of cells are externally oriented; sometimes called *eversion*.

inviable gametes   Gametes incapable of taking part in the process of fertilization.

isidium (plural, **isidia**)   Rigid protuberance of upper part of lichen thallus, which may break off and serve for vegetative reproduction.

isodiametric   Having equal diameters; used to describe cell shape essentially equal in length and width.

isogamous Having isogamy.

isogamy Fusion of gametes that are the same size and are morphologically alike.

isokont Having flagella of equal length.

isomorphic Morphologically alike.

karyogamy Fusion of two sex nuclei following fusion of protoplasts (plasmogamy).

keel Canal or cleft in valve of some pennate diatoms.

key A device to aid in the identification of an object by presenting mutually exclusive descriptive phrases with which to compare the object in question.

key classification A key.

kinetochore A region connecting the two chromatids of a chromosome; having a layered appearance at the ultrastructural level.

kombu Edible product of some Laminariales (brown algae).

lamella (plural, **lamellae**) Plate, or layer: submicroscopic structure of chloroplast membranes (thylakoids) containing pigments; in bryophytes, refers to thin sheets of flaplike plates of tissue on the dorsal surface of the thallus or leaves; in basidiomycetes, the gills of a mushroom.

lamina (plural, **laminae**) Leaf or blade: in brown algae, expanded leaflike part of thallus.

laminaran Storage polysaccharide composed of β-1,3, 1,6 glucoside linkages, characteristic of the brown algae.

laminate Flattened and broadly expanded; leaflike.

lanceolate Narrow and tapering toward each end.

laver General name given to edible dried preparation made from algae such as *Ulva* (green laver, Chlorophyceae) and *Porphyra* (purple laver, Bangiophyceae).

lenticular Lens-shaped (a double convex).

leptoid Cell specialized for conducting metabolites in bryophytes.

leucosin See **chrysolaminaran**.

limited growth Determinate growth; having a fixed, definite limit.

Linnaean hierarchy The arrangement of organisms into taxa, each subordinate to the one above it, with kingdom as the highest taxon and species as the lowest.

lipid Any group of compounds comprising fats; composed of glycerol and fatty acids.

locule Compartment, cavity, or chamber: in ascomycetes, stromatic chambers containing asci.

locus (plural, **loci**) A particular location on a chromosome of a gene.

lorica (plural, **loricae**) Surrounding case that is separate from protoplast in some algae.

lutein Xanthophyll pigment.

lysotrophy Nutritional type in which food is digested external to cell or thallus, the digestion products then absorbed.

Ma Symbol used in geochronology signifying the age of rocks $\times$ 10^6.

macrocyst Cyst developed by aggregation of myxamoebae in dictyosteliomycetes; within it, karyogamy and meiosis occur, and myxamoebae are released upon germination.

mamillose Conspicuously bulging.

mannan Polysaccharide material composed of mannose units occurring in cell walls of some algae and yeasts.

mannitol Saccharide alcohol; part of the polysaccharide laminaran.

mannoglycerate Saccharide storage product in some red algae.

mastigoneme Hairlike thread or process occurring along the length of some flagella; also known as "flimmer" or *tinsel*.

mating type In algae and fungi, the term used to designate a particular genotype with respect to compatibility in sexual reproduction; used when gametes are identical in appearance and referred to as plus ($+$) and minus ($-$) rather than male and female.

medulla Innermost region of thallus in lichens and in some brown and red algae.

meiosis Reduction division in which the number of chromosomes is reduced from the diploid ($2n$) to the haploid ($1n$) state.

meiosporangium Structure in which spores are produced by meiosis (reduction division).

meiospore Spore produced by meiosis, with a reduction in chromosome number from diploid to haploid (spores usually produced in fours).

meristem Tissue concerned with formation of new cells.

meristematic Having a meristem or the characteristics of a meristem.

meristoderm Outer meristematic cell layer (epidermis) of Laminariales (brown algae).

mesokaryote Term for dinoflagellate nuclear structure in which histone proteins are lacking; also used to denote dinoflagellate evolutionary position between prokaryotes and eukaryotes.

mesosome Invagination of plasma membrane in prokaryotic cells; generally bears respiratory enzymes.

microfossils Microscopic fossils, including spores, small algae, fungi, etc.

microtubule Elongate, proteinaceous, cytoplasmic unit, about 25 nm in diameter; involved in maintaining cell shape; basic component of eukaryotic flagella.

mitochondrion (plural, **mitochondria**) Cell organelle in which cellular respiration occurs.

mitosis Equational division in which chromosome number in new cells is same as parental number.

mitosporangium Structure producing spores by mitosis (equational division).

mitospore Spore produced by mitosis and having same chromosome number as spore mother cell.

mixotroph Photoautotroph capable of utilizing organic compounds in environment; may also be termed *facultative heterotroph* or *facultative parasite*.

moniliform Beadlike; sometimes used to describe chromatin in nucleus in dinoflagellates.

monoecious Condition in seed plants in which separate pollen-bearing and ovule-bearing cones or flowers are both borne on the same plant.

monoicous Condition of bryophyte gametophyte in which the gametophore bears both kinds of sex organs; bisexual.

monokaryotic Hyphal condition in which the compartments contain a single haploid nucleus (e.g., the primary mycelium of basidiomycetes).

monophyletic Evolving from a single ancestral stock.

monophyly Evolution of a taxon from one evolutionary line.

monopodial Having one main axis of growth.

monosiphonous Having a uniseriate row of cells; not surrounded by pericentral cells (Florideophyceae).

monosporangium Vegetative cell that metamorphoses to produce a single spore; characteristic of some red algae.

monospore Single spore produced by metamorphosis of single vegetative cell, the monosporangium; characteristic of some red algae.

monostromatic Having thallus one cell thick.

mordanted Treated with a substance (mordant) which forms an insoluble compound with a stain or dye and causes its fixation in cells or tissues.

mucilage canal Elongate cells in the cortex of Laminariales (brown algae) which may conduct mucilaginous materials.

mucilage hairs Specialized mucilage-producing hairs in bryophytes, commonest near leaf axils and growing points of the gametophore.

mucilaginous Jellylike; slimy.

mucopeptide Complex proteinaceous material giving prokaryotic cells rigidity (see also **peptidoglycan**).

multiaxial Having a main (central) axis composed of many parallel or almost parallel filaments.

multiple alleles The presence of several alleles for any one locus within a population.

mutualistic Refers to symbiotic associations in which both partners benefit.

multiseriate Having many rows of cells.

multistratose Having many layers.

mutation Change in genetic composition; change in gene.

mycologist One who specializes in the study of fungi.

mycelium (plural, **mycelia**) Mass of hyphae; the thallus of a fungus.

mycobiont Fungal partner, or component of a lichen.

mycolaminaran Storage product produced by oomycetous fungi— so called because of its similarity to laminaran.

mycorrhiza (plural, **mycorrhizae**) Symbiotic relationship of fungus and root or rootlike structure.

mycoses Fungus diseases of humans or animals (e.g., ringworm, athlete's foot).

myxamoeba (plural, **myxamoebae**) Naked amoeboid cell characteristic of myxomycetes and dictyosteliomycetes.

nannoplankton Plankton with dimensions less than 75 μm.

natural classification A classification utilizing as many characteristics as possible.

natural selection A natural process that results in a change in gene frequency in a population as a result of different degrees of reproduction and survival among individuals of a population.

natural taxon A taxon composed of organisms that are more closely related to each other than they are to organisms in another taxon at the same level.

neck Slender part of archegonium through which male gamete travels to reach the female gamete.

neck cells Cells comprising the narrow tube or neck of the archegonium.

neck canal cells Inner row of

cells in neck region of archegonium; at maturity these cells disintegrate.

neutral spore In brown algae, term sometimes used to describe the mitospore produced in a plurilocular mitosporangium.

nitrogen fixation Conversion of free atmospheric nitrogen to bound organic forms.

nitrogen-fixing Refers to prokaryotes having the ability to fix atmospheric nitrogen.

node Point on an axis where one or more parts are attached.

nomenclature The application of appropriate names to plant taxa following the International Rules of Botanical Nomenclature.

nori Edible dried preparation of *Porphyra* (Bangiophyceae); also known as *purple laver*.

nuclear body Area in bacterial or cyanophyte cell containing DNA and associated cytoplasm: in bacteria, the area of cytoplasm is devoid of ribosomes (see also **nucleoid**).

nucleus (plural, **nuclei**) DNA-containing organelle in eukaryotic cells.

nucleoid Bacterial "nucleus," consisting of a circular strand of DNA and its associated cytoplasm.

nucule In charophytes, female reproductive structure (gametangium) including oogonium and surrounding sterile vegetative cells; in fossil charophytes called *gyrogonites*.

numerical taxonomy The erection of classifications by numerical analysis of characteristics; taximetrics.

numerical phyletics The erection of an inferred phylogeny of extant organisms utilizing numerical analysis of characteristics.

nurse cells In some hepatics, sterile cells among the spores that

lack any special wall thickenings and frequently disappear before the spores mature.

nutritive cell In some Florideophyceae, a special cell in the carpogonial branch that fuses with the carpogonium following fertilization.

obligate parasite Term generally used for an organism that occurs only as a parasite.

odonthalan Mucilaginous galactan occurring in cell walls and intercellular spaces of some red algae.

ontogenetic Refers to the life history or development of an individual organism.

ontogeny Development of an organism or structure in its various stages from initiation to maturity.

oogamous Having oogamy.

oogamy Production of gametes in which the female is large and nonmotile, and the male is small and either motile or nonmotile.

oogonium (plural, **oogonia**) Female gametangium consisting of a single cell (occurring in fungi and algae).

oosphere Egg or female gamete produced in an oogonium (oomycetes).

oospore Thick-walled resting spore of oomycetes developing from a fertilized oosphere.

operculate Refers to a sporangium or ascus that opens by a small lid or cover.

opercular cell A cell that forms a lid or cover.

operculum Lid or cover: in fungi, part of a cell wall; in the Bryidae, a multicellular tissue in the sporangium.

order A level in the Linnaean hierarchy within a class that includes closely related families.

organ Distinct and differentiated part of multicellular organism composed of tissues.

organelle Membrane-bound structure within the eukaryotic cell (e.g., nucleus, chloroplast).

osmophilic bodies Lipid-containing granules; appear dark when treated with osmium fixation.

ostiole Opening or pore.

overturn Complete mixing of a body of water from surface to bottom resulting from several ecological factors.

paleobotanist One who studies fossil plants and their relationships to environment and age of rocks.

palynologist One who studies living and fossil spores, pollen grains, algal cysts, etc.

papilla (plural, **papillae**) Blunt projection or protuberance.

papillate Having papillae.

papillose With small warts; descriptive of the sculpturing of spores and cell walls.

paramylon Polysaccharide storage product composed of β-1,3 glucoside linkages, occurring in euglenophytes.

paraphyllium (plural, **paraphyllia**) Filamentous or leaflike, chlorophyll-containing outgrowth on stem near leaf base, occurring in some pleurocarpous Bryidae and some Hepaticae.

paraphysis (plural, **paraphyses**) Sterile hair or thread: in the ascomycetes, sterile hypha in the hymenium; in bryophytes and brown algae, unicellular or multicellular hair associated with the sporangia or sex organs.

parasexuality (parasexual cycle) Mechanism of recombination of hereditary material based on the mitotic cycle rather than the sexual cycle; occurs in some higher fungi.

parasite   Organism that derives its nutrients and hence energy from a living host.

paraspore   In some Florideophyceae, spore produced in a parasporangium; believed to be a mitospore.

parenchyma   Tissue composed of living, thin-walled, randomly arranged cells.

parenchymatous   Composed of living, thin-walled, randomly arranged cells.

parietal   Peripheral in position; for example, in some algae the chloroplast is located near the periphery of the cell.

parthenogenesis   The development of a gamete into a new individual without fertilization.

parthenogenetic development   Production of a new plant from an unfertilized egg.

partial veil   Membrane extending from the cap margin to the stipe and covering the gill or pore layer (Agaricales).

partial wall   Membranous layer covering the developing hymenium in some basidiomycetes.

pathogenic   Disease causing.

peat   Deposits of incompletely decomposed plant material, primarily *Sphagnum* (Musci).

pectin (pectic)   Polysaccharide material in cell wall and middle lamella.

pellicle   Proteinaceous, articulating strips of material enclosing the protoplast in the euglenophytes (under the plasma membrane).

peltate   Shield-shaped.

pennate   Common name for diatoms generally of rectangular shape with markings in parallel rows.

penultimate   Next to the last.

peptidoglycan   Rigid compo-
nents of cell walls in prokaryotes; composed of cross-linked polysaccharides and polypeptide chains; also called *murein* and *mucopeptide.*

perennating   Causing or permitting to be perennial.

perennial   Refers to a plant that continues to live for several years.

perfect stage   Sexual stage in fungi.

perianth   Protective organs around reproductive structures: in Jungermannials (Hepaticae), sheath of fused leaves surrounding archegonia and developing sporophyte.

pericarp   In Florideophyceae, urn-shaped gametophyte tissue surrounding the carposporophyte (sometimes collectively referred to as *cystocarp*).

pericentral cell   Cell that occurs outside the central axis (as in *Polysiphonia*, Florideophyceae).

perichaetial shoot (perichaetium)   Composite structure, comprising the archegonia and surrounding or subtending leaves of some Hepaticae and most Musci.

perichaetium   Enlarged leaves surrounding archegonia of bryophytes.

periclinal   Parallel to the circumference of the surface.

peridiole   Lenticular body in which basidiospores are formed in the Nidulariales (basidiomycetes).

peridium   Membranous covering or outer sterile layer of sporangium (myxomycetes) and some basidiocarps (gasteromycetes).

perigonial leaves   Leaves surrounding the antheridia of some Musci and Hepaticae; also known as *perigonial bracts* in Jungermanniales.

perigonium (plural, perigonia)   Antheridium plus associated perigonial leaves or bracts in Hepaticae and Musci.

perigynium (plural, perigynia)   In Hepaticae, a sleevelike extension of stem or thallus tissue that surrounds the archegonia.

periphysis (plural, periphyses)   Sterile hyphae lining the ostiolar canal of perithecia in some ascomycetes.

periplast   Proteinaceous layer surrounding the protoplast of cryptophytes; inside plasma membrane.

peristome   Sporangium mouth of Musci.

peristome teeth   Multicellular, toothlike structures ringing mouth (peristome) of sporangium of Musci.

perithecium   Ascocarp in which the hymenium is completely enclosed at maturity of the asci and ascospores, except for a small opening, or ostiole; generally urn-shaped.

phaeophyte tannin   Tanninlike substance in refractile granules near the nucleus in some brown algae; originally thought to be a polysaccharide and termed *fucosan.*

phagocyte   Specialized cell in the body's defensive system which is capable of ingesting and destroying invading baterial cells or other foreign particles.

phagocytosis   Ingestion of foreign particles or cells, often by specialized cells (phagocytes) within the animal body; the foreign material (antigen) is first acted upon by antibodies.

phagotroph(ic)   Organism that ingests solid food particles.

phagotrophically   Functioning as a phagotroph; ingesting solid food particles.

phenetic   Referring to the relationships of organisms based on phenotypic characteristics.

phenetic classification (system)   A classification based on phenotypic characteristics of extant organisms; the arrangement does not reflect phylogeny, because the evolutionary (fossil) history is unknown.

phenetic series An arrangement of taxa demonstrating the degree of relationship believed to exist among them, based on similarity or dissimilarity of characteristics, and without knowledge of their origins, ancestors, or descendants.

phenotype (phenotypic) The external, visible appearance or physical expression of a genetic trait.

phenotypic plasticity Different expressions of one genotype when grown under different environmental conditions.

phialides Specialized conidium-producing cells; conidia are produced successively at the open end of the phialide (ascomycetes and Fungi Imperfecti).

phloem Food-conducting tissue of vascular plants.

photic zone Depth of water through which light penetrates.

photoautotroph Autotrophic organism that derives energy for metabolism from visible light.

phototactic Responding to light by moving toward or away from the stimulus; a locomotory response of motile cells or organisms.

phototropism (phototropic) Growth response toward (positive) or away from (negative) a unilateral light stimulus.

phycobilin Water-soluble, proteinaceous pigment, similar to bile pigment, occurring in bluegreen algae, red algae, and cryptophytes; phycobiliprotein.

phycobiliprotein Water-soluble red or blue pigment, similar to bile pigment, occurring in bluegreen

algae, red algae, and cryptophytes, serving as an accessory pigment; phycobilin.

phycobilisomes Granules containing biliproteins attached to thylakoid membranes in Rhodophyta and Cyanophyta.

phycobiont Algal partner or component of a lichen.

phycocolloid Complex colloidal substance produced by algae, especially some brown and red algae (e.g., agar, algin, carrageenan).

phycocyanin Blue phycobilin pigment occurring in bluegreen algae, red algae, and cryptophytes.

phycoerythrin Red phycobilin pigment occurring in bluegreen algae, red algae, and cryptophytes.

phycologist One who studies algae.

phyllid Flattened, leaflike appendage in bryophytes.

phylogenetic classification A classification based on the phylogeny of the organisms classified; also, a classification that reflects the phylogeny of the organisms classified.

phylogeneticist One who studies phylogeny.

phylogeny An evolutionary history of an organism or group of organisms; includes both living (extant) and extinct (fossil) representatives.

phylogenetic tree A pictorial representation of a real or inferred phylogeny.

phylum (plural, **phyla**) One of the main categories used in classification of organisms, often restricted to the animal kingdom (see **division**).

physode See **fucosan vesicles.**

phytoplankton Free-floating or weakly swimming microscopic aquatic plant life.

pileate Having a pileus.

pileus Cap or structure bearing hymenium in some ascomycetes and basidiomycetes.

pilus (plural, **pili**) In gram-negative bacteria, proteinaceous filaments protruding through cell wall; smaller than flagella, the filaments are adhesive, and one type is important in conjugation.

pinnate Branching at right angles from a central axis, and all in one plane (as in a feather).

pit connection In red algae, "plugged" connection between cells; also referred to as *pit plugs.*

plankton Aquatic organisms, usually microscopic, that are free-floating or weakly swimming; position determined primarily by water currents.

planogamete Motile gamete, generally by means of one or more flagella.

planospore Motile spore with one or more flagella; also termed *zoospore.*

planozygote Zygote motile by means of flagella.

plasma membrane Double-layered membrane surrounding the protoplast of cell.

plasmodesma (plural, **plasmodesmata**) In some algae and bryophytes, a fine cytoplasmic connection that passes through the cell wall and joins the protoplasts of adjacent cells.

plasmids In bacteria, minute, self-replicating, extrachromosomal strands of DNA; they contribute to the genotype.

plasmodiocarp Sessile sporangium developing from main plasmodial branches in myxomycetes.

plasmodium (plural, **plasmodia**) Naked, acellular, assimilative stage in myxomycetes.

plasmogamy Fusion of cytoplasm of gametes prior to nuclear fusion, or karyogamy; in higher fungi, fusion of cytoplasm and establishment of the dikaryotic stage.

plastid Organelle in the cytoplasm of a cell in which carbohydrate metabolism occurs, or a storage organelle, or a colored organelle.

pleomorphism Having more than one form or shape.

pleurocarpous Growth form in Bryidae in which gametophore is multibranched and creeping, with sporophyte borne on short, lateral branch.

plurilocular Having many chambers; used to describe gametangia and mitosporangia occurring in brown algae.

pneumatocyst Hollow area of stripe of some brown algae containing gas, which helps keep them afloat.

podetium (plural **podetia**) Stiff, erect, secondary branch of the thallus bearing the apothecia in some lichens (especially reindeer lichens).

polar ring A short hollow cylinder located at each pole of the spindle of a dividing nucleus (in Florideophyceae).

polarity Condition in which one end of a cell of a developing organism is differentiated from the other, usually in both structure and function.

polyglucan granules Carbohydrate storage product in bluegreen algae.

polyhedral body Crystalline body in some bluegreen algae and bacteria; shown in some bacterial cells to contain enzymes involved in carbon dioxide fixation.

polyphosphate granule Phosphate storage granules in bluegreen algae.

polyphylogeny (polyphyletic) Evolution from more than one ancestral stock.

polyphyly Evolution of a taxon from more than one evolutionary line.

polyploid Having multiple genomes.

polyploidy Having more than two complete sets of chromosomes.

polysaccharide Organic compound composed of a large number of sugar (saccharide) units attached to one another.

polysiphonous In Florideophyceae, composed of several filaments in tiers of parallel, vertically elongated cells (e.g., *Polysiphonia*).

polyspore Type of mitospore in which more than four spores are produced from one spore mother cell (occurs in some Florideophyceae).

polystromatic Composed of several (many) cell layers.

population The members of one species that occupy one site at one time.

porphyran Mucilaginous saccharide occurring in cell walls and intercellular spaces of some red algae.

postical Branches arising in plane of amphigastria (some Hepaticae).

postfertilization In red algae, refers to stages of development immediately following fertilization.

prealga A phagocytic amoeboid prokaryote preceding the first true eukaryote in the (hypothetical) evolution of eukaryotes from prokaryotes.

primary mycelium In basidiomycetes, the mycelium developing from a basidiospore; that is, the haploid or monokaryotic phase.

primary pit connection In red

algae, the plugged aperture that remains following division of a cell.

primary plasmodium Plasmodium developing in root hair following infection by haploid swarm cell in *Plasmodiophora*.

primary producers (primary food producers) Photosynthetic organisms that are able to convert light energy into chemical energy for use by themselves and by other organisms.

primary protonema The chlorophyllose protonema produced by a germinating moss spore; sometimes treated as synonymous with *chloronema*.

primitive Ancient or ancestral; refers to any characteristic in a taxon or lineage that is ancestral, and has not changed with evolution.

primordium Beginning of an organ; the very earliest cellular level at which an organ can be discerned.

proboscis In the Fucales, an external, membranous structure closely associated with the anterior flagellum.

progametangial Refers to a progametangium.

progametangium (plural, **progametangia**) In Mucorales, the fertile branch tip in conjugation (the gametangium develops by deposition of a wall in the progametangium).

prokaryotic Refers to organisms (**prokaryotes**) that have nuclear bodies and lack membrane-bound organelles (nuclear membranes, chromosomes, etc.) in their cells.

propagula (plural, **propagulae**) Propagules: in bryophytes, vegetative reproductive structure produced by the gametophore, which is basically a miniature gametophore.

propagule The part of a plant

that propagates it; either a vegetative structure or a spore.

prophage "Dormant" stage of bacterial virus in which the viral genome is integrated into, and replicates with, that of the host; if the prophage becomes active, or vegetative, viral particles are produced and the host cell is destroyed.

protomitochondria Stage in evolution of mitochondria postulated for certain bacteria that contributed to the formation of the first eukaryotic cell by endosymbiosis.

protonema (plural, **protonemata**) Filamentous gametophyte stage of charophytes and many bryophytes; usually results from spore germination.

protoplastid Photosynthetic bodies postulated for early eukaryotic cells, representing blue-green algae that were engulfed during endosymbiosis.

protozoa Unicellular animals, usually microscopic.

proximal(ly) The part nearest the point of attachment.

pseudocapillitium Threadlike plasmodial vestiges in fructification of some myxomycetes.

pseudoelater Sterile structures among meiospores in sporangium of Anthocerotae.

pseudoparaphyllia Reduced, chlorophyllose, leaflike or filamentous structures on the stems of some mosses, arising at branch primordia.

pseudoparenchyma(tous) In algae, mass of densely packed filaments, randomly arranged, which may lose their individuality and resemble parenchyma tissue; in fungi, tissue in which hyphal cells become inflated, isodiametric, and adherent laterally to similar cells of adjacent hyphae.

pseudoplasmodium See **grex**.

pseudopodium False foot: in Andreaeidae and Sphagnidae (Musci), leafless gametophytic tissue acting as a seta, raising the sporangium above the main part of the gametophore.

pseudoraphe Clear area on valve between rows of striae or costae in some pennate diatoms.

punctum (plural, **puncta**) Fine pore in the siliceous walls of diatoms; may contain finer pores, but not as complicated as areola.

purple laver Common name given to species of the red alga *Porphyra*, especially when dried (see also **nori**).

pycnidium (plural, **pycnidia**) Flask-shaped structure in which conidia are formed in some ascomycetes and Fungi Imperfecti.

pycniospore Sporelike, uninucleate cell produced in pycnia of rusts and functioning in fertilization (i.e., dikaryotization); also called a *spermatium*.

pycnium Flasklike to variously shaped subepidermal structure wherein spermatizing agents, the pycniospores or spermatia, develop (Uredinales); also called a spermagonium.

pyrenoid Organelle associated with chloroplasts of some algae and the Anthocerotae; usually a center for storage-product formation.

pyriform Pear-shaped.

raphe Unsilicified longitudinal groove or split in cell wall of some pennate diatoms.

receptive hypha In some Ascomycotina and Uredinales, specialized filament to which the spermatangial nucleus is transferred.

reciprocal parasitism Partnership between two dissimilar organisms in which both benefit; also referred to as *symbiosis*.

recombination The bringing together of alleles as a result of sexual reproduction following meiosis.

refractile Capable of reflecting.

repetitive germination Germination of a spore by production of a new spore essentially identical to the first.

reservoir Enlarged posterior part of groove in some motile cells, such as euglenophytes.

resistant sporangium Sporelike body, often functioning in overwintering; gives rise to zoospores upon germination.

resupinate Flat or spread on the substrate with hymenium on outer side.

refringent Refractive; refract (reflect).

reticulate Netlike.

reticulum Network.

retort cell Flask-shaped cell with an apical pore, occurring on the stem of some Sphagnidae (Musci).

rhizine Bundle of hyphae that attaches the lichen thallus to the substrate.

rhizoid Unicellular or multicellular rootlike filament that attaches some plants to the substrate; in chytrids, enucleate, rootlike extension from cyst.

rhizoidal Having the characteristics of a rhizoid.

rhizome Underground stem; in bryophytes, either vertical or horizontal.

rhizoplast Threadlike organelle connecting basal body to nucleus in some motile cells.

rhizopodial Morphological type or growth form in which the cell is somewhat amoeboid.

ribosome Granular structural element, often associated with endoplasmic reticulum of the cytoplasm and sites of protein synthesis; composed of RNA and protein.

saccate   Saclike.

saccharide   Sugar units.

saprobe (saprobic)   Heterotrophic organism deriving its source of energy from dead organisms; also termed *saprophyte*.

saprophyte   See **saprobe**.

sargasterol   A sterol occurring in brown algae.

sclerotium (plural, **sclerotia**)   In Myxomycota, a hard plasmodial resting stage; in Eumycota, a resting body composed of a hardened mass of hyphae, frequently rounded.

secondary mycelium   In basidiomycetes, dikaryotic mycelial stage initiated by fusion of primary hyphae or spermatia and receptive hyphae.

secondary plasmodium   Plasmodium, possibly dikaryotic, developing in root cortex following fusion of motile cells in *Plasmodiophora*.

secondary pit connection   In red algae, the plugged aperture remaining following nuclear transfer between one vegetative cell and another.

secondary protonema   In mosses, erect, chlorophyll-rich branches that arise from the creeping protonema.

secondary spore   In basidiomycetes, a spore similar to the basidiospore in its development but produced by the germinating spore rather than on the basidium.

secretory cell   In brown algae, small cells having a secretory function, surrounding mucilage canals.

segregation   Separation of homologous chromosomes, and hence linkage groups of genes, at the time of meiosis.

selection   A natural process

which results in certain individuals leaving more offspring, thereby changing the nature of the population in which the individuals occur.

separation disc   Breakage area in a filament formed by death of a cell (bluegreen algae).

septate   Divided by a partition.

septum (plural, **septa**)   A transverse wall, generally perpendicular to the length of a filament.

sessile   Without a stalk.

seta   Sporophyte stalk in bryophytes.

serological   Refers to a reaction between substances (antibodies) formed in the body and foreign substances (antigens; e.g., components of bacterial cells).

sheath   Covering external to cell wall.

sieve cell   Type of phloem cell characteristic of nonflowering vascular plants.

sieve element   One cell in a series constituting a sieve tube.

sieve plate area   Area in wall of sieve element or sieve cell with fine pores; occupied by connecting protoplasmic strands; also termed *sieve plate* or *sieve area*.

sieve tube   In Laminariales (brown algae), a conducting structure composed of tubelike series of sieve elements with sieve plate areas in common end walls.

sinistral   Having left-hand helical arrangement of sheath cells around oogonium (charophytes).

siphonous   Refers to morphological type of growth form that is nonseptate, multinucleate, and often elongate; group or series of multinucleate green algae (Bryopsidophyceae).

sirenin   A substance produced by female gametes that attracts the male gametes in *Allomyces*.

soredia   Mass of algal cells surrounded by fungus hyphae, extruded through upper or outer cortex of lichen and functioning in vegetative reproduction.

sorocarp   The simple fruiting body of dictyosteliomycetes; lacks a containing membrane and is often of irregular shape.

sorophore   Stalk holding the sorus in the dictyosteliomycetes.

sorus (plural, **sori**)   Cluster of spores or spores together with spore-producing structures; may include associated sterile elements.

spawning   In invertebrates and fishes especially, the process of liberating or depositing eggs freely into the water or on the surface of substrates in large numbers.

special-purpose classification   A classification based on only a few characteristics and designed to fulfill a special or particular purpose.

specialized   The state of a characteristic that is derived or highly modified in a taxon or lineage.

speciation   The creation of new species in natural habitats by mutation, genetic recombination, and natural selection.

species   Taxonomic unit in Linnaean hierarchy within a genus, in which the organisms included have one or more distinctive characteristics and generally interbreed freely.

spermagonium   Flask-shaped structure producing the small, sporelike spermatia in some ascomycetes and in the Uredinales (basidiomycetes); see **pycnium**.

spermatangium (plural, **spermatangia**)   Structure that produces a spermatium in red algae.

spermatium (plural, **spermatia**) Nonmotile cell functioning as male gamete in some ascomycetes, Uredinales (basidiomycetes), and red algae.

spheroidal Approaching the form of a sphere.

spirillum (plural, **spirilla**) Helical or coiled morphological form of bacterial cell; also termed *spiril*.

sporangiophore Special branch bearing sporangia.

sporangiospore Spore produced by cleavage in sporangium.

sporangium (plural, **sporangia**) Structure in which spores are produced: unicellular in algae, fungi, and bacteria; in bryophytes, multicellular with outer sterile layer of protective cells.

spore General name for reproductive structure, usually unicellular, but multicellular in some fungi and a few bryophytes.

sporeling Young plant produced by germination of a spore.

sporic meiosis Pattern of alternation of generations in which spores are formed by meiosis, and the spores develop into multicellular gametophytes before gametes are produced; occurs in many algae, all bryophytes, and vascular plants.

sporocarp Many-celled structure bearing spores; a fruiting body in fungi.

sporophyll Leaflike appendage bearing sporangia.

sporophyte Spore-producing plant (generally diploid and producing meiospores).

squamule Small, loosely attached lobe in squamulose lichen.

squamulose Lichen growth form similar to foliose type but with numerous, small, loosely attached thallus lobes, or *squamules*.

stalk cell Cell in the fertile axis of some red algae (e.g., *Polysiphonia*), to which the tetrasporangium is attached.

starch Storage polysaccharide composed of α-1,4, 1,6 glucoside linkages (amylose and amylopectin), characteristic of green algae, pyrrhophytes, and cryptophytes.

statospore In chrysophytes, type of ornamented resting spore, usually formed endogeously; may be a zygote.

stellate Star-shaped.

stem-calyptra In some hepatics, a protective sheath that is formed from both stem tissue and archegonial cells; sheathes the young sporophyte.

stephanokont Having an anterior ring of flagella (in some Chlorophyta).

stereids Thick-walled, elongate supportive cells in the gametophores and seta of mosses.

sterigma (plural, **sterigmata**) Minute spore-bearing process (basidiomycetes).

sterol Type of organic compound present in some plants, possibly as a storage product, such as ergosterol, fucosterol, sargasterol, and sitosterol.

stipe Stalk lacking vascular tissue; may be unicellular or multicellular.

stipitate Having a stipe or special stalk.

stolon Aerial runner: in zygomycetes, aerial hyphae, usually bearing rhizoids and sporangiophores at points of contact with the substrate.

stria (plural, **striae**) Fine row of puncta in wall of a diatom, appearing as a line.

stroma (plural, **stromata**) A compact mass of fungus cells, or of mixed host and fungal cells, in or on which spores or sporocarps are formed; a granular matrix surrounding the thylakoids within the chloroplast.

stromatolite Finely layered calcareous formations produced by some bluegreen algae; the formations are usually extensive.

submarginal Near or under the margin.

substrate (substratum) Foundation: underlying surface providing point of attachment or host for plant.

subtidal Refers to that portion of the continental shelf below the lowest low-tide level (never exposed).

succubous Leaf insertion in Jungermanniales (Hepaticae) in which the lower margin of a leaf lies on top of the upper margin of the leaf directly below it on the same side of the stem.

sulcus Longitudinal furrow: in dinoflagellates, longitudinal posterior groove containing the trailing flagellum.

supporting cell Specialized cell from which carpogonial branch arises in some Florideophyceae.

sushi A type of Japanese "sandwich" made with red seaweed (*Porphyra*).

swarm cell Flagellated cell resulting from spore germination in myxomycetes, also called *swarmer*.

swarmer Motile, flagellated reproductive cell: in bacteria, released by certain filamentous forms (e.g., *Sphaerotillus*); in algae, motile cell of unknown origin; see also **swarm cell.**

symbiosis Partnership between two dissimilar organisms, from which both benefit; also referred to as *reciprocal parasitism*.

sympleisiomorphy The generalized traits of a group.

synapomorphy Shared derived trait in members of a taxon.

synaptonemal complex Structure present during meiosis that links paired, homologous chromosomes.

syngamy   Fusion of gametes; fertilization.

tactic   Refers to directed movements of motile cells in response to stimuli (e.g., aerotaxis, phototaxis, etc.).

taximetrics   See **numerical taxonomy.**

taxon (plural, **taxa**)   General term that can be applied to any group of plants or entities.

teliospore   Thick-walled resting spore that bears the basidium in some rusts and smuts (basidiomycetes).

telium (plural, **telia**)   Structure producing teliospores in some rusts and smuts (basidiomycetes).

tentacle   Organelle of unknown structure in *Noctiluca* (dinoflagellates).

terete   Cylindrical and tapering.

tetrasporangium (plural, **tetrasporangia**)   Meiosporangium in Florideophyceae in which four spores are produced.

tetraspore   Meiospore produced in Florideophyceae (red algae).

tetrasporine group   Series of uninucleate green algae with a nonlinear or linear arrangement of nonmotile cells.

tetrasporophyte   Plant producing tetraspores, usually free-living, diploid plant (Florideophyceae).

thallose   Having a simple plant body without differentiation into leaves or leaflike structures; type of growth form in some Hepaticae.

thallus   A plant body not differentiated into roots, stems, and leaves.

theca (plural, **thecae**)   Case; cell wall of diatoms and dinoflagellates.

thermophilic   Refers to plants capable of growth at high temperatures above 40 °C, as in hot springs (optimum 40–50 °C for many thermophilic fungi; up to about 90 °C for thermophilic bacteria and 73 °C for bluegreen algae).

thylakoid   Photosynthetic membranes: in eukaryotic cells, grouped within chloroplast envelope.

thylakoid membrane   Cytoplasmic membrane in prokaryotes containing photosynthetic and respiratory pigments and ribosomes.

tinsel flagellum   Flagellum with many fine, tubular hairs, or *mastigonemes*, in one or two rows along the length of the flagellum.

tissue   Group of cells organized into a structural and functional unit.

transduction   Transfer of genetic material from one bacterial cell to another by bacterial viruses (bacteriophages).

transformation   The incorporation of genetic material of dead cells from the medium into the genetic make-up of a living cell, as in some bacteria.

transition zone   Intercalary meristem between lamina and stipe in some brown algae.

trichoblast   Simple or branched, often colorless, hairlike branch in some Florideophyceae.

trichocine group   Series of uninucleate green algae with linear (filamentous) or parenchymatous arrangement of cells.

trichocyst   Cytoplasmic organelle in some flagellated algae (cryptophytes, raphidophytes, and dinoflagellates) that can be released upon being disturbed; also known as *ejectosome*.

trichogyne   Receptive, hairlike extension of female gametangium

in Florideophyceae and ascomycetes.

trichome   Linear row of cells in the bluegreen algae, exclusive of any sheath material.

trichothallic growth   Intercalary growth at base of hairlike, uniseriate filament in brown algae.

trigone   Conspicuous corner-thickening in cell walls of leaves of Hepaticae.

trilete   A three-armed scar in the shape of a Y on some meiospores.

triploid   Polyploid having three times the haploid chromosome number.

true branching   Branching in a filamentous form produced by a change in direction of cell division.

tuber   Underground storage cells: in the bryophytes, group of nutrient-rich cells that can remain dormant during the unfavorable season, permitting the gametophore to survive and reproduce itself vegetatively.

turbinate   Shaped like a top.

umbonate   Bearing a convex elevation in the center.

unarmored   Lacking specific articulated plates or armor, as in some dinoflagellates.

uniaxial   Having a main (central) axis consisting of a single filament of usually large cells.

unilocular   Having one chamber; usually refers to the meiosporangium in the brown algae.

uniseriate   Having a single linear row of cells.

unisexual   Having only one type of sexual structure (either male or female) produced by any one individual.

unistratose   Having one layer.

universal veil   Membrane covering the developing basidiocarp in the Agaricales (basidiomycetes).

unlimited growth   Indeterminate growth; growth which occurs indefinitely, or without limit.

unnatural taxon   A taxon that includes some organisms that are more closely related to organisms in another taxon at the same rank.

upgrade theory   The hypothesis that the thalloid habit evolved first in bryophytes, with the leafy habit being derived later (opposite to *downgrade theory*).

uredium (plural, **uredia**)   Structure producing uredospores in some rusts (basidiomycetes).

uredospore   Dikaryotic repeating spore in some rusts (basidiomycetes).

vacuole   A membrane-bound cavity in the protoplasm of a cell, containing cell sap.

valve   In diatoms, each half of the silicified portion of the cell; in hepatics, parts resulting from bending outward of sporangium wall when sporangium opens by means of regular longitudinal splits.

valve view   Surface view of diatom cell.

variation   The presence of more that one form or type in a population.

vascular system   A plant conductive system composed of xylem (that transports water) and phloem (that transports metabolites).

vector   Organism that transmits a disease from one plant or animal to another (e.g., the mosquito that transmits malaria).

vegetative   Refers to the nonreproductive or fertile phase of plant structure and growth.

vegetative reproduction   Asexual reproduction; progeny have same genetic constitution as parent.

venter   Lower, swollen, egg-containing portion of archegonium.

vesicle   Membrane forming enclosed sacs.

vibrio   Short, curved, rod-shaped bacterial cell.

violaxanthin   Xanthophyll pigment.

volutin   Stored food substance in bacteria, often appearing as granules.

volva   Cuplike fragment of universal veil at base of stipe of some Agaricales (basidiomycetes).

volvocine group   Series of uninucleate green algae with flagellated vegetative cells.

whiplash flagellum   Smooth-surfaced flagellum, generally lacking tubular hairs (although other hairs can be present); may have distal thin region.

xanthophyll   General name for group of yellow, carotenoid pigments composed of oxygenated hydrocarbons.

xenogenous   Refers to the origin of eukaryotic organelles from outside the cell; mitochondria and chloroplasts derived from the engulfment of bacteria and bluegreen algae, respectively (see **autogenous**).

xylan   Water-soluble polysaccharide xylose units occurring in cell walls of some red algae and green algae.

zeaxanthin   Xanthophyll pigment.

zooplankton   Free-floating or weakly swimming aquatic animal life.

zoospore   Spore motile by means of one or more flagella; also termed *planospore*.

zooxanthellae   Algal cells (often yellow) living symbiotically in cells of certain invertebrate animals; algae known to be members of the dinoflagellates, cryptophytes, and xanthophytes.

zygospore   Thick-walled resting spore resulting from the fusion of gametangia (conjugation) in zygomycetes.

zygote   Product of syngamy; diploid cell resulting from fusion of two haploid gametes.

zygotic meiosis   The pattern of alternation of generations in which meiosis occurs directly in the zygote, restoring the haploid condition; only diploid stage is zygote.

INDEX

peat, 474, 478 (*see also Sphagnum*)
peatland development, 478
pectic acid, brown algae, 356
pectic substances, 311, 321
pectin, green algae, 254; brown algae, 355; red algae, 400
Pediastrum, 261, 269, **275**
Pedinomonas, **303**, 306
Pelagophycus, 356, 391, 392
Pellia neesiana, **459**
pellicle, 345, **347**, 348, 349
Pelvetiopsis, 356, 361, **363**
penicillin, 36
Penicillium, **132**, 173, 189, 191, **196**, 241; *P. chrysogenum*, 191
Peniophora, 228, 229, **231**
pennate diatoms, **322**, **324**, 325, 326, 330
peptidoglycan, 20, 28, 72; bacteria, 33, 36, 68, 72; bluegreen algae, 72
Peranema, **347**, 349
percentage G + C, 136
Percursaria, 261, **264**, **277**
perennial, brown algae, 377
perianth, bryophytes, **452**, 453
pericarp, red algae, 421, 427, 429, **430**, **432**
pericentral cells, red algae, 418, **432**, **433**
perichaetial leaves, bryophytes, 473, 474
perichaetium, bryophytes, 446
periclinal divisions, brown algae, 367
peridial layers, higher fungi, **236**, **247**
Peridinium, 341, **345**, **346**, 353
peridioles, 234, **245**
peridium, 233, 234, **236**, 242; higher fungi, 233, 234, 236, **236**, **242**; slime molds, 95, 96, 97, 100
perigonia, bryophytes, 451, **452**
perigynium, bryophytes, 456, **459**
periplast, 350
peristome, Bryidae, 488, 495, 496; bryophytes, 449; Buxbaumiidae, **487**; Polytrichidae, **484**, 485; Tetraphidae, 481, 485
peristome teeth, bryophytes, 473, **482**, **484**, **486**, **487**
perithecia, 181, 185, 189, 193, **193**, 201, 203, 211
Peronospora, 145, 146, **146**, **164**
Peronosporales, 137, 139, 142, 145, 146
Pestalotia, 237, **248**
Petalonia, 356, 367
Petalophyllum ralfsii, 459
Petrocelis, 425
Peziza, 209, **211**
Pezizales, 203, 209
pH, lower fungi, 130
Phacus, **347**, 349, **349**
Phaeoceros, 468; *P. laevis*, 468
Phaeocystis, **333**
Phaeophyceae, 355, 361, 391 (*see also* brown algae)
Phaeophycidae, 355, 356, 361
Phaeophyta, **9**, 28, 308, 317, 335, 355, 369, 391 (*see also* brown algae); classification, 356; phenetic relationships, **520**; phenetic series, **520**; relationships, 389
phaeophyte tannin, brown algae, 358
Phaeostrophion, 356, 367, 369, **371**
Phaeothamnion, **320**, 321
phage, 44; bacterial, **55**
phagocytosis, 32
phagotrophy, 123; chrysophytes, 321; euglenoids, 349; dinoflagellates, 345; lower fungi, 131; slime molds, 90, 93, 107, 109, 114, 118, 123
Phallales, 234
Phallus, 234, **244**
pharmaceutical preparations, 395
phenetic relationships, chrysophytes, 309; Chlorophyta, 306, **525**; phaeophyte orders, **520**
phenetic schemes, 520; fungi, 516
phenetic series, 513, **517**, 521; green algae, 306; red algae, 520
phenetics, 11

phenotype, 3
phenotypic plasticity, 2
phialides, higher fungi, 189, 241
Philonotis fontana, **494**
phloem, 373
Phoma, 237, **247**
phosphorescence, dinoflagellates, 343
photoautotrophs, 505
photoautotrophy, bacteria, 31, 36, 54; bluegreen algae, 83, 84; pseudomonads, 54
photosynthate, 27, 326
photosynthesis, 26, 143, 250; bacteria, 22, 88; bluegreen algae, 83, 85, 88; eukaryotes, 23, 26; sulfur bacteria, 55, 57
photosynthetic lamella, 20 (*see also* thylakoids); brown algae, 356; red algae, 401
photosynthetic sulfur bacteria, 55, 57
phototropic, lower fungi, 130, 165
phototropic response, higher fungi, 196
phycobilin pigments, red algae, 403, 429, 434
phycobiliproteins, 26, 28, 351, 353 (*see also* biliproteins); bluegreen algae, 72, 83, 85; cryptophytes, 351, 353
phycobilisomes, bluegreen algae, 72, **78**, 88; red algae, 404
phycobiont, 25, 185, 187, 305
phycocolloids, 27, 28; brown algae, 355
phycocyanin, 27
phycocyanobilin, red algae, 403
phycoerythrin, 27, 404
phycoerythrobilin, red algae, 403
Phycomyces, 163, **164**, 165
phyllids, bryophytes, 443
Phyllospadix, 407
Phyllothallia nivicola, **456**
phylogenetic systematics, 11
phylogeny, 504–530; algae, 516–520; Bryophyta, 527–529; Charophyceae, 521; Chlorophyta, **304**, 306, 513, 516, 520–524; Chrysophyta, 513, 516, 521; Codiales, 521; Corallinaceae, 513; Cryptophyta, 516, 521, 524; Cyanophyta, 516; dasyclad genera, **515**; diatoms, **526**; Euglenophyta, 524; evolution, eukaryotes, 506:
  land plants, **304**, 306, 525–527
evolutionary patterns, 513; fungi, 513–516; green algae, **304**, 306, 513, 516, 520, 521; Phaeophyta, 360, 361, 520, 521; Pyrrophyta, 524; red algae, 431; Rhodophyta, 431, 520
Physarales, 100
Physarum, **95**, **96**, **97**, 100; *P. bivalve*, **100**; *P. polycephalum*, 100
Physcia, 207, **207**
physiology, higher fungi, 187
physodes, brown algae, 358
Phytophthora, 142, 143, **144**, 145; *P. infestans*, 143, **144**, 145; *P. palmivora*, **144**
phytoplankton, 305, 331, 343, 394
phytosterols, red algae, 405
pigmentation, algae, 26, 28, 340; Bryopsidophyceae, 291; Chrysophyta, 309; Jungermanniales, 451; *Sphagnum*, 478; red algae, 403
pigments, 26, 27 (*see also* pigmentation)
Pilayella, 356, **358**, **360**, 363; flagella, ultrastructure, **359**, **360**
pileus, 209, 229, 230, 231
pili, 47; bacteria, 37, 40; bipolar, **42**
pilins, bacteria, 40
Pilobolus, 163, 165, 167, 168; sporangiophores, **163**
Pinus, **226**
pit connection, formation, red algae, **418**; ultrastructure, **401**
pit connections, red algae, **401**, 406, 439
Plagiochila, **453**; *P. splenoides*, **452**; *P. tridenticulata*, **448**; *P. virginica*, **448**; *P. yokogurensis fragilifolia*, **448**
plankton, 305, 317, 321, 331, 337, 343
planktonic animals, 148, **348**
planospore, **21**, 130 (*see also* zoospores)
planozygote, dinoflagellates, 341

plant diseases, bacteria, 71
plant parasites, 203
plant pathology, 145
plasma membrane, **16**, 19; bacteria, 32, 36, 37, 40; lower fungi, 126; slime molds, 91, 92
plasmid, 44; bacteria, 37
plasmodesmata, 369; brown algae, **385**; green algae, 279
plasmodiocarps, 96, 100
Plasmodiophora, 118; life history, **116**; *P. brassicae*, 108, **116**, **117**, 118
Plasmodiophorales, 108
Plasmodiophoromycetes, 90, 91, 108–110, 118; life history, 109; secondary plasmodium, 109; spores, 109; swarm cells, 109
plasmodium; slime molds, 90, 91, 92, 93, 94, 96, 100, 108, 112, 118
plasmogamy, 25, 26; higher fungi, 175, 178, 182, 187, 189; slime molds, 109
Plasmopara, 145; *P. viticola*, 145, **146**
plastids, 26, 109 (*see also* chloroplasts); green algae, 279
Platymonas, **303**
Plectomycetes, 189
Pleodorina, 260, 261, **267**
Pleospora, 185, 209, **213**
Pleosporales, 209
Pleurocapsales, 78
pleurocarpous, bryophytes, 471
Pleurochloris, **310**, 315, 317
Pleurochrysis, **330**, 333
pleuropneumonia, bovine, 71
Pleuropneumonia, 71
Pleurotus, 231, **231**, **238**
plurilocular, brown algae, **365**, 373, **387**
pneumatocysts, brown algae, 373, **380**
pneumonia, 44
pneumonic plague, 51
Poaceae, 10
Podosphaera, 191, **200**
Pogonatum, **483**
Pohlia annotina, 447
polar flagellum, **42**
polar ring, red algae, 401
polarity, *Fucus*, 369, 383
pollen, 148
pollution, 85, 331
poly-beta-hydroxybutyrate, 61
poly-beta-hydroxybutyric acid, bacteria, 36, 55, 61, 63
polyglucan granules, bacteria, 74, 75, 77; bluegreen algae, 75
polyhedral bodies, bacteria, 74, 75
polyhedral granules, bluegreen algae, 75
Polykrikos, **341**, **343**
Polyneura, **424**
polyolefins, red algae, 405
polyphosphate granules, 74, 75; bluegreen algae, 75
polyphyly, 11–12
polyploids, 5–6
polyploidy, 5; brown algae, 360
Polyporus, 229, **233**
polysaccharides, 27; green algae, 281; brown algae, 358; lower fungi, 125; red algae, 405
Polysiphonia, 415, 418, 423, 425, 439; fertile trichoblast, 421; life history, **430**; reproduction, **430**, **432**
polysiphonous, red algae, 418, **432**, **433**
Polysphondylium, 105, 107, **115**
polyspores, red algae, 423
Polytrichidae, 483–486; leaves, 483, **484**; protonema, 486; sporangium, **484**, 485; stem structure, **472**, **484**, 485; vascular system, **472**, **484**, 485
Polytrichum, 485; *P. commune*, **472**, **485**
Pongola rock series, 505
population genetics, 1
populations, 1
pore, higher fungi, **174**
poroid hymenium, **235**
Porphyra, **401**, **404**, 407, 409, **410**, **412**, **413**,

Vascular Plants

SUBDIVISION	CLASS	SUBCLASS	ORDER	FAMILY
Rhyniophytina (*Rhyniophytes*)	Rhyniopsida Trimerophytopsida		Rhyniales Trimerophytales	
Zosterophyllophytina (*Zosterophyllophytes*)	Zosterophyllopsida		Zosterophyllales	
Psilotophytina (*Psilotophytes*)	Psilotopsida		Psilotales	
Lycophytina (*Lycopods*)	Lycopsida		Lepidodendrales Pleuromeiales Isoetales Selaginellales Lycopodiales	
Sphenophytina (*Articulates*)	Sphenopsida		Hyeniales Sphenophyllales Calamitales Equisetales	
Pterophytina (*Pterophytes*) (*Preferns*) { (*True Ferns*) {	Cladoxylopsida Coenopteridopsida Filicopsida		Cladoxylales Coenopteridales Ophioglossales Marattiales Osmundales Schizaeales Pteridales Dicksoniales Hymenophyllales Gleicheniales Cyatheales Aspidiales Blechnales Matoniales Polypodiales Marsileales Salviniales	
		Ophioglossidae Marattiidae Osmundidae Filicidae		
		Marsileidae Salviniidae		
	Progymnospermopsida (*Progymnosperms*)		Aneurophytales Archaeopteridales	
(*Gymnosperms*)	Pteridospermopsida (*Seed Ferns*)		Pteridospermales	Lyginopteridaceae, Medullosaceae, Callistophytaceae, Peltaspermaceae, Corystospermaceae, Caytoniaceae
	Cycadopsida (*Cycadophytes*)		Cycadeoidales	Cycadeoidaceae Williamsoniaceae
			Cycadales	
	Ginkgopsida (*Ginkgos*)		Ginkgoales	
	Coniferopsida (*Coniferophytes*)		Cordaitales Voltziales Coniferales	Lebachiaceae, Voltziaceae Pinaceae, Cupressaceae, Podocarpaceae, Araucariaceae, Taxaceae, Cephalotaxaceae
	Gnetopsida (*Gnetophytes*)		Ephedrales Welwitschiales Gnetales	
(*Flowering Plants*)	Magnoliopsida (*Angiosperms*)	Magnoliidae (*Dicots*)	Magnoliales	Magnoliaceae, Degeneriaceae, Annonaceae
			Laurales	Calycanthaceae, Lauraceae
			Piperales	Piperaceae
			Aristolochiales	Aristolochiaceae
			Rafflesiales	Rafflesiaceae
			Nymphaeales	Nymphaeaceae, Ceratophyllaceae
			Illiciales	Illiciaceae, Schizandraceae
			Nelumbonales	Nelumbonaceae
			Ranunculales	Ranunculaceae, Lardizabalaceae, Berberidaceae
			Papaverales	Papaveraceae, Fumariaceae
			Sarraceniales	Sarraceniaceae
			Trochodendrales	Trochodendraceae
			Cercidiphyllales	Cercidiphyllaceae
			Eupteleales	Eupteleaceae
			Didymelales	Didymelaceae
			Hamamelidales	Hamamelidaceae, Platanaceae
			Eucommiales	Eucommiaceae
			Urticales	Ulmaceae, Moraceae, Cannabaceae, Urticaceae